Lecture Notes in Computer Science 1233

Edited by G. Goos, J. Hartmanis and J. van Leeuwen

Advisory Board: W. Brauer   D. Gries   J. Stoer

W0055462

Springer-Verlag Berlin Heidelberg GmbH

Walter Fumy   (Ed.)

Advances in Cryptology — EUROCRYPT '97

International Conference on the Theory and
Application of Cryptographic Techniques
Konstanz, Germany, May 11-15, 1997
Proceedings

 Springer

Series Editors

Gerhard Goos, Karlsruhe University, Germany

Juris Hartmanis, Cornell University, NY, USA

Jan van Leeuwen, Utrecht University, The Netherlands

Volume Editor

Walter Fumy
Siemens AG, Corporate Technology
Otto-Hahn-Ring 6, D-81730 Munich, Germany
E-mail: walter.fumy@mchp.siemens.de

Cataloging-in-Publication data applied for

Die Deutsche Bibliothek - CIP-Einheitsaufnahme

Advances in cryptology : proceedings / EUROCRYPT '96,
International Conference on the Theory and Application of
Cryptographic Techniques, Konstanz, Germany, May 11 - 15, 1997.
Walter Fumy (ed.). - Berlin ; Heidelberg ; New York ; Barcelona ;
Budapest ; Hong Kong ; London ; Milan ; Paris ; Santa Clara ;
Singapore ; Tokyo : Springer, 1997
   (Lecture notes in computer science ; Vol. 1233)

CR Subject Classification (1991): E.3.4, G.2.1, D.4.6, F.2.1-2, C.2, J.1, K.6.5

ISSN 0302-9743
ISBN 978-3-540-62975-7   ISBN 978-3-540-69053-5 (eBook)
DOI 10.1007/978-3-540-69053-5

© Springer-Verlag Berlin Heidelberg 1997

Originally published by Springer-Verlag Berlin Heidelberg New York in 1997.

Typesetting: Camera-ready by author
SPIN 10548767   06/3142 – 5 4 3 2 1 0   Printed on acid-free paper

Preface

EUROCRYPT '97, the 15th annual EUROCRYPT conference on the theory and application of cryptographic techniques, was organized and sponsored by the International Association for Cryptologic Research (IACR). The IACR organizes two series of international conferences each year, the EUROCRYPT meeting in Europe and CRYPTO in the United States.

The history of EUROCRYPT started 15 years ago in Germany with the Burg Feuerstein Workshop (see Springer LNCS 149 for the proceedings). It was due to Thomas Beth's initiative and hard work that the 76 participants from 14 countries gathered in Burg Feuerstein for the first open meeting in Europe devoted to modern cryptography. I am proud to have been one of the participants and still fondly remember my first encounters with some of the celebrities in cryptography.

Since those early days the conference has been held in a different location in Europe each year (Udine, Paris, Linz, Linköping, Amsterdam, Davos, Houthalen, Aarhus, Brighton, Balantonfüred, Lofthus, Perugia, Saint-Malo, Saragossa) and it has enjoyed a steady growth. Since the second conference (Udine, 1983) the IACR has been involved, since the Paris meeting in 1984, the name EUROCRYPT has been used. For its 15th anniversary, EUROCRYPT finally returned to Germany.

The scientific program for EUROCRYPT '97 was put together by a 18-member program committee which considered 104 high-quality submissions. These proceedings contain the revised versions of the 34 papers that were accepted for presentation. In addition, there were two invited talks by Ernst Bovelander and by Gerhard Frey.

A successful EUROCRYPT conference requires the combined efforts of many people. First, I would like to thank the members of the program committee, who devoted a tremendous amount of time and energy to reading the papers and making the difficult selection. They are: Michael Burmester, Hans Dobbertin, Marc Girault, Shafi Goldwasser, Alain P. Hiltgen, Don B. Johnson, Pil Joong Lee, Tsutomu Matsumoto, David Naccache, Kaisa Nyberg, Paul Van Oorschot, Torben P. Pedersen, Josef Pieprzyk, Bart Preneel, Rainer Rueppel, Claus Schnorr, and William Wolfowicz.

In addition, I gratefully acknowledge the support to the program committee by the following experts: Albrecht Beutelspacher, Simon R. Blackburn, Carlo Blundo, Antoon Bosselaers, Odoardo Brugia, Marco Bucci, Anne Canteaut, Chris Charnes, Ivan Damgård, Yvo Desmedt, Erik De Win, Markus Dichtl, Michele Elia, Piero Filipponi, Marc Fischlin, Roger Fischlin, Steven Galbraith,

Oded Goldreich, Dieter Gollmann, Shai Halevi, Helena Handschuh, Erwin Hess, Stanislaw Jarecki, Joe Kilian, Lars R. Knudsen, Xuejia Lai, Françoise Levy-dit-Vehel, Keith M. Martin, Willi Meier, Alfred Menezes, Renato Menicocci, Daniele Micciancio, Preda Mihailescu, Thomas Mittelholzer, Sean Murphy, Pascal Paillier, Birgit Pfitzmann, Tal Rabin, David M'Raihi, Vincent Rijmen, Ron Rivest, Rei Safavi-Naini, Jacques Traoré, and Peter Wild. I apologize to those whose names have inadvertently escaped this list.

I also thank Alfred Büllesbach, Roland Müller, Roland Nehl, and Susanne Röhrig for taking the resposibility to organize EUROCRYPT '97, and Christina Strobel for her help with the proceedings.

Finally, I would like to thank the authors of all submissions (including those whose papers could not be accepted because of the large number of high-quality submissions received) for their hard work and cooperation.

March 1997   Walter Fumy

EUROCRYPT '97

May 11-15, 1997, Konstanz, Germany

Sponsored by the

International Association for Cryptologic Research (IACR)

General Chairmen

Roland Nehl, Deutsche Telekom, Germany
Alfred Buellesbach, debis Systemhaus, Germany

Program Chairman

Walter Fumy , Siemens AG, Germany

Program Committee

Michael Burmester .... University of London, U.K.
Hans Dobbertin .... BSI, Germany
Marc Girault .... SEPT, France
Shafi Goldwasser .... MIT, USA
Alain P. Hiltgen .... Crypto AG, Switzerland
Don B. Johnson .... Certicom, USA
Pil Joong Lee .... Postech, Korea
Tsutomu Matsumoto .... Yokohama National University, Japan
David Naccache .... Gemplus, France
Kaisa Nyberg .... Finnish Defence Forces, Finland
Paul Van Oorschot .... Entrust Technologies, Canada
Torben P. Pedersen .... Cryptomathic, Denmark
Josef Pieprzyk .... University of Wollongong, Australia
Bart Preneel .... K.U. Leuven, Belgium
Rainer Rueppel .... R3 Security Engineering, Switzerland
Claus Schnorr .... University of Frankfurt, Germany
William Wolfowicz .... Fondazione Ugo Bordoni, Italy

Contents

Two Attacks on Reduced IDEA
(Extended Abstract)

Johan Borst[*1], Lars R. Knudsen[2], Vincent Rijmen[2**]

[1] T.U. Eindhoven, Discr. Math., P.O. Box 513, NL-5600 MB Eindhoven,
borst@win.tue.nl
[2] K.U. Leuven, Dept. Elektrotechniek-ESAT, Kard. Mercierlaan 94, B-3001 Heverlee,
{lars.knudsen,vincent.rijmen}@esat.kuleuven.ac.be

Abstract. In 1991 Lai, Massey and Murphy introduced the IPES (Improved Proposed Encryption Standard), later renamed IDEA (International Data Encryption Algorithm). In this paper we give two new attacks on a reduced number of rounds of IDEA. A truncated differential attack on IDEA reduced to 3.5 rounds and a differential-linear attack on IDEA reduced to 3 rounds. The truncated differential attack contains a novel method for determining the secret key.

1 Introduction

The block cipher IDEA (International Data Encryption Algorithm) was proposed by X. Lai and J. Massey in [11] as a strengthened version of PES (for Proposed Encryption Standard) proposed by the same authors in [10]. The blocks are 64 bits and the keys are 128 bits. Both ciphers are based on the design concept of "mixing operations from different algebraic groups". IDEA was developed to increase the security against differential cryptanalysis. In [9] it was argued that for 3 rounds of IDEA there are no useful differentials and concluded that IDEA is resistant against a differential attack after 4 of its 8 rounds.

IDEA is an iterated cipher consisting of 8 rounds followed by an output transformation. We count the output transformation as an extra half round. The complete first round and the output transformation are depicted in the computational graph shown in Figure 1. The two multiplications and the two additions in the middle of the figure are called the MA-structure. The key schedule takes as input a 128 bit key and returns 52 subkeys, each of 16 bits.

W. Meier cryptanalysed 2 rounds of IDEA in a differential-like attack using a partial distributive law [14]. J. Daemen found large classes of weak keys for IDEA [4] and also described an attack on 2.5 rounds of IDEA for all keys in [3].

Differential cryptanalysis was introduced by Biham and Shamir in [1]. In an attack on an iterated cipher one considers plaintext pairs P, P^* of a certain difference and the corresponding ciphertexts C and C^*. The main tool in the

---

[*] The work of the first author was done while visiting K.U. Leuven.
[**] F.W.O. research assistant, sponsored by Funds for Scientific Research-Flanders (Belgium)

W. Fumy (Ed.): Advances in Cryptology - EUROCRYPT '97, LNCS 1233, pp. 1-13, 1997.
© Springer-Verlag Berlin Heidelberg 1997

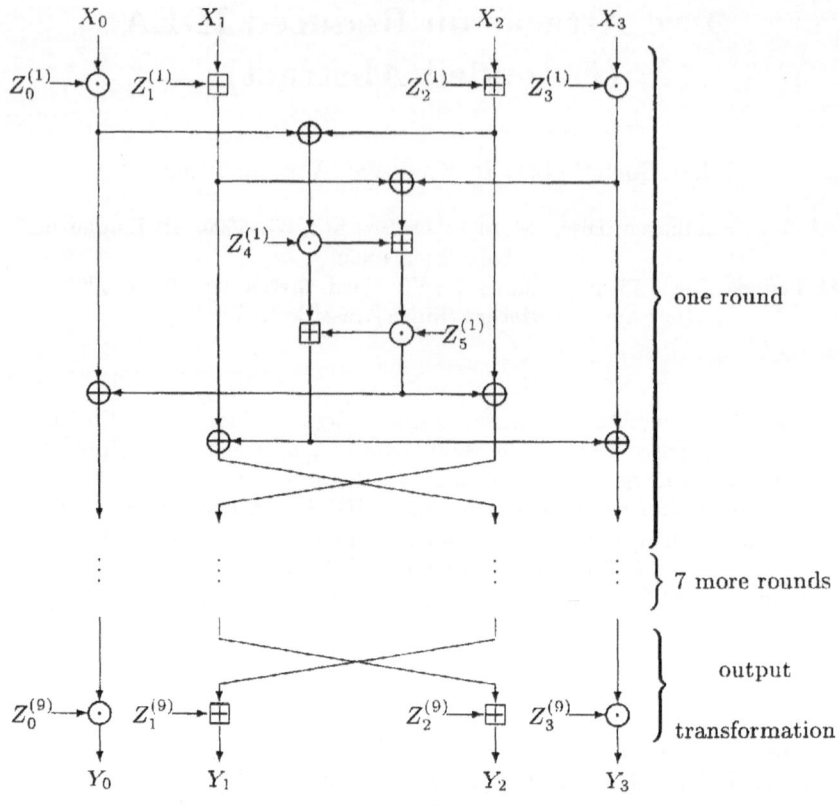

X_i : 16-bit plaintext subblock
Y_i : 16-bit ciphertext subblock
$Z_i^{(r)}$: 16-bit key subblock
$\oplus$: bit-by-bit exclusive-OR of 16-bit subblocks
$\boxplus$: addition modulo 2^{16} of 16-bit integers
$\odot$: multiplication modulo $2^{16} + 1$ of 16-bit integers
with the zero subblock corresponding to 2^{16}

Fig. 1. Computational graph for the encryption process of the IDEA cipher.

differential attack is the characteristic, a list of the expected differences in the ciphertexts after each round of the cipher. Lai and Massey introduced the notions of differentials in [11, 9]. Later in [6] Knudsen extended the notions of differentials to that of *truncated differentials*, where only subsets of the differences are predicted. A *right* pair is a pair of plaintexts, for which the ciphertext differences follow the differential. In a differential attack an attacker needs to get at least one right pair. However, an attacker might not be able to determine which pairs are right pairs from the differences in the ciphertexts, but if the characteristic or differential predicts also the differences in (parts of) the ciphertexts, often an attacker can discard pairs, which are not right pairs. A *wrong* pair is a pair of plaintexts, for which the differences in the ciphertexts do not follow the differential, but which looks like a right pair to the attacker.

In the linear attack [12] by Matsui one considers linear combinations of some bits of the plaintext, the ciphertext and the key, and defines linear characteristics. Nyberg introduced the *linear hull* [15], the analogue to differentials in differential attacks. In [5] Hellman and Langford combined the differential and the linear attack to the *differential-linear* attack, and applied it to 8 rounds of the DES.

In this paper we give two new attacks on IDEA. In Section 2 the differential attack using truncated differentials is described, which can be used to break 3.5 rounds of IDEA. In Section 3 the differential-linear attack is described, which can be used to break 3 rounds of IDEA and Section 4 gives concluding remarks. Full versions of the attacks in this paper are described in [2, 8].

2 Truncated Differential Attack

In this section we describe a differential attack on 3.5 rounds of IDEA using truncated differentials. We define the difference of two bit strings A and A^* of the same length as

$$\Delta A = A \oplus A^*. \tag{1}$$

For differential cryptanalysis of IDEA with other definitions of difference, we refer to [11, 14]. Under the definition of difference (1) IDEA is not a Markov cipher [11]. Also, as we will see, the probabilities of the differentials used depend very much on the key used in the encryptions. Thus, the hypothesis of stochastic equivalence [11], i.e., that the average probability of a differential taken over all keys is approximately the same as the probability for a fixed key for virtually all keys, does not hold for IDEA with difference (1).

Consider the following one-round differential for IDEA.

$$(a, b, c, d) \xrightarrow{p_1} (e, f, g, h) \xrightarrow{(e \oplus g, f \oplus h) \xrightarrow{p_2} (k, l)} (c \oplus l, g \oplus l, f \oplus k, h \oplus k)$$

(a, b, c, d) denotes the four-word input difference and (e, f, g, h) denotes the difference after the key addition. This transition has probability p_1. With probability p_2 the input difference $(e \oplus g, f \oplus h)$ to the MA-structure leads to an output difference (k, l). The output difference of the round is given as $(e \oplus l, g \oplus l, f \oplus k, h \oplus k)$.

The 3-round truncated differential used in our attack on IDEA is:

$$(A, 0, B, 0) \xrightarrow{2^{-16}} (C, 0, C, 0) \xrightarrow{(0,0) \xrightarrow{1} (0,0)} (C, C, 0, 0)$$

$$(C, C, 0, 0) \xrightarrow{1} (D, E, 0, 0) \xrightarrow{(D,E) \xrightarrow{2^{-32}} (E,D)} (0, D, 0, E)$$

$$(0, D, 0, E) \xrightarrow{2^{-16}} (0, F, 0, F) \xrightarrow{(0,0) \xrightarrow{1} (0,0)} (0, 0, F, F)$$

$$(0, 0, F, F) \xrightarrow{1} (0, G, 0, H)$$

where the words A to H represent any values. The average probability of the truncated differential is 2^{-64}. This probability is computed over all choices of the inputs to a round and to the MA-structure and over all choices of the round keys and where we have also assumed that the MA-structure acts like a random function.

This differential has a mirror image with the same probability:

$$(0, A, 0, B) \xrightarrow{2^{-16}} (0, C, 0, C) \xrightarrow{(0,0) \xrightarrow{1} (0,0)} (0, 0, C, C)$$

$$(0, 0, C, C) \xrightarrow{1} (0, 0, D, E) \xrightarrow{(D,E)^{2^{-32}}(E,D)} (D, 0, E, 0)$$

$$(D, 0, E, 0) \xrightarrow{2^{-16}} (F, 0, F, 0) \xrightarrow{(0,0) \xrightarrow{1} (0,0)} (F, F, 0, 0)$$

$$(F, F, 0, 0) \xrightarrow{1} (G, 0, H, 0)$$

These differentials are called truncated differentials, since we predict only two of the four words, the zeros, of the differences after each round.

We consider the attack also for reduced versions of IDEA, that operate on four nibbles, IDEA(16) and on four bytes IDEA(32), respectively, instead of four 16-bit words [9]. These reductions allow us to actually implement the attack and experimentally verify our results. The above differentials are defined similarly for the reduced versions. The average probabilities are 2^{-16} respectively 2^{-32}.

2.1 Description of the attack

First the attack on IDEA (full block length) is described. A structure of plain-texts consists of 2^{32} texts: p_1 and p_3 are fixed, p_0 and p_2 take on all possible values. We can use every combination of two texts as a pair. This means we generate $2^{32} \cdot (2^{32} - 1)/2 \approx 2^{63}$ pairs from a structure. For every structure the expected number of right pairs is 0.5. The differential requires that Δc_0 and Δc_2 are equal to zero, and only such pairs are considered. On the average only one out of 2^{32} pairs will survive this test. For each surviving pair do the following: for all possible keys $Z_0^{(1)}, Z_2^{(1)}$ check whether

$$(p_0 \odot Z_0^{(1)}) \oplus (p_0^* \odot Z_0^{(1)}) = (p_2 \boxplus Z_2^{(1)}) \oplus (p_2^* \boxplus Z_2^{(1)}) . \tag{2}$$

On the average, this holds for 2^{16} values of $(Z_0^{(1)}, Z_2^{(1)})$. Similarly we check for which keys in the output transformation, it holds that

$$(c_1 \odot (Z_1^{(4)})^{-1}) \oplus (c_1^* \odot (Z_1^{(4)})^{-1}) = (c_3 \boxminus Z_3^{(4)}) \oplus (c_3^* \boxminus Z_3^{(4)}) . \tag{3}$$

Note that for a right pair these tests are successful for the correct value of the key. In total it can be expected that each pair suggests 2^{32} 64-bit key values and therefore every structure will suggest 2^{63} keys. Therefore every value of the key will be suggested 0.5 times per used structure and, as indicated above, every structure will suggest the correct value of the key 0.5 times. One might expect that among all the key values suggested by wrong pairs is also the correct value of the key. However, a wrong pair in the above attack will not suggest the correct value of the key. For a non-discarded pair of plaintexts and their ciphertexts a key will be suggested if the tests (2) and (3) succeed. For the correct value of the key this means that the input difference to the second round will be $(\tilde{C}, \tilde{C}, 0, 0)$. The output difference of the third round will be $(0, 0, \tilde{F}, \tilde{F})$, and the input difference of the third round will be $(0, \tilde{D}, 0, \tilde{E})$. Thus, the difference in the second round after the key addition will be $(D', E', 0, 0)$ and the output difference of the round is $(0, \tilde{D}, 0, \tilde{E})$. But this implies that $D' = \tilde{D}$ and $E' = \tilde{E}$, because of the structure of the round function of IDEA. It follows that if the correct value of the key is suggested for a pair of plaintexts, this must be a right pair. Summing up, for every structure in the attack there will be 0.5 right pairs, which suggest the correct value of the key, and 2^{31} wrong pairs, which on the average suggest a wrong value of the key 0.5 times. Thus, for the above attack the traditional method of Biham-Shamir [1] will not work, the S/N-ratio is 1, meaning that the correct value of the key cannot be distinguished from any other value of the key.

However, as we will see, the probability of the above differentials used in the attack depends very much on the secret key. For some keys the probability is less than the average probability and for other keys it is larger. We extend the key search method of a differential attack to the cases where the probability of the differential for the correct value of the secret key is different from the average probability over all keys. The bigger this difference the faster the attack. If the difference is big enough and if we assume that wrong values of the secret key is suggested randomly and uniformly by the attack, the correct value of the key will be found using sufficiently many plaintext pairs. This is a novel approach in differential attacks and reminiscent of the key search method in a linear attack [12].

For the actual attack, there is an overlap between the key bits we count on in the first round and the bits we count on in the last round. Furthermore, because of the absence of a carry bit after the highest order bit of the modular addition, we are unable to distinguish keys that differ only in these bits, so we will regard these two values of the key as one. These two observations are very important to reduce the memory requirements when we implement the attack. Using the first differential above 14 key bits overlap and two bits are indistinguishable for IDEA, which means that we would search for only 48 bit key values. For the reduced versions of IDEA we implemented key schedules, such that relatively as many key bits overlap. For IDEA(32) and IDEA(16) seven and three bits overlap, respectively. This means that in these cases we search for only 23 bit and 11 bit key values. To find other key bits a similar attack with the second differential above can be executed.

2.2 Experimental verification

We implemented the attack using the first differential on IDEA(16). First we calculated the probability of the differential for all keys by exhaustive search. Table 1 shows these probabilities for different classes of keys. The average probability over all keys was estimated to $2^{-16.5}$. The key-dependency of the proba-

#Keys/All keys	Probability
13%	0
12%	$0 < p \le 2^{-18}$
21%	$2^{-18} < p \le 2^{-17}$
30%	$2^{-17} < p \le 2^{-16}$
14%	$2^{-16} < p \le 2^{-15}$
10%	$2^{-15} < p \le 1$

Table 1. Probability of the used differential for IDEA(16) with 3.5 rounds for classes of the secret key.

#Keys/All keys	# Structures	# Chosen plaintexts
25%	16	2^{12}
40%	32	2^{13}
51%	64	2^{14}
59%	128	2^{15}
67%	256	2^{16}

Table 2. Average number of chosen plaintexts needed in the attack on IDEA(16) with 3.5 rounds in 1000 attacks.

bilities stems mostly from the second round of the differential, where a difference (D, E) in the inputs to the MA-structure must result in difference (E, D) in the outputs of the MA-structure. Of most interest are the classes of keys that deviate most from the average probability. It is interesting to see that for about 1 in every 8 possible values of the secret key the probability of the used differential is zero. The numbers in Table 1 also indicate that the attack will not work for some classes of keys, namely the classes of keys for which the probabilities are too close to the average probability over all choices of the keys.

In Table 2 we list the results of 1000 implementations of our attack on IDEA(16) for increasing number of chosen plaintexts. We used key rankings as in [13] and tested whether the correct value of the key was among the eight least and eight most suggested values, thus the attack returns 16 suggestions for 11 bits of the secret key. As seen, using all plaintexts the correct value of the key is among those 16 values in about 67% of all cases. Note that there are a total of 2^{16} plaintexts of IDEA(16) and that an exhaustive search for the key will take the time of about 2^{32} encryptions.

Next we implemented attacks on IDEA(32). First we estimated the probabilities of the used differentials for different classes of keys. The result follows from Table 3. Based on the results of 160 experiments with random keys, we estimated the average probability over all keys to $2^{-32.7}$. Note that this is slightly less than first estimation made in the beginning of this section. This difference

#Keys/All keys	Probability
14%	$0 \leq p \leq 2^{-35}$
10%	$2^{-35} < p \leq 2^{-33.0}$
31%	$2^{-33.0} < p \leq 2^{-32.5}$
45%	$2^{-32.5} < p$

Table 3. Probability of the used differential for IDEA(32) for classes of the secret key.

#Keys/All keys	# Structures	# Chosen plaintexts
1%	16	2^{20}
7%	64	2^{22}
15%	128	2^{23}
31%	256	2^{24}
54%	512	2^{25}
65%	1024	2^{26}
83%	2048	2^{27}

Table 4. Average number of chosen plaintexts needed in the attack on IDEA(32) with 3.5 rounds in 100 attacks.

is caused by the fact that the MA-structure is not a random mapping. We implemented the attack for 100 different randomly chosen keys using up to 2048 structures. The results are given in Table 4. Using the above results on reduced versions of IDEA, we estimate the number of chosen plaintexts needed in our attack on IDEA. From Table 2 it follows that one finds 25% and 51% of the keys using $2^{3n/4}$ respectively $2^{7n/8}$ chosen plaintexts for $n = 16$ for IDEA(16). From Table 4 it follows that one finds 1% and more than 83% of the keys using $2^{5n/8}$ respectively $2^{7n/8}$ chosen plaintexts for $n = 32$ for IDEA(32). As can be seen the number of keys we can recover increases for larger block sizes with relatively the same amount of data. We predict that a similar increase will occur for the attack on IDEA. Next we consider the workload and the amount of memory needed. One needs enough memory to store one structure. Once one structure has been analysed it is thrown away and a new structure analysed. Thus, the memory requirement for the attack on IDEA is 2^{32} words of each 64 bits. The workload is the estimated number of operations needed to perform the attack, measured as the number of encryptions of the cipher. The 2^{32} ciphertexts in a structure are hashed on the values of c_0 and c_2, since for a right pair the pairs of these values are equal. The workload of the hashing and storing of the ciphertexts is small compared to the time of the rest of the attack. For each pair that survives the filtering process we try all possible 2^{16} values of the affected keys of each side of Eq. (2). These tests can be sped up by pre-calculating a table to avoid the expensive multiplication operation. This table would be of size 2^{32} 16-bit

#Keys/All keys	# Structures	# Chosen plaintexts	Workload
>1%	2^8	2^{40}	2^{51}
>31%	2^{16}	2^{48}	2^{59}
>83%	2^{24}	2^{56}	2^{67}

Table 5. Estimated number of chosen plaintexts needed in the attack on IDEA with 3.5 rounds with 2^{32} words of memory.

words. We estimate that a multiplication takes the equivalent of 3.5 additions, and that an addition, an exclusive-or and a table-lookup take about the same time. The workload is about 2^{12} encryptions of IDEA with 3.5 rounds for every analysed pair. Totally, the workload is about 2^{43} encryptions for every structure. Because of the overlap of key bits in this first round test with the key bits in the output transformation, the second part of the key search, i.e. using Equation (3), is much faster than the first and can be ignored in the workload estimation.

The estimated number of chosen plaintexts and the workload for our attack on IDEA is given in Table 5. Note that an exhaustive search for the key of IDEA takes the time of about 2^{128} encryptions of IDEA. Finally we discuss how to find additional key bits. The attack outlined above finds 48 bits of the 128 bit key of IDEA. However, once these key bits have been found, one can do a similar attack using the second truncated differential. As noted earlier the key-dependency of the probability of the first differential comes mostly from the second round of the differentials. Since the second round is the same for the two differentials, one can expect that for a fixed key the probabilities of the two differentials are very close. After doing the attack with the second differential one has all 64 key bits in the beginning of the first round and all 64 key bits of the output transformation. Subsequently, one can do similar attacks on a further reduced version of IDEA.

3   A Differential-linear Attack

In this section we give a differential-linear attack on IDEA reduced to 3 rounds. We will use the notation $P = (p_0, p_1, p_2, p_3)$, $C = (c_0, c_1, c_2, c_3)$ to describe plaintexts, ciphertexts and their 16-bit subblocks. The version we look at is 3 rounds without the output transformation and where we omit the swapping of the second and third ciphertext blocks. We will write $A[i]$ to indicate the ith bit of A, where $A[0]$ is the least significant bit (LSB) of A and $A[15]$ denotes the most significant bit (MSB) for a 16-bit word A. These indices will be omitted whenever the context makes it clear which bit(s) we are considering. With $A[i, \dots, j]$ we will indicate the row of bits $A[i] \dots A[j]$. Also, we define some special 16-bit symbols μ_i for $i = 0, \dots, 15$, where $\mu_i[i] = 1$ and $\mu_i[j] = 0$ for $j \neq i$.

9

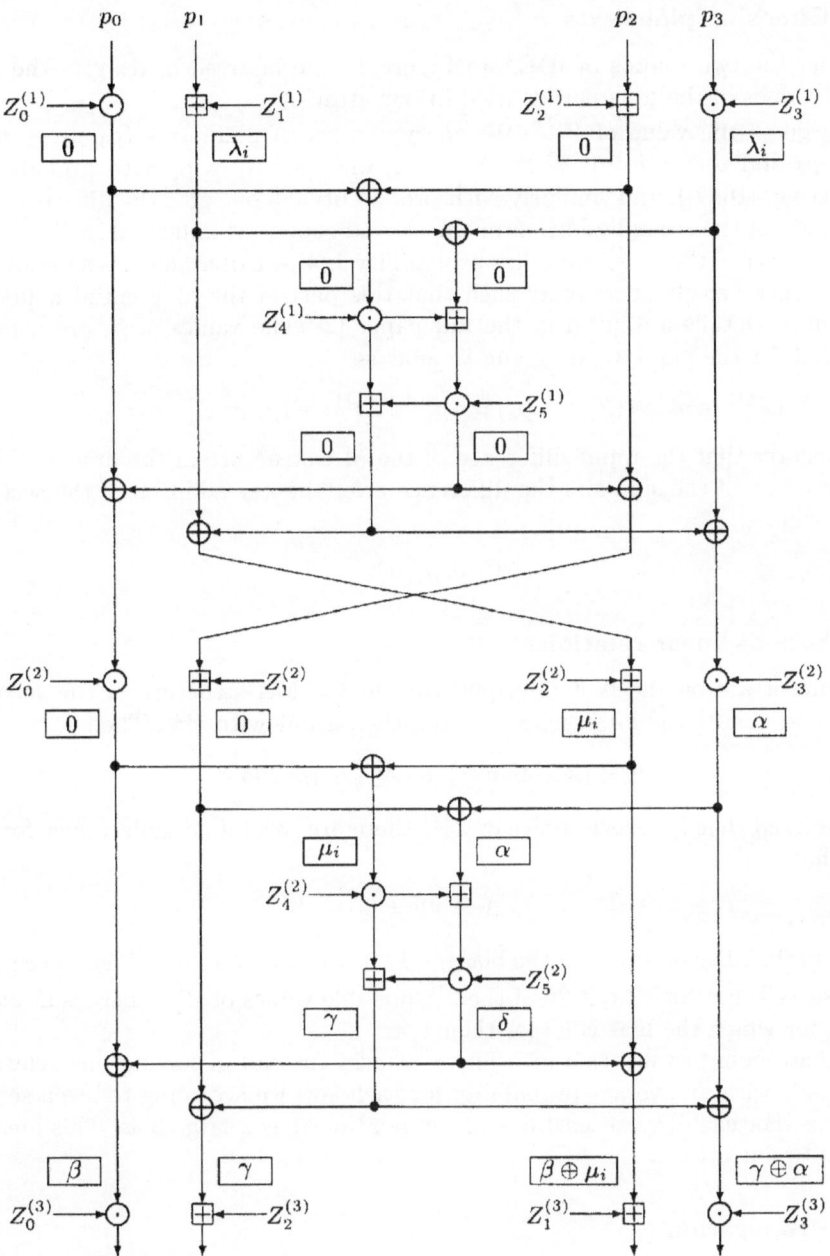

Fig. 2. Two rounds of IDEA.

3.1 Choosing plaintexts

Consider the two rounds of IDEA in Figure 2. The inserted boxes give the expected values of the differentials used in our attack.

We guess the value of $Z_3^{(1)}$. We encrypt a set of plaintexts (p_0, p_1, p_2, p_3), where p_0 and p_2 are fixed. With $\Delta p_1 = \mu_i$ one gets $\lambda_i = \mu_i$ with probability 0.5 (see e.g. [16, 7]), and similarly with probability 0.5 one gets the difference μ_i in outputs of the second addition in the second round, as indicated in Figure 2, thus this part of the differential has probability $1/4$. A closer analysis shows that one can pick six plaintext pairs such that this part of the differential holds at least once. Details are given in the full paper [2]. The values of p_3 are chosen such that for the pairs we are going to analyse

$$(Z_3^{(1)} \odot p_3^*) \oplus (Z_3^{(1)} \odot p_3) = (p_1^* \boxplus Z_1^{(1)}) \oplus (p_1 \boxplus Z_1^{(1)}) = \lambda_i \, .$$

This ensures that the input difference of the MA-structure in the first round is zero. For one of the six pairs the difference after the key addition of the second round will be

$$(0, 0, \mu_i, \alpha). \tag{4}$$

3.2 Sets of linear relations

We concentrate on the first multiplication in the MA-structure of the second round and denote the input with $p^{(2)}$ and the output with $r^{(2)}$. Then

$$\Delta r^{(2)} = (Z_4^{(2)} \odot p^{(2)}) \oplus (Z_4^{(2)} \odot (p^{(2)} \oplus \mu_i)) \, .$$

We observed that for every choice of $Z_4^{(2)}$ there are several possible values for μ_i such that

$$\Delta r^{(2)}[0] = 0 \tag{5}$$

with a probability p, such that the bias $|p - 1/2| > 0.166$ over all $p^{(2)}$. Furthermore we observed that for all but 26 of the 2^{16} possible values of $Z_4^{(2)}$ there is at least one μ_i for which the bias is larger then $1/4$.

We are going to use this in a linear attack. Instead of having one relation that holds with an average probability for each key, we are going to use a set of relations. For each key at least one of the relations has a large bias. This idea is central to our attack.

3.3 Propagation

From now on, we only consider the least significant bits of the various 16-bit intermediate results. For these bits the modular addition reduces to an exclusive-or. Denote by $t^{(2)}$ and $t^{(3)}$ the outputs of the second multiplication in the MA-structure of the second and third round, respectively. Using (4) we get for the difference after the second round

$$(\Delta t^{(2)}, \mu_i \oplus \Delta t^{(2)}, \Delta r^{(2)} \oplus \Delta t^{(2)}, \alpha \oplus \Delta r^{(2)} \oplus \Delta t^{(2)}).$$

Because the ciphertext (c_0, c_1, c_2, c_3) equals the output of the third round, we can calculate

$$\Delta t^{(3)} = \Delta c_2 \oplus \Delta r^{(2)} \oplus \Delta t^{(2)}$$
$$\Delta r^{(3)} = \Delta t^{(3)} \oplus \Delta c_1 \oplus \mu_i \oplus \Delta t^{(2)} = \Delta c_1 \oplus \Delta c_2 \oplus \mu_i \oplus \Delta r^{(2)},$$

where $r^{(3)}$ is defined in a similar way as $r^{(2)}$. In other words, we are able to predict the least significant bit of the output difference of the first multiplication of the MA-structure of the last round. The inputs of this multiplication are the subkey $Z_4^{(3)}$ and an intermediate result that equals $c_0 \oplus c_2$. For every ciphertext pair we can calculate $c_0 \oplus c_2$ and predict $\Delta r^{(3)}$ with a high probability. We keep a counter for every possible value of $Z_4^{(3)}$ and increment the counters of the key values that are compatible with the calculated $c_0 \oplus c_2$ and $\Delta r^{(3)}$.

Note that we don't know for which μ_i (5) holds with large probability. Therefore we have to repeat the attack for different values of μ_i. Also we guessed the value of $Z_3^{(1)}$. Our experiments suggest that for wrong guesses of $Z_3^{(1)}$ the algorithm fails to suggest a specific value for $Z_4^{(3)}$. Thus we can recognize wrong guesses. With this algorithm it is impossible to distinguish between the correct subkey values and their additive inverses modulo $2^{16} + 1$.

When $Z_3^{(1)}$ is guessed correctly, tests have shown that we need at most $9000 < 2^{14}$ pairs to determine $Z_4^{(3)}$. On the average we guess correctly after 2^{15} trials, therefore we need about 2^{29} plaintext pairs. Examining one plaintext pair takes a few exclusive-or operations and 2^{16} table look-ups, one for each value of $Z_4^{(3)}$. Since we examine 16 differentials, our attack needs totally about 2^{20} simple operations, i.e., addition or exclusive-or, for each pair and the total workload is therefore about 2^{49} simple operations. Using the estimate of Section 2 that an exclusive-or takes the same time as an addition and a multiplication takes 3.5 times as much time as either of them, the workload is about equal to $0.75 \cdot 2^{44}$ encryptions with 3 rounds of IDEA.

3.4 Finding additional key bits

In this paragraph we will describe how to find the subkeys $Z_0^{(3)}$ and $Z_5^{(3)}$ (or their additive inverses modulo $2^{16} + 1$). For this a method will be used similar to the main one described in [3]. First we will give a definition of compatibility.

Definition 1. A word A is said to be compatible with B modulo N if there exists a pair of words C, C^* with $C \oplus C^* = A$ and $C - C^* \pmod{N} = B$.

It is easy to see that a word A is compatible to at most 2^k words modulo N, where k is the Hamming weight of A. The probability that a randomly chosen word with Hamming weight k and another one are compatible modulo 2^{16} is therefore smaller or equal to 2^{k-16}.

For this part of the attack we will consider only the plaintext pairs that we already constructed with the correct guess for $Z_3^{(1)}$ (or its additive inverse modulo $2^{16} + 1$) that yield μ_{15} after the key addition of the second round. The difference after the second round is $(\beta, \beta \oplus \mu_i, \gamma, \gamma \oplus \alpha)$, see Figure 2. Like α, γ and δ the difference β is unknown. However, since $Z_4^{(3)}$ is known (or $2^{16} + 1 - Z_4^{(3)}$), when also $Z_5^{(3)}$ and $Z_0^{(3)}$ would be known, we would be able to calculate β for each pair and the intermediate values $(q_0^{(3)}, q_1^{(3)}, q_2^{(3)}, q_3^{(3)})$ before the MA-structure of the last round. Then $\beta \oplus \mu_{15}$ must be compatible modulo 2^{16} with $(q_1^{(3)} \boxplus Z_1^{(3)}) \boxminus (q_1^{(3)*} \boxplus Z_1^{(3)}) = q_1^{(3)} \boxminus q_1^{(3)*}$. To find $Z_0^{(3)}$ and $Z_5^{(3)}$ we simply guess their values and for each guess check this compatibility requirement. It can be shown [3] that the expected number of pairs needed to eliminate a wrong guess for a pair $Z_0^{(3)}, Z_5^{(3)}$ is approximately equal to 1 divided by the probability that a random 16-bit word is compatible modulo 2^{16} to another one. Tests have shown that this number is between 1 and 5.

As in the previous section this search method doesn't make a distinction between $Z_0^{(3)}, Z_5^{(3)}$ and their additive inverses modulo $2^{16} + 1$. It takes two multiplications with $Z_5^{(3)}$ and $Z_4^{(3)}$ to find $\Delta q_0^{(3)}$ and $\Delta q_1^{(3)}$, but as $Z_4^{(3)}$ is fixed, multiplications with this key are many times the same. Then it takes one multiplication with $(Z_0^{(3)})^{-1}$ to find β. So finding $Z_0^{(3)}$ and $Z_5^{(3)}$ takes at most 2^{33} multiplications modulo $2^{16} + 1$. According to the estimates earlier made this is about equal to $1.5 \cdot 2^{29}$ encryptions with 3 rounds of IDEA.

Finally, one can find the remaining key bits by doing additional attacks using similar characteristics as the above. The attack will have a better performance, since many key bits are already known.

4 Conclusions

We have presented two attacks on IDEA with a reduced number of rounds. The first attack finds the secret key of 3.5 rounds of IDEA in more than 86% of all cases using an estimated number of 2^{56} chosen plaintexts and a workload of about 2^{67} encryptions of 3.5 rounds of IDEA. With 2^{40} chosen plaintexts the attack works for 1% of all keys. The second attack finds the secret key of 3 rounds of IDEA. It needs at most 2^{29} chosen pairs of plaintext and a workload of about 2^{44} encryptions with 3 rounds of IDEA.

Although our attacks make use of some sophisticated techniques, the efficiencies, in particular the workloads, of the algorithms probably can be improved greatly. Further we think that similar attacks can be successful against more rounds of IDEA, but it is questionable if in this way anything substantial can be achieved against the full 8.5-rounds version of IDEA.

References

1. E. Biham and A. Shamir. *Differential Cryptanalysis of the Data Encryption Standard*. Springer Verlag, 1993.
2. J. Borst. *Differential-Linear Cryptanalysis of IDEA*. Technical Report ESAT-COSIC Report 96-2, Department of Electrical Engineering, Katholieke Universiteit Leuven, Febr. 1997.

3. J. Daemen, R. Govaerts, and J. Vandewalle. Cryptanalysis of 2,5 rounds of IDEA. Technical Report ESAT-COSIC Report 94-1, Department of Electrical Engineering, Katholieke Universiteit Leuven, March 1994.

4. J. Daemen, R. Govaerts, and J. Vandewalle. Weak keys for IDEA. In T. Helleseth, editor, *Advances in Cryptology - Proc. Eurocrypt'93, LNCS 773*, pages 224–231. Springer Verlag, 1994.

5. M.E. Hellman and S. K. Langford. Differential linear cryptanalysis. In Y. G. Desmedt, editor, *Advances in Cryptology - Proc. Crypto'94, LNCS 839*, pages 26–39. Springer Verlag, 1994.

6. L.R. Knudsen. Truncated and higher order differentials. In B. Preneel, editor, *Fast Software Encryption - Second International Workshop, Leuven, Belgium, LNCS 1008*, pages 196–211. Springer Verlag, 1995.

7. L.R. Knudsen and W. Meier. Improved differential attack on RC5. In Neal Koblitz, editor, *Advances in Cryptology - Proc. Crypto'96, LNCS 1109*, pages 216–228. Springer Verlag, 1996.

8. L.R. Knudsen and V. Rijmen. *Truncated Differentials of IDEA*. Technical Report ESAT-COSIC Report 97-1, Department of Electrical Engineering, Katholieke Universiteit Leuven, Febr. 1997.

9. X. Lai. *On the Design and Security of Block Ciphers*. PhD thesis, ETH, Zürich, Switzerland, 1992.

10. X. Lai and J.L. Massey. A proposal for a new block encryption standard. In I.B. Damgård, editor, *Advances in Cryptology - Proc. Eurocrypt'90, LNCS 473*, pages 389–404. Springer Verlag, 1991.

11. X. Lai, J.L. Massey, and S. Murphy. Markov ciphers and differential cryptanalysis. In D.W. Davies, editor, *Advances in Cryptology - Proc. Eurocrypt'91, LNCS 547*, pages 17–38. Springer Verlag, 1992.

12. M. Matsui. Linear cryptanalysis method for DES cipher. In T. Helleseth, editor, *Advances in Cryptology - Proc. Eurocrypt'93, LNCS 765*, pages 386–397. Springer Verlag, 1993.

13. M. Matsui. The first experimental cryptanalysis of the Data Encryption Standard. In Y. G. Desmedt, editor, *Advances in Cryptology - Proc. Crypto'94, LNCS 839*, pages 1–11. Springer Verlag, 1994.

14. W. Meier. On the security of the IDEA block cipher. In T. Helleseth, editor, *Advances in Cryptology - Eurocrypt'93, LNCS 765*, pages 371–385. Springer Verlag, 1993.

15. K. Nyberg. Linear approximations of block ciphers. In A. De Santis, editor, *Advances in Cryptology - Proc. Eurocrypt'94, LNCS 950*, pages 439–444. Springer Verlag, 1994.

16. R.A. Rueppel. *Analysis and Design of Stream Ciphers*. Springer Verlag, 1986.

Combinatorial Properties of Basic Encryption Operations

Extended Abstract

Thilo Zieschang
ARCOR
Kölner Strasse 12, 65760 Eschborn, Germany
email: zieschang@acm.org

Abstract. The basic ingredients of modern fast software block encryption schemes are computer instructions like SHIFT, ADD, XOR etc. We analyze the algebraic structure of different combinations of those cryptographic primitives from a purely combinatorial point of view. Different subsets of such operations will yield an interesting variety of different permutation groups, e.g. semidirect products, affine linear groups, wreath products, and symmetric groups. As we will show, a simple pair of a SHIFT and an ADD operation is already powerful enough to generate every possible encryption function on its set of input blocks. On the other hand, any possible combination of SHIFT and XOR operations can only produce a subset of at most $n2^n$ functions within the symmetric group of order $n!$. The present results are useful in theory at first. Their cryptographic applications can be found in providing practical tools for the analysis of the algebraic structure of new block encryption schemes and evaluation of their subroutines.

1 Introduction

One of the main goals in secret key cryptography is the development of design criteria for good block ciphers. Several necessary conditions are known which have to be fulfilled by every secure cipher. To support fast software encryption, many of the recently developed block ciphers are compositions exclusively of efficient, basic computer instructions like SHIFT, XOR, and ADD, for example. Further, arithmetic operations like exponentiation or multiplication in Fermat prime moduli have been used. Examples of such encryption schemes are IDEA, SAFER, RC5 and others. A lot of trial and error is involved in the development of new block ciphers. Different subsets and combinations of the above basic operations result in completely different levels of security. Besides such properties as confusion, diffusion, avalanche, nonlinearity, resistance against several known cryptanalytic attacks, etc., it is an important criterion that a given encryption algorithm realizes a large variety of different permutations among its binary input vectors. The importance of those combinatorial properties on the security of block ciphers had been pointed out by several authors. Nevertheless, few systematic work has been done in this area so far. Thus, it is important to analyze the combinatorial structure of those basic cryptographic functions which are the underlying components of most encryption schemes. We will study different combinations of basic operations with respect to the above security measure. By analyzing their group generating properties we answer the question how many different encryption functions can be realized by given subsets of those operations, and which is their corresponding cycle structure. Another important question in terms of resistance against cryptanalytic

W. Fumy (Ed.): Advances in Cryptology - EUROCRYPT '97, LNCS 1233, pp. 14-26, 1997.
© Springer-Verlag Berlin Heidelberg 1997

attacks is whether the generated permutation group is a primitive or an imprimitive group. Suppose we have a pair (P,C) of plaintext and ciphertext, $C = E_K(P)$, and assume that the encryption algorithm E generates an imprimitive group G. Let Δ_i and Δ_j be blocks for G with $P \in \Delta_i$ and $C \in \Delta_j$. Then, according to the definition of imprimitivity, every other plaintext P' which is also contained in Δ_i will automatically be mapped into Δ_j. Since this can considerably reduce the search amount, this property would severely undermine the security of an encryption scheme. Thus it is necessary to verify whether a given block cipher generates an imprimitive or a primitive group.

It is the intention of this present work to provide useful results and proofs that can be employed in the analysis of combinatorial properties of block ciphers. Note that combinatorial results like those below are designed to simplify the determination of the specific group generated by a given algorithm. In practice, however, we have to make use of additional randomness assumptions to consider the data-dependence of our encryption algorithm appropriately.

As one can see, a rich algebraic structure is involved when combining different sets of computer instructions. Among others, the following groups of permutations do occur: symmetric groups of different degree, semilinear products, wreath products, and affine general linear groups.

2 Basic Operations

First, we have to identify a set of simple instructions that are bijective functions on the set of n-bit binary vectors. Subsequently, we will refer to such functions as *basic operations*. Consider the addition of two variables, for example. If we add always some fixed value c, we get a function ADD_c, which operates on single input variables. All basic operations can be regarded as permutations, which allows us in the present context to speak about their cycle structure, element order, and so on. Since we study CPU registers, their length n to be used below will often be a power of two.

It is an interesting question whether it is feasible to realize every possible permutation as a product of simple basic operations? As we will see - the answer is yes. This is even possible if we restrict ourselves to products consisting of only two different, fixed basic operations.

2.1 Elementary Properties of Individual Basic Operations

In this section we will introduce some of the individual basic operations which have to be studied in more detail.

XOR_c   This function operating on binary vectors of length n works as follows. The input value is XOR'ed with some arbitrary, but constant value c. If c is not equal to the all-zero string, then we get an involution. One of the transpositions, say (a,b), in the cycle decomposition of XOR_c can be constructed as desired, by choosing an appropriate c value. We can simply take c = (a XOR b), so that a maps on b. By choice of $2^n - 1$ different fixed values c, this yields $2^n - 1$ different fixed point-free involutions.

Note that XOR_c is always an even permutation, since the number of transpositions in its cycle decomposition is equal to 2^{n-1}.
Example: let $n = 4$ and suppose that $c = [1010]$. We identify the set of 4-bit strings with the corresponding decimal numbers 0, 1, ..., 15. This yields the following permutation:
XOR_c $\equiv (0, 10)(1, 11)(2, 8)(3, 9)(4, 14)(5, 15)(6, 12)(7, 13)$.

SHIFT_k   We will understand this function as a logical shift (rotation) to the right by k positions. The order of this permutation is equal to $n/\gcd(k,n)$. Further, the permutation is not fixed point-free and contains cycles of different lengths dividing n. Shifting by a different number of positions gives n-1 nontrivial permutations of this type. We can determine the fixed points of SHIFT_k explicitly. Let the variable X represent an arbitrary bit string of some fixed length. If $X = 101$, for example, then XXXX stands for a string of the form 101101101101. The function SHIFT_k has exactly $2^{\gcd(k,n)}$ fixed points of form XX..X, where X is an arbitrary string of length gcd(k,n).
It is easy to determine the cycle structure of a SHIFT_k permutation. Since SHIFT_k = $(\text{SHIFT\_1})^k$, it is sufficient to examine SHIFT_1. Let $[a_1, a_2, ..., a_n] = [X_1, X_2, ..., X_b]$ be a partition of the input vector into b subblocks of equal length m, bm = n, such that those subblocks do not allow further subdivision of the form $X_i = YY$, where Y is a subblock of length m/2. Then, in the cycle decomposition of SHIFT_1, the element $[a_1, a_2, ..., a_n]$ is contained in a cycle of length m. This implies the following numbers of cycles. The number of elements contained in n-cycles corresponds with the number of n-bit vectors $[X_1, X_2]$ with $X_1 \neq X_2$. This is equal to $2^n - 2^{n/2} = (2^{n/2}-1)2^{n/2}$. Hence, SHIFT_1 possesses $(2^{n/2} - 1)2^{n/2}/n$ cycles of length n. In general, the number of elements contained in n/m-cycles, m | n and m < n, corresponds with the number of n-bit vectors $[(X_1 X_2)^m]$ with $X_1 \neq X_2$. Hence, SHIFT_1 contains $(2^{n/m} - 2^{n/2m})m/n$ cycles of length n/m. As we can see from the above, for every given b, the number of cycles of length b is a multiple of two, which shows that SHIFT_1, and hence SHIFT_k = $(\text{SHIFT\_1})^k$, is always an even permutation.
Example: let $n = 4$. Again, we identify the set of 4-bit strings with the corresponding decimal numbers 0, 1, ..., 15. For $k = 1$, for example, we get the permutation SHIFT_1 $\equiv$ (0)(1, 2, 4, 8)(3, 6, 12, 9)(5, 10)(7, 14, 13, 11)(15).

ADD_c   This function represents addition modulo 2^n of the constant value c, ignoring overflow. ADD_1, for example, results in the 2^n-cycle (0, 1, 2, .., 2^n-1). ADD_c, $0 < c < 2^n$, is always a fixed point-free, regular permutation, which means that it consists of k cycles of length m, $m \neq 1$, such that km = n. The cycle length m is equal to $m = 2^n/\gcd(c,2^n)$. This shows that ADD_c is not always an even permutation. ADD_c is an odd permutation if and only if the resulting cycle decomposition yields a 2^n-cycle, which is the case for those c with $\gcd(c,2^n) = 1$. Hence, ADD_c is even if and only if c is even.
Example: let $n = 4$ and suppose that $c = 5$. This yields the permutation ADD_5 $\equiv$ (0, 5, 10, 15, 4, 9, 14, 3, 8, 13, 2, 7, 12, 1, 6, 11).

MUL_c   This operation means multiplication modulo 2^n (ignoring overflow) by some constant c, where c is odd and $0 < c < 2^n$. MUL_c is a bijective mapping, if and only if, gcd(c , 2^n) = 1, which yields the above condition on possible values for c. Zero and 2^{n-1} are always fixed points; odd numbers are always mapped on odd, and even numbers are mapped on even numbers.

Example: let n = 4 and c = 5, then we get the function MUL_5 with the following cycle structure: MUL_5 $\cong$ (0)(1, 5, 9, 13)(2, 10)(3, 15, 11, 7)(4)(6, 14)(8)(12).

The length of every cycle in the permutation MUL_c divides 2^{n-2}, since $(Z/2^nZ)^* \cong$ $(Z/2Z) \times (Z/2^{n-2}Z)$. MUL_c produces odd permutations as well as even permutations, depending on c; more exactly, the following holds.

Lemma: MUL_c is an even permutation even if and only if $c \equiv 1 \bmod 4$.

Proof: Every element $a \in (Z/2^nZ)^*$ has one of two possible forms: either $a = 4m - 1$ or a = 4m' + 1, with m, m'$\in$ N. In both cases, the square a^2 of a has the same form, $a^2 = 4k + 1$, where k = (4m - 2)m or k = (4m' + 2)m', respectively. Hence, if $b \in (Z/2^nZ)^*$ is a quadratic residue, then it follows that $b \equiv 1 \bmod 4$. Since $(Z/2^nZ)^*$ has exactly 2^{n-2} quadratic residues, as well as it has exactly 2^{n-2} elements b with $b \equiv 1 \bmod 4$, and since further every quadratic residue necessarily yields an even permutation, this already proves the assertion.

MULT_c   For values c such that gcd(c, 2^n+1) = 1, MULT_c can be understood as the (bijective) function which multiplies the input variable by the constant c, modulo 2^n+1. To see how this works, we give a short example. While computing with our binary strings, we will not identify them with the set of numbers {0, 1, ..., 2^n-1}. Instead we will take the set {1, 2, ..., 2^n}, thus replacing zero by 2^n. If 2^n+1 is prime, then MULT_c is a fixed point-free, regular permutation. This can be seen as follows. If MULT_c were not regular, then an appropriate power of MULT_c would have some fixed points. This, however, is impossible, since nonzero multiplication in a prime modulus is invertible, which implies that ca $\neq$ a for every c $\neq$ 1 and a $\neq$ 0. The operation MULT_c can efficiently be executed by a method called „low-high multiplication", as described in Lai [Lai], page 35.

Lemma: The basic operation MULT_c, considered as a permutation on the set of n-bit binary vectors, is even if and only if c is a quadratic residue modulo 2^n+1.

Proof: The set {MULT_c | 1 $\leq$ c $\leq$ 2^n} forms a group G of order 2^n containing both, odd and even permutations. This follows, since G always contains an odd cycle MULT_d of length 2^n, where d is a generator of the multiplicative group of the prime field $Z/(2^n+1)Z$. Each permutation group containing odd elements has an equal number of odd and even elements.

If c is a quadratic residue modulo $2^n + 1$, then MULT_c must be an even permutation. But $(Z/(2^n + 1)Z)^*$ has exactly 2^{n-1} quadratic residues, hence each quadratic nonresidue e $\in (Z/(2^n + 1)Z)^*$ must yield an odd permutation MULT_e. This proves the assertion.

Example: let n = 2, then 2^n+1 = 5 is prime; as a result, the following permutations do occur: MULT_2 $\cong$ (1, 2, 4, 3), MULT_3 $\cong$ (1, 3, 4, 2), MULT_4 $\cong$ (1, 4)(2, 3).

2.2 Closure and Generating Properties of Individual Functions

We have the following generator properties of individual basic operations. Here, <M> denotes the group generated by the set M.

<{XOR_c | c $\in$ {0, 1}^{n}}> = <{XOR_c$_j$ | 1 $\leq$ j $\leq$ n} >, where c$_j$:= [b$_1$b$_2$...b$_n$], with b$_i$ = δ_{ij}. The resulting group is an elementary abelian 2-group of order 2^n, which is isomorphic to (Z$_2$)^n. Every minimal generating set for this group comprises of n elements.

<{SHIFT_k | 0 $\leq$ k < n}> = <SHIFT_1>. The generated group is isomorphic to the cyclic group of order n.

<{ADD_c | 0 $\leq$ c < 2^n}> = <ADD_1>. The generated group is isomorphic to the cyclic group of order 2^n.

<{MULT_c | 0 $\leq$ c < 2^n}>, where p := 2^n + 1 is prime. Then the multiplicative group (Z$_p$)$^{'}$ is cyclic and we find a generating element g such that <{MULT_c | 0 $\leq$ c < 2^n}> = <MULT_g>. Hence, this group is isomorphic to a cyclic group of order 2^n. If 2^n + 1 is not prime, then the structure of the multiplicative group of Z/(2^n + 1)Z varies in accordance with n. In practice, however, we use the operation MULT only in Fermat prime moduli, which is the case for n = 8 and n = 16, for example.

<{MUL_c | c odd, 0 $\leq$ c < 2^n}> = <MUL_a, MUL_-1>, where a is an element of order 2^{n-2}. The multiplicative group of Z/(2^{n}Z) is an abelian group of order 2^{n-1}. This group is isomorphic to Z/(2^{n-2}Z) × Z/(2Z).

3 Combination of Basic Operations

In the above sections we have studied several types of basic operations that can be used in the construction of fast software encryption schemes. Now the question arises which specific subsets of those basic operations should be combined together to achieve a large number of encryption functions, i.e. permutations of binary n-bit vectors, with presumably high structural complexity? Therefore, we have to analyze the group generating properties of different combinations of those basic operations. In the present paper, due to its space limitations, we restricted ourselves to a collection of some interesting pairs of generators, resulting in different permutation groups.

As we will see, there are surprisingly big differences among slightly different mixtures. For example, consider two different cryptosystems operating on binary input vectors of length n, the first consisting of a combination of SHIFT and XOR operations, while the second system is using a combination of SHIFT and ADD operations. Then the first scheme will only be able to produce at most n2^n different, simple-structured „encryptions", while the second scheme has the power to generate the whole amount of 2^n! possible encryption functions.

If we further mention the result that, taken two arbitrary permutations from the symmetric group of degree n, it is most probable that those two elements already generate the whole alternating or symmetric group (probability $\rightarrow$ 1 for n $\rightarrow \infty$), then it is even more surprising that the groups generated by different pairs of basic operations are much smaller in most cases.

3.1 {XOR, SHIFT} : Semidirect Products

Theorem: The group G generated by the set S:= {SHIFT_d, XOR_c | $0 \le d < n$, $c \in \{0,1\}^n$} of all SHIFT and XOR operations is isomorphic to a semidirect product G = NU of an elementary abelian 2-group N of order 2^n by a cyclic group U of order n. The order of G is equal to $n2^n$ and G can be generated by n + 1 elements. If $n = 2^k$ is a power of 2, then G is a nonabelian 2-group of order 2^{k+n}. The maximal required word length with generating set S is equal to 2.

Proof: Suppose we have a finite sequence $a_1 a_2 ... a_m$ of XOR and SHIFT operations. Then it is always possible to replace this sequence by a single XOR operation, followed by a single SHIFT operation (or, alternatively, first SHIFT and afterwards XOR). The reason for this is the following obvious exchange property:
XOR_e $\cdot$ SHIFT_d = SHIFT_d $\cdot$ XOR_c, where c := SHIFT_-d(e).
Hence, we only have to enumerate all $n2^n$ possible words SHIFT_d $\cdot$ XOR_c to get all elements of the group generated by the set {SHIFT_d, XOR_c | $0 \le d < n$, $c \in \{0,1\}^n$}. The cardinality of this group is equal to $n2^n$: Suppose that SHIFT_d_1 $\cdot$ XOR_c_1 = SHIFT_d_2 $\cdot$ XOR_c_2. Then left multiplication of both sides by SHIFT_-d_1 yields: XOR_c_1 = SHIFT_$(d_2 - d_1)$ $\cdot$ XOR_c_2. If we apply both sides to the register A := c_2, then we get $(c_1$ XOR $c_2) = 0$. It follows that $c_1 = c_2$, which further implies that $d_1 = d_2$. We have shown that the group generated by XOR and SHIFT operations is of order $n2^n$.
Next we have to determine the structure of this group. The set N := {XOR_c | $c \in \{0,1\}^n$} of XOR operations is a subgroup of the alternating group of degree 2^n and is isomorphic to an elementary abelian 2-group, N $\cong (Z_2)^n$. The set U := {SHIFT_d | $0 \le d < n$} is isomorphic to a cyclic subgroup of order n. Further, let XOR_c $\in$ N and SHIFT_d $\in$ U be two arbitrary elements of the above sets, then
SHIFT_d $\cdot$ XOR_c $\cdot$ (SHIFT_d)$^{-1}$ = SHIFT_d $\cdot$ XOR_c $\cdot$ SHIFT_-d = XOR_e $\in$ N, with e := SHIFT_-d(c). Hence, N is a normal subgroup of the product group UN = NU. Since N $\cap$ U = {id}, this shows that the set {SHIFT_d, XOR_c | $0 \le d < n$, $c \in \{0,1\}^n$} = NU is a semidirect product of N by U whose order is $n2^n$.
If the register length $n = 2^k$ is a power of two, the order of G is also a power of two and G is a nonabelian 2-group of order 2^{k+n}. G can be generated by n+1 elements: one SHIFT and n XOR operations, as shown in the previous section. If we take the generating set {SHIFT_d, XOR_c | $0 \le d < n$, $c \in \{0,1\}^n$} for G, then the diameter of the corresponding Cayley graph is equal to 2.

3.2 {ADD, SHIFT} : Symmetric Groups of Degree 2^n

Theorem: The group G of degree 2^n, $n \in N$, generated by the set {ADD_c, SHIFT_d | 0 $\leq$ c < 2^n, 0 $\leq$ d < n} of all ADD and SHIFT operations is isomorphic to the symmetric group of degree 2^n.

Proof: Our strategy is to construct a specific group element g $\in$ G whose existence within primitive groups of sufficiently large degree forces the group to be isomorphic to the alternating or symmetric group of corresponding degree. Therefore, our first task is to prove that the generated group G is not imprimitive. Suppose that G is imprimitive and $\Delta \subset \Omega := \{0, 1, 2, ..., 2^{n-1}\}$ is a nontrivial block for G (i.e. $1 < |\Delta| < 2^n$) containing zero, $0 \in \Delta$. If Δ contains an odd element a, $0 < a < 2^n$, then a = ADD_a(0) and therefore Δ = ADD_a(Δ). Further, this implies that Δ = (ADD_a)^i(Δ) for every $0 \leq i < n$. But for odd a we have <{(ADD_a)^i | $0 \leq i < n$}> = <ADD_1>. Hence, it follows that Ω = <ADD_1>(0) $\subseteq \Delta$, which shows that in this case $\Delta = \Omega$, a contradiction. Hence, Δ contains only even numbers. Let k $\in \Delta$ be such an even number, k $\neq$ 0. Then the binary n-bit register representing the integer k contains at least one bit equal to 1. By an appropriate power SHIFT_c of SHIFT_1, this „1" can be shifted by c positions to the least significant bit of our register, the result of this operation thus representing an odd number. Since 0 is a fixed point of SHIFT_c, we see that SHIFT_c(Δ) = Δ. But then we also have SHIFT_c(k) $\in \Delta$, which is a contradiction, since, by construction, SHIFT_c(k) is odd. Hence, the group G must be primitive.

The second step of our proof is the construction of an appropriate group element g, from which we can deduce that G contains the alternating group of degree 2^n. Here, we will show that the primitive group G of degree 2^n always contains a 2^{n-1}-cycle g. Let s be a cycle, s = (s_1, s_2, s_3, ..., s_r), and let t be an arbitrary permutation. Then one can verify that it always holds the equation tst^{-1} = (t(s_1), t(s_2), t(s_3), ..., t(s_r)); the conjugate of s has the same cycle structure than s itself. It is obvious that the same then holds for arbitrary permutations s, s consisting of two or more nontrivial cycles.
Now consider the 2^n-cycle ADD_1 = (0, 1, 2, 3, ..., 2^n - 1) $\in$ G. By the above observation we see that a := SHIFT_1 $\cdot$ ADD_1 $\cdot$ SHIFT_-1 =
= (SHIFT_1(0), SHIFT_1(1), SHIFT_1(2), SHIFT_1(3), ..., SHIFT_1(2^n - 1)).
For $0 \leq m \leq 2^{n-1}$-1 we have SHIFT_1(m) = 2m, since the rightmost bit in the binary representation of m is zero. Hence, in this case the logical shift SHIFT_1 is equivalent to an arithmetic shift. For $2^{n-1} \leq m \leq 2^n$ -1 we can write m = 2^{n-1} + m', $0 \leq$ m' $\leq 2^{n-1}$-1. This allows us to see that for such m we get SHIFT_1(m) = SHIFT_1(2^{n-1}) + SHIFT_1(m') = 1 + 2m'. Therefore, the above permutation is equal to a = (0, 2, 4, 6,...,2^n - 2, 1, 3, 5,..., 2^n - 1). From this, by setting e := 2^{n-1}, we get a^e = (0, 1)(2, 3)(4, 5)...(2^n - 2, 2^n - 1). Finally, we compute g := ADD_-1 $\cdot$ a^e = (2^n - 1, 2^n - 3, 2^n - 5, ..., 3, 1). This element g fixes every even number and, as desired, is a cycle of length 2^{n-1}. Since every primitive permutation group of degree 2^n containing a cycle of length 2^{n-1} is (at least) (2^{n-1}+1)-fold transitive ([Wiel], Theorem 13.8), we see that for register length n, n > 2, the group G must be 9-fold transitive. But (see the classification of finite simple groups and [Came], for example) we know that in this case G must be isomorphic to either the alternating or symmetric group of corresponding degree. Since G contains odd

elements (take the 2^n-cycle ADD_1, for example), G is isomorphic to the symmetric group of degree 2^n. Also, for n = 2, we get G = <ADD_1, SHIFT_1> = <(0,1,2,3), (1,2)> $\cong S_4$.

3.3 {MUL, SHIFT} : Symmetric Groups of Degree 2^n - 1

Since 0 is a fixed point for MUL as well as for SHIFT operations, it is reasonable to consider zero as an exceptional value here. Therefore, we will subsequently consider the operation of G = <{MUL_c, SHIFT_d | 0 < c < 2^n, c odd, 0 ≤ d < n}> on the set of register variables $\Omega' = \Omega - \{0\} = \{1, 2, 3, ..., 2^n - 1\}$.

Theorem: The group G = <{MUL_c, SHIFT_d | 0 < c < 2^n, c odd, 0 ≤ d < n}> of degree 2^n- 1, n $\in$ N, generated by all MUL and SHIFT operations acting on the set $\Omega' = \Omega - \{0\}$ of order 2^n - 1, is isomorphic to the symmetric group of degree 2^n - 1.

Proof: Our strategy is as follows. First we show that G is transitive, primitive, and then, by a quite technical procedure, we construct an element in G which is a 3-cycle.
Let x, y $\in \Omega'$ be two arbitrary points. First we show that there exists a permutation $\pi \in$ G such that $\pi(x) = y$.
Case 1: x and y are odd. By MUL_y, since y is odd, we can map 1 onto y. In the same way we can map x onto 1 by taking the inverse of MUL_x, i.e. multiplication by x^{-1}. Combining both operations yields MUL_x^{-1}y(x) = y, as desired.
Case 2: x even, y odd. Since x $\neq$ 0, we can shift x by an appropriate number k of positions, such that the least significant bit of x' := SHIFT_k(x) is equal to „1", thus x' is odd. Now we can apply case (1) above on x' and map x' onto y. This shows that a combination of one SHIFT followed by one MUL operation can map x onto y.
Case 3: x odd, y even. By simply taking inverse permutations, this is the same as in (2).
Case 4: x and y are even. This is an immediate consequence of cases (2) and (3).
Hence, the group G acts transitive on Ω'. However, we note that, in the general case, we need a sequence of not only two but three basic operations to map two arbitrary even numbers onto each other. The resulting operation has the form SHIFT · MUL · SHIFT.
Now we have to show that the above transitive operation is primitive. Let Δ be a block for G which contains -1. If we assume $|\Delta| > 1$, then Δ contains an element c having zeros and ones in its binary representation. Since SHIFT_1(Δ) = Δ, this shows that Δ contains both, even as well as odd elements: simply shift c by an appropriate number of positions.
Consider the permutation MUL_$(2^{n-1} + 1) \in$ G. MUL_$(2^{n-1} + 1)$ is an involution that fixes every even element and consists of transpositions of the form (x, x + 2^{n-1} mod 2^n). This can be seen as follows. Let x = 2y + 1 be odd. Then MUL_$(2^{n-1} + 1)(x) = (2^{n-1} + 1)(2y + 1) = 2^ny + 2^{n-1} + 2y + 1 = x + 2^{n-1}$ (mod 2^n). Further, $(2^{n-1} + 1)^2 = 1$ (mod 2^n), which shows that the order of MUL_$(2^{n-1} + 1)$ is equal to two. As we have noted, Δ contains even elements, and therefore MUL_$(2^{n-1} + 1)(\Delta) = \Delta$. It follows that, for every odd element $[b_{n-1}, b_{n-2}, ..., b_1, 1]$ in Δ, we also have $[1 - b_{n-1}, b_{n-2}, ..., b_1, 1] \in \Delta$.
Suppose we have a number in Δ whose binary representation has the form $[..., b_{i+2}, 1, b_i, ...]$, then we can shift the „1" to the rightmost position, replace b_i by 1 - b_i, shift back

and finally get the element $[..., b_{i+2}, 1, 1 - b_i, ...] \in \Delta$. Thus, we can construct every binary n-bit vector which is different from the all-zero vector by starting from $-1 = [1, 1, ..., 1]$ and then replacing 1's and 0's accordingly. This is always possible if we start constructing each substring of adjacent zero positions by beginning replacement with the rightmost position within this substring. As we have shown, the block Δ must contain the whole set $\Omega' = \{1, 2, 3, ..., 2^{n-1}\}$. Hence, the group G acts primitive on Ω'.

We proceed by shortly explaining the action of some specific permutations in G, which we will combine subsequently to yield the appropriate 3-cycle, forcing our group to be either alternating or symmetric.

(i) Structure of SHIFT_-1:

The general structure is as follows: SHIFT_-1 maps

$x \rightarrow x/2$, x even; $x \rightarrow (x - 1)/2 + 2^{n-1}$, x odd.

Example n = 4: SHIFT_-1 $\equiv$ (0)(1,8,4,2)(3,9,12,6)(5,10)(7,11,13,14)(15)

(ii) Structure of MUL_$(2^{n-1} + 1)$:

The general structure is as follows: MUL_$(2^{n-1} + 1)$ maps

$x \rightarrow x$, x even; $x \rightarrow x + 2^{n-1} \mod 2^n$, x odd;

Example n = 4: (0)(2)(4)(6)(8)(10)(12)(14)(1,9)(3,11)(5,13)(7,15)

(iii) Structure of A := (MUL_$(2^{n-1} + 1))^{SHIFT\_-1}$:

Let s be a cycle, $s = (s_1, s_2, s_3, ..., s_k)$, and let t be an arbitrary permutation. Then $tst^{-1} = (t(s_1), t(s_2), t(s_3), ..., t(s_k))$. Hence we can see that, by the above analysis of the structure of MUL_$(2^{n-1} + 1)$ and SHIFT_-1, we get the following cycle structure of the permutation A:= (MUL_$(2^{n-1} + 1))^{SHIFT\_-1}$:

The fixed points of A comprise of the set $\{x/2, 0 \leq x < 2^n$, x even$\} = \{0, 1, 2, 3,..., 2^{n-1} - 1\}$. The transpositions in A have the form $(x, x + 2^{n-1})^{SHIFT\_-1}$, where $1 \leq x < 2^{n-1}$, x odd, which yields transpositions of the following form:

$((x - 1)/2 + 2^{n-1}, (x + 2^{n-1} - 1)/2 + 2^{n-1}) = ((x - 1)/2 + 2^{n-1}, (x - 1)/2 + 2^{n-1} + 2^{n-2})$, where $1 \leq x < 2^{n-1}$, x odd. As we can see, the permutation A does not move any element of size smaller than 2^{n-1}. Example n = 4: A = (8,12)(9,13)(10,14)(11,15).

(iv) Structure of B := A^{MUL_-1}:

First, notice that MUL_-1 consists of two fixed points, namely 0 and 2^{n-1}, together with transpositions of the form $(x, 2^n - x)$, for $x = 1, 2, ..., 2^{n-1} - 1$.

Thus, under conjugation by MUL_-1, the fixed points of A move into the following set of fixed points of B = A^{MUL_-1}: $\{0\} \cup \{2^n - x, 1 \leq x \leq 2^{n-1} - 1\} = \{0, 2^{n-1} + 1, 2^{n-1} + 2, 2^{n-1} + 3, ..., 2^n - 1\}$. Hence, the permutation B does not move any element of size larger than 2^{n-1}. Example n = 4: B = (1,5)(2,6)(3,7)(4,8).

We combine the above observations and consider the product of A and B. Both permutations, A and B, are involutions. The only element which is going to be moved by A as well as by B is the element 2^{n-1}. Therefore, the permutation AB $\in$ G consists of the union of those transpositions of A and B which do not contain the element 2^{n-1}, together with exactly one further cycle of length 3. This 3-cycle $(x, y, 2^{n-1})$ stems from the product of a transposition $(x, 2^{n-1})$ in A with a transposition $(y, 2^{n-1})$ in B. The square C := $(AB)^2 \in$ G is a 3-cycle. Since the group G is primitive, this proves that G contains the alternating group of degree $2^n - 1$. Further, the group G contains odd permutations (take SHIFT_1, for example). Therefore, G is isomorphic to the symmetric group of degree $2^n - 1$.

3.4 Groups of Larger Subsets of Basic Operations

In the above sections we studied the properties of each individual basic operation, as well as the combinatorial properties of different pairs of possible combinations of those basic operations. With this knowledge about specific groups, each generated by two types of generating functions, it is relatively easy to determine the algebraic structure of larger generating subsets. If, for example, we have a pair $\{A, B\}$ of basic operations, such that $\{A, B\}$ already generates the whole symmetric group of degree 2^n, then obviously every larger set $\{A, B, C, ...\}$ of basic operations forms also a generating set for this group. This applies, for example, to the systems $\{XOR, SHIFT, ADD\}$, $\{MUL, SHIFT, ADD\}$, $\{XOR, SHIFT, MUL\}$. As a funny property in this context, we can observe the following. Every addition ADD_c, $0 \leq c < 2^n$, $n \in N$, can be executed as a sequence of multiplications modulo 2^n and multiplications modulo $2^n + 1$, i.e. $ADD \subset$ $<MUL, MULT>$. By this property we get additional results on subsets of basic operations which can generate the complete symmetric group of degree 2^n. This holds, for example, for the generating set $\{MULT, SHIFT, MUL\}$.

4 MIX-2 - A Simple System Generating the Symmetric Group

4.1 Description of MIX-2

Based upon our analysis of basic encryption operations in the previous section, we will present a simple three-line algorithm named *MIX-2* which is able to produce every possible permutation on its set of input vectors, i.e. MIX-2 can generate the symmetric group. MIX-2 uses only two different operations, namely SHIFT_a and ADD_b. Here, a and b are odd. The register length n depends on the computer; for example $n = 64$. We will deduce that the round function of MIX-2, as well as the full-round MIX-2 scheme do generate the whole symmetric group of degree 2^n.

Let r denote the number of rounds, M the input vector of length n, and let $(b_1, b_2, ..., b_r)$ be the first r bits of the key K. Then the complete MIX-2 algorithm can be described as follows.

```
FOR (i = 0; i < r; ++i)
      IF b_i   ADD_b(M);
      ELSE  SHIFT_a(M);
```

As our key K we take a binary k-bit vector $K = (b_1, b_2, ..., b_k)$, k being (preferrably) a power of two, for example $k = 256$. The number r of rounds is equal to $k - (n + \log_2 (n) - 2)$. If $n = 64$, for example, then we get $r = 256 - 64 - 6 + 2 = 188$. Here, the bit sequence $(b_{r+1}, b_{r+2}, ..., b_{r+n-1}, 1)$ provides the binary representation of the odd number b. Similarly, the sequence $(b_{r+n}, b_{r+n+1}, ..., b_k, 1)$ represents the odd number a, $0 \leq a < 2^n$. Now we can show the generator property of MIX-2.

Theorem: The full-round MIX-2, as well as its set of round functions, can generate the symmetric group of degree n, where n is the length of the input vectors.

Proof: As we have shown, the set of all SHIFT and ADD operations {SHIFT_x, ADD_y | $0 \leq x < n, 0 \leq y < 2^n$} generates the symmetric group of degree 2^n. But for every odd a and odd b, we have <SHIFT_a> = <{SHIFT_x| $0 \leq x < n$} and <ADD_b> = <{ ADD_y | $0 \leq y < 2^n$}>. Hence, the two basic operations SHIFT_a and ADD_b, and therefore the round functions of MIX-2, do generate the symmetric group.

Now we have to study the resulting permutations of MIX-2 after r rounds. Take key K_1 := $(1_1, 1_2, ..., 1_r, ...)$, then MIX-2 using this key corresponds with a simple ADD_rb operation. By the above definition it follows that r is odd, hence rb is odd and <ADD_rb> = <{ ADD_y | $0 \leq y < 2^n$}>. Similiarly, by taking key K_0 = $(0_1, 0_2, ..., 0_r, ...)$, we see that SHIFT_ra generates all SHIFT operations. Thus we have shown that MIX-2 generates the symmetric group.

4.2 Poor Combinatorial Behaviour Through Different Functions in MIX-2

In the above definition of MIX-2, we could have chosen other pairs of basic operations as well. Suppose we replace the addition modulo 2^n, ADD_b, by the similiar operation XOR_b (in this case, the constant b does not have to be odd). The difference between the XOR_b and the ADD_b operation consists, simply speaking, only in the consideration of carry bits. Some cryptographic properties of those carry bits have already been studied in another, non-combinatorial setting, by W. Meier and O. Staffelbach [Staf/Mei]. It is the propagation of those carry bits that saves the above algorithm from complete nonsense. The result of the replacement of ADD by XOR is disastrous. Independent from the actual number of rounds r, the resulting function

```
FOR (i = 0; i < r; ++i)
      IF b_i   XOR_b(M);
      ELSE  SHIFT_a(M);
```

shrinks and is equivalent to a „2-round version"
```
XOR_b'(M);
SHIFT_a'(M);
```

of the previous algorithm, where the corresponding key K = $(b_1, b_2, ...)$ can be chosen as K = (1, 0, ...), for example.

Exactly the same problem occurs if, alternatively, we keep the ADD operator, but replace the SHIFT_a operator by MUL_a, i.e. multiplication modulo 2^n. (Here we have to adjust key lengths appropriately, since the odd constant a now satisfies $0 < a < 2^n$.) Again, the complete algorithm

```
FOR (i = 0; i < r; ++i)
      IF b_i   ADD_b(M);
      ELSE  MUL_a(M);
```

is reducible to
ADD_b'(M);
MUL_a'(M);

since the affine general linear group $AGL(1, 2^n)$ generated by those operations is a semidirect product $AGL(1, 2^n) = GH$ of a group G by a group H, where G is generated by the set $\{ADD\_x \mid 0 \le x < 2^n, x \text{ odd}\}$ and H is generated by the set $\{MUL\_y \mid 0 \le y < 2^n, y \text{ odd}\}$. The shrinking property follows, similiarly as in the previous case of <XOR, SHIFT>, from the fact that $GH = HG$.

Another weakness occurs if we consider the pairs {XOR, ADD} or {XOR, MUL} as ingredients to the above MIX-2 system. Both resulting permutation groups are imprimitive in the sense that the set of odd input vectors and the set of even input vectors (i.e. least significant bit equal to 1 or 0, respectively) are permuted among themselves. Suppose we know one plaintext-ciphertext pair. Then, for example, if both text strings are even, we can deduce that every even (odd) plaintext will be mapped into an even (odd) ciphertext. For a cryptosystem, of course, this property would be intolerable. Note also that the multiplication operator MUL_a(.) always preserves the least significant bit of its operand. This fact, together with its relatively slow running time, is an additional disadvantage of the MUL operation.

Note that there exist examples in the literature, one of which had been designed by S. Murphy, K. Paterson, and P. Wild [Mur/Pate/Wil], to demonstrate the existence of a weak cipher that nevertheless can generate the symmetric group. In their block encryption scheme, however, one of the bijections in the round function consists of a trivial permutation θ whose cycle decomposition is $\theta = (0)(2^{n-1})(1, 2, ..., 2^{n-1} - 1, 2^{n-1} + 1, ..., 2^n -3, 2^n - 2, 2^{n-1})$. Thus, in all but a few cases, θ just increments its input by one, which is, of course, no very helpful encryption step. As the authors show, their cryptosystem, whose second encryption component is an XOR with the secret round key, can easily be broken by a known plaintext attack. We remark that MIX-2 is just a good example demonstrating a very simple algorithm having the potential to generate the whole symmetric group. We did not call it a cryptosystem and do not assume that MIX-2, without further ingredients and modifications, is of any use as a strong block cipher. Nevertheless, it is not *that* easy to break MIX-2, in comparison with its quite trivial structure. The above provides another reason to recommend the implementation of subroutines having strong combinatorial properties.

5 Conclusions

As we have shown, care has to be taken in the right combination of basic computer instructions to strengthen the combinatorial properties of an encryption algorithm. Since most schemes involve data-dependence of their basic encryption operations, some randomness assumptions have to be accepted in practice and further research in this direction should be undertaken to facilitate the determination of the group generated by a given block encryption scheme. We analyzed the algebraic structure of different combinations of some basic encryption operations from a purely combinatorial point of view. Different subsets of such operations yield an interesting variety of different permutation groups, e.g. semidirect products, affine linear groups, wreath products, and symmetric groups. Even though the above results are useful in theory at first, their cryptographic applications can be found in providing practical tools for the analysis of the algebraic structure of block encryption schemes and combinatorial evaluation of their subroutines.

6 Bibliography

[Came] P. J. Cameron, *Finite permutation groups and finite simple groups*, Bull. London Math. Soc. 13 (1981), pages 1-22.

[CampK/Wien] K. Campbell, and M. Wiener, *DES is not a group*, Proc. of Crypto'92, pages 512-520.

[Cop/Gro] D. Coppersmith, and E. Grossman, *Generators for certain alternating groups with applications to cryptography*, SIAM Journal Appl. Math., 29(4), pages 624-627, 1975.

[Eve/Gol2] S. Even, and O. Goldreich, *DES-like functions can generate the alternating group*, IEEE Transaction on Inf. Theory, IT-29(6), pages 863-865, 1983.

[Hup] B. Huppert, *Endliche Gruppen 1*, 2. Nachdruck der 1.Auflage, Springer-Verlag Berlin, Heidelberg, New York 1967.

[Hup/Bla] B. Huppert, and N. Blackburn, *Finite groups 3*, Springer-Verlag Berlin, Heidelberg, New York 1982.

[Isa/Zie] I. M. Isaacs, and T. E. Zieschang, *Generating symmetric groups*, American Math. Mon., Oct. 1995, pages 734-739.

[Mur/Pate/Wil] S. Murphy, K. Paterson, and P. Wild, *A weak cipher that generates the symmetric group*, J.Cryptology (1994) 7, pages 61 - 65.

[Pie/Zha] J. Pieprzyk, and X. Zhang, *Permutation generators of alternating groups*, Auscrypt'90, pages 237-244.

[Staf/Mei] O. Staffelbach, and W. Meier, *Cryptographic Significance of the Carry for Ciphers Based on Integer Addition*, Crypto'90, pages 601 - 614.

[Wer] R. Wernsdorf, *The 1-round functions of DES generate the alternating group*, Proc. of Eurocrypt'92, pages 99-112

[Wiel] H. Wielandt, *Finite permutation groups*, Academic 1964.

A New Public-Key Cryptosystem

David Naccache

Gemplus Card International

1 place de la Méditerranée

Sarcelles CEDEX, F-95206, France

100142.3240@compuserve.com

Jacques Stern

Ecole Normale Supérieure

45 rue d'Ulm

Paris CEDEX 5, F-75230, France

jacques.stern@ens.fr

Abstract. This paper describes a new public-key cryptosystem where the ciphertext is obtained by multiplying the public-keys indexed by the message bits and the cleartext is recovered by factoring the ciphertext raised to a secret power. Encryption requires four multiplications / byte and decryption is roughly equivalent to the generation of an RSA signature.

1 Introduction

It is striking to observe that two decades after the discovery of public-key cryptography, the cryptographer's toolbox contains only a dozen of asymmetric encryption schemes. This rarity and the fact that today's most popular schemes had so far defied all complexity classification attempts strongly motivates the design of new asymmetric cryptosystems.

Interestingly, the cryptographic community has been relatively more successful in the related field of identification, where a user attempts to convince another entity of his identity by means of an on-line communication. For example, there have been several attempts to build identification protocols based on simple operations (see [19, 21, 22, 16]). Although the devising of new public key cryptosystems appears much more difficult (since it deals with trapdoor functions rather than simple one-way functions) we feel that research in this direction is still in order : simple yet efficient constructions may have been overlooked and, in a way, the present paper is an example of such a situation.

As observed by [18], most asymmetric encryption schemes present the following common design morphology :

• Start with an intractable problem P and find an easy instance $P[\text{easy}] \in P$ which should be solvable in polynomial space and time.

• Shuffle or scramble $P[\text{easy}]$ until the resulting problem $P[\text{shuffle}]$ does not resemble $P[\text{easy}]$ any more and becomes indistinguishable from P.

• Publish $P[\text{shuffle}]$ and describe how it should be used for encryption. The information s by the means of which $P[\text{shuffle}]$ is reduced to $P[\text{easy}]$ is kept as a secret trapdoor.

• Construct the cryptosystem in such a way that decryption is essentially different for the cryptanalyst and the legitimate receiver. Whilst the former must solve $P[\text{shuffle}]$, the latter may use s and solve only $P[\text{easy}]$.

W. Fumy (Ed.): Advances in Cryptology - EUROCRYPT '97, LNCS 1233, pp. 27-36, 1997.

Roughly at the same time when RSA was discovered [17], knapsack encryption was introduced by Merkle and Hellman [11]. It used the knapsack problem where $P[\text{easy}]$ was superincreasing and shuffling was a linear operation modulo some large integer. As is well known, the knapsack cryptosystem was broken by Shamir. A variant of the knapsack system was proposed by Chor and Rivest [4] where shuffling was more elaborate since it was based on computing discrete logarithms in finite fields. Later on, building on Chor and Rivest's work, Lenstra [10] introduced the powerline system which, instead of computing logarithms, used directly the multiplicative structure of the field. For the sake of accurate paternity respect, let us stress that the construction presented in this paper uses a multiplicative version of the basic (additive) knapsack problem by combining two old, and once well-known, techniques : the multiplicative Merkle-Hellman knapsack [11] and Pohlig-Hellman's secret-key cryptosystem [15]. The new scheme therefore relates to Merkle-Hellman's cryptosystem very much the same way as the powerline system is related to the Chor-Rivest scheme. Actually, we were not aware of [10] and it is through a note by Paul Camion [3] that we understood that we had found a missing species.

The scheme presented in this article is based on the following problem :

P : given p, c and a set $\{v_i\}$, find a binary vector x such that

$$c = \prod_{i=0}^{n} v_i^{x_i} \bmod p$$

It is easy to observe that if the v_i-s are relatively prime and much smaller than p, P can be solved in polynomial time by factoring c :

$P[\text{easy}]$ is an instance of P where $p > \prod_{i=0}^{n} v_i$ and $\gcd(v_i, v_j) = 1$ for $i \neq j$.

The scrambled $P[\text{shuffle}]$ is obtained by extracting a secret (s-th) modular root of each v_i in $P[\text{easy}]$. By raising a product of such roots to the s-th power, each v_i shrinks back to its original size and x can be found by factoring.

The following sections describe how to use P for public-key encryption.

2 The new scheme

Let p be a large public prime and denote by n the largest integer such that :

$$p > \prod_{i=0}^{n} p_i \text{ where } p_i \text{ is the } i\text{-th prime (start from } p_0 = 2)$$

The secret-key $s < p - 1$ is a random integer such that $\gcd(p-1, s) = 1$ and the public-keys are the $n+1$ roots generated à la Pohlig-Hellman [15] :

$$v_i = \sqrt[s]{p_i} \bmod p$$

$$m = \sum_{i=0}^{n} 2^i m_i \in \mathcal{M} \text{ is encrypted as } c = \prod_{i=0}^{n} v_i^{m_i} \bmod p \text{ and recovered by :}$$

$$m = \sum_{i=0}^{n} \frac{2^i}{p_i - 1} \times \left(\gcd(p_i, c^s \bmod p) - 1 \right)$$

Naturally, as in all knapsack-type systems, the v_is can be permuted and re-indexed for increased security.

2.1 a small example

key generation for $n = 7$ The prime $p = 9700247 > 2 \times 3 \times 5 \times 7 \times 11 \times 13 \times 17 \times 19$ and the secret $s = 5642069$ yield the v-list :

$$v_0 = \sqrt[s]{2} \bmod p = 8567078 \qquad v_4 = \sqrt[s]{11} \bmod p = 8643477$$
$$v_1 = \sqrt[s]{3} \bmod p = 5509479 \qquad v_5 = \sqrt[s]{13} \bmod p = 6404090$$
$$v_2 = \sqrt[s]{5} \bmod p = 2006538 \qquad v_6 = \sqrt[s]{17} \bmod p = 1424105$$
$$v_3 = \sqrt[s]{7} \bmod p = 4340987 \qquad v_7 = \sqrt[s]{19} \bmod p = 7671241$$

encryption of $m = 202 = 11001010_2$

$$c = v_7^1 \times v_6^1 \times v_5^0 \times v_4^0 \times v_3^1 \times v_2^0 \times v_1^1 \times v_0^0 \bmod p = 7202882$$

decryption by exponentiation, we retrieve :

$$c^s \bmod p = 7202882^{5642069} \bmod 9700247 = 6783$$

whereby :

$$6783 = 19^1 \times 17^1 \times 13^0 \times 11^0 \times 7^1 \times 5^0 \times 3^1 \times 2^0 \quad \rightarrow \quad m = 11001010_2$$

information rate The information rate of our scheme (number of cleartext bits packed into each ciphertext bit) is sub-optimal since, in this example :

$$\mathcal{I} = \log{(m)} / \log{(c)} = \frac{8}{24} \simeq 33.33\% < 1$$

2.2 p as a function of n

Evaluating the growth of p and n is important for comparing and understanding the characteristics on the new scheme since message-space mainly depends on n while computational complexity is proportional to the square of p's size.

Lemma 1 *Asymptotically :*

$$p \, e^{\mathrm{li}(n)} \sim n! \log^n(n) \quad where \quad \mathrm{li}(n) = \int_2^n \frac{dx}{\log(x)} \sim \frac{n}{\log(n)}$$

whereas interpolation for $128 \le n \le 418$ and $989 < \log p < 4096$ yields :

$$|1000 \log p + 144525 - n\,(8062.11 + 6.74\,n) + 4.26337\,(n/10)^3\,| < 1012$$

The following table summarises the relation between p and n for five frequent sizes of p :

size of p	n	p_n	$\mathcal{M}$	size of the v-list	$\mathcal{I}$
512 bits	74	379	75 bits	4,800 bytes	14.6 %
640 bits	88	461	89 bits	7,120 bytes	13.9 %
768 bits	103	569	104 bits	9,984 bytes	13.5 %
1,024 bits	130	739	131 bits	16,768 bytes	12.8 %
2,048 bits	232	1471	233 bits	59,648 bytes	11.4 %

Although, as explained in the next sub-section, the first three instances (512, 640 and 768) are only given for illustrative purpose.

2.3 The size of p

$\mathcal{M}$ must be sufficiently large (we recommend *at least $n \geq 160$*) to prevent birthday-search [20] through two lists of $2^{n/2}$ elements to find a couple of sets such that :

$$\prod_{i \in set[1]} v_i = \left(\prod_{i \in set[2]} v_i \right)^{-1} c \bmod p$$

$\mathcal{M}$ and $\mathcal{I}$ can be increased by combining the following strategies :
Represent m in a non-binary base $(m = \sum_{i=0}^{n} r^i m_i, 0 \leq m_i < r)$ and let

$$p > \prod_{i=0}^{n} p_i^{r-1}$$

Encryption and decryption become :

$$c = \prod_{i=0}^{n} v_i^{m_i} \bmod p \text{ and } m = \sum_{i=0}^{n} \frac{r^i}{\log(p_i)} \times \log \gcd(p_i^{r-1}, c^s \bmod p)$$

size of p	n	p_n	r	$\mathcal{M}$	size of the v-list	$\mathcal{I}$
1,024 bits	74	379	3	119 bits	9,600 bytes	11.6 %
2,048 bits	130	739	3	208 bits	33,536 bytes	10.2 %
2,048 bits	93	491	4	188 bits	24,064 bytes	9.2 %
2,048 bits	47	223	8	144 bits	12,288 bytes	7.0 %
2,048 bits	39	173	10	133 bits	10,240 bytes	6.5 %

Let $p < \prod_{i=0}^{n} p_i$ but restrict $\sum_{i=0}^{n} m_i = w$ so that $\forall m \in \mathcal{M}, \prod_{i=0}^{n} p_i^{m_i} < p$.

This variant implies a non-standard coding (constant-weight messages are rather suited to random-challenge identification and less for encryption) but results in drastically smaller v-lists :

size of p	n	p_n	w	$\mathcal{M}$	size of the v-list	$\mathcal{I}$
512 bits	131	743	55	125 bits	8,448 bytes	24.4 %
512 bits	271	1747	47	176 bits	17,408 bytes	34.4 %
768 bits	199	1223	76	187 bits	19,200 bytes	24.3 %
768 bits	274	1777	71	222 bits	26,400 bytes	28.9 %
1,024 bits	419	2903	89	308 bits	53,760 bytes	30.1 %
1,024 bits	479	3413	87	323 bits	61,440 bytes	31.5 %

Note that it is also possible to require that $\sum_{i=0}^{n} m_i \leq w$ but this complicates coding and has a very limited effect on $\mathcal{I}$.

2.4 The arithmetic properties of p

The mulipliticative property of the Legendre symbol yields :

$$\prod_{i \in A} (-1)^{m_i} = \left(\frac{c}{p} \right) \text{ where } A = \{0 \leq i \leq n, \ p_i \in NQR_p\}$$

Even if the leakage of the bit :

$$b = \sum_{i \in A} m_i \mod 2$$

is not serious in itself, it may become dangerous in some specific scenari; typically, if the same m is sent to several users, relations of the form

$$b_j = \sum_{i \in set[j]} m_i \mod 2$$

can be collected and m reconstructed by linear algebra.

A trivial countermeasure would be to restrict $p_i \in QR_p$ (in this case, s can also be even)[1] but one may proceed in a more elegant way by specifying $p_0 = 2 \in NQR_p$ and *simila similibus curantur*, let

$$m_0 = \sum_{i \in A-\{0\}} m_i \mod 2$$

cancel b.

---

[1] there are exactly 54 one-byte primes, 43 nine-bit primes and 75 ten-bit primes. If one has to discard half of them, and if one wants to have a sub-minimal 160-bit message space, 50 of the primes will be eleven-bit numbers and key generation will only be possible in the lucky event where the quadratic residues have an uneven distribution and concentrate on small values.

Other small factors of $p-1$ produce similar phenomena. If q is such a factor, then, by raising the ciphertext to the power $(p-1)/q$, one ends up with an element of a multiplicative sub-group of order q. Since q is small, discrete logarithms can be computed in this sub-group and yield a linear equation modulo q where the message bits are the unknowns. Leakage through other factors of $p-1$ is avoided by using a safe prime *i.e.* a prime p such that $(p-1)/2$ is prime as well.

3 Some applications

3.1 Processing encrypted data

A major weakness of software encryption is that while being processed, data are in a vulnerable state. For being modified, information must be deciphered and re-encrypted again. Unfortunately, while in clear, secrets are exposed to a wide gamut of threats ranging from scanning by hostile TSR-programs to interception in residual electromagnetic radiation.

The new cryptosystem seems interesting for processing encrypted data as it allows to modify (only) m_k by multiplying (or dividing) c by v_k. If $m_k = 1$, an additional multiplication by v_k is likely to have no effect on the cleartext[2] but if $m_k = 0$, modular division (by v_k) will destroy the whole plaintext.

3.2 Incremental encryption

Similarly, the sender can pre-encrypt a chunk of m and complete c later. This feature can be used in group-encryption protocols where each participant adds an encrypted chunk to a common ciphertext without gaining knowledge about the chucks encrypted by his peers (again, each chunk should be sufficiently big to avoid exhaustive search and properly protected against modular division).

When protection against active attacks is needed (that is, when the peers are malicious active adversaries), this feature can be inhibited by using a part of m as a (sufficiently big) CRC or by pre-encrypting m with a conventional block-cipher keyed with some public constant.

3.3 Batch encryption

Surprisingly, encrypting a pair of random message-blocks (here $m[1]$ and $m[2]$) requires only 75% of the multiplications needed for two sequential encryptions $(i = 1, 2)$:

$$c[i] = \mathsf{encrypt}(m[i] \oplus m[1] \wedge m[2]) \times \mathsf{encrypt}(m[1] \wedge m[2]) \bmod p$$

Although this strategy can be generalised to more than two blocks by building an intersection tree, accurate evaluation indicates that bookkeeping quickly costs the gain.

---

[2] the probability that $p_k \prod_{i=0}^{n} p_i^{m_i} < p$ is very close to one if m is uniformly distributed.

4 Implementation

In order to fit into a 68HC05-based ST16CF54 smart-card (4,096 EEPROM bytes, 16,384 ROM bytes and 352 RAM bytes), key storage was replaced by a command that re-computes the v-list upon request (re-computation and transmission take 310 ms per v_i but have to be done only once after reset). The p-list is compressed into a string of 48 bytes (in our implementation, $n = 74$) which k-th bit equals one if and only if k is prime. p_i is extracted by scanning this string until i ones were read (p_i is then the value of the scan-counter). To speed-up decryption (215 ms plus 33 ms for DES pre-encryption), our 824-byte program uses a composite p (four 256-bit factors) and sub-contracts all base-conversion operations ($r = 3$) to the smart-card reader. Benchmarks were done with a 5 MHz oscillator and ISO 7816-3 T=0 transmission at 115,200 bauds.

As strange as it may appear, the PC encrypts RSA-compatible ciphertexts without using a public exponent. Publishing $e = 1/s \bmod \phi(p)$ will make the computation of the v-list public but result in a standard RSA with a particular message format.

Although we see no immediate objection to restrict s to 160 bits, we recommend to avoid doing so before a reasonable scrutiny period (in particular, using a short s with a composite p seems related to [24, 23]) and enforce, in general, the following recommendations :

• As for any block cipher, too short messages (≤ 64 bits) should not be encrypted, unless concatenated to an appropriate randomiser [6].

• As for RSA and DSA [9], correct implementation must hide the correlation between processing time and the weights of m and s.

• To avoid oracle attacks [1], we recommend to reject all decrypted messages that, when re-encrypted *by the receiver* do not re-yield c.

• Since the p-list is not necessary for encryption, we recommend to keep it secret in practice but assume its knowledge as a weakened target for the sake of academic research.

Unlike RSA, our scheme is not patented; hardware and software implementing the cryptosystem can therefore be freely used and disseminated.

5 Challenge

It is a tradition in the cryptographic community to offer cash rewards for successful cryptanalysis. More than a simple motivation means, such rewards also express the designers' confidence in their own schemes. As an incentive to the analysis of the new scheme, we therefore offer (as a souvenir from Eurocrypt'97) DM 1024 to whoever will decrypt :

c = 9D581F9E996C5D0878DC92BF5D5A8D2177B8B853E6697007
47D2C1411FAC6346045C76596193DE57A3996F04395E7BD44780
157CE4497E506DA61F09B73BAF3286272AC1625A5D989749BD38
46B634819BD26DF278CF6CD9157B891C629D3ECB49CB6E18D57E
4D9D4B70DA14738E1654F7466B48A0FCF96E0A7CBEF7A7A05DDA$_{16}$

p = EB17673456CF46F2F819B1FB5B15D330FCF1BB063E6C5DBB
A2A675D1639F0AF897C6CF04B3DEE33EBA5795C4A2E7EEF7CD28
5721B97F184159987F91DDC9C8270E5D36B2562F23B3881DD795
FB53634679944F3F11027B1D90BB8D3767151069626420E64E02
029BE0FA5ECEFC6987C72C10451CC033FFD77A78E8B8B2A60623$_{16}$

where $r = 4$, $n = 74$ and the coding convention is space $= 0$, a $= 1$, b $= 2, \cdots, $ z $= 26$. The challenger should be the first to decrypt at least 50% of c (the v-list is available by email) and publish the cryptanalysis method which must be different than computing the discrete logarithm of one of the v_i-s but the authors are ready to carefully evaluate *ad valorem* any feedback they get.

6 Further research

Since a first (informal) presentation of the scheme, several researchers began to investigate its different aspects and compare its features to RSA [5, 12, 2].

Elliptic curving the scheme is still an open problem (since elliptic curves are Abelian groups and not Euclidean domains, gcds can not be computed). Provable security, strategies for reducing the size of the public-key or signing with the scheme are also important for increasing the *practical* usefulness of the new cryptosystem.

A general knapsack taxonomy also seems in order. The idea of multiplicative knapsack is roughly 20 years old and was first proposed in the open literature by Merkle and Hellman [11] in their original paper. As, observed by Desmedt in his 1986 survey [7], encryption in the multiplicative Merkle-Hellman knapsack is actually additive. It is in fact the decryption which is multiplicative. The scheme presented here is in this respect thoroughly multiplicative. It should also be noted that Merkle-Hellman's knapsack was (partially) cryptanalyzed in by Odlyzko [13] but all our attempts to extend this attack to the new scheme failed.

As a final conclusion, although our scheme seems practical and simple, it can hardly compete with RSA on concrete commercial platforms as its public keys are typically eighty times bigger than RSA ones. Nevertheless, the new concept appears to be a promising starting-point for improvements and further research.

7 Acknowledgements

The authors thank Yvo Desmedt, Philippe Hoogvorst, David Kravitz and Ronald Rivest and Eurocrypt's referees for helpful comments and discussions.

References

1. R. Anderson, *Robustness principles for public-key protocols*, LNCS, Advances in Cryptology, Proceedings of Crypto'95, Springer-Verlag, pp. 236–247, 1995.
2. R. Anderson & S. Vaudenay, *Minding your p's and q's*, LNCS, Advances in Cryptology, Proceedings of Asiacrypt'96, Springer-Velrag, pp. 26–35, 1996.

3. P. Camion, *An example of implementation in a Galois field and more on the Naccache-Stern public-key cryptosystem*, manuscript, October 27–29, 1995.

4. B. Chor & R. Rivest, *A knapsack-type public key cryptosystem based on arithmetic on finite fields*, IEEE Transactions on Information Theory, vol. IT 34, 1988, pp. 901–909.

5. T. Cusick, *A comparison of RSA and the Naccache-Stern public-key cryptosystem*, manuscript, October 31, 1995.

6. D. Denning (Robling), *Cryptography and data security*, Addison-Wesley Publishing Company, p. 148, 1983.

7. Y. Desmedt, *What happened with knapsack cryptographic schemes*, Performance limits in communication - theory and practice, NATO ASI series E : Applied sciences, vol. 142, Kluwer Academic Publishers, pp. 113-134, 1988.

8. W. Diffie & M. Hellman, *New directions in cryptography*, IEEE Transactions on Information Theory, vol. IT 22 n° 6, pp. 644–654, 1976.

9. P. Kocher, *Timing attacks in implementations of Diffie-Hellman, RSA, DSS and other systems*, LNCS, Advances in Cryptology, Proceedings of Crypto'96, Springer-Verlag, pp. 104–113, 1996.

10. H. Lenstra, *On the Chor-Rivest knapsack cryptosystem*, Journal of Cryptology, vol. 3, pp. 149–155, 1991.

11. R. Merkle & M. Hellman, *Hiding information and signatures in trapdoor knapsacks*, IEEE Transactions on Information Theory, vol. IT 24 n° 5, pp. 525–530, 1978.

12. M. Naor, *A proposal for a new public-key by Naccache and Stern*, presented at the Weizmann Institute Theory of Computation Seminar, November 19, 1995.

13. A. Odlyzko, *Cryptanalytic attacks on the multiplicative knapsack cryptosystem and on Shamir's fast signature scheme*, IEEE Transactions on Information Theory, vol. IT 30, pp. 594–601, 1984.

14. H. Petersen, *On the cardinality of bounded subset products* , Technical report TR-95-16-E, University of Technology Chemnitz-Zwickau, 1995.

15. S. Pohlig & M. Hellman, *An improved algorithm for computing logarithms over $GF(q)$ and its cryptographic significance*, IEEE Transactions on Information Theory, vol. 24, pp. 106–110, 1978.

16. D. Pointcheval, *A new identification scheme based on the perceptrons problem*, LNCS, Advances in Cryptology, Proceedings of Eurocrypt'94, Springer-Verlag, pp. 318–328, 1995.

17. R. Rivest, A. Shamir & L. Adleman, *A method for obtaining digital signatures and public-key cryptosystems*, CACM, vol. 21, n°. 2, pp. 120-126, 1978.

18. A. Salomaa, *Public-key cryptography*, EATCS Monographs on theoretical computer science, vol. 23, Springer-Verlag, page 66, 1990.

19. A. Shamir, *An efficient identification scheme based on permuted kernels*, LNCS, Advances in Cryptology, Proceedings of Crypto'89, Springer-Verlag, pp. 606–609.

20. G. Simmons, *Contemporary cryptology : The science of information integrity*, IEEE Press, pp. 257–258, 1992.

21. J. Stern, *A new identification scheme based on syndrome decoding*, LNCS, Advances in Cryptology, Proceedings of Crypto'93, Springer-Verlag, pp. 13–21, 1994.

22. J. Stern, *Designing identification schemes with keys of short size*, LNCS, Advances in Cryptology, Proceedings of Crypto'94, Springer-Verlag, pp. 164–173, 1994.

23. P. van Oorschot & M. Wiener, *On Diffie-Hellman key agreement with short exponents*, LNCS, Advances in Cryptology, Proceedings of Eurocrypt'96, Springer-Verlag, pp. 332–343, 1996.

24. M. Wiener, *Cryptanalysis of short RSA secret exponents*, IEEE Transactions on Information Theory, vol. 36, n°. 3, pp. 553–558, 1990.

On the Importance of Checking Cryptographic Protocols for Faults

(Extended abstract)

Dan Boneh
dabo@bellcore.com

Richard A. DeMillo
rad@bellcore.com

Richard J. Lipton*
lipton@bellcore.com

Math and Cryptography Research Group, Bellcore,
445 South Street, Morristown NJ 07960

Abstract. We present a theoretical model for breaking various cryptographic schemes by taking advantage of random hardware faults. We show how to attack certain implementations of RSA and Rabin signatures. We also show how various authentication protocols, such as Fiat-Shamir and Schnorr, can be broken using hardware faults.

1 Introduction

Direct attacks on the famous RSA cryptosystem seem to require that one factor the modulus. Therefore, it is interesting to ask whether there are attacks that avoid this. The answer is yes: the first was the recent attack based on timing [4]. It was observed that a few bits could be obtained from the *time* that operations took. This would allow one to break the system without factoring.

We have a new type of attack that also avoids directly factoring the modulus. We essentially use the fact that from time to time the hardware performing the computations *may* introduce errors. There are several models that may enable a malicious adversary to collect and possibly cause faults. We give a high level description:

Transient faults Consider a certification authority (CA) that is constantly generating certificates and sending them out to clients. Due to random transient hardware faults the CA might generate faulty certificates on rare occasions. If a faulty certificate is ever sent to a client, we show that in some cases that client can break the CA's system and generate fake certificates. Note that on many systems, a client is alerted when a faulty certificate is received.

Latent faults Latent faults are hardware or software bugs that are difficult to catch. As an example, consider the Intel floating point division bug. Such bugs may also cause a CA to generate faulty certificates from time to time.

Induced faults When an adversary has physical access to a device she may try to purposely induce hardware faults. For instance, one may attempt to attack

---

* Also at Princeton University. Supported in part by NSF CCR–9304718.

W. Fumy (Ed.): Advances in Cryptology - EUROCRYPT '97, LNCS 1233, pp. 37-51, 1997.
© Springer-Verlag Berlin Heidelberg 1997

a tamper-resistant device by deliberately causing it to malfunction. We show that the erroneous values computed by the device enable the adversary to extract the secret stored on it.

We consider a fault model in which faults are transient. That is, the hardware fault only affects the current data, but not subsequent data. For instance, a bit stored in a register might spontaneously flip. Or a certain gate may spontaneously produce an incorrect value. Note that the change is totally silent: the hardware and the system have no clue that the change has taken place. We assume that the probability of such faults is small so that only a small number of them occur during the computation.

Our attack is effective against several cryptographic schemes such as the RSA system and Rabin signatures [10]. The attack also applies to several authentication schemes such as Fiat-Shamir [5] and Schnorr [11]. As expected, the attack itself depends on the exact implementation of each of these schemes. For an implementation of RSA based on the Chinese remainder theorem we show that given *one* faulty version of an RSA signature one can efficiently factor the RSA modulus with high probability. The same approach can also be used to break Rabin's signature scheme. In Section 6 we show that hardware faults can be used to break other implementations of the RSA system though many more faulty values are required.

In Section 4 we show that the Fiat-Shamir identification scheme [5] is vulnerable to our hardware faults attack. Given a few faulty values an adversary can completely recover the private key of the party trying to authenticate itself. In Section 5 we obtain the same result for Schnorr's identification protocol [11]. Both schemes are suitable for use on smart cards.

It is important to emphasize that the attack described in this paper is currently theoretical. We are not aware of any published results physically experimenting with this type of attack. The purpose of these results is to demonstrate the danger that hardware faults pose to various cryptographic protocols. The conclusion one may draw from these results is the importance of verifying the correctness of a computation for *security* reasons. For instance, a smart card using RSA to generate signatures should check that the correct signature has indeed been produced. The same applies to a certification authority using RSA to generate certificates. In protocols where the device has to keep some state (such as in identification protocols) our results show the importance of protecting the registers storing the state information by adding error detection bits (e.g. CRC). We discuss these points in more detail at the end of the paper.

We note that FIPS [6] publication 140-1 suggests that hardware faults may compromise the security of a module. Our results explicitly demonstrate the extent of the damage caused by such faults.

2 Chinese remainder based implementations

2.1 The RSA system

In this section we consider a system using RSA to generate signatures in a naive way. Let $N = pq$ be a product of two large prime integers. To sign a message x using RSA the system computes $x^s \bmod N$ where s is a secret exponent. Here the message x is assumed to be an integer in the range 1 to N (usually one first hashes the message to an integer in that range). The security of the system relies on the fact that factoring the modulus N is hard. In fact, if the factors of N are known then one can easily break the system, i.e., sign arbitrary documents without prior knowledge of the secret exponent.

The computationally expensive part of signing using RSA is the modular exponentiation of the input x. For efficiency some implementations exponentiate as follows: using repeated squaring they first compute $E_1 = x^s \bmod p$ and $E_2 = x^s \bmod q$. They then use the Chinese remainder theorem (CRT) to compute the signature $E = x^s \bmod N$. We explain this last step in more detail. Let a, b be two precomputed integers satisfying:

$$\begin{cases} a \equiv 1 \pmod{p} \\ a \equiv 0 \pmod{q} \end{cases} \quad \text{and} \quad \begin{cases} b \equiv 0 \pmod{p} \\ b \equiv 1 \pmod{q} \end{cases}$$

Such integers always exist and can be easily found given p and q. It now follows that

$$E = aE_1 + bE_2 \pmod{N}$$

Thus, the signature E is computed by forming a linear combination of E_1 and E_2. This exponentiation algorithm is more efficient than using repeated squaring modulo N since the numbers involved are smaller.

2.2 RSA's vulnerability to hardware faults

Our simple attack on RSA signatures using the above implementation enables us to factor the modulus N. Once the modulus is factored the system is considered to be broken. Our attack is based on obtaining two signatures of the same message. One signature is the correct one; the other is a faulty signature. At the end of the section we describe an improvement due to Arjen Lenstra [9] that factors the modulus using just a single faulty signature of a known message M.

Let M be a message and let $E = M^s \bmod N$ be the correct signature of the message. Let $\hat{E}$ be a faulty signature. Recall that E and $\hat{E}$ are computed as

$$E = aE_1 + bE_2 \pmod{N} \quad \text{and} \quad \hat{E} = a\hat{E}_1 + b\hat{E}_2 \pmod{N}$$

Suppose that by some miraculous event a hardware fault occurs only during the computation of *one* of $\hat{E}_1, \hat{E}_2$. Without loss of generality, suppose a hardware fault occurs during the computation of $\hat{E}_1$ but no fault occurs during the computation of $\hat{E}_2$, i.e. $\hat{E}_2 = E_2$. Observe that

$$E - \hat{E} = (aE_1 + bE_2) - (a\hat{E}_1 + b\hat{E}_2) = a(E_1 - \hat{E}_1)$$

Now, if $E_1 - \hat{E}_1$ is not divisible by p then

$$\gcd(E - \hat{E}, N) = \gcd(a(E_1 - \hat{E}_1), N) = q$$

and so N can be easily factored. Notice that if the factors of N are originally chosen at random then it is extremely unlikely that p divides $E_1 - \hat{E}_1$. After all, $E_1 - \hat{E}_1$ can have at most $\log N$ factors.

To summarize, using one faulty signature and one correct one the modulus used in the RSA system can be efficiently factored. We note that the above attack works under a very general fault model. It makes no difference what type of fault or how many faults occur in the computation of E_1. All we rely on is the fact that faults occur in the computation modulo only one of the primes.

Arjen Lenstra [9] observed that, in fact, one faulty signature of a known message M is sufficient. Let $E = M^s \bmod N$. Let $\hat{E}$ be a faulty signature obtained under the same fault as above, that is $E \equiv \hat{E} \bmod q$ but $E \not\equiv \hat{E} \bmod p$. It now follows that

$$\gcd(M - \hat{E}^e, N) = q$$

where e is the public exponent used to verify the signature, i.e. $E^e = M \bmod N$. Thus, using the fact that the message M is known it became possible to factor the modulus given only one faulty signature. This is of interest since most implementations of RSA signatures avoid signing the same message twice using some padding technique. Lenstra's improvement shows that as long as the entire signed message is known, even such RSA/CRT systems are vulnerable to the hardware faults attack.

The attack on Chinese remainder theorem implementations applies to other cryptosystems as well. For instance, the same attack applies to Rabin's signature scheme [10]. A Rabin signature of a number $x \bmod N$ is the modular square root of x. The extraction of square roots modulo a composite makes use of CRT and is therefore vulnerable to the attack described above.

3 Register faults

From here on our attacks are based on a specific fault model which we call *register faults*. Consider a tamper-resistant device. We view the device as composed of some circuitry and a small amount of memory. The circuitry is responsible for performing the arithmetic operations. The memory (registers plus a small on chip RAM) is used to store temporary values.

Our fault model assumes that the circuitry contains no faults. On the other hand, a value stored in a register may be corrupted. With low probability, one (or a few) of the bits of the value stored in some register may flip. We will need this event to occur with sufficiently low probability so that there is some likelihood of the fault occurring exactly once throughout the computation. As before, all errors are transient and the hardware has no clue that the change has taken place.

4 The Fiat-Shamir identification scheme

The Fiat-Shamir [5] identification scheme is an efficient method enabling one party, Alice, to authenticate it's identity to another party, Bob. They first agree on an n-bit modulus N which is a product of two large primes and a security parameter t. Alice's secret key is a set of invertible elements $s_1, \ldots, s_t \bmod N$. Her public key is the square of these numbers $v_1 = s_1^2, \ldots, v_t = s_t^2 \pmod{N}$. To authenticate herself to Bob they engage in the following protocol:

1. Alice picks a random $r \in \mathbf{Z}_N^*$ and sends $r^2 \bmod N$ to Bob.
2. Bob picks a random subset $S \subseteq \{1, \ldots, t\}$ and sends the subset to Alice.
3. Alice computes $y = r \cdot \prod_{i \in S} s_i \bmod N$ and sends y to Bob.
4. Bob verifies Alice's identity by checking that $y^2 = r^2 \cdot \prod_{i \in S} v_i \pmod{N}$.

For the purpose of authentication one may implement Alice's role in a tamper resistant device. The device contains the secret information and is used by Alice to authenticate herself to various parties. We show that using register faults one can extract the secret $s_1, \ldots, s_t$ from the device. We use register faults that occur while the device is waiting for a challenge from the outside world.

Theorem 1. *Let N be an n-bit modulus and t the predetermined security parameter of the Fiat-Shamir protocol. Given t faulty runs of the protocol one can recover the secret $s_1, \ldots, s_t$ in the time it takes to perform $O(nt + t^2)$ modular multiplications.*

Proof. Suppose that due to a miraculous fault, one of the bits of the register holding the value r is flipped while the device is waiting for Bob to send it the set S. In this case, Bob receives the correct value $r^2 \bmod N$, however y is computed incorrectly by the device. Due to the fault, the device outputs:

$$\hat{y} = (r + E) \cdot \prod_{i \in S} s_i$$

where E is the value added to the register as a result of the fault. Since the fault is a single bit flip we know that $E = \pm 2^i$ for some $i = 0, \ldots, n - 1$. Observe that Bob knows the value $\prod_{i \in S} v_i$ and he can therefore compute

$$(r + E)^2 = \frac{\hat{y}^2}{\prod_{i \in S} v_i} \pmod{N}$$

Since there are only n possible values for E Bob can guess its value. When E is guessed correctly Bob can recover r since

$$(r + E)^2 - r^2 = 2E \cdot r + E^2 \pmod{N}$$

and this linear equation in r can be easily solved. Bob's ability to discover the secret random value r is the main observation which enables him to break the system. Using the value of r and E Bob can compute:

$$\prod_{i \in S} s_i = \frac{\hat{y}}{r + E} \pmod{N}$$

To summarize, Bob can compute the value $\prod_{i \in S} s_i$ by guessing the fault value E and using the formula:

$$\prod_{i \in S} s_i = \frac{2E \cdot \hat{y}}{\frac{\hat{y}^2}{\prod_{i \in S} v_i} - r^2 + E^2} \quad (\text{mod } N)$$

We now argue that Bob can verify that the fault value E was guessed correctly. Let T be the hypothesized value of $\prod_{i \in S} s_i$ obtained from the above formula. To test if T is correct Bob can verify that the relation $T^2 = \prod_{i \in S} v_i$ mod N holds. Usually only one of the possible values for E will satisfy the relation. In such a case Bob correctly obtains the value of $\prod_{i \in S} s_i$.

Even in the unlikely event that two values E, E' satisfy the relation, Bob can still break the system. If there are two possible values E, E' generating two values $T, T', T \neq T'$ satisfying the relation then clearly $T^2 = (T')^2$ mod N. If $T \neq -T'$ mod N then Bob can already factor N. Suppose $T = -T'$ mod N. Then since one of T or T' must equal $\prod_{i \in S} s_i$ (one of E, E' is the correct fault value) it follows that Bob now knows $\prod_{i \in S} s_i$ mod N up to sign. For our purposes this is good enough.

The testing method above enables Bob to check whether a certain value of E is the correct one. By testing all n possible values of E until the correct one is found Bob can compute $\prod_{i \in S} s_i$. Consequently, to correctly determine the value of $\prod_{i \in S} s_i$ for one set S requires $O(n + t)$ modular multiplications. For t sets we need $O(nt + t^2)$ modular multiplications.

Observe that once Bob has a method for computing $\prod_{i \in S} s_i$ for various sets S of his choice, he can easily find $s_1, \ldots, s_t$. The simplest approach is for Bob to construct $\prod_{i \in S} s_i$ for singleton sets, i.e. sets S containing a single element. If $S = \{k\}$ then $\prod_{i \in S} s_i = s_k$ and hence the s_i's are immediately found. However, it is possible that the device might refuse to accept singleton sets S. In this case Bob can still find the s_i's as follows. We represent a set $S \subseteq \{1, \ldots, t\}$ by its characteristic vector $U \in \{0,1\}^t$, i.e. $U_i = 1$ if $i \in S$ and $U_i = 0$ otherwise. Bob picks sets $S_1, \ldots, S_t$ such that the corresponding set of characteristic vectors $U_1, \ldots, U_t$ form a $t \times t$ full rank matrix over $\mathbf{Z}_2$. Bob then uses the method described above to construct the values $T_i = \prod_{i \in S_i} s_i$ for each of the sets $S_1, \ldots, S_t$. To determine s_1 Bob constructs elements $a_1, \ldots, a_t \in \{0,1\}$ such that

$$a_1 U_1 + \ldots + a_t U_t = (1, 0, 0, \ldots, 0) \quad (\text{mod } 2)$$

These elements can be efficiently constructed since the vectors $U_1, \ldots, U_t$ are linearly independent over $\mathbf{Z}_2$. When all computations are done over the integers we obtain that

$$a_1 U_1 + \ldots + a_t U_t = (2b_1 + 1, 2b_2, 2b_3, \ldots, 2b_t)$$

for some known integers $b_1, \ldots, b_t$. Bob can now compute s_1 using the formula

$$s_1 = \frac{T_1^{a_1} \cdots T_t^{a_t}}{v_1^{b_1} \cdots v_t^{b_t}} \quad (\text{mod } N)$$

Recall that the values $v_i = s_i^2 \pmod{N}$ are publicly available. The values $s_2, \ldots, s_t$ can be constructed using the same procedure. This phase of the algorithm requires $O(t^2)$ modular multiplications.

To summarize, the entire algorithm above made use of t faults and made $O(nt + t^2)$ modular multiplications.   □

We emphasize that the faults occur while the device is waiting for a challenge from the outside world. Consequently, the adversary knows at exactly what time the register faults must be induced.

We described the algorithm above for the case where a register fault causes a *single* bit flip. More generally, the algorithm can be made to handle a small number of bit flips per register fault. However, finding the correct fault value E becomes harder. If a single register fault causes c bits in the register to flip then the running time of the algorithm becomes $O(n^c t)$ modular multiplications.

4.1 A modification of the Fiat-Shamir scheme

One may suspect that our attack on the Fiat-Shamir scheme is successful due to the fact that the scheme is based on squaring. Recall that Bob was able to compute the random value r chosen by the device since he was given r^2 and $(r + E)^2$ where E is the fault value. One may try to modify the scheme and use higher powers. We show that our techniques can be used to break this modified scheme as well.

The modified scheme uses some publicly known exponent e instead of squaring. As before, Alice's secret key is a set of invertible elements $s_1, \ldots, s_t \bmod N$. Her public key the set of numbers $v_1 = s_1^e, \ldots, v_t = s_t^e \bmod N$. To authenticate herself to Bob they engage in the following protocol:

1. Alice picks a random r and sends $r^e \bmod N$ to Bob.
2. Bob picks a random subset $S \subseteq \{1, \ldots, t\}$ and sends the subset to Alice.
3. Alice computes $y = r \cdot \prod_{i \in S} s_i \bmod N$ and sends y to Bob.
4. Bob verifies Alice's identity by checking that $y^e = r^e \cdot \prod_{i \in S} v_i \pmod{N}$.

When $e = 2$ this protocol reduces to the original Fiat-Shamir protocol. Using the methods described in the previous section Bob can obtain the values $L_1 = r^e \bmod N$ and $L_2 = (r + E)^e \bmod N$. As before we may assume that Bob guessed the value of E correctly. Given these two values Bob can recover r by observing that r is a common root of the two polynomials

$$x^e = L_1 \pmod{N} \quad \text{and} \quad (x + E)^e = L_2 \pmod{N}$$

Furthermore, r is very likely to be the only common root of the two polynomials. Consequently, when the exponent e is polynomial in n Bob can recover r by computing the GCD of the two polynomials. Once Bob has a method for computing r he can recover the secrets $s_1, \ldots, s_t$ as discussed in the previous section.

5 Attacking Schnorr's identification scheme

The security of Schnorr's identification scheme [11] is based on the hardness of computing discrete log modulo a prime. Alice and Bob first agree on a prime p and a generator g of $\mathbf{Z}_p^*$. Alice chooses a secret integer s and publishes $y = g^s \bmod p$ as her public key. To authenticate herself to Bob, Alice engages in the following protocol:

1. Alice picks a random integer $r \in [0, p)$ and sends $z = g^r \bmod p$ to Bob.
2. Bob picks a random integer $t \in [0, T]$ and sends t to Alice. Here $T < p$ is some upper bound chosen ahead of time.
3. Alice sends $u = r + t \cdot s \bmod p - 1$ to Bob.
4. Bob verifies that $g^u = z \cdot y^t \bmod p$.

For the purpose of authentication one may implement Alice's role in a tamper resistant device. The device contains the secret information s and is used by Alice to authenticate herself to various parties. We show that using register faults one can extract the secret s from the device. In this section $\log x$ denotes logarithm of x to the base e.

Theorem 2. *Let p be an n-bit prime. Given $n \log 4n$ faulty runs of the protocol one can recover the secret s with probability at least $\frac{1}{2}$ in the time it takes to perform $O(n^2 \log n)$ modular multiplications.*

Proof. Bob wishing to extract the secret information stored in the device first picks a random challenge $t \in \mathbf{Z}_{p-1}^*$. The same challenge will be used in all invocations of the protocol. Since the device cannot possibly store all challenges given to it thus far, it cannot possibly know that Bob is always providing the same challenge t. The attack enables Bob to determine the value $t \cdot s \bmod p - 1$ from which the secret value s can be easily found. For simplicity we set $x = ts \bmod p - 1$ and assume that $g^x \bmod p$ is known to Bob.

Suppose that due to a miraculous fault, one of the bits of the register holding the value r is flipped while the device is waiting for Bob to send it the challenge t. More precisely, when the third phase of the protocol is executed the device finds $\hat{r} = r \pm 2^i$ in the register holding r. Consequently, the device will output $\hat{u} = \hat{r} + x \bmod p - 1$. Suppose $\hat{r} = r + 2^i$. Bob can determine the value of i (the fault position) by trying all possible values $i = 0, \ldots, n - 1$ until an i satisfying

$$g^{\hat{u}} = g^{2^i} g^r g^x \pmod{p}$$

is found. Assuming a single bit flip, there is exactly one such i. The above identity proves to Bob that $\hat{r} = r + 2^i$ showing that the i'th bit of r flipped from a 0 to a 1. Consequently, Bob now knows that indeed that i'th bit of r must be 0. Similar logic can be used to handle the case where $\hat{r} = r - 2^i$. In this case Bob can deduce that the i'th bit of r is 1.

More abstractly, Bob is given $x + r^{(1)}, \ldots, x + r^{(k)} \bmod p - 1$ for random values $r^{(1)}, \ldots, r^{(k)}$ (recall $k = n \log 4n$). Furthermore, Bob knows the value of some bit of each of $r^{(1)}, \ldots, r^{(k)}$. Obtaining this information requires $O(n^2 \log n)$ modular multiplications since for each of the k faults one must test all n possible values of i. Each test requires a constant number of modular multiplications.

We claim that using this information Bob can recover x in time $O(n^2)$. We assume the k faults occur at uniformly and independently chosen locations in the register r. The probability that at least one fault occurs in every bit position of the register r is at least $1 - n\left(1 - \frac{1}{n}\right)^k \geq 1 - n \cdot e^{-\log 4n} = \frac{3}{4}$. In other words, with probability at least $\frac{3}{4}$, for every $0 \leq i < n$ there exists an $r^{(i)}$ among $r^{(1)}, \ldots, r^{(k)}$ such that the i'th bit of $r^{(i)}$ is known to Bob (we regard the first bit as the LSB).

To recover x Bob first guesses the $\log 8n$ most significant bits of x. Later we show that Bob can verify whether his guess is correct. Bob tries all possible $\log 8n$ bit strings until the correct one is found. Let X be the integer that matches x on the most significant $\log 8n$ bits and is zero on all other bits. For now we assume that Bob correctly guessed the value of X. Bob recovers the rest of x starting with the LSB. Inductively suppose Bob already knows bits $x_{i-1} \ldots x_1 x_0$ of x (Initially $i = 0$). Let $Y = \sum_{j=0}^{i-1} 2^j x_j$. To determine bit x_i Bob uses $r^{(i)}$, of which he knows the i'th bit and the value of $x + r^{(i)}$. Let b be the i'th bit of $r^{(i)}$. Then

$$x_i = b \oplus i\text{'th bit}(x + r^{(i)} - Y - X \bmod p - 1)$$

assuming no wrap around, i.e., $0 \leq x + r^{(i)} - Y - X < p - 1$. By construction we know that $0 \leq x - X - Y < p/8n$. Hence, wrap around will occur only if $r^{(i)} > (1 - \frac{1}{8n})p$. Since the r's are independently and uniformly chosen in the range $[0, p)$ the probability that this doesn't happen in all n iterations of the algorithm is $(1 - \frac{1}{8n})^n > \frac{3}{4}$.

To summarize, we see that for the algorithm to run correctly two events must simultaneously occur. First, all bits of r must be "covered" by faults. Second, all the r_i must be less than $(1 - \frac{1}{8n})p$. Since each event occurs with probability at least $\frac{3}{4}$, both events happen simultaneously with probability at least $\frac{1}{2}$. Consequently, with probability at least $\frac{1}{2}$, once X is guessed correctly the algorithm runs in linear time and outputs the correct value of x. Of course, once a candidate x is found it can be easily verified using the public data. There are $O(n)$ possible values for X and hence the running time of this step is $O(n^2)$. Since the first part of the algorithm requires $O(n^2 \log n)$ modular multiplications it dominates in the overall running time. $\qquad\square$

We note that the attack also works when a register fault induces multiple bit flips in the register r (i.e. $\hat{r} = r + \sum_{j=1}^c 2^{i_j}$). As long as the number of bit flips c is constant, their exact location can be found in polynomial time. We also note that the faults we use occur while the device is waiting for a challenge from the outside world. Consequently, the adversary knows at exactly what time the faults should be induced.

6 Breaking other implementations of RSA

In Section 2.1 we observed that CRT based implementations of RSA can be easily broken in the presence of hardware faults. In this section we show that using register faults it is possible to break other implementations of RSA as well. Let N be an n-bit RSA composite and s a secret exponent. The exponentiation function $x \longrightarrow x^s \bmod N$ can be computed using either one of the following two algorithms (we let $s = s_{n-1}s_{n-2}\ldots s_1 s_0$ be the binary representation of s):

- Algorithm I
  init $y \leftarrow x$; $z \leftarrow 1$.
  main For $k = 0, \ldots, n-1$.
      If $s_k = 1$ then $z \leftarrow z \cdot y \pmod{N}$.
      $y \leftarrow y^2 \pmod{N}$.
  Output z.
- Algorithm II
  init $z \leftarrow 1$.
  main For $k = n - 1$ down to 0.
      If $s_k = 1$ then $z \leftarrow z^2 \cdot x \pmod{N}$.
      Otherwise, $z \leftarrow z^2 \pmod{N}$.
  Output z.

For both algorithms given several faulty values one can recover the secret exponent in polynomial time. Here by faulty values we mean values obtained in the presence of register faults. The attack only uses erroneous signatures of *randomly* chosen messages; the attacker need not obtain the correct signature of any of the messages. Furthermore, an attacker need not obtain multiple signatures of the same message. The following result was the starting point of our research on fault based cryptanalysis:

Theorem 3. *Let N be an n-bit RSA modulus. For any $1 \le m \le n$, given $(n/m)\log 2n$ faults, the secret exponent s can be extracted from a device implementing the first exponentiation algorithm with probability at least $\frac{1}{2}$ in the time it takes to perform $O((2^m n^3 \log^2 n)/m^2)$ RSA encryptions.*

Proof. We use the following type of faults: let $M \in \mathbb{Z}_N$ be a message to be signed. Suppose that at a single random point during the computation of $M^s \bmod N$ a register fault occurs. More precisely, at a random point in the computation one of the bits of the register z is flipped. We denote the resulting erroneous signature by $\hat{E}$. We intend to show that an ensemble of such erroneous signatures enables one to recover the secret exponent s. Even if other types of faulty signatures are added to the ensemble, they do not confuse our algorithm.

Let $l = (n/m)\log 2n$ and let $M_1, \ldots, M_l \in \mathbb{Z}_N$ be a set of random messages. Set $E_i = M_i^s \bmod N$ to be the correct signature of M_i. Let $\hat{E}_i$ be an erroneous signature of M_i. We are given $\hat{E}_i$ but do not know the value of E_i. A register fault occurs at exactly one point during the computation of $\hat{E}_i$. Let k_i be the value of k (recall k is the counter in algorithm I) at the point at which the fault occurs. Thus, for each faulty signature, $\hat{E}_i$, there is a corresponding k_i indicating the time at which the fault occurs. We may sort the messages so that $0 \leq k_1 \leq k_2 \leq \ldots \leq k_l < n$. The time at which the faults occur is chosen uniformly (among the n iterations) and independently at random. It follows that given l such faults, with probability at least half, $k_{i+1} - k_i < m$ for all $i = 1, \ldots, l - 1$. To see this observe that the probability that no fault occurs in a specific interval of width m is $\left(\frac{n-m}{n}\right)^l < 1/2n$. Since there are at most n such intervals the probability that all of them contain a fault is at least $1 - n \cdot \frac{1}{2n} = \frac{1}{2}$. Note that since we do not know where the faults occur, the values k_i are unknown to us.

Let $s = s_{n-1} \ldots s_1 s_0$ be the bits of the secret exponent s. We recover a block of these bits at a time starting with the MSBs. Suppose we already know bits $s_{n-1} \ldots s_{k_i}$ for some i. Initially $i = l + 1$ indicating that no bits are known. We show how to recover bits $s_{k_i-1} s_{k_i-2} \ldots s_{k_{i-1}}$. We intend to try all possible bit vectors until the correct one is found. Since even the length of the block we are looking for is unknown, we have to try all possible lengths. The algorithm works as follows:

1. For all lengths $r = 1, 2, 3 \ldots$ do:
2. For all candidate r-bit vectors $u_{k_i-1} u_{k_i-2} \ldots u_{k_i-r}$ do:
3. Set $w = \sum_{j=k_i}^{n-1} s_j 2^j + \sum_{j=k_i-r}^{k_i-1} u_j 2^j$. In other words, w matches the bits of s and u at all known bit positions and is zero everywhere else.
4. Test if the current candidate bit vector is correct by checking if one of the erroneous signatures $\hat{E}_j$, $j = 1, \ldots, l$ satisfies

$$\exists b \in \{0, \ldots, n\} \text{ s.t. } \left(\hat{E}_j \pm 2^b M_j^w\right)^e = M_j \pmod{N}$$

Recall that e is the public signature verification exponent. The $\pm$ means that the condition is satisfied if it holds with either a plus or minus.

5. If a signature satisfying the above condition is found output $u_{k_i-1} u_{k_i-2} \ldots u_{k_i-r}$ and stop . At this point we know that $k_{i-1} = k_i - r$ and $s_{k_i-1} s_{k_i-2} \ldots s_{k_{i-1}} = u_{k_i-1} u_{k_i-2} \ldots u_{k_i-r}$.

We show that the condition at step (4) is satisfied by the correct candidate $u_{k_i-1} u_{k_i-2} \ldots u_{k_{i-1}}$. To see this recall that $\hat{E}_{i-1}$ is obtained from a fault at the k_{i-1}'st iteration. That is, at the k_{i-1}'st iteration the value of z was changed to $\hat{z} \leftarrow z \pm 2^b$ for some b. Notice that at this point $E_{i-1} = z M_{i-1}^w$. From that point on no fault occurred and therefore the signature $\hat{E}_{i-1}$ satisfies

$$\hat{E}_{i-1} = \hat{z} M_{i-1}^w = E_{i-1} \pm 2^b M_{i-1}^w \pmod{N}$$

When in step (4) the signature $\hat{E}_{i-1}$ is corrected it properly verifies when raised to the public exponent e. Consequently, when the correct candidate is tested, the faulty signature $\hat{E}_{i-1}$ guarantees that it is accepted.

To bound the running time of the algorithm we bound the number of times the condition of step (4) is executed. One must try all possible candidate bit vectors u against all possible error locations b and erroneous signatures $\hat{E}_j$. Consequently, the number of times the condition is tested is at most

$$n \cdot l \cdot \left[\sum_{r=1}^{n-k_l} 2^r + \sum_{r=1}^{k_l - k_{l-1}} 2^r + \ldots + \sum_{r=1}^{k_2 - k_1} 2^r + \sum_{r=1}^{k_1} 2^r \right] \leq nl \left[l \cdot \sum_{r=1}^{m} 2^r \right] \leq O(nl^2 2^m)$$

Hence, the algorithm runs in the time it takes to perform $O((2^m n^3 \log^2 n)/m^2)$ RSA encryptions.

We still need to show that a wrong candidate will not pass the test of step (4) with high probability. Suppose some signature $\hat{E}_v$ incorrectly causes the wrong candidate u' to be accepted at some point in the algorithm. That is, $\hat{E}_v \pm 2^b M_v^w = E_v \bmod N$ even though $\hat{E}_v$ was generated by a different fault (here w is defined as in step (3) using the bits of u'). We know that $\hat{E}_v = E_v \pm 2^{b_1} M_v^{w_1}$ for some b_1, w_1 with $w_1 \neq w$. Therefore,

$$E_v \pm 2^{b_1} M_v^{w_1} \pm 2^b M_v^w = E_v \pmod{N}$$

In other words, M_v is a root of a polynomial of the form $a_1 x^{w_1} + a_2 x^w = 0 \bmod N$ for some a_1, a_2, w_1, w. To bound the number of roots write $\varphi(N) = \prod_{i=1}^{\eta} q_i^{\nu_i}$ and $\gcd(w_1 - w, \varphi(N)) = \prod_{i=1}^{\theta} q_i^{\mu_i}$ where the q_i are distinct primes. The number of roots is upper bounded by $\alpha \stackrel{def}{=} \prod_{i=1}^{\theta} q_i^{\nu_i}$ (this is the maximum number of roots of a polynomial of the form $x^{w_1 - w} = a_3 \bmod N$). Observe that α is a function of w and w_1. Since the message M_v is chosen independently of the fault location (i.e. independently of b_1, w_1) it follows that M_v is a root with probability at most α/N. Consequently, the probability that a specific $\hat{E}_v$ causes a specific wrong candidate u' to be accepted is bounded by α/N.

Define $\bar{\alpha}$ to be the maximum value of α over all possible values of w, w_1 (note that there are l possible values for w_1 and $O(2^m l)$ possible values for w). Let B be the number of times the equality test at step (4) is invoked, i.e. $B = O(nl^2 2^m)$. Then the probability that throughout the algorithm a wrong candidate is ever accepted is bounded by $B\bar{\alpha}/N$. We argue that with high probability (over the fault locations) $\bar{\alpha} < N/nB$. This will prove that a wrong candidate is never accepted with probability at least $1 - \frac{1}{n}$ (over the random messages M_v). This will complete the proof of the theorem.

Suppose that over the random choice of the secret exponent s, and the random choice of the fault location k_i we have that $\Pr[\bar{\alpha} > N/nB] > 1/n^c$ for some fixed $c \geq 1$. We show that in this case there is an efficient algorithm for factoring N. This will prove that we may indeed assume that $\bar{\alpha} < N/nB$ with probability bigger than $1 - \frac{1}{n^c}$ for all $c \geq 1$ (since otherwise N can already be factored).

The factoring algorithm works as follows. It picks a random exponent s and random messages $M_1, \ldots, M_l \in \mathbb{Z}_N$. It then computes erroneous signatures $\hat{E}_i$ of the M_i by using the first exponentiation algorithm to compute $M_i^s \bmod N$ and deliberately simulating a random register fault at a random iteration. By assumption, with probability at least $1/n^c$ we have $\bar{\alpha} > N/nB$. Here the values w, w_1, α and $\bar{\alpha}$ are defined as above using the simulated faults. Since $\bar{\alpha} > N/nB$

there exist some w, w_1 for which $\alpha > N/nB$. By definition of α it follows that $\varphi(N)$ divides $t(w_1 - w)^n$ for some integer $0 < t \le nB$. To see this observe that α divides $(w_1 - w)^n$ and $\alpha = \varphi(N)/t$ for some $0 < t \le nB$. These values w, w_1, t can be found using exhaustive search since there are only $O(l \cdot 2^m l \cdot nB) = (n2^m)^{O(1)}$ possibilities. Once a multiple of $\varphi(N)$ is constructed, namely $t(w_1 - w)^n$, the modulus N can be efficiently factored. By repeating this process n^c times we factor N with constant probability. The total running time of the algorithm is polynomial in n and 2^m.

$\square$

If one allows the algorithm to obtain both the erroneous and correct signature of each message M_i then the running time of the algorithm can be improved. The test at step (4) can be simplified to

$$\exists b \in \{0, \ldots, n\} \quad \text{s.t.} \quad \hat{E}_j \pm 2^b M_j^w = E_j \pmod{N}$$

thus saving the need for an RSA encryption on every invocation of the test.

7 Defending against an attack based on hardware faults

One can envision several methods of protection against the type of attack discussed in the paper. The simplest method is for the device to check the output of the computation before releasing it. Though this extra verification step may reduce system performance, our attack suggests that it is crucial for security reasons. In some systems verifying a computation can be done efficiently (e.g. verifying an RSA signature when the public exponent is 3). In other systems verification appears to be costly (e.g. DSS).

Our attack on authentication protocols such as the Fiat-Shamir scheme uses a register fault which occurs while the device is waiting for a response from the outside world. One can not protect against this type of a fault by simply verifying the computation. As far as the device is concerned, it computed the correct output given the input stored in its memory. Therefore, to protect multi-round authentication schemes one must ensure that the internal state of the device can not be affected. Consequently, our attack suggests that for security reasons devices must protect internal memory by adding some error detection bits (e.g. CRC).

Another way to prevent our attack on RSA signatures is the use of random padding. See for instance the system suggested by Bellare and Rogaway [1]. In such schemes the signer appends random bits to the message to be signed. To verify the RSA signature the verifier raises the signature to the power of the public exponent and verifies that the message is indeed a part of the resulting value. The random padding ensures that the signer never signs the same message twice. Furthermore, given an erroneous signature the verifier does not know the full plain-text which was signed. Consequently, our attack cannot be applied to such a system.

8 Summary and open problems

We described a general attack which makes use of hardware faults. The attack applies to several cryptosystems. We showed that encryption schemes using Chinese remainder, e.g. RSA and Rabin signatures, are especially vulnerable to this kind of attack. Other implementations of RSA are also vulnerable though many more faults are necessary. The idea of using hardware faults to attack cryptographic protocols applies to authentication schemes as well. For instance, we explained how the Fiat-Shamir and Schnorr identification protocols may be broken using hardware faults. The same applies to the Guillou-Quisquater identification scheme [8] though we do not give the details here. Recently several symmetric ciphers such as DES have also been analyzed for their ability to withstand a faults based attack [2].

Verifying the computation and protecting internal storage using error detection bits defeats attacks based on hardware faults. We hope that this paper demonstrates that these measures are necessary for *security reasons*. Methods of program checking [3] may come in useful when verifying computations in cryptographic protocols. Specifically, a recent result of Frankel, Gemmel and Yung [7] could prove useful in this context.

An obvious open problem is whether the attacks described in this paper can be improved. That is, can one mount a successful attack using fewer faults? To make the problem crisp we pose the following concrete question: can a general implementation of RSA be broken using significantly fewer faults than n, say $\sqrt{n}$? (here n is the size of the modulus). Such a result would significantly improve our Theorem 3. Ideally we would like to break a general implementation of RSA using only a constant number of erroneous encryptions.

Acknowledgements

We are grateful to Arjen Lenstra for his many helpful comments. We also thank R. Venkatesan for his help in working out some preliminary details of Differential Fault Analysis of DES in parallel to Biham and Shamir.

References

1. M. Bellare, P. Rogaway, "The exact security of digital signatures - How to sign with RSA and Rabin", in Proc. Eurocrypt 96, pp. 399–416.
2. E. Biham, A. Shamir, "A New Cryptanalytic Attack on DES: Differential Fault Analysis", Manuscript.
3. M. Blum, H. Wasserman, "Program result checking", proc. FOCS 94, pp. 382–392.
4. P. Kocher, "Timing attacks on implementations of Diffie-Hellman, RSA, DSS, and other systems", Proc. of Cyrpto 96, pp. 104–113.
5. U. Feige, A. Fiat, A. Shamir, "Zero knowledge proofs of identity", Proc. of STOC 87.

6. Federal Information Processing Standards, "Security requirements for cryptographic modules", FIPS publication 140-1, http://www.nist.gov/itl/csl/fips/fip140-1.txt.

7. Y. Frankel, P. Gemmell, M. Yung, "Witness based cryptographic program checking and robust function sharing", proc. STOC 96, pp. 499–508.

8. L. Guillou, J. Quisquater, "A practical zero knowledge protocol fitted to security microprocessor minimizing both transmission and memory", in Proc. Eurocrypt 88, pp. 123–128

9. A.K. Lenstra, Memo on RSA signature generation in the presence of faults, manuscript, Sept. 28, 1996. Available from the author.

10. M. Rabin, "Digital signatures and public key functions as intractable as factorization", MIT Laboratory for computer science, Technical report MIT/LCS/TR-212, Jan. 1979.

11. C. Schnorr, "Efficient signature generation by smart cards", J. Cryptology, Vol. 4, (1991), pp. 161–174.

Lattice Attacks on NTRU

Don Coppersmith* Adi Shamir**

Abstract. NTRU is a new public key cryptosystem proposed at Crypto 96 by Hoffstein, Pipher and Silverman from the Mathematics department of Brown University. It attracted considerable attention, and is being advertised over the Internet by NTRU Cryptosystems. Its security is based on the difficulty of analyzing the result of polynomial arithmetic modulo two unrelated moduli, and its correctness is based on clustering properties of the sums of random variables. In this paper, we apply new lattice basis reduction techniques to cryptanalyze the scheme, to discover either the original secret key, or an alternative secret key which is equally useful in decoding the ciphertexts.

1 Introduction

NTRU [1] was proposed at the rump session of Crypto 96, as a fast public-key encryption system. The authors explored several potential attacks against the scheme, but concluded that they are extremely unlikely to succeed. In particular, they considered the standard lattice-based attack and showed that the attackers could not expect to find the secret key by computing the shortest vector in this lattice with the LLL [3] algorithm, since the secret key was surrounded by a "cloud" of exponentially many unrelated lattice vectors.

In this paper we present another lattice-based attack, which should either find the original secret key $\mathbf{f}$ or an alternative key $\mathbf{f}'$ which can be used in place of $\mathbf{f}$ to decrypt ciphertexts with only slightly higher computational complexity. We construct a lattice L, each of whose elements corresponds to a potential decrypting key $\mathbf{f}'$; the effectiveness of $\mathbf{f}'$ for decrypting is directly related to the length of the corresponding lattice element. If we find any vector $\mathbf{f}'$ as short as $\mathbf{f}$, we can decrypt easily. If, instead, we find several vectors $\mathbf{f}'^{(i)}$, each being 2 or 3 times the length of $\mathbf{f}$, then we can obtain partial decryptions from each potential key $\mathbf{f}'^{(i)}$ and piece them together to form a total decryption.

The paper is organized as follows. Section 2 gives some notation, and introduces a norm which will be useful to our analysis. In Section 3 we sketch the NTRU cryptographic system. In Section 4 we describe the lattice L. Section 5 relates the probability of success to the lengths of the recovered short vectors in the lattice.

---

* IBM Research, Yorktown Heights, NY, USA; copper@watson.ibm.com
** Dept. Computer Science, The Weizmann Institute of Science, Rehovot 76100, Israel; shamir@wisdom.weizmann.ac.il

W. Fumy (Ed.): Advances in Cryptology - EUROCRYPT '97, LNCS 1233, pp. 52-61, 1997.
© Springer-Verlag Berlin Heidelberg 1997

2 Notation

We denote the integers by $\mathbf{Z}$ and the integers modulo q by $\mathbf{Z}_q$. N is a positive integer. We will identify the vector space $\mathbf{Z}^N$ (respectively $\mathbf{Z}_q^N$) with the ring of polynomials $\mathbf{Z}[X]/(X^N - 1)$ (resp. $\mathbf{Z}_q[X]/(X^N - 1)$), by

$$\mathbf{f} = (f_0, f_1, \ldots, f_{N-1})^T = \sum f_i X^i.$$

A boldface letter $\mathbf{f}$ represents a vector. The convolution of two vectors is given by $\mathbf{f} * \mathbf{g}$ where

$$(\mathbf{f} * \mathbf{g})_k = \sum_{\substack{i+j=k \pmod{N}}} f_i g_j;$$

this is the ordinary polynomial product in $\mathbf{Z}_q[X]/(X^N - 1)$, and is both commutative and associative. The vector of all 1's is denoted by $\mathbf{1}$. The matrix of all 1's is J, and the identity matrix is I. The symbol p_q^{-1} denotes a multiplicative inverse of p modulo q.

2.1 Approximations to a norm

For $\mathbf{x} \in \mathbf{Z}^N$ define

$$\bar{\mathbf{x}} = \tfrac{1}{N} \sum_{i=0}^{N-1} x_i$$
$$|\mathbf{x}|_\perp = \left(\sum_{i=0}^{N-1} (x_i - \bar{\mathbf{x}})^2 \right)^{1/2}$$

So $|\mathbf{x}|_\perp$ is the standard deviation of the entries of $\mathbf{x}$, scaled by $\sqrt{N}$. This norm is invariant under the operation of adding $t\mathbf{1}$ to $\mathbf{x}$, that is, adding t to each entry x_i. It is the L^2 norm of the projection of $\mathbf{x}$ orthogonal to the vector $\mathbf{1}$, hence the $\perp$ symbol.

In some circumstances we will use the *approximation*

$$|\mathbf{x} * \mathbf{y}|_\perp \approx |\mathbf{x}|_\perp \, |\mathbf{y}|_\perp . \tag{1}$$

Indeed, letting $x_i = \bar{\mathbf{x}} + w_i$ and $y_j = \bar{\mathbf{y}} + z_j$ we find

$$|\mathbf{x} * \mathbf{y}|_\perp^2 = \sum_k [(\mathbf{x} * \mathbf{y})_k - \overline{\mathbf{x} * \mathbf{y}}]^2$$
$$= \sum_k (\mathbf{w} * \mathbf{z})_k^2$$
$$= \sum_k \left(\sum_i w_i z_{k-i} \right) \left(\sum_j w_j z_{k-j} \right),$$

with indices being considered modulo N. For each product $w_i z_{k-i} w_j z_{k-j}$ counted in the sum, the difference $j - i$ between the $\mathbf{w}$-indices is the same as the difference $(k - i) - (k - j)$ between the $\mathbf{z}$-indices. Letting $d = i - j$ denote this common difference, and setting $\ell = k - j$, we rearrange the sum as:

$$|\mathbf{x} * \mathbf{y}|_\perp^2 = \sum_d \left(\sum_i w_i w_{i+d} \right) \left(\sum_\ell z_\ell z_{\ell+d} \right)$$
$$= \left(\sum_i w_i^2 \right) \left(\sum_\ell z_\ell^2 \right) + \sum_{d \neq 0} \left(\sum_i w_i w_{i+d} \right) \left(\sum_\ell z_\ell z_{\ell+d} \right)$$
$$= |\mathbf{x}|_\perp^2 \, |\mathbf{y}|_\perp^2 + \sum_{d \neq 0} \left(\sum_i w_i w_{i+d} \right) \left(\sum_\ell z_\ell z_{\ell+d} \right)$$

Now let us assume that $\mathbf{w}$ and $\mathbf{z}$ behave like random vectors. For each of the $N-1$ terms corresponding to nonzero d, the autocorrelation coefficient $\sum_i w_i w_{i+d}$ should be smaller than the corresponding sum with $d = 0$, namely $\sum_i w_i^2$, by a factor of about $1/\sqrt{N}$, and similarly with the autocorrelation coefficient $\sum_\ell z_\ell z_{\ell+d}$, so that the product should be smaller by a factor of $1/N$. Further, these terms come in with random sign, so that some cancellation should occur. So, in the random case, we can assume that the second sum (over nonzero values of d) is much smaller than the first term, corresponding to $d = 0$. This leads us to the approximation (1):

$$|\mathbf{x} * \mathbf{y}|_\perp^2 = |\mathbf{x}|_\perp^2 |\mathbf{y}|_\perp^2 + \text{smaller terms}$$

$$|\mathbf{x} * \mathbf{y}|_\perp = |\mathbf{x}|_\perp |\mathbf{y}|_\perp + \text{smaller terms}$$

3 The NTRU system

We sketch the NTRU system, as developed in [1]. We give sample parameters, based on the authors' original recommendations, to aid the reader's intuition, but with the caution that these parameters can be modified in future versions of the NTRU system.

Public parameters include three positive integers, (N, p, q), with p and q relatively prime. For example we might have $N = 167$, $p = 15$, $q = 1024$. Part of the public key is a vector $\mathbf{h} \in \mathbf{Z}_q^N$. The space of allowable plaintext messages $\mathbf{m}$ is $S_{\mathbf{m}} = \{0, 1, \ldots, p - 1\}^N$.

There are additionally spaces

$$S_\phi, S_{\mathbf{f}}, S_{\mathbf{g}} \subseteq \mathbf{Z}_q^N$$

of allowable values of vectors ϕ, $\mathbf{f}$ and $\mathbf{g}$, to be described in the next few paragraphs. For example, we might have each of $S_\phi = S_{\mathbf{f}} = S_{\mathbf{g}}$ being the collection of all $\binom{N}{d}$ N-vectors with $d = 71$ entries of 1 and $N - d = 96$ entries of 0.

The private key contains vectors $\mathbf{f} \in S_{\mathbf{f}}$ and $\mathbf{g} \in S_{\mathbf{g}}$ related to the public key $\mathbf{h}$, and integers s, t, t_1, t_2 which need not be kept secret. The values $\mathbf{f}$ and $\mathbf{g}$ satisfy

$$p\mathbf{g} = \mathbf{f} * \mathbf{h} \pmod{q}. \tag{2}$$

The private key also includes a vector $\mathbf{f}_p^{-1}$, calculated from $\mathbf{f}$, satisfying

$$\mathbf{f} * \mathbf{f}_p^{-1} = (1, 0, 0, \ldots, 0)^T \pmod{p}.$$

This product corresponds to the polynomial 1.

Encryption: To encrypt the plaintext $\mathbf{m}$, the encryptor randomly selects $\phi \in S_\phi$ and computes the ciphertext

$$\mathbf{e} = \phi * \mathbf{h} + \mathbf{m} \pmod{q}.$$

A different random choice of ϕ is made for each plaintext $\mathbf{m}$.

Decryption: The decryptor computes

$$a = f * e = (f * h) * \phi + f * m \pmod q,$$

and adjusts the entries by

$$b_k = a_k + t - \begin{cases} 0 & \text{if } a_k < s \\ q & \text{if } a_k \geq s. \end{cases}$$

Notice that

$$(f * h) * \phi = pg * \phi \pmod q.$$

Parameters are chosen so that both $t_1 1 + pg * \phi$ and $t_2 1 + f * m$ are "small enough": the entries of the non-modular expression

$$b = t1 + pg * \phi + f * m$$

are guaranteed to lie between $-q/2$ and $q/2$ most of the time. If in fact all entries lie in that range, the decryptor can switch from computation modulo q to computation modulo p, and calculate

$$b * f_p^{-1} \pmod p.$$

This removes dependence on the unknown ϕ, and recovers m.

We estimate the bound on the elements of b which still make this computation possible. Using the approximation (1) we can say that

$$|t_1 1 + pg * \phi|_\perp \approx p |g|_\perp |\phi|_\perp$$

where $|\phi|_\perp$ is the norm of a typical element of S_ϕ. (We can arrange things so that all such ϕ have the same norm.) Similarly

$$|t_2 1 + f * m|_\perp \approx |f|_\perp |m|_\perp.$$

Making the second assumption that the two vectors $t_1 1 + pg * \phi$ and $t_2 1 + f * m$ are nearly orthogonal, we would obtain

$$|b|_\perp^2 = |t1 + pg * \phi + f * m|_\perp^2 \approx |t_1 1 + pg * \phi|_\perp^2 + |t_2 1 + f * m|_\perp^2$$
$$\approx p^2 |g|_\perp^2 |\phi|_\perp^2 + |f|_\perp^2 |m|_\perp^2,$$

which we choose to write as

$$|b|_\perp^2 \approx \left(p^2 |\phi|_\perp^2\right) |g|_\perp^2 + \left(|m|_\perp^2\right) |f|_\perp^2. \tag{3}$$

Make the third assumption that the entries of b are *normally distributed*, with mean near 0 (this governed our choice of t) and standard deviation $\sigma \approx |b|_\perp/\sqrt{N}$. The decoding procedure will fail if any of the N entries b_i exceeds $q/2$ in absolute value.

In the table below, we see the effect of letting $q/2$ be a reasonable multiple (3,4,5 or 6) of the standard deviation σ. The second column gives the probability that an individual term $|b_i|$ will exceed $q/2$ (and hence be misinterpreted), and the third column gives the probability that at least one of the $N = 167$ terms $|b_i|$ exceeds $q/2$ (and hence the decryption is incorrect).

$(q/2)/\sigma$	Individual failure $\rho = \text{Prob}\{|b_i| > q/2\}$	Failure among 167 entries $1 - (1 - \rho)^N$
3	2.70×10^{-3}	3.63×10^{-1}
4	6.33×10^{-5}	1.05×10^{-2}
5	5.73×10^{-7}	9.57×10^{-5}
6	1.97×10^{-9}	3.30×10^{-7}

So if $q/2 = 5\sigma$ (that is, $\sigma = q/10$) the procedure will correctly decode most messages. We would want to arrange parameters so that $\sigma < q/10$, and a smaller value of σ would ensure higher reliability.

Remark: We are essentially using an estimate on the L^2 norm of $\mathbf{b}$ to produce an estimate of its L^∞ norm; the L^2 bound is relatively easy to estimate, but the L^∞ bound is what is required for error-free decoding.

4 The lattice

We have seen that reliability of decoding is directly related to the ratio of $\sigma \approx |\mathbf{b}|_\perp /\sqrt{N}$ to q. In turn, equation (3) gives an estimate of $|\mathbf{b}|_\perp$ in terms of $|\mathbf{f}|_\perp$ and $|\mathbf{g}|_\perp$, where $p\mathbf{g} = \mathbf{f} * \mathbf{h}$ (mod q).

Let us consider an alternate N-vector $\mathbf{f}'$ which the cryptanalyst can use in place of the correct value $\mathbf{f}$. Calculate from equation (2) a value $\mathbf{g}'$, and from equation (3) an estimate of $|\mathbf{b}'|_\perp$. If this value $|\mathbf{b}'|_\perp$ is comparable to $|\mathbf{b}|_\perp$ (smaller or not much larger), then the cryptanalyst will be able to mimic the legitimate decoder, using $\mathbf{f}'$ and $\mathbf{g}'$, to decode the message.

Thus we find the system of equations

$$p\mathbf{g}' = \mathbf{f}' * \mathbf{h} \quad (\text{mod } q) \tag{4}$$

$$|\mathbf{b}'|_\perp^2 \approx \left(p^2 |\phi|_\perp^2\right) |\mathbf{g}'|_\perp^2 + \left(|\mathbf{m}|_\perp^2\right) |\mathbf{f}'|_\perp^2 . \tag{5}$$

Consider $|\phi|_\perp$ and $|\mathbf{m}|_\perp$ to be held constant at their "typical" values. Setting

$$\lambda = \frac{|\mathbf{m}|_\perp}{p\,|\phi|_\perp},$$

we are left with

$$\sigma'^2 = \frac{|\mathbf{b}'|_\perp^2}{N} \approx \left(\frac{p^2 |\phi|_\perp^2}{N}\right) \left(|\mathbf{g}'|_\perp^2 + \lambda^2 |\mathbf{f}'|_\perp^2\right)$$

It is a simple matter to build a lattice L, whose elements correspond to choices of $\mathbf{f}'$ and corresponding $\mathbf{g}'$, and with the squared norm of the elements being

$$|\mathbf{g}'|_\perp^2 + \lambda^2 |\mathbf{f}'|_\perp^2 .$$

Start with a $2n \times 2n$ matrix L':

$$L' = \begin{bmatrix} \lambda I & 0 \\ H & qI \end{bmatrix}.$$

Here I is the $N \times N$ identity matrix, and H is the circulant matrix whose columns are circularly shifted versions of the vector $\mathbf{h}p_q^{-1}$ (mod q); recall p_q^{-1} is an integer satisfying $p_q^{-1}p = 1$ (mod q).

A vector in the column span of L' will be of the form

$$\mathbf{v}'_{\mathbf{f}',\mathbf{x}} = L' \begin{bmatrix} \mathbf{f}' \\ \mathbf{x} \end{bmatrix} = \begin{bmatrix} \lambda\mathbf{f}' \\ \mathbf{g}' \end{bmatrix},$$

where $\mathbf{g}'$ satisfies $p\mathbf{g}' = \mathbf{f}' * \mathbf{h}$ (mod q), and $\mathbf{x}$ is an arbitrary integer vector representing multiples of q.

The presence of H in the lower left of L' insures the relation $\mathbf{g}' = p_q^{-1}\mathbf{h} * \mathbf{f}'$ (mod q), and the block qI serves to perform the reduction modulo q.

The vectors $\lambda\mathbf{f}'$ and $\mathbf{g}'$ will generally have nonzero mean, but we are interested in the orthogonal norms $|\mathbf{f}'|_\perp$ and $|\mathbf{g}'|_\perp$. To this end, subtract from each column vector $\mathbf{v}$ in the top half of L' the constant vector $(\bar{\mathbf{v}})\mathbf{1}$ so that the result has zero mean; similarly each vector $\mathbf{w}$ in the bottom half of L' is replaced by $\mathbf{w} - (\bar{\mathbf{w}})\mathbf{1}$. Our new matrix L is then

$$L = \begin{bmatrix} \lambda I - (\lambda/N)J & 0 \\ H - \alpha J & qI - (q/N)J \end{bmatrix},$$

where J is the matrix of all 1's, and α is a suitably chosen scalar.

Remark: L has only $2N - 2$ independent vectors, because

$$L \begin{bmatrix} \mathbf{1} \\ \mathbf{0} \end{bmatrix} = L \begin{bmatrix} \mathbf{0} \\ \mathbf{1} \end{bmatrix} = \mathbf{0}.$$

Now a typical vector is

$$\mathbf{v}_{\mathbf{f}',\mathbf{x}} = L \begin{bmatrix} \mathbf{f}' \\ \mathbf{x} \end{bmatrix} = \begin{bmatrix} \lambda(\mathbf{f}' - (\overline{\mathbf{f}'})\mathbf{1}) \\ \mathbf{g}' - (\overline{\mathbf{g}'})\mathbf{1} \end{bmatrix},$$

and the square of its L^2 norm is

$$|\mathbf{v}_{\mathbf{f}',\mathbf{x}}|^2 = \lambda^2 |\mathbf{f}'|_\perp^2 + |\mathbf{g}'|_\perp^2 = \left(\frac{1}{p^2|\phi|_\perp^2}\right)\left[|\mathbf{m}|_\perp^2 |\mathbf{f}'|_\perp^2 + p^2 |\phi|_\perp^2 |\mathbf{g}'|_\perp^2\right]$$
$$= \left(\frac{1}{p^2|\phi|_\perp^2}\right) |\mathbf{b}'|^2$$

Thus the norm of the lattice element $\mathbf{v}_{\mathbf{f}',\mathbf{x}}$ is directly related to the suitability of $\mathbf{f}'$ as a decrypting key.

Remark: We also need $\mathbf{f}'$ to be invertible modulo p, so that $\mathbf{f}'^{-1}_p$ can be used in the decrypting process. This seems to be a weak requirement.

For a given vector $\mathbf{f}'$, select $\mathbf{x}$ to minimize this norm, and define

$$n_{\mathbf{f}'} = (p|\phi|_\perp)\min_{\mathbf{x}} |\mathbf{v}_{\mathbf{f}',\mathbf{x}}| = |\mathbf{b}'|.$$

5   Lengths of suitable vectors

We have seen that the correct key $\mathbf{f}$ should have

$$n_{\mathbf{f}} < q/10$$

in order to insure that messages are decoded correctly at least 0.9999 of the time.

If the lattice basis reduction finds a vector $\mathbf{f}'$ with, say, $n_{\mathbf{f}} = q/4$, then the cryptanalyst can still gain much useful information. The entries b'_k of the recovered vector $\mathbf{b}$ are likely to be contained in the interval

$$[-3\sigma, +3\sigma] = [-3q/4, 3q/4],$$

since there are only 167 entries and the probability of any given entry lying outside the 3σ interval is about 0.0026. Any entry b'_k in the interval $[q/4, 3q/4]$ (mod q) is unreliable, because it could represent either b'_k or $b'_k - q$ and still lie within the range $[-3\sigma, 3\sigma]$. But entries b'_k in the intervals $[0, q/4) \cup (3q/4, q)$ (mod q) are reliable; one can assume that they represent integers in the range $(-q/4, q/4)$ with no aliasing. We expect a fraction 0.68 of all b'_k to lie in this reliable range. Each represents knowledge of a linear relation among the message components m_i (mod p), namely

$$b'_k = \sum_i m_i f'_{k-i} \quad (\text{mod } p).$$

If we find two such vectors $\mathbf{f}'^{(1)}$ and $\mathbf{f}'^{(2)}$, each yielding about $0.68N$ linear relations (modulo p) among the N entries m_i, then we can solve the resulting system of linear equations to recover the message $\mathbf{m}$.

If the recovered vectors $\mathbf{f}'$ are somewhat longer, say

$$n_{\mathbf{f}'^{(i)}} \approx 4 \times n_{\mathbf{f}}$$

then we may have to work with faulty partial information: a few of the estimate integers b'_k might be incorrect, leading to a few incorrect linear equations among a collection of mostly correct ones. Then we will have to resort to techniques from error-correcting codes to discover the incorrect equations among the correct ones.

So our success depends on the success of lattice basis reduction methods in finding relatively short vectors in the lattice. If we find a vector as short as $\mathbf{f}$:

$$n_{\mathbf{f}'} \leq n_{\mathbf{f}}$$

then clearly we can use $\mathbf{f}'$ as a decrypting key. If we find two vectors not much longer than $\mathbf{f}$:

$$n_{\mathbf{f}'^{(1)}} = n_{\mathbf{f}'^{(2)}} \leq 2.5 \times n_{\mathbf{f}}$$

then each will give us partial information, and we can combine this information via linear algebra to recover $\mathbf{m}$. If we find several vectors somewhat longer yet,

$$n_{\mathbf{f}'^{(i)}} \approx 4 \times n_{\mathbf{f}}$$

then we still have a chance, if error-correcting techniques can be applied.

The Lovasz lattice basis reduction methods [3] are only guaranteed to find a vector whose length satisfies

$$n_{\mathbf{f}'} < 2^{N/2} n_{\mathbf{f}}$$

which is clearly insufficient. Schnorr [4], [5] has improved the original methods by using block techniques; he can find shorter vectors, at a higher computational price, than LLL. But it is still not guaranteed to find vectors as short as

$$n_{\mathbf{f}'} \approx 4 n_{\mathbf{f}}.$$

To summarize: if there are many vectors $\mathbf{f}'$ with $n_{\mathbf{f}'} \leq n_{\mathbf{f}}$ then we are likely to stumble across one and be able to decrypt. If $\mathbf{f}$ is much shorter than all other vectors, then we are likely to find $\mathbf{f}$. The only hope for the scheme to remain secure is for many vectors to satisfy, say, $n_{\mathbf{f}'} = 10 \times n_{\mathbf{f}}$ and hope that the lattice basis reduction methods fail to find $\mathbf{f}$ among the sea of $\mathbf{f}'$. With any improvements in the technology of lattice basis reduction, this temporary security would vanish.

6 Other comments

The lattice used in our main attack contains linear combinations of the columns of the circulant matrix H and appropriate multiples of the identity matrix I. An alternative lattice attack is to consider the dual lattice which characterizes all the integral solutions of the following homogeneous equation $H * \mathbf{f} = p\mathbf{g} + q\mathbf{k}$, where $\mathbf{f}$, $\mathbf{g}$ and $\mathbf{k}$ are three vectors with integral unknowns, and p, q are the two moduli. This lattice is closely related to that described in Section 4, except for the difference between $\mathbf{x}$ and $|\mathbf{x}|_\perp$; it is hoped that this alternative description might help the reader's intuition.

We consider the set of all the column vectors $\begin{bmatrix} \mathbf{f} \\ \mathbf{g} \end{bmatrix}$ of $2n$ integers which make the n entries in $\mathbf{k}$ integral. It is easy to show that it forms a lattice since its discrete and closed under addition. This lattice has full dimension $2n$ (except in degenerate cases), and we can find the $2n$ basis vectors in two groups of n. In each group we combine the n column vectors into a matrix, and denote the resultant $n \times n$ matrices F G and K:

1. Find a basis for the homogeneous case in which $K = 0$. The resultant equation is $H * F = pG$, which can be solved by $F = pI$ and $G = H$ since $HpI = pH$.

2. Find a basis for the inhomogeneous case in which $K = I$. The resultant equation is $H * F = pG + qI$. To solve it, we assume that H is invertible modulo p, and find two integral matrices B, C satisfying $H * B = I + pC$ (that is, B is the inverse of H modulo p, and C is the matrix of multiples of the modulus p in the modular reductions.) Then $F = qB$ and $G = qC$ is a solution since $H * F = qH * B = qI + pqC$, and $pG + qI = pqC + qI$.

We now combine the two cases into a single $2n \times 2n$ matrix A whose columns generate the lattice of $\begin{bmatrix} \mathbf{f} \\ \mathbf{g} \end{bmatrix}$ vectors. The matrix A is:

$$A = \begin{bmatrix} pI & qB \\ H & qC \end{bmatrix}.$$

The small column vector we are looking for in this lattice has entries of zero and one in the top half, and around two or three in the bottom half. We believe that for the recommended parameters of the NTRU cryptosystem, the LLL algorithm will be able to find the original secret key $\mathbf{f}$ as the first half of such an unusually short lattice vector.

7 Extensions

We understand that the authors of NTRU, after learning the details of our attack, are continuing their research into related schemes [2].

One direction of their research involves schemes similar to NTRU but with larger parameters. The expense, for the designers of the system, comes with larger public keys and more time-consuming encryption. The added security comes from the notion that in a lattice of higher dimension (several hundred) it will be computationally harder for the opponent to find high-quality vectors. To maintain this security, one must keep ahead of advances in lattice basis reduction techniques.

Another direction of their research involves extensions to noncommutative groups. Instead of using a group algebra over $\mathbf{Z}_N$ (that is, the ring $\mathbf{Z}_q[X]/(X^N - 1)$), one would use a group algebra over a noncommutative group. At the time of this writing we have not had sufficient time to analyze these proposed extensions, but we hope to be able to comment on the noncommutative version in the final version of the paper.

8 Acknowledgments

We thank Claus Schnorr for insight into lattice basis reduction methods.

References

1. J. Hoffstein, J. Pipher and J. H. Silverman, "NTRU: A new high speed public key cryptosystem," Manuscript, August 30, 1996; presented at rump session of Crypto 96.
2. J. Hoffstein, J. Pipher and J. H. Silverman, private communications, October 1996 and January 1997.

3. A. K. Lenstra, H. W. Lenstra and L. Lovasz, "Factoring Polynomials with Integer Coefficients," Matematische Annalen **261** (1982), 513–534.
4. C. P. Schnorr, "A hierarchy of polynomial time lattice basis reduction algorithms," Theoretical Computer Science **53** (1987), 201-224.
5. C. P. Schnorr, "Block reduced lattice bases and successive minima," Combinatorics, Probability and Computing **3** (1994), 507-522.

Kleptography:
Using Cryptography Against Cryptography

Adam Young* and Moti Yung**

Abstract. The notion of a Secretly Embedded Trapdoor with Universal Protection (SETUP) has been recently introduced. In this paper we extend the study of stealing information securely and subliminally from black-box cryptosystems. The SETUP mechanisms presented here, in contrast with previous ones, leak secret key information without using an explicit subliminal channel. This extends this area of threats, which we call "kleptography".

We introduce new definitions of SETUP attacks (strong, regular, and weak SETUPs) and the notion of m out of n leakage bandwidth. We show a strong attack which is based on the discrete logarithm problem. We then show how to use this setup to compromise the Diffie-Hellman key exchange protocol. We also strengthen the previous SETUP against RSA. The strong attacks employ the discrete logarithm as a one-way function (assuring what is called "forward secrecy"), public-key cryptography, and a technique which we call probabilistic bias removal.

Key words: cryptanalytic attacks, kleptography, leakage bandwidth, Discrete Log, Diffie-Hellman, RSA, design and manufacturing of cryptographic devices and software, black-box devices, subliminal channels, information hiding, SETUP mechanisms, randomness, pseudorandomness.

1 Introduction

Numerous problems and subtleties exist when constructing a cryptosystem for use, since designing and manufacturing secure cryptosystems is a demanding task. Some of these problems are immediate and known, yet they require diligent engineering. Other issues are more involved or are yet unknown.

One area where problems have been recognized is in the information-hiding aspect of cryptosystems, and in particular the existence of "subliminal channels" in cryptosystems. Subliminal channels can be used to convey information in the output of a cryptosystem in a way that is different from the intended output. This notion was put forth by Simmons [Sim85, Sim94]. Other works on subliminal channels are [Des90] which showed an RSA channel and [KL95] which showed how to hide a shadow public key inside a key distribution method. The usage of subliminal channels expose information universally (to anyone).

* Dept. of Computer Science, Columbia University Email: ayoung@cs.columbia.edu.
** CertCo, NY, USA. Email: moti@cs.columbia.edu, moti@certco.com

W. Fumy (Ed.): Advances in Cryptology - EUROCRYPT '97, LNCS 1233, pp. 62-74, 1997.
© Springer-Verlag Berlin Heidelberg 1997

Recently, it was shown that a cryptosystem, when implemented as a black-box (i.e., when the user has only input/output access to the hardware or software cryptographic facility), can be designed such that it gives a unique advantage to the attacker. This is accomplished using SETUP mechanisms [YY96]. SE-TUPs are are unnoticeable in black-box environments and they resist reverse-engineering as well (the device may still use a strong random source).

Indeed, black-box cryptography is both endorsed and employed by the U.S. government ("trusted" hardware devices). It is also in use in the private sector (e.g., embedded cryptography in devices like cellular phones). It is often the case that crucial cryptographic key management functions are implemented in hardware and that companies that produce commercial software implementations of cryptographic systems do not publicize source code to protect proprietary information. Even when specifications are available, users rarely check the validity or compliance of the available implementation against the specifications.

Previously, the SETUP threat employed subliminal channels and combined subliminal channels with public key cryptography (with a private key known only to the attacker). In this paper we present various "kleptographic threats". Kleptography, in turn, is defined as the "study of stealing information securely and subliminally"; we limit ourselves to the context of cryptographic systems. The kleptographic attacker can steal the secrets securely, and in an exclusive and subliminal manner.

The attack that we present involves public-key cryptography and strong one-way functions, and is in the same spirit as SETUP attacks (avoiding trivial attacks on the pseudorandomness of the device and similar simple attacks where reverse engineering implies knowledge of the future states of the device). What is new in this work is that we show how to implement SETUPs without using explicit subliminal channels. Rather than employing an "information leaking channel," the implementation, in conjunction with the internal cryptographic tools, *generates* opportunities for leaking information.

In this paper:

1. We refine the notion of a SETUP [YY96] and define the notions of weak, regular, and strong SETUPs.
2. We expand the range of setup attacks that can be carried out on cryptographic devices. We define (m, n)-leakage bandwidth.
3. We present a setup mechanism that employs the discrete logarithm problem; (previously, only weak attacks were known on discrete log systems). We show how it can be embedded within a device that conducts Diffie-Hellman key exchanges.
4. The mechanism is used to strengthen the SETUP in RSA keys (presented in [YY96]), so that after reverse-engineering of the RSA key generation device, one cannot tell whether the past keys that were generated were generated by a kleptographic mechanism or by a regular one.
5. A key technique that is presented is "probabilistic bias removal". Bias removal simply eliminates the biases of a distribution caused by the algebra employed by the setup mechanism within a cryptographic device.

2 Definitions and Background

A Secretly Embedded Trapdoor with Universal Protection is an algorithm that can be embedded within a cryptosystem to leak encrypted secret key information to the attacker in the output of that cryptosystem [YY96]. This encryption is performed by a PKC function E that is contained within the cryptosystem. E may be a probabilistic public key encryption function [GM84]. The outcome is a strong 'encryption' that is leaked in a fashion that is noticeable only to the owner of the private portion of E. The following is the definition of a (regular) setup, which is based on the definition from [YY96].

Definition 1. Assume that C is a black-box cryptosystem with a publicly known specification. A (regular) SETUP mechanism is an algorithmic modification made to C to get C' such that:

1. The input of C' agrees with the public specifications of the input of C.
2. C' computes efficiently using the attacker's public encryption function E (and possibly other functions as well), contained within C'.
3. The attacker's private decryption function D is not contained within C' and is known only by the attacker.
4. The output of C' agrees with the public specifications of the output of C. At the same time, it contains published bits (of the user's secret key) which are easily derivable by the attacker (the output can be generated during key-generation or during system operation like message sending).
5. Furthermore, the output of C and C' are polynomially indistinguishable (as in [GM84]) to everyone except the attacker.
6. After the discovery of the specifics of the setup algorithm and after discovering its presence in the implementation (e.g. reverse-engineering of hardware tamper-proof device), users (except the attacker) cannot determine past (or future) keys.

2.1 Weak SETUP

Definition 2. A *weak setup* is a regular setup except that the output of C and C' are polynomially indistinguishable to everyone except the attacker and the owner of the device who is in control (knowledge) of his or her own private key (i.e., requirement 5 above is changed).

It may seem that a weak setup is cryptographicaly insecure. Indeed it is in the sense that it can be detected in poly-time by the owner of the device (but not compromised by anyone). Note however, that the user (owner) must first assume that the device in question contains a SETUP, and must also know exactly how to test the black-box device for the presence of it. Weak SETUPs are sufficient for the case where the end users are in collaboration. An example of this (as shown in [YY96]) is the prisoner's dilemma of Gus Simmons [Sim85]. In this scenario, Alice is in prison and wants to leave. She contaminates her own cryptosystem with a weak SETUP so as to leak her private key to Bob through digital signatures. After securely leaking her private key she can send data to him subliminally through digital signatures.

2.2 Strong SETUP

The key aspect of a regular setup is that we assume that the users do not have access to the actual implementation of C. This is in fact necessary for polynomial indistinguishability. Now assume that devices/implementations sometimes use the contaminated algorithm (namely, the setup) and sometimes use the uncontaminated (setup-free) version. Now we can make an interesting strengthening.

Definition 3. A *strong setup* is a regular setup, but in addition we assume that the users are able to hold and fully reverse-engineer the device after its past usage and before its future usage. They are able to analyze the actual implementation of C' and deploy the device. However, the users still cannot steal previously generated/future generated keys, and if the setup is not always applied to future keys, then setup-free keys and setup keys remain polynomially indistinguishable.

A strong setup is a much more powerful notion than a regular setup. To exemplify, consider the following problem. Suppose that we are given a cryptographic device such that with 50% probability it uses the setup mechanism, and with 50% probability it behaves normally (based on a random bit, say). The claim is that if the setup is a strong setup, then a user who is handed the output of such a device cannot tell with probability greater than 50% whether or not the output contains hidden secret key information. The obvious assumption being made here is that the user did not observe the computation (randomness) that yielded the output in question, but otherwise he can observe the device's algorithms and control.

This notion is useful to an attacker as protection against the threat of "key revocation", since even if the device is reverse-engineered, previously generated setup keys are indistinguishable from normal ones. Furthermore, the decision as to which keys to steal may be dictated by a secret policy used or given as input at the time of stealing. Mathematically, the notion gives an extra challenge beyond the polynomial indistinguishability based on the public key of the attacker and the pseudorandomness which is protected by the device. In fact, the involvement of a hard to invert one-way function and the notion of "forward secrecy" seems to be needed ("forward secrecy" is the notion that applies to key distributions – it requires that compromising the long lived key should not give away previous session keys distributed using the compromised long-lived key). The strong setup further requires that the distributions of the cryptosystem and the setup one be "indistinguishable" – even when given the public keys and tools embedded inside the black-box device. A weak setup in ElGamal signature was presented in [YY96], which is all they achieved based on algebraic properties of the discrete logarithm problem.

2.3 Leakage Bandwidth

We now define the notion of leakage bandwidth in cryptosystems. It defines what can be leaked in cryptographic systems (e.g., key generation or key exchange) that are repetitively invoked.

Definition 4. A (m, n)-leakage scheme is a SETUP mechanism that leaks m keys/secret messages over n keys/messages that are output by the cryptographic device $(m \leq n)$.

The discrete log attack that we present is a (1,2)-leakage scheme where in two key generations we are able to leak one key to the attacker. We will show how this scheme can be extended to become a $(m, m + 1)$-leakage scheme.

3 Discrete Log based SETUP against Diffie-Hellman

Previously, the underlying strategy was to somehow modify a cryptosystem to 'display' the public key encrypted ciphertext of secret key information in the output of the cryptographic device. Such modifications are difficult to come by, since the modification must not interfere with the normal operation of the device, and the SETUPs output must also be embedded in the normal output of the device. Hence, the data that is output by the device is dual in nature. A subliminal channel is the traditional vehicle for leaking such data, since a channel has a known bandwidth and does not interfere with the expected operation of a device. What we will now present is a different approach to leaking data securely.

We will now briefly review the Diffie-Hellman key exchange protocol [DH76]. Alice and Bob want to agree on a secret key using an insecure communication channel. Diffie-Hellman uses the parameters p, which is a large prime, and g which is a generator modulo p. These parameters are public. To establish a secret key k, they do the following. Alice generates a value a randomly, where $a < p - 1$. Bob generates a value b in the same fashion. Alice sends Bob $A = g^a \bmod p$ and Bob sends Alice $B = g^b \bmod p$. They can both compute k, where $k = A^b = B^a \pmod{p}$.

The primary attack that is presented in this paper introduces a setup for Diffie-Hellman. Let p is a large strong prime and g is a generator mod p. The user's private key is x where x is less than $p - 1$ (as in ElGamal scheme [ElG85]). The user's public key is (y, g, p) where $y = g^x \bmod p$. To encrypt a message m $(m < p)$, k is chosen randomly such that $k < p - 1$. We then compute $r = g^k \bmod p$, and $s = y^k m \bmod p$. The ciphertext of m is the pair (r, s). To recover m, we compute $s/r^x \bmod p$.

3.1 Discrete Log Attack

Suppose that the only information that we are allowed to display is $g^c \bmod p$ for some $c < p - 1$ (as in Diffie-Hellman). The question is, how can we leak

c efficiently? The following is a way to leak a value, call it c_2, over the single message $m_1 = g^{c_1} \bmod p$, such that the subsequent message $m_2 = g^{c_2} \bmod p$ is compromised. In this attack we assume that the device is free to choose the exponents used. Let the attacker's private key be X, and let the corresponding public key be Y. Let W be a fixed odd integer, and let H be a cryptographically strong hash function. WLOG, assume that H generates values less than $\phi(p)$. The following algorithm describes the operation of the Diffie-Hellman device when it is used two times.

1. For the first usage, $c_1 \in Z_{p-1}$ is chosen uniformly at random
2. The device outputs $m_1 = g^{c_1} \bmod p$.
3. c_1 is stored in non-volatile memory for the next time the device is used.
4. For the second usage $t, \in \{0, 1\}$ is chosen uniformly at random.
5. $z = g^{c_1 - Wt} Y^{-ac_1 - b} \bmod p$.
6. $c_2 = H(z)$
7. The device outputs $m_2 = g^{c_2} \bmod p$.

The attacker need only passively tap the communications line, and obtain m_1 and m_2, in order to calculate c_2. The value for c_2 is found as follows.

1. $r = m_1{}^a g^b \bmod p$
2. $z_1 = m_1 / r^X \bmod p$
3. if $m_2 = g^{H(z_1)} \bmod p$ then output $H(z_1)$
4. $z_2 = z_1 / g^W$
5. if $m_2 = g^{H(z_2)} \bmod p$ then output $H(z_2)$

The value c_2 can be used by the attacker to determine the key from the second DH key exchange. Note that only the attacker can perform these computations since only the attacker knows X. The reason for using W will become clear in the next section.

What is strange about the above setup mechanism is that we didn't choose c_2 randomly and then public key encrypt it. Instead, we designated $g^{c_1} \bmod p$ to be the ElGamal encryption of something, and then calculated what that something was. Note that $g^{c_1} \bmod p$ acts as both the first and second parts of the ElGamal encryption of z. So, we are doing ElGamal encryptions (r, s), where $r = s$. This is made possible due to fact that the device is free to choose its own random parameters. Hence, it is possible to leak an exponent efficiently using exponentiated values $g^c \bmod p$ alone. The discrete log attack, in effect, securely discloses a pseudo-random value c_2 to the attacker and then deliberately uses it in a subsequent message. We call z a *hidden field element* with respect to Y, since it is an element of Z_p that can only be recovered using the trapdoor information in Y (or at least as conjectured). As described, this is a (1,2)-leakage system (note that we never said we could choose our messages explicitly!).

In order for c_2 to be able to take on any value less than $p - 1$ we assume that $g_1 = g^{-Xb-W}$, $g_2 = g^{-Xb}$, and $g_3 = g^{1-aX}$ are generators mod p.

Claim 1 *z is uniformly distributed in Z_p.*

Proof. We have the equation $Y^{ac_1+b}g^{Wt}z = g^{c_1} \bmod p$. Solving for z, we get $z = g^{-Xb-Wt}g^{(1-aX)c_1} = g_i g_3^{c_1} \bmod p$, where i is 1 or 2. But $g_i = g_3^u \bmod p$, for some integer u. So, $z = g_3^{c_1+u} \bmod p$. Since c_1 is chosen uniformly at random, the claim holds. QED.

If H is a pseudo-random function [GGM86], then c_2 can take on any value less than $p-1$ as desired. Note that the attack works when $(p-1)/2$ is a prime (and it also works when it is composite).

3.2 Security of Discrete Log SETUP Mechanism

There are two issues to consider with respect to the discrete log attack. It must be intractable for people other than the attacker to recover c_2. It must also be intractable for people other than the attacker to detect that this SETUP Mechanism is in use. We consider these in turn.

Claim 2 *The Discrete Log SETUP is secure iff the DH problem is secure.*

Proof. Suppose we have an oracle A that solves the DH problem. So, $A(g^u, g^v) = g^{uv}$. Let $f = g^{c_1}/A(g^{ac_1+b}, Y)$. Clearly, f or f/g^W is z. From z we can readily obtain c_2. Suppose we have an oracle B that breaks the Discrete Log SETUP mechanism where $B(y, m_1) = (z_1, z_2)$. Here z_1 corresponds to $t = 0$ and z_2 corresponds to $t = 1$. We have g^u and g^v and wish to find g^{uv}. We can use B to solve the DH problem as follows. We run $B(g^v, g^u)$ and take z_1 of the output. We then calculate $f = g^u(g^v)^{-b}z_1$. It follows that $z = f^{1/a} \bmod p$. QED.

We have shown that the setup is secure in the sense that a user, not knowing the random choices of exponents of the device, cannot determine the second exponent c_2. It remains to show that users cannot detect the presence of the Discrete Log SETUP.

Claim 3 *Assuming H is a pseudorandom function, and that the device design is publicly scrutinizable, the outputs of C and C' are polynomially-indistinguishable.*

Proof. We know that z is uniformly distributed in Z_p from Claim 1. Therefore, since H is a pseudo-random function (whose domain is Z_{p-1}), c_2 is distributed uniformly in Z_{p-1}. So, the exponentiated values that are output by C and C' have polynomially indistinguishable probability distributions. QED.

From Claims 2 and 3 it follows that

Theorem 1 *The Discrete Log problem has a strong setup implementation, assuming DH is hard.*

It remains to explain why the value W was used in the setup mechanism. This is used as a precaution in the case that H were found to be invertible. This precautionary mechanism is intended to further insure undetectability for black-box implementations. So, suppose that the device is a black-box, the choice of exponents are made available to the user, and H is invertible. Furthermore, suppose that the outcome of t is always zero (i.e., W isn't used). WLOG, assume that $a = 1$ and $b = 0$ is publicly known. The user can detect the presence of the setup probabilistically as follows. The user generates several Diffie-Hellman values, and corresponding exponents. Consider one such pair of exponents c_1 and c_2. Since H is invertible, the user can calculate z. But, the user does not know Y since the device is a black-box. The user hypothesizes that the attacker's private key X is odd (so g^{1-x} isn't a generator). If this is the case, then the user would expect that if c_1 is even, then g^{c_1}/z would be a residue mod p. If c_1 where odd, then the user would expect that g^{c_1}/z would be a non-residue mod p. Now suppose that the user hypothesizes that X is even. Then if c_1 is odd or even, g^{c_1}/z is always a residue. Hence, under these circumstances, the user can detect the presence of the setup on a probabilistic basis by looking for quadratic residues (or non-residues) modulo p.

3.3 Strong Setup in Diffie-Hellman

The discrete log setup attack can be used to implement a strong setup in Diffie-Hellman, so long as the device does not output the exponents it chooses to the user. Implementing the attack is straightforward. The attacker includes his or her Y within Alice's device. The attacker then need only passively tap the communications line. It is assumed that g and p remain fixed.

Theorem 2 *The Diffie-Hellman key exchange has a $(l, l+1)$- leakage bandwidth SETUP implementation.*

We need to show that we can increase the bandwidth of the attack dramatically. We can do so by chaining together the values that are leaked. We calculate $c_3 = H(z)$ using the equation $Y^{ac_2+b}g^{Wt}z = g^{c_2} \bmod p$. The value of $g^{c_3} \bmod p$ is then used in the next key exchange. We continue this process, say l times. This permits the leakage l contiguous Diffie-Hellman keys. After l times a new c_1 is chosen entirely random, thus insuring that all such contaminated devices behave differently. Thus, the attack can be expanded to become a $(l,l+1)$-leakage setup. Note that this attack requires the storage of a small amount of state information to work.

4 Probabilistic Bias Removal Method (PBRM)

Consider the following effective, albeit trivial, setup attack on a hybrid cryptosystem based on RSA and IDEA (PAP is a kleptographic version of PGP where 'Good' is changed to 'Awful' [YY96]). This version contains the attacker's 512 bit RSA public key and requires that its users use 1024 bit public keys. PAP operates as follows. After the user has given PAP his or her own public and private keys, PAP recovers the users prime p, where $n = pq$. PAP then divides p

into two equal length bit-strings and then probabilistically encrypts both using the attacker's public key. The result is two ciphertext bit-strings each of which is 512 bits in length. Since the key size of IDEA is 128 bits, PAP proceeds to leak these bit-strings by using them as the next eight symmetric keys used by the program. This constitutes a (1,8)-leakage setup attack.

If the attacker succeeds in retrieving enough of these session keys (e.g., by convincing the user to e-mail him stuff), then he can compute the user's private key. If the user suspects his PGP is really PAP, then he cannot simply encrypt his prime p and compare, since the encryptions were probabilistic. However, if the user generates enough symmetric keys using PAP he can detect the contamination. The method for doing so was noted by [Sch] in regards to the version of PAP in [YY96]. Note that each of the two ciphertext bit-strings that are leaked are each less than the attacker's public modulus N. The output of the device is therefore biased towards outputting session keys, which when concatenated in sets of four, are less than N, whereas the values should be uniformly distributed in $\{0,1\}^{512}$. This is in fact a very general problem in kleptography, since it is public key encrypted values that are publicly displayed.

An abstract version of the 'biasing problem' can be stated as follows. We are given a value x that is uniformly distributed in $[1..R]$, and we want a value x' that is uniformly distributed in $[1..S]$, where $R > S/2$. Furthermore, we require that x be easily obtainable from x'. We will now describe our Probabilistic Bias Removal Method (PBRM) which accomplishes this. Assume that we have access to an unbiased coin. We flip the coin and obtain either heads or tails. If $x \leq S - R$ and we get heads then we set $x' = x$. But, if $x \leq S - R$ and we get tails then we set $x' = S - x$. If $x > S - R$ and we get heads, then $x' = x$. If $x > S - R$ and we gets tails then we repeat the entire algorithm from the beginning. It is clear that x is readily obtainable from x', since $x = x'$ unless $x' > R$, in which case $x = S - x'$.

Claim 4 x' *is uniformly distributed in* S.

Proof. x is chosen uniformly at random from $[1..R]$. So, the probability that a particular x is chosen is $1/R$. In the case that $x \leq S - R$, x' will be set to x with probability $1/2R$ and x' will be set to $S - x$ with probability $1/2R$. Thus the values of x' at the beginning and ending of the range of S are uniformly distributed. It remains to show that the values in the middle have the same probability of occurring. If $x > S - R$, then x' will be set to x with probability $1/2R$. If the toss comes out tails, then the experiment is repeated. QED.

Note that in the version of PAP presented above, if we take $R = Z_N^*$, the values 1, p, and q are not in R. Such minute discrepancies can be ignored however.

5   Strong Setup in RSA Key Generation

In [YY96] a setup for RSA [RSA78] key generation was proposed. This setup constitutes a regular setup but can be modified to be a strong setup. To see why the previous attack does not constitute a strong setup, consider the following. The user knows his public modulus n, his public exponent e, and his private exponent d. From these he can factor n and recover the secret primes p and q. If the user knows exactly how the attack is implemented (i.e., the attacker's public key, the fixed symmetric key, etc.), then he can detect the mechanism based on p and n. The user simply encrypts p in the same way as the mechanism would and compares the result to the upper order bits of n. If they match, then he has successfully distinguished C' from C in poly-time. However, the setup as stated is a regular setup, since knowledge of the fixed symmetric key is needed to detect any possible bias in the output.

We will now describe a modification to the setup based on the discrete log attack that constitutes a strong setup. This version of PAP contains the attacker's ElGamal public key (Y, g, P). P is the same size as the prime p being generated. The attacker keeps his private key X secret. Let $a = G(b, c)$ denote a pseudo-random function G that when applied to the data b using the key c produces a value a. Let M be the number of bits in the representation of P. Finally, let K be a fixed symmetric key which need not be secret that is included within the device. Below is the pseudo-code for the setup attack.

1. choose c_1 randomly where $c_1 < P - 1$
2. solve for z in $Y^{ac_1+b} g^{Wt} z = g^{c_1} \bmod P$ (the discrete log attack)
3. remove the bias of z to get z' using the PBRM (assuming that the input of H needs to be distributed uniformly in some domain larger than P), goto step 1 if repeat is necessary
4. set $z'' = H(z')$
5. set lowest order bit of z'' to 1 (so z'' is odd)
6. set $p = z'' + num$ where num is the smallest positive integer that makes p prime (increment in steps of 2 and check odd values for primality. We assume that $num \leq B_1$ where B_1 is some constant)
7. apply PBRM to $g^{c_1} \bmod P$ to get a value v, repeat step 6 as necessary
8. for $(i = 0; i < B_2; i++)$ do steps 8 through 12
9. $U = G(v, K + i)$
10. choose the value RND uniformly at random from $\{0, 1\}^M$
11. Let $[U][RND]$ be the concatenation of the bit-strings U and RND
12. solve for q in the equation $[U][RND] = pq + r$
13. if q is prime then set $n = [U][RND] - r$ and goto step 14
14. goto step 1
15. calculate the RSA exponents e and d

To find out if a given public key was created using PAP, the attacker does the following. He first sets U to be the upper order bits of the victim's public modulus n such that there are M bits to the right of this value. He then decrypts U using $K + i$ and where i ranges from 0 to $B_2 - 1$. If any of the resulting values is greater than or equal to N, then a toss of tails occurred in the last application of the PBRM, so the correct value for $g^{c_1} \bmod P$ needs to be calculated. The attacker then decrypts all of the values for $g^{c_1} \bmod P$ using his private key to get the set of possible values for z. Since the PBRM was used, there are at most two possible values z' for each z. For each z', we compute $z'' = H(z')$ and set the lowest order bit of z'' to one. We then increment in steps of two to get the set of candidate values for p. Like before, we increment in steps of two to check only odd values. The number of candidate values are limited by the value B_1. If any of one of the resulting values divides n, then the attacker has successfully factored the victim's modulus. If a factor isn't found, then the attacker decrypts $U + 1$ and proceeds as before. Note that since the PAP ignores the remainder upon dividing [U][RND] by p, it is possible that a borrow bit modified the upper order bits of n. It is for this reason that the attacker must try $U + 1$ as well. If by then, a factor isn't found, the attacker concludes that his version of PAP was not used to generate the public key.

A few explanations for why PAP operates in this way are in order. PAP applies bias removal to $g^{c_1} \bmod P$ to prevent statistical detection of the setup mechanism. We assume that H is a pseudo-random function, so we did not apply bias removal to z''. Note that the transformation $z'' = H(z')$ insures that p can have a value larger than the attacker's public modulus. The reason for encrypting $g^{c_1} \bmod P$ using G is to take advantage of the pseudo-randomness and to avoid the overhead of excessive modular arithmetic, the amount of which is dictated by the prime number theorem. Hence, this step is essential to ensure a good probability of finding a valid p and q. We implemented the strong setup for RSA. See the appendix for an analysis of its performance.

We would like to briefly add that the setup attacks on DSA and Kerberos given in [YY96] can be readily modified to become strong setups. This can be accomplished by leaking probabilistic public key encrypted data, where the PBRM has been applied to the ciphertext that results from the probabilistic encryption. The probabilistic encryptions prevent the user from detecting the contamination by re-encrypting the secret information (which he knows) and comparing.

5.1 Security of Strong RSA Key Setup

By making certain reasonable cryptographic assumptions, the values for p and q that are chosen by PAP are random.

Lemma 5. *Assuming that p and the upper order bits [U] of [U][RND] are random, q is random in the set of M-bit primes.*

Claim 5 *Assuming the design of PAP is publicly available, the output of C and C' are polynomially indistinguishable.*

Proof. PAP does not make its choices of exponents c_1 known. Hence, Claim 2 applies, and PAP is secure iff the DH problem is hard. Clearly the upper order bits [U] are chosen randomly in PAP. Since p is found from the strong one-way hash (and pseudo-random function [GGM86]) of z', it follows from lemma 5 that the probability distributions of C and C' are polynomially indistinguishable. QED.

It follows that

Theorem 3 *RSA has a strong setup as long as the DH problem is hard.*

As a side note, this setup can be modified to accommodate the generation of strong primes.

6   Conclusion

We found kleptographic attacks against systems that do not have explicit subliminal channels. The stealing was made more effective by repetitive correlated usage, and by increasing the leakage bandwidth through chaining. It was demonstrated that repeated use of a cryptosystem may generate "implicit channels" for attacks. Chaining, in turn, increases the applicability of stealing via SETUP mechanisms. We also refined and strengthened the notion of SETUP attacks.

Acknowledgments:
We would like to acknowledge Jo Schueth for pointing out the statistical attack on the RSA key setup and Hari Sundaram for improving the efficiency of the PBRM recovery algorithm.

References

[Des90]   Yvo Desmedt. Abuses in Cryptography and How to Fight Them. In *Advances in Cryptology—CRYPTO '88*, pages 375–389, Berlin, 1990. Springer-Verlag.

[DH76]   W. Diffie, M. Hellman. New Directions in Cryptography. In *IEEE Trans. on Information Theory*, 22(6), pages 644-654, 1976.

[ElG85]   T. ElGamal. A Public-Key Cryptosystem and a Signature Scheme Based on Discrete Logarithms. In *Advances in Cryptology—CRYPTO '84*, pages 10–18, Berlin, 1985. Springer-Verlag.

[GGM86]   O. Goldreich, S. Goldwasser, and S. Micali, How to Construct Random Functions. *J. of the ACM*, 33(4), pp 210-217, 1986.

[GM84]   S. Goldwasser and S. Micali, Probabilistic Encryption. *J. Comp. Sys. Sci.* 28, pp 270-299, 1984.

[KL95]   J. Kilian and F.T. Leighton. Fair Cryptosystems Revisited. In *Advances in Cryptology—CRYPTO '95*, pages 208–221, Berlin, 1995. Springer-Verlag.

[RSA78]   R. Rivest, A. Shamir, L. Adleman. A method for obtaining Digital Signatures and Public-Key Cryptosystems. In *Communications of the ACM*, volume 21, n. 2, pages 120–126, 1978.

[Sch]   Jo Schueth, public communication (sci.crypt).

[Sim85]   G. J. Simmons. The Subliminal Channel and Digital Signatures. In *Advances in Cryptology—EUROCRYPT '84*, pages 51–57, Berlin, 1985. Springer-Verlag.

[Sim94]   G. J. Simmons. Subliminal Channels: Past and Present. In *European Trans. on Telecommunication*, 5(4), 1994, pages 459–473.

[YY96]   A. Young, M. Yung. The Dark Side of Black-Box Cryptography. In *Advances in Cryptology—CRYPTO '96*, pages 89–103, Springer-Verlag.

A   Performance: Strong RSA SETUP

We demonstrated the practicality of the attack by implementing it and noticing that it performs reasonably well (takes longer in general but sometimes it is faster than a comparable setup-free version).

Our program was written in ANSI C and was linked with the GNU MP library version 1.3.2. Our program generates a 512 bit RSA public/private key pair using the strong setup mechanism described in this paper. Our implementation uses truerand of D. Mitchell and M. Blaze as a source of true randomness (it is part of AT&T CryptoLib by J. Lacy, D. Mitchell, W. Schell). These physically random values are used as seeds for a pseudo-random number generator. We chose to use Wheeler and Needham's TEA as our pseudo-random function (any other block cipher like DES will do). We used the probabilistic primality test from Knuth to test the random values. We chose B_1 equal to 256. The value for B_2 was also 256.

Table 1

512 bit RSA key generation times in seconds

Trial	SETUP gen	SETUP decr
1	404	93
2	35	15
3	63	52
4	104	120
5	17	176
6	150	262
7	172	131
8	334	153
9	132	264
10	133	116
Average	154.4	138.2

The SETUP gen column lists the SETUP key generation times. The SETUP decr column lists the amount of time required to derive a private key from the corresponding public key. We note that the times reported may potentially be decreased by doing the following. By simply hashing the pseudorandomly calculated value instead of applying the PBRM and then hashing, it is likely that the key generation times would be shorter. This would of course be done at the expense of not suppling the hash function with inputs that are uniformly distributed. What we see is variability in the timing; it may be possible therefore, to modify a system like PGP to contain a strong RSA SETUP mechanism such that it can't be detected by noticing a "substantial" delay in the key generation times.

Fast and Secure Immunization Against Adaptive Man-in-the-Middle Impersonation

Ronald Cramer (ETH Zurich *)   and
Ivan Damgård (Aarhus University ** & BRICS ***)

Abstract. We present a simple method for constructing identification schemes resilient against impersonation and man-in-the-middle attacks. Though zero-knowledge or witness hiding protocols are known to withstand attacks of the first kind, all such protocols previously proposed suffer from a weakness observed by Bengio et al. : a malicious verifier may simply act as a moderator between the prover and yet another verifier, thus enabling the malicious verifier to pass as the prover.
We exhibit a general class of identification schemes that can be efficiently and securely tranformed into identification schemes withstanding an adaptive man-in-the-middle attacker. The complexity of the resulting (witness hiding) schemes is roughly twice that of the originals. Basically, any three-move, public coin identification scheme that is zero knowledge against the honest verifier and that is secure against passive impersonation attacks, is eligible for our transformation. This indicates that we need only seemlingly weak cryptographic intractability assumptions to construct a practical identification scheme resisting adative man-in-the-middle impersonation attacks. Moreover, the required primitive protocols can efficiently be constructed under the factoring or discrete logarithm assumptions.

1   Introduction

An (public key) identification scheme (see for instance [9]) is an (interactive) protocol by means of which one party (the prover) proves its identity to another party (the verifier). Securing log-in procedures is a main application of such schemes. An identification scheme consists of an algorithm to generate public-key/private-key pairs, and a protocol for the prover and the verifier. The collection of eligible key-pairs is chosen such that it is infeasible to compute a corresponding private key when only the public key is observed. Typically, the protocol's purpose is to show that the prover "knows" the private key that corresponds to the prover's public key. Most known identification schemes take the

---

* Inst. for Theoretical Comp. Sc., ETH Zurich, CH-8092 Zurich, Switzerland. Email: cramer@inf.ethz.ch. Research done while employed at CWI, Amsterdam, The Netherlands.
** Maths. & Comp. Sc. Dept., Ny Munkegade, Aarhus, Denmark. Email: ivan@daimi.aau.dk
*** Basic Research in Computer Science, Center of the Danish National Research Foundation

W. Fumy (Ed.): Advances in Cryptology - EUROCRYPT '97, LNCS 1233, pp. 75-87, 1997.
© Springer-Verlag Berlin Heidelberg 1997

form of three move interactive where the verifier is required to send a random bitstring as a challenge. For such methods to be secure, the verifier must not be able to extract this private key from the prover. Formally, this notion of security is captured by considering *adaptive impersonation attacks*. The (probabilistic polynomial time) attacker is given a prover, who has access to a key-pair as produced by the key-generation algorithm, as a black-box. Thus, the attacker only sees the prover's outputs as dictated by the identification protocol and not any of its internal coinflips, private inputs, etc. Next, the attacker is allowed to query the black-box a polynomial number of times, playing the role of a (malicious) verifier. This means that the attacker is allowed to choose the challenges in any way thought suitable to extract information about the private key. In particular, the choice of any next challenge may depend on the entire history of the attack and public key. Next, the attacker is denied any further access to this black-box prover. The identification scheme is called secure against adaptive impersonation attacks if the attacker is still unable to impersonate the prover (execute the prover's part of the protocol, facing an honest verifier).

In [4] a weakness of identification schemes proposed until then was exposed. There, the authors explained how a malicious *man-in-the-middle* $\tilde{V}$ may abuse his conversations with an honest prover P to misrepresent himself as P to yet another verifier V. The attack is not by cryptographic ingenuity. But, simply pretending to be a verifier himself, $\tilde{V}$ actually forwards V's challenges to P and forwards P's replies to V. Thus, while P is under the impression that he is identifying himself to $\tilde{V}$, he is actually identifying himself to V, to the possible advantage of $\tilde{V}$. A remedy suggested in [4] has the prover and verifier (rather the devices that represent them) isolate themselves physically from the outside world. A Faraday's cage could be a suitable implementation. However, for identification over networks, for instance, this measure seems not to be useful.

We present a simple method to construct identification schemes resilient against adaptive impersonation *and* man-in-the-middle attacks. Though zero-knowledge [13] or witness hiding protocols [10] are known to withstand attacks of the first kind, all such protocols previously proposed suffer from the weakness observed by Bengio e.a. [4], since a malicious verifier may simply act as a moderator between the prover and yet another verifier, thus enabling the malicious verifier to pass as the prover. Using a three-move public coin protocol that is collision intractable (without knowing the private key, it is infeasible to pass the protocol) and honest verifier zero knowledge we build a witness-hiding identification scheme that differs from previous proposals in that an execution of a given proof of identity can only be unambiguously appreciated by the intended verifier. This is achieved by having the prover direct the protocol to the intended verifier's public key. It is consequently shown that resilience against man-in-the-middle-attacks follows from this approach. Note that the required primitive protocol corresponds to an identification scheme secure against passive impersonation and honest verifiers. Directing a proof to an intended verifier has been considered by other researchers in a different context, as we will explain later. Our contribution is to provide a general, secure and efficient immunization against adaptive man-in-the-middle impersonation attacks in identification schemes. Furthermore, we want the immunization to work even if the the orginal identification scheme satisfies only weak security properties.

Example schemes that satisfy our requirements include Schnorr's scheme based on discrete logarithms [18] or Guillou-Quisquater's scheme based on RSA [15]. But more generally, any one-way group homomorphism or any pair of claw-free trapdoor permutations gives rise to the desired building block. If we would take, for example, Schnorr's scheme [18] as input to our constructions, the resulting identification scheme would have twice the complexity (in terms of computation and communication) of [18]. But we are then able to prove that our scheme is witness-hiding and resilient against man-in-the-middle attacks if computing discrete logarithms is hard.

Conceptually, our method to disable man-in-the-middle attacks is as follows. Let X and Y be two players, where X wishes to identify himself to Y. Suppose now that we have an efficient method by which X could take Y's public key, and his own key-pair (his public key and secret key), and securely prove the statement "I know X's secret key or I know Y's secret key". If this protocol is *witness indistinguishable* (no information is released as to which is the case), only Y can be sure he is talking to X rather than anyone else. For, any other verifier Z would only know that he is talking to X or Y. Thus, if X directs his proof to Y as outlined above, the proof is unambiguous only to Y.

So why would this help against man-in-the-middle attacks? By the symmetry of the statement proved and by the asserted witness-indistinguishability of the proof, if Y could abuse his conversation with X to pass as X at Z as the man-in-the-middle would do, he must be able to do so without talking to X. Thus the man-in-the-middle attack reduces to a cryptographic attack. But now we invoke the witness-indistinguishability again to show that if Y's attack would succeed, he could compute X's secret key. This then contradicts our assumption that it is hard to compute the secret key from a random public key. We stress that this approach makes sense only if the keys are sufficiently indepedently generated. In the extreme case that two verifier keys are identical, it is clear that man-in-the-middle attacks are still feasible. More generally, a proof of security will fail if there is dependendence among these keys: if one is chosen as a clever function of the other (such as a random and secret power of a given key based on discrete logarithms), proof given to one verifier may still be "diverted" to another verifier. In Sections 6 and 7 we discuss this matter in detail and give examples of how proper key-generation can be enforced.

We note that the same basic idea of proving one of two statements in order to direct a proof to one specific verifier was found independently by Jacobson, Impagliazzo and Sako in [16]. Their main motivation was to make undeniable signature schemes more secure and non-interactive. Their method for building a verifier designated protocol uses a trapdoor bit commitment scheme. In comparison, our method shows that if you start with a protocol of a certain form, then a separate trapdoor bit commitment is not needed. On the other hand, their methods works for some protocols that are not of the form we consider. We also note that, in a different context, Chaum [5] proposed using trapdoor commitment schemes to ensure that only a particular verifier can appreciate a given proof. Dolev, Dwork and Naor [8] have introduced *non-malleable cryptography*, a theoretical primitive that includes prevention of man-in-the-middle attacks in a number of scenarios, and have proposed protocols that work under general cryptographic assumptions.

It is not so much the concept explained above that we advocate as the most significant contribution here. We would like to stress that the concept has been applied implicitly before, prior to [16]. [16] is the first paper applying the ideas to verifier-directed proofs, however. We know of at least one example, namely the protocol of Feige and Shamir [12] for bounded round general zero knowledge proofs. There, the prover commits to a witness for the NP-statement to be proved using an unconditionally hiding *trapdoor commitment scheme*, an instance of which is generated by the verifier. Indeed, the proof conducted there can be seen as showing that the NP-statement is true, or that the prover knows the verifier's trapdoor! To get the designated verifier proofs for general languages, postulated in [16] but not given, we can use the result of [12] and make sure that verifiers' instances of the trapdoor commitment scheme are independently generated.

In our setting, we restrict ourselves to the problem of identification, and attempt to formulate a very efficient solution to the problem of identification in the presence of an *adaptive man-in-the-middle attacker*. Moreover, we are only interested in solutions that allow for some well-defined and accepted cryptographic intractability assumption to be reduced to the security of the identification scheme.

It is interesting to note that our results apply to a general class of identification schemes which in their normal mode of operation need only satisfy seemingly weak security properties. Namely, zero knowledge with respect to the honest verifier and collision intractability (that is, the scheme is secure against passive impersonation attacks). As a result of our simple and efficient transformation, we obtain the required security level, namely *security against adaptive man-in-the-middle attackers*.

Technically speaking, our approach is close to the ones taken in [7,6]. However, it is not clear from those papers (which may partly be seen as investigations into witness hiding) how we can efficiently obtain security against adaptive man-in-the-middle attackers in our context. Please note that such was neither clear from [16], since there the focus is on undeniable signatures. Although it appears to be true that their approach using trapdoor bitcommitments has a wider applicability than that, their approach does not indicate that immunization of an identification scheme against man-in-the-middle attackers, can be done efficiently and securely *even if* the given scheme is only weakly secure in normal mode of operation, as we discussed above.

Please note that digital signatures also lead to identification schemes secure against impersonation and man-in-the middle attacks. The prover would simply sign a message consisting of the concatenation of a random challenge (supplied by the verifier) and the verifier's public key. Although we feel that our schemes could compare favorably in terms of practical value to even such solutions, we like to point out that we aim for a practical identification scheme that is proven secure if some standard cryptographic intractability assumption holds. Seen in this light, digital signatures, for example, with such proven security, i.e. signatures secure against adaptively chosen message attacks, still come at too high a price in this context. Nevertheless, it may be reasonable here to use them for key-certification. Note that in this signature based approach, the prover (in this case the signer) leaves a trace: the verifier can later prove to a third party that he talked to the prover. In some cases this is undesirable as it might damage the privacy of the prover. This problem is not present in our approach: because the verifier could (using his own secret key) simulate the protocol perfectly, he cannot use a transcript of the protocol to convince a third party.

If one aims at practical value *and* proven security (relative to a plausible assumption), it may be true that our proposal for identification schemes secure against impersonation and man-in-the-middle attacks comes close to what one could reasonably achieve in this area, due to its conceptual simplicity and efficient implementation.

This work is organized as follows. First, we define a general class of "weak" identification schemes in Section 2, to be used later as the building block for our transformation. The existence of our building blocks is discussed in Section 5. The main result and its proof of security are given in Sections 3 and 4. Section 6 discusses in detail the key-generation requirements. Finally, we give an application to *access-control* in Section 7.

2 Model

We define the basic ingredients to our results.

Σ-Protocols Let (A, B) be a three move protocol where the prover A speaks first. The verifier B is required to send random bits only. A and B are probabilistic polynomial time (PPT) machines. The protocol (A, B) resembles a proof of knowledge for a binary relation R (see for instance [9] for details), in that the prover can always make the verifier accept on common input x, if the prover knows w such that $(x, w) \in R$. By running (probabilistic) polynomial time algorithm $a(\cdot)$ on x and his secret witness w, the prover A computes his initial message a. After having received the initial message, the verifier B chooses a bitstring $c \in \{0, 1\}^{t_B}$ uniformly at random, and sends it as a challenge to A. The challenge length t_B is assumed to depend only on the binary length of the common input x (and the protocol (A, B) of course). The prover completes the conversation by running (probabilistic) polynomial time algorithm $z(\cdot)$ on x, w, a, c, thereby possibly re-using the random bits used in the computation of the initial message. The resulting response z is submitted to the verifier. By invoking the (probabilistic) polynomial time procedure ϕ, the verifier tests the validity of the conversation. We call such a protocol (A, B) with the properties described above a Σ-protocol [1] for relation R.

Furthermore, we introduce the following terminology and notation. A sequence (x, a, c, z) is called an accepting conversation if and only if $\phi(x, a, c, z) = accept$. A pair of accepting conversations (x, a, c, z) and (x, a, c', z') with $c \neq c'$, is called a *collision*. When a verifier B follows the protocol, i.e. chooses the challenge indeed at random, that verifier is called *honest*. For an arbitrary prover A^*, (A^*, B) denotes the interaction between A^* and the honest verifier B, on some given common input.

Required Security Properties First, we need the protocol to satisfy a weak form of knowledge-soundness.

Definition 1. Let k be a security parameter for protocol (A, B). Suppose we are given a PPT generator G for relation R that on input 1^k produces $(x, w) \in R$, such that no PPT algorithm E, given x as input, can generate two accepting conversations (a, c, r), (a, c', r') with $c \neq c'$ (a "collision for x"), except with negligible probability of success (probability taken over the coinflips of E and G). Then (A, B) is called *collision intractable* over G.

Note that we don't require that a witness can be extracted from a successful prover. Thus, the protocol need not be a proof of knowledge. The property implies that, given as input a random instance x only, it is infeasible to construct a successful prover for that instance. In particular it follows from our assumptions that it must be hard to compute a witness w from a given x (when x is generated according to G). By a standard rewinding argument (see Bellare and Goldreich [3]), we have the following.

Proposition 2. *Let a Σ-protocol (A, B) for relation R be given, and let $x \in \{0,1\}^*$. Suppose that A^* is an arbitrary PPT prover such that (A^*, B) succeeds with probability ϵ, on common input x. Let $T_{A^*}(x)$ be A^*'s running time and suppose that $\epsilon > 1/2^{t_B}$. Then there exists a probabilistic algorithm Ext that outputs two accepting conversations x, a, c, z and x, a, c', z' with $c \neq c'$ (that is, a collision), with expected running time polynomial in $T_{A^*}(x)$ and $1/(\epsilon - 1/2^{t_B})$. Ext is allowed to run A^* as a rewindable blackbox. The probability is taken over the coin tosses of Ext and A^*.*

Next, we will assume the protocol (A, B) to be honest verifier zero-knowledge, that is, we only demand that conversations with the honest verifier can be simulated (perfectly).

---

[1] Of course, there is nothing new about three move, public coin protocols as such in cryptography, but we have decided to give them a name, derived from *zig-zag* and *Merlin-Artur* (see [2])

Definition 3. Let $(x, w) \in R$. Let a prover A and a verifier B execute (A, B), both following the protocol. Let x be the common input and let w be private input to the prover. Suppose we are given a probabilistic polynomial time algorithm M with the following properties.

1. On input x, M outputs an accepting conversation.
2. The distribution of the conversations generated by A and B is equal to $M(x)$.

Then (A, B) is said to satisfy *honest verifier zero knowledge*, with *simulator M*.

Relation with Identification Schemes We can view a Σ-protocol (A, B) for relation R as an identification scheme by identifying a public/private key-pair with a pair $(x, w) \in R$, as generated by some given generator G.

It is easy to see that such a protocol constitutes an identification scheme *secure against passive attacks*, if (A, B) is collision-intractable over G and if the length t_B of the challenges is large enough, say linear in the security parameter. Indeed, by Proposition 2, we can extract collisions with non-negligible probability from a passive attacker (that is, one which is given the public x only) having non-negligible probability of success. But this would contradict our assumption that (A, B) is collision-intractable over G.

Adding honest verifier zero knowledge to our requirements, makes sure that the resulting scheme is *secure against random challenge atacks*. By this we mean that even an attacker which is allowed to query a prover on *random* challenges, cannot *later* pose as that prover. Note that we use here the previous observation that collision-intractability implies security against passive attacks.

Security against adaptive attacks means that even though the attacker is allowed to query a prover on *any* challenge of his choice and in an adaptive fashion, it can still not later pose as that prover. This is basically the notion of security from [11].

The *adaptive man-in-the-middle attacker*, is one which has "adaptive access" to a prover X as well. Additionally however, the attacker is allowed to pose as any verifier $\tilde{Y}$ out of a given set V of verifiers, and have X identify itself to this verifier. The attacker's goal is to make an honest verifier Y, with $Y \notin V$, accept X, possibly running executions of X's identification to any $\tilde{Y} \in V$ online. If this is infeasible for any PPT attacker, we say that the identification scheme is *secure against adaptive man-in-the-middle impersonation*. Note that our definition combines the notions of security from Feige *et al.* [11] and Bengio *et al.* [4].

Our purpose is to transform identification schemes that are only secure against random challenge attacks into ones that withstand even adaptive man-in-the-middle impersonation, which seems to be the most desirable security level for public key identification schemes.

3   Main Result

Let (A, B) be a collision-intractable Σ-protocol for relation R and generator G. Suppose that (A, B) is honest verifier zero-knowledge, with simulator M, and that the challenge length t_B is linear in the security parameter k. Thus, by the remarks above, (A, B) constitutes an identification scheme secure against random challenge attacks. Our purpose is to transform (A, B) into a new identification scheme which is secure against adaptive man-in-the middle impersonation. This transformation works as follows.

Key Generation   A keypair $(x, w) \in R$, consisting of a public key x and a secret key w, for participant X is generated as

$$(x, w) \leftarrow G(1^k)$$

for an appropriate security parameter k. The public key x is placed in X's public directory. The secret key w is held privately.

Identification of X to Y   Here, participant X will identify itself to participant Y. Let their respective public keys be x and y, and let X's secret key be w. The claimed identification protocol runs as follows.

Move 1: X computes $a \leftarrow a(x, w)$ and $(b, d, s) \leftarrow M(y)$. Then X sends the pair (a, b) to Y.

Move 2: Y selects C uniformly at random from $\{0, 1\}^{t_B}$ and sends C as a challenge to X.

Move 3: X puts $c \leftarrow C \oplus d$ and computes $z \leftarrow z(x, w, a, c)$, and sends z, d, s to Y. Finally, Y checks the conversation by verifying whether $\phi(x, a, C \oplus d, z) \stackrel{?}{=} accept$ and $\phi(y, b, d, s) \stackrel{?}{=} accept$. If these verifications are satisfied, X is accepted by Y.

Please note that the secret key of the verifier Y is not used during the identification. One can imagine a scenario where the set of provers is disjoint from the set of verifiers. In this case, no storage of secret data is required at the verifier's side.

From a technical point of view the protocol above is quite similar to that given in Corollary 13 from [7] (while collision-intractability and honest verifier zero knowledge as a building block is taken from [6]). That result may be viewed as a way to transform identification schemes secure against random challenge attacks into ones that withstand adaptive challenge attacks only.

The cryptographic assumptions needed here are potentially weaker. But most importantly, here we show how the protocol from Corollary 13 [7] can be "rearranged" so as to withstand even man-in-the-middle attackers. Thus from the point of view of functionality, the protocol presented here is superior. Another difference is that here the length of the public key is invariant under the transformation.

4 Security Analysis

We give proof of security under the assumption that the participants' keys are generated as prescribed in the Key Generation protocol. In Section 6 we explain in detail why this assumption is needed and we also propose ways of enforcing this. An application where this condition is satisfied in a natural way is presented in Section 7.

Before we give the proof, we 'd like to point out that an execution of the protocol from Section 3 *leaves no trace*, in the sense that a verifier Y cannot later prove to a third that X identified itself to Y earlier. This follows from the symmetry of the protocol: Y can generate the conversations of the identification of X to Y with exactly the same distribution on its own.

Theorem 4. *Let (A, B) be a collision intractable, honest verifier zero knowledge Σ-protocol for relation R and generator G. Assume that the challenge length t_B is linear in the security parameter k. Then the identification scheme based on (A, B) from Section 3 is secure against adaptive man-in-the-middle imperson-ation.*

Proof. The idea is as follows. First we generate public key x' according to G, and discard the corresponding secret key. We show that, if the protocol were not witness hiding or were not resilient against man-in-the-middle attacks, there exists an efficient algorithm that takes x' as input and outputs a collision for x' in the protocol (A, B). But this would then contradict (A, B)'s collision-intractability.

The following game is easily be seen as modelling the situation. Let m be polynomial in the security parameter k. We generate m public keys with known secret keys by running G m times. We flip a coin b. If $b = 0$, then we put $x \leftarrow x'$ and assign the m key pairs to $Y_1 \ldots Y_m$. If $b = 1$, we select j at random from $\{1, \ldots, m\}$, and put $y_j \leftarrow x'$, and assign the m key pairs to X, $Y_1, \ldots, Y_{j-1}, Y_{j+1}, \ldots Y_m$.

The game consists of two stages.

1. The attacker gets the following prover as a black-box. We define P as the prover who gets x and all public keys y_i as input, plus the secret keys as generated above. P can perform the identification protocol for all pairs (x, y_i). The attacker is allowed to play with P (as a blackbox, but not rewindable) for a polynomial amount of time. Then, the attacker gives us a number $j' \in \{1, \ldots, n\}$, and hands back P. This models the idea that before the real attack, the attacker may try to extract as much information as needed for winning in the second stage.

2. With probability $(m + 1)/(2m)$, the attacker chose $j' = j$ such that P was not given the secret key for y_j in the beginning or was not given the secret key for x. Let's assume that this event happens (If not, we re-run the previous stage). Next, the attacker gets as input the secret keys for all public keys y_i with $i \neq j$. This models the idea that (possibly via a man-in-the-middle

attack), the attacker tries to pass as X to any other verifier intended by X. To make the proof easier, we just give the attacker the secret keys which allow him to perfectly simulate X's behaviour at any other site than Y_j, rather than giving him X as a blackbox: if he can't do it *with* the secret keys, than he certainly can't when he is given X as a blackbox who only identifies himself at Y_i with $i \neq j$. The attacker wins the game, if he can pass the protocol against the honest verifier on input (x, y_j).

Let's assume that the attacker won with probability $\epsilon > 2^{-t_B}$ (recall that t_B is assumed to be of linear size in k). Then, by Proposition 2, we can extract a collsion for y_j or for x from the attacker (running it as a rewindable blackbox) with expected time polynomial in the running time of the attacker and $1/(\epsilon - 1/2^{t_B})$. Thus, if ϵ is non-negligible, then we can extract a collision from the attacker in expected polynomial time. But, this is a collision for key x' with probability $1/2$, since the attacker cannot distinguish between the cases $b = 0$ and $b = 1$ by witness indistinguishability of the protocol (which follows by the properties of the simulator M). This contradicts the assumption that (A, B) is collision-intractable over G.

5 Existence

The following theorem can be derived from the results in [6], and gives an indication of the generality of our primitive.

Theorem 5. *Suppose that a family of claw-free pairs of trapdoor permutations exists, or that a family of one-way group homomorphisms exists. Then there exists a Σ-protocol for relation R, with generator G, that is collision-intractable and honest verifier zero knowledge and that has a challenge length linear in the security parameter.*

If based on claw-free pairs of trapdoor permutations, we can always efficientlyenforce the challenge length of (A, B) to be linear in the security parameter, while keeping the size of the initial message, the reply and the length of the common string constant in length. For one-way group homomorphisms, we can do something similar, under the condition that for each such homomorphism f, there exists a (large) prime v with the following property: for each y in the range of f, it is easy to compute a preimage x of y^v (using multiplicative notation for the group operation in the range). Two important examples of such families of one-way group homomorphisms can be constructed under the factoring and discrete logarithm assumptions. We give no further details of the general construction here.

A particularly efficient implementation, for example, is obtained when (A, B), for instance, is Schnorr's protocol [18] or Guillou-Quisquater's [15]. The following example is based on Schnorr's identification protocol. Let G_q be a group of prime order q such that computing discrete logarithms in G_q is hard. Let g be a fixed member of G_q.

Key Generation A keypair, consisting of a public key and a secret key, for participant X is generated as

$$(x = g^w, w)$$

where w is chosen at random from $\mathbb{Z}_q$. The public key x is placed in X's public directory. The secret key w is held privately.

Identification of X to Y Here, participant X will identify itself to participant Y. Let their respective public keys be x and y, and let X's secret key be w. The claimed identification protocol withstanding adaptive man-in-the-middle impersonation runs as follows.

Move 1: X computes $a \leftarrow g^u$ and $b \leftarrow g^s y^{-d}$, where u, s and d are chosen at random from $\mathbb{Z}_q$. Then X sends the pair (a, b) to Y.

Move 2: Y selects C at random from $\mathbb{Z}_q$ and sends C as a challenge to X.

Move 3: X puts $c \leftarrow C + d \bmod q$ and computes $z \leftarrow cw + u \bmod q$, and sends z, d, s to Y. Finally, Y checks whether $g^z \stackrel{?}{=} ax^c$ and $g^s \stackrel{?}{=} by^d$, where c is defined as $C + d \bmod q$ If these verifications are satisfied, X is accepted by Y.

6 A Note on Key-Generation

Using our example based on discrete logarithms from Section 5, we explain why it is important that key-generation takes place as demanded; if key-generation is not taken care of as required, the following attack could be mounted against the scheme. Let's assume that some malicious party $\tilde{Y}$ wishes to be accepted as any prover X by some verifier Y. Let x and y denote their respective public keys.

The attacker $\tilde{Y}$ proceeds by selecting $\alpha, \beta \in \mathbb{Z}_q$, computing $\tilde{x} \leftarrow g^\alpha y^\beta$, and defining $\tilde{x}$ as its public key. Whenever any prover X identifies itself to $\tilde{Y}$, the latter can easily divert the communication to Y and be accepted as X as follows:

Move 1: Prover X identifies itself to $\tilde{Y}$ and the attacker $\tilde{Y}$ claims to be X to verifier Y. The attacker $\tilde{Y}$ proceeds as follows. Receive a and b from X. Compute $\tilde{b} \leftarrow b^{1/\beta}$ Forward a and $\tilde{b}$ to Y.

Move 2: Receive Y's challenge C, and forward it to X.

Move 3: Receive X's replies z, d and s. Compute $\tilde{s} \leftarrow (s - \alpha d)/\beta \bmod q$, and forward z, $\tilde{s}$ and d to Y, who checks that $g^z \stackrel{?}{=} ax^c$ and $g^s \stackrel{?}{=} \tilde{b}y^d$, where c is defined as $C + d \bmod q$. As a result, $\tilde{Y}$ is accepted as X by Y.

A simple way to enforce proper key-generation, is by having a *trusted registration authority*. This authority need only be active during registration of the public keys, and participants basically have to proof knowledge of their secret key before the public key can be registered. Some care must be taken however, because a man-in-the-middle attacker may also try to abuse an interactive key-generation protocol for the purpose of later misrepresenting himself. One possible solution is the following. Let X be a participant who wishes to have a public key registered. Then the authority computes $g_* \leftarrow g^{w'}$, where w' is chosen at random from $\mathbb{Z}_q$, and sends g_* to X. Next, X chooses w'' at random from $\mathbb{Z}_q$, computes $x \leftarrow g_*^{w''}$ and proves knowledge of w'' with respect to g_*, using a suitable interactive (zero-knowledge) protocol for instance. Finally, the authority registers x as X's public key and sends w' to X, who computes the secret key as $w \leftarrow w'w'' \bmod q$.

7 An Application

In this section, we give an example where the conditions on key-generation are satisfied in a natural way. Imagine an organization with m sites to which restricted access is applicable. Some n officials are granted access to some of these sites. When an accessor presents himself at one of these sites, his access rights are checked by verifying his identity. These sites may vary from buildings, specific sections of buildings, or even databases or computer systems. The organization keeps a central list of the identities of the officials and their specific access rights. It is assumed that each site has access to this list, either by having a copy of the list at hand, or by consulting the central database.

Let $X_1, \ldots, X_n$ be the collection of participants. The collection of sites with restricted access is denoted $Y_1, \ldots, Y_m$. The organization generates a keyset (x_i, w_i) for each participant X_i, as described in the Key Generation protocol in Section 3. Each participant X_i is given a tamperresistant smartcard S_i, capable of performing our protocols. The keyset is securely loaded into the cards. Now, for each site Y_j, the organization generates a keyset (y_j, v_j). The secret key v_j is destroyed. We assume that each site is represented too by some device capable of performing the protocols. For each site, the organization prepares a list of the public keys of the officials that are granted access to this site. This list is made available to the site. Please note that the devices for the sites need not store any secret information. One only has to make sure that the data they store is authentic and cannot be modified by unauthorized parties.

When participant X_i wishes to exercise his right of access to site Y_j, he lets his smartcard simply perform the identification protocol with site Y_j as the verifier, on common input (x_i, y_j). By the security properties of the identification scheme, the resulting protocol is secure against adaptive impersonation attacks, but furthermore, no adversary can by means of a man-in-the-middle attack, divert the communication to a different site Y_t, and pass there as X_i, even if X_i has the right of access at site Y_t.

References

1. M. Abadi, E. Allender, A. Broder, J. Feigenbaum and L. Hemachandra: *On Generating Solved Instances of Computational Problems*, Proceedings of Crypto '88, Springer Verlag LNCS, vol. 403, pp. 297–310.

2. L. Babai and S. Moran: *Arthur- Merlin Games: A Randomized Proof System and a Hierarchy of Complexity Classes*, JCSS, vol. 36, pp. 254–276, 1988.

3. M. Bellare and O. Goldreich: *On Defining Proofs of Knowledge*, Proceedings of Crypto '92, Springer Verlag LNCS, vol. 740, pp. 390–420.

4. S. Bengio, G. Brassard, Y. Desmedt, C. Goutier and J.J. Quisquater: *Secure Implementation of Identification Systems*, Journal of Cryptology, 1991 (4): 175–183.

5. D. Chaum: *Provers Can Limit the Number of Verifiers, unpublished.*

6. R. Cramer and I. Damgård: *Secure Signature Schemes based on Interactive Protocols*, Proceedings of Crypto '95, Springer Verlag LNCS, vol. 963, pp. 297–310.

7. R. Cramer, I. Damgård and B. Schoenmakers: *Proofs of Partial Knowledge and Simplified Design of Witness Hiding Protocols*, Proceedings of Crypto '94, Springer verlag LNCS, vol. 839, pp. 174–187.

8. D. Dolev, C. Dwork and M. Naor: *Non-malleable cryptography*, Proceedings of STOC '91, pp. 542–552.

9. A. Fiat and A. Shamir: *How to Prove Yourself: Practical Solutions to Identification and Signature Problems*, Proceedings of Crypto '86, Springer Verlag LNCS, vol. 263, pp. 186–194

10. U. Feige, A. Shamir: *Witness Indistinguishable and Witness Hiding Protocols*, Proceedings of STOC '90, pp. 416–426.

11. U. Feige, A. Fiat and A. Shamir: *Zero-Knowledge Proofs of Identity*, Journal of Cryptology 1 (1988) 77–94.

12. U. Feige and A. Shamir: *Zero-Knowledge Proofs of Knowledge in Two Rounds*, Proceedings of Crypto '89, Springer Verlag LNCS, vol. 435, pp. 526–544.

13. S. Goldwasser, S. Micali and C. Rackoff: *The Knowledge Complexity of Interactive Proof Systems*, SIAM J.Computing, Vol. 18, pp. 186-208, 1989.

14. *Efficient Identification Schemes Secure against Impersonation and Man-in-the-Middle Attacks*, preprint, October 1995.

15. L. Guillou, J.J. Quisquater: *A Practical Zero-Knowledge Protocol fitted to Security Microprocessor Minimizing both Transmission and Memory*, Proceedings of Eurocrypt '88, Springer Verlag LNCS, vol. 330, pp. 123–128.

16. M. Jacobson, R. Impagliazzo and K. Sako: *Designated Verifier Proofs and their Applications*, Proc. of Eurocrypt '96, Springer Verlag LNCS, vol. 1070, pp. 143–154.

17. T. Okamoto: *Provably Secure and Practical Identification Schemes and Corresponding Signature Schemes*, Proceedings of Crypto '92, Springer Verlag LNCS, vol. 740, pp. 31–53.

18. C. P. Schnorr: *Efficient Signature Generation by Smart Cards*, Journal of Cryptology, 4 (3): 161–174, 1991.

Anonymous Fingerprinting

Birgit Pfitzmann[1] *, Michael Waidner[2]

[1] Universität Dortmund, Informatik 6, D-44221 Dortmund, Germany;
email pfitzb@ls6.informatik.uni-dortmund.de
[2] IBM Zürich Research Laboratory, Säumerstrasse 4, CH-8803 Rüschlikon, Switzerland;
email wmi@zurich.ibm.com

Abstract. Fingerprinting schemes deter people from illegally redistributing digital data by enabling the original merchant of the data to identify the original buyer of a redistributed copy. Recently, asymmetric fingerprinting schemes were introduced. Here, only the buyer knows the fingerprinted copy after a sale, and if the merchant finds this copy somewhere, he obtains a proof that it was the copy of this particular buyer.

A problem with all previous fingerprinting schemes arises in the context of electronic marketplaces where untraceable electronic cash offers buyers privacy similar to that when buying books or music in normal shops with normal cash. Now buyers would have to identify themselves solely for the purpose of fingerprinting. To remedy this, we introduce and construct anonymous asymmetric fingerprinting schemes, where buyers can buy information anonymously, but can nevertheless be identified if they redistribute this information illegally.

A subresult of independent interest is an asymmetric fingerprinting protocol with reasonable collusion tolerance and 2-party trials, which have several practical advantages over the previous 3-party trials. Our results can also be applied to so-called traitor tracing, the equivalent of fingerprinting for broadcast encryption.

1 Introduction

Fingerprinting schemes are cryptologic mechanisms for the copyright protection of digital data. They do not rely on tamper-resistance, i.e., it is assumed that the buyers obtain the data digitally and can in principle copy them. Buyers who abuse this possibility by illegitimately redistributing the data are called traitors. Fingerprinting schemes discourage traitors by enabling the original merchant of the data to identify the traitor who originally bought the copy.

1.1 Known Classes of Fingerprinting Schemes

Conventional fingerprinting schemes, called symmetric here, essentially work as follows: The merchant prepares a slightly different "copy" of the data item for each buyer. If he finds a redistributed data item, he finds out to which of the copies sold it corresponds. This concept was introduced in [W83]. Examples of how one can make imperceptible differences in copies and more references can be found in [ZK95, BRD95, CKLS96]. Fingerprinting became a cryptologic topic with the problem of collusion tolerance: What if several traitors collude and compare their copies to find and then eliminate differences? This problem was first considered in [BMP86]; solutions that can tolerate larger collusions were presented in [BS95].

---

* The work of this author was done at the University of Hildesheim and supported by the DFG (German Research Foundation).

In these symmetric schemes, the merchant finds out the identity of a traitor, but cannot convince any third party of this treachery because he does not find anything in the redistributed copy that he could not have made up himself. In contrast, in asymmetric schemes, introduced in [PS96], the merchant obtains a proof of the treachery. For this, fingerprinting must be an interactive protocol between the buyer and the merchant where the buyer also inputs a secret and the merchant does not see the fingerprinted copy that this buyer obtains. Only if he finds this copy after a redistribution, he can extract the proof. The same collusion tolerance as in the symmetric schemes in [BS95] was achieved for asymmetric fingerprinting in [PW97, BM97].

So-called traitor tracing is the equivalent of fingerprinting for cryptologic keys. It was introduced in [CFN94] for broadcast encryption, i.e., for situations where the real data, e.g., a Pay-TV movie, are broadcast in encrypted form, and only the keys needed to decrypt the data are sold. Now a different personal key is sold to each buyer; the encryption scheme is adapted so that all the personal keys can be used to decrypt the same ciphertext. The schemes in [CFN94] already achieve good collusion tolerance. (Actually, these techniques were the basis for collusion-tolerant normal fingerprinting in [BS95].) Asymmetric traitor tracing, introduced in [P96], analogous to asymmetric fingerprinting, guarantees that the merchant obtains a proof of treachery if he finds a redistributed key. Reasonable collusion tolerance for asymmetric traitor tracing was also achieved in [PW97].

One type of scheme that so far only exists for traitor tracing [PW97], but not for normal fingerprinting, is an asymmetric scheme with reasonable collusion tolerance and 2-party trials. A 2-party trial means that the merchant can simply take his proof and convince any arbiter with it, whereas in a 3-party trial, the buyer also has to take part. One advantage of 2-party trials is that one need not find the buyer to carry out the trial. However, this advantage is minor because in a real trial, the buyer would have to be notified anyway and non-technical points would have to be discussed, e.g., whether someone could have stolen the data item from an honest buyer. More importantly, in a 3-party trial, the buyer also still has to find some secrets, which means that she should not have forgotten the password needed to use them or died without leaving it to someone else. Additionally, one has to take care with multiple trials about the same data item because the buyer might have to divulge something about her secrets in each trial. Finally, 2-party trials are much easier to use as subprotocols in other schemes, as we will see below.

1.2 Anonymous Fingerprinting

Electronic marketplaces are supposed to offer similar privacy as current marketplaces. Thus it should be possible to buy cheap objects like books, pictures, and pieces of music anonymously. This becomes even more important if one buys individual articles of what would have been a book or a magazine on paper because the choice of articles gives a lot of information about a person's lifestyle, habits, etc. For such purposes, anonymous networks, anonymous cash-like payment systems, and even protocols for anonymous, but secure exchange of payment and goods exist, see, e.g., [C81, C85, BP90] for early examples and [B94] for an efficient anonymous off-line payment system with identification of double-spenders.

It would be a pity if all this anonymity were destroyed just because the buyers had to identify themselves for the purpose of fingerprinting or traitor tracing. However, this undesirable situation would occur with all previous symmetric and asymmetric fingerprinting schemes: The buyer has to identify herself for (key) fingerprinting during a purchase, and thus for each particular data item bought, e.g., one picture in fingerprinting or one Pay-TV movie in traitor tracing.

The goal in this paper is therefore to carry out fingerprinting anonymously, but nevertheless to enable the merchant to identify traitors later. This possibility of identification will *only* exist for traitors, whereas honest buyers will remain anonymous. All our schemes will be asymmetric, i.e., the merchant can also convince any third party that a particular person was a traitor.

1.3 Our Results

In Section 2, we introduce the exact model of anonymous fingerprinting and discuss some variants. In Section 3, we present a construction framework for anonymous fingerprinting that makes certain assumptions about an underlying fingerprinting scheme. In Section 4, we show how this framework can be instantiated with some existing fingerprinting and traitor tracing schemes, and why a gap remains. In Section 5, we fill this gap by constructing a scheme for collusion-tolerant asymmetric fingerprinting with 2-party trials, using Reed-Solomon-codes for low-rate error-and-erasure decoding. This scheme is of interest in its own right, too. The complexity of our constructions is still rather high; we mainly regard them as constructive proofs of existence.

2 Precise Model

We assume that at the start of our scheme, each buyer already has a key pair (sk_B, pk_B) of a digital signature scheme, so that the public key can serve as a digital identity. Thus we can require a buyer to sign something under her identity in a protocol.

For modularity, we also require buyers to register specifically for the fingerprinting scheme under their digital identity. This allows us to make the protocols of the fingerprinting scheme concrete, without fixing how the validity of the initial digital identity is verified. In some situations, this registration could be joined with the initial establishment of the digital identity. The parties where registration can be done are called registration centers. A reasonable choice is the buyer's bank, in particular if the fingerprinted data are paid with anonymous digital cash, because the buyer has to register with a bank anyway and will only be anonymous among this bank's clients. We do *not* require the registration centers to be particularly trusted by any other party; in the strongest of our models, the only bad thing a registration center can successfully do is to refuse registration.

Thus we have four types of parties: Merchants, buyers, registration centers, and arbiters who should be convinced in trials. Technically, the role of arbiter should not be restricted, i.e., it should be possible to convince anyone as long as they know a few specific public keys. We can still get quite a number of different definitions, depending on how active the registration centers and arbiters have to be, and whether the merchants and buyers have to trust the registration centers for any or many requirements. We are primarily interested in cryptologic solutions with minimal trust (where a cheating registration center can only refuse registration), but we also mention weaker models.

We only present a detailed definition for fingerprinting schemes, not for traitor tracing. We follow the style of [PS96], but introduce somewhat less explicit notation for brevity.

Definition 1 (Components of anonymous fingerprinting). An anonymous fingerprinting scheme consists of seven protocols. Each interactive algorithm for a party in a protocol is polynomial-time and may be probabilistic, and it may produce an output *failed* to indicate that the protocol could not be finished in the normal way. Security parameters k for computational security, σ for error probabilities in information-theoretic properties, and *coll_size* for the maximum number of colluding traitors are common inputs.

- *Registration center key distribution*: A registration center generates a key pair (sk_{RC}, pk_{RC}), typically of an underlying signature scheme, and distributes pk_{RC} reliably to all merchants, arbiters, and the buyers that might register at this center.

- *Registration* is a two-party protocol between a buyer and a registration center. The common inputs are the buyer's digital identity pk_B, the registration center's public key pk_{RC}, and possibly an upper bound N_B on the number of purchases that the buyer can make based on one such registration. The registration center's secret input is its secret key. We call the outputs the registration center's and the buyer's registration records.

- *Data initialization* is an algorithm the merchant carries out for each data item to be sold. He inputs the data item and possibly an upper bound N_M on the number of times he will sell it. (This protocol could be included into the first sale, i.e., the first execution of fingerprinting, but it is often useful to consider common precomputations separately.) The output is called the merchant's initial data record.

- *Fingerprinting* is a two-party protocol between a merchant and an anonymous buyer. The merchant secretly inputs the data item and the corresponding initial data record, and, not necessarily secretly, the public key of the registration center with whom the buyer registered. The buyer inputs her registration record or an update of it, and both input a common text that describes what this purchase is about.

  The output for the merchant is called a purchase record. The main output for the buyer is the fingerprinted data item; she may also obtain an update on her registration record (e.g., a purchase counter is increased and in schemes with 3-party trials, evidence is stored).

- *Identification* is either an algorithm for the merchant alone or a two-party protocol between the merchant and the registration center. The merchant's input is a redistributed data item whose original buyer he wants to identify, the original version of this data item, the initial data record, and all the purchase records for this data item. If the registration center takes part, its input is its registration records.

  The output for the merchant should be the identity of a buyer, the text used in the particular purchase, and another string called *proof*.

- *Enforced identification*. For cases where the registration center is needed in identification, but refuses to cooperate, there must be a 3-party version of identification that includes an arbiter. The merchant should get the same outputs as in identification, and the arbiter either obtains the output *ok* or *center_guilty*, which denotes that the arbiter has noticed misbehavior by the registration center.

- *Trial* is a two- to four-party protocol between at least the merchant and an arbiter, and possibly a buyer and a registration center. The common inputs are the identity of the accused buyer and the text denoting the disputed purchase. The merchant also inputs the string *proof* obtained in identification. If the registration center takes part, it inputs the registration record of this buyer, and if the buyer takes part, she inputs her current updated registration record (typically just the evidence from the disputed purchase).

The main output is the arbiter's result. It may be *guilty*, which means that the arbiter finds the buyer a traitor, or *not_guilty*, which means that he rejects the accusation. In some systems, the output can also be *center_guilty*, which means that no decision between the merchant and the buyer could be reached because of wrong behaviour of the registration center. ♦

In the following, we describe the security requirements on such a scheme. All should also be fulfilled under active attacks. Generally, an active attack means that the attackers can influence the sequence of protocols the honest users carry out and the user inputs (e.g., the texts), obtain some outputs from the users (e.g., whether a protocol failed or not), and behave maliciously during the protocol executions.

Definition 2 (Effectiveness).

- Correct case. Registration and data initialization should end successfully, i.e., not with the output *failed*, if the parties in the given protocol execution are honest. Similarly, fingerprinting should end successfully if the merchant, the buyer, and the buyer's registration center are honest, and the fingerprinted data item should be sufficiently similar to the data item input by the merchant. Similarity can be formalized by a given relation as in [PS96].

- No jamming by registration center. Even for a cheating registration center, it is infeasible to carry out registration with a buyer such that it ends successfully, but nevertheless an execution of fingerprinting between this buyer and an honest merchant will fail later. ♦

The second property is one of those that define minimal trust in the registration center. Of course, it cannot be avoided that a cheating registration center refuses or messes up registration altogether. However, if the buyer notices this by the output *failed*, it is no problem: She can register at another center. It would only be a problem if fingerprinting failed later and the buyer and the merchant would not know whether to blame each other or the center. The name "jamming" was taken from the consideration of similar frauds by arbiters in arbitrated authentication schemes in [DY91].

Definition 3 (Integrity).

- Security for the merchant. For any algorithm $\tilde{B}$ of the cheating buyers that buys at most *coll_size* copies of a certain data item (i.e., engages in at most *coll_size* executions of fingerprinting for it) and then produces another copy sufficiently similar to the original for the merchant to feel cheated, the merchant will successfully identify a buyer, i.e., obtain a valid digital identity as an output in identification, together with a text used and a string *proof*, and then win a trial with any honest arbiter. Similarity is defined by a second given relation as in [PS96], and $\tilde{B}$ may carry out any other transactions, such as additional registrations and buying other data items, in between as part of its active attack.

  This should hold even if the registration centers are cheating, i.e., $\tilde{B}$ also comprises them. In this case, the protocol for enforced identification may be needed if normal identification failed, and the output for the arbiter in either this protocol or the trial may be *center_guilty*, instead of *guilty* in the trial.[1]

- Protecting the merchant from making wrong accusations. As the merchant will usually damage his reputation if he accuses a buyer and then loses the trial, we require that this does not happen to honest merchants. Thus, even if there are more than *coll_size* traitors, it should be infeasible for the other participants to make up a data item such that identification succeeds, and then a trial with an honest arbiter leads to the output *not_guilty*.

- Security for the buyer. Honest buyers are not found guilty in trials. More precisely, if a buyer only takes part in the prescribed protocols and keeps their results secret (in particular, the data item bought), then, no matter what the other parties do, an honest arbiter will not obtain the output *guilty* in a trial where he entered the identity of this buyer. Even if the other parties can adaptively obtain some data items this buyer bought, selected by the texts used in the corresponding execution of fingerprinting, the buyer will not be found guilty for any other texts.

- Security for registration centers. In schemes with strong security for the merchant, i.e., where an arbiter may decide *center_guilty*, honest registration centers require that honest arbiters never decide this about them.   ◆

In a weaker version of security for the merchant, the requirement would only hold if at least the registration centers the dishonest buyers registered with are honest. A similar weak version of the security for the buyer is not desirable because being wrongly found guilty as a traitor is a fate much worse than losing some revenue.

Finally, we come to the privacy requirements. We only make them explicitly for buyers, corresponding to the usual model of payer anonymity in digital payment systems. However, the identity of the merchant is not needed anyway, neither above nor in other types of fingerprinting.

Definition 4 (Anonymity). Nothing about the purchase behaviour of honest buyers becomes known to any other party, except, if the registration center cooperates, for facts that can simply be derived from the knowledge of who registered and for what number of purchases, N_B, and at what time protocols are executed. This should even hold for the remaining purchases if the other parties can adaptively obtain some data items this buyer bought.   ◆

The exception cannot be avoided. For instance, if the first person who registers buys something before anybody else registers, the merchant and the registration center together naturally know who it was. Furthermore, the definition assumes, like that of anonymous payment systems, that the underlying communication does not identify the buyers. The definition is otherwise very strict. For instance, it implies that the merchant cannot learn whether a particular buyer bought a particular data item by accusing her unjustly of redistribution.

---

[1] A stronger requirement that it is always a buyer who is identified would not make much sense: If a registration center colludes with some traitors, it can be regarded as one of them; actually, identifying a cheating registration center is more important than identifying a normal buyer and the merchant is more likely to get compensation.

We could also define weaker versions of anonymity, in particular k-out-of-n traceability and linkability. Similar models have been considered with payment systems, often without distinguishing them. Some types of fingerprinting with weak anonymity can be implemented quite easily and without any real additional cryptology, but we omit these constructions in favor of stronger ones.

3 Construction Framework for Full Anonymity

During fingerprinting, the buyer has to input identifying information that will be embedded into the data item; we call it *emb*. The merchant must be convinced that this information is correct, but without learning more about it. Hence a construction has to address two major issues:

- Relating the identifying information *emb* to the public key of the registration center, so that the merchant has a starting point for the verification that does not identify the buyer and does not make purchases linkable, together with a minimum-knowledge verification procedure.

- A mechanism for the merchant to extract *emb* from a redistributed data item. This is not trivial because in most non-anonymous schemes, information is not simply "extracted" from the data item found, but derived in combination with other information or in interaction with an accused buyer, each of which is more complicated here.

In this section, we show a construction framework that includes a solution to the first issue, but assumes a subprotocol that solves the second issue.

Construction 1 (Framework for anonymous fingerprinting). We only show those protocols where anything interesting happens at this level of abstraction.

- In registration, the buyer selects a pseudonym, i.e., a key pair (sk_B^*, pk_B^*) of a signature scheme, and signs under her normal identity that she will be responsible for this pseudonym. She obtains a certificate $cert_B$ from the registration center, i.e., a signature with sk_{RC} on pk_B^*. Intuitively, this certificate means that the registration center declares that it knows the identity of the buyer who chose this pseudonym.

- In fingerprinting, the anonymous buyer secretly computes a signature on the text identifying the purchase, $sig := sign(sk_B^*, text)$. The entire value to be embedded is $emb := (text, sig, pk_B^*, cert_B)$. This buyer hides this value in a commitment (see [BCC88]), sends the commitment to the merchant, and proves the validity of the hidden signature and certificate in zero-knowledge.

  Instead of embedding *emb* directly, the buyer can encrypt it, send the ciphertext to the merchant, and commit to and embed the key, which may be much shorter. The zero-knowledge proof now refers to the value obtained by decrypting the given ciphertext with the hidden key.

- In identification, the merchant extracts *emb*. He sends $proof_1 := (text, sig, pk_B^*)$, which proves that the owner of this pseudonym has redistributed the data item corresponding to *text*, to the registration center and asks for identification. If the registration center refuses, the merchant shows $proof_1$ to an arbiter, together with $cert_B$ to prove that the registration center knows the corresponding identity. Thus, in enforced identification, the registration center either has to identify or will be found guilty. The registration center also has to send the buyer's signature that she is responsible for this pseudonym. This signature and $proof_1$ constitute *proof*. The merchant verifies all the values before making an accusation.

In the version with encryption, the merchant tries to decrypt the ciphertexts from all the purchase records for this data item. He verifies the resulting cleartexts as above, and uses the first that fulfils the criterion.

- In a trial, the arbiter first verifies the accused buyer's signature that she is responsible for the pseudonym pk_B^*, and then that *sig* is a valid signature on *text* corresponding to this pseudonym.

Theorem 1. If all the underlying primitives are secure, the construction framework yields a provably secure anonymous fingerprinting protocol. ◆

The proof is quite straightforward and omitted here. The security assumptions about the underlying scheme for embedding are (a) for the security and anonymity of the buyer, that it does not leak information about *emb*, and (b) for the security of the merchant, that extracting will in fact recover the embedded value if there are at most *coll_size* traitors. The security of the zero-knowledge proof scheme must be assumed in the same kind of composition that we allow for our protocols.

4   Instantiation with Known Fingerprinting Schemes

We now identify existing fingerprinting schemes that offer the combination of embedding and extracting needed in Construction 1. We also describe some details of other fingerprinting schemes because they help understanding the new construction in Section 5.

For the cryptologic aspects of fingerprinting, it is typically assumed (starting with [BMP86, BS95]) that a marking scheme is given, i.e., a data-type-dependent scheme for hiding individual bits in data items. Each mark is a part of the data item for which 2 versions exist. In data initialization, the merchant probabilistically selects a tuple of l marks for the given data item. Each fingerprinted data item can now be described by a binary codeword of length l: the i-th bit denotes which version of the data is used in the i-th mark. It is assumed that traitors can only notice and delete marks by comparing their copies. More precisely, the Marking Assumption [BS95] states that if the codewords of all traitors agree in the i-th bit, any redistributed copy they make will correspond to a word with the same i-th bit.

A consequence of the Marking Assumption is that in any redistributed data item produced by at most *coll_size* traitors, the merchant will find a word that has at least $l / coll\_size$ bits in common with the codeword of at least one traitor. (If the traitors delete a mark they have identified, instead of using one of the 2 versions, the merchant arbitrarily sets the corresponding bit in the word to 0 or 1.) The merchant now has to derive some real information; this can be seen as a problem of error correction for far more errors than correct symbols. We now consider how different fingerprinting schemes deal with this problem, and whether they offer the direct extraction we need (+/–):

+ Symmetric schemes with (almost) no collusion tolerance: If there is no collusion at all, the marking assumption implies that the whole codeword of the traitor remains intact. Hence it can simply be extracted. Some schemes do not assume traitors to be clever and hope that the majority of one word will still be intact, so that a normal error-correcting code can be used.

– Symmetric collusion-tolerant schemes [BMP86, BS95]: Essentially, the merchant looks through the list of the codewords he has used and checks which of them has $l/coll\_size$ symbols in common with the redistributed word. (In fact, a somewhat more complicated code and comparison is used to make it provably unlikely that an honest participant's codeword also has so many symbols in common with the redistributed word.) These schemes cannot be used for embedding and extracting a significant amount of information because then the merchant would not know the codewords that were used, and a list of all possible ones would be exponentially long.

– Asymmetric schemes with 3-party trials also had to address the problem that the codewords used cannot be known to the merchant entirely because parts of them are needed to make up *proof*, the proof of redistribution, when they are found. The basic idea in [PW97, BM97] is to make one half of the codeword known to the merchant in fingerprinting and to keep the other half secret. In identification, the merchant first searches a list of the known halves to identify a buyer, whom he accuses. He only has the other half, which should contain *proof*, with a large number of errors, too many for efficient decoding. Thus the accused buyer is now asked to show the real *proof*, and the arbiter compares if it has enough symbols in common with what the merchant found.

However, this three-party idea cannot be used in the anonymous case because the merchant does not know whom to accuse before he has found the correct secret, and one cannot ask many buyers to divulge theirs. More technically, we see that *proof* is not actually extracted.

+ Asymmetric collusion-tolerant traitor tracing with 2-party trials [PW97, Section 4] (based on ideas from [CFN94]): A code is used where some parts of the codeword must be taken from one traitor as a whole. The entire secret that will be the main part of *proof* is used as many such parts, so that it will come through at least once.

This scheme can be used for embedding and extracting arbitrary values *emb*: These values are treated just as the main part of the proof was treated above. In the notation of [PW97] for readers familiar with it: *emb* is used as the second-level codewords instead of rid_B. All parts of the scheme that do not deal with embedding and extracting, i.e., the one-way image of rid_B and its signing and verification, are omitted.

For fingerprinting, there seems to be no idea yet how to glue large parts together so that they have to be taken from one traitor as a whole, as in traitor tracing. However, in the following, we will use much smaller parts that will be correct as a whole, and apply error-and-erasure-correcting codes.

5 Collusion-Tolerant Asymmetric Fingerprinting with 2-Party Trials

5.1 Ideas

Recall the basic idea from [BS95] to achieve a certain level of collusion tolerance among a large number of participants: A concatenated code (called nested in [B83]) is used where the outer words are of length l over the alphabet $\{1, ..., q\}$, and the inner code, which is used to encode each symbol of an outer codeword, is a fixed binary code Γ_0 of length $d(q-1)$, where l, d, and q are three parameters that we adapt to our purposes below.

The important property of Γ_0 is that it has a decoding procedure that guarantees that, except with exponentially small probability, an outer symbol that appeared in the codeword of at least one traitor will be extracted in each position. The precise error probability is $2^{-\sigma}$ for all l outer symbols together if d is chosen as $2q^2(\log_2(2ql) + \sigma)$.

Thus the symbols of the outer codeword are blocks that have to be taken from one traitor as a whole, as desired in the construction idea in Section 4. However, they can only encode a very small number of bits because the inner code is essentially unary. Thus we proceed in a more complicated way to put several such small pieces together again, i.e., to try and find a certain number that come from the same traitor. For this, we will link known and secret halfsymbols (in contrast to the unconnected known and secret symbols of the words in [PW97, BM97]), so that symbols that disagree on the known halfsymbols can be excluded right away. This leaves us with many erasures, but hopefully few errors, and thus we can hope for efficient decoding. We will do this with Reed-Solomon codes, but we first present the rest of the construction without fixing the code.

5.2 Construction with Generic Code

The following construction is only a scheme for embedding and extracting data. It can either be used in Construction 1 to obtain an anonymous collusion-tolerant fingerprinting scheme, or as a normal collusion-tolerant asymmetric fingerprinting scheme with 2-party trials. For the latter, the values *emb* are selected and treated like the values id_{sym} in Construction 1 of [PS96]: In fingerprinting, the buyer randomly chooses *emb* and gives the merchant a one-way image *im* of it, together with a signature. Later, knowing the preimage *emb* of *im* proves that the merchant found the redistributed data.

Note that in both these applications, Construction 2 and the surrounding scheme are coupled over a *secret* value, *emb*, that must be the same in both schemes, i.e., the same commitment must be used.

We denote the binary length of the values to be embedded as a function $len(k)$ of the computational security parameter because they are usually cryptologic secrets. The following construction is in terms of four parameters l, d, q_1, and q_2, which will be chosen as polynomial functions of the given parameters k, σ, *coll_size*, and N_M. Here, l and d will be used for a concatenated code exactly as explained in Section 5.1, and the parameter q for that code will be $q_1 q_2$. We assume that q_1 and q_2 are small powers of 2, say $q_i = 2^{\kappa_i}$. Thus each symbol of the outer code can be represented as the concatenation of two short strings of length κ_1 and κ_2.

We also need an error-and-erasure-correcting code *EECC* of the same length l over an alphabet of size q_2 and of sufficient dimension *dim* to encode the values to be embedded, i.e., $\kappa_2 dim \geq len(k)$. The precise error-and-erasure-correcting properties needed are discussed below.

Construction 2 (Embedding and extracting).

- *Data initialization.* The merchant chooses marks for the data item using the underlying marking scheme. Furthermore, for each of the l positions of the outer code, he chooses a substitution $subst_i$ randomly, i.e., a permutation of the alphabet $\{1, ..., q\}$. Recall that the alphabet is small enough for a random permutation to be represented as a table.

- *Embedding*: The merchant's secret inputs are the data item and the initial data record. The commitment that fixes the value emb_B that will be embedded for the current buyer is a common input.[2] The buyer's secret input is emb_B and the auxiliary data needed to open the commitment.

  - The merchant secretly selects κ_1 random bits for each of the l symbols of the outer codeword. We call them halfsymbols and denote the choice as

    $$halfword\_search_B := (halfsym\_search_{B,1}, \ldots, halfsym\_search_{B,l}).$$

  - Now emb_B is encoded with the error-and-erasure-correcting code $EECC$ into l halfsymbols of κ_2 bits each. We call them $halfsym\_emb_{B,1}, \ldots, halfsym\_emb_{B,l}$. The buyer can do this alone if she hides the result in commitments again and proves in zero-knowledge that the computation was correct.

  - The halfsymbols from the merchant and the buyer are mixed into symbols by the operation

    $$sym_{B,i} := subst_i(halfsym\_search_{B,i} \parallel halfsym\_emb_{B,i}),$$

    where $subst_i$ is the substitution chosen in data initialization for this symbol position. We will see below why this encryption is necessary for the security of the merchant. This step and the following one require secure 2-party computation because secrets from both parties are used. The outer codeword of this buyer is

    $$word_B := (sym_{B,1}, \ldots, sym_{B,l}).$$

  - Each outer symbol $sym_{B,i}$ is encoded using the inner code Γ_0, and the resulting word is used to fingerprint the data item. The result is only output to the buyer.

- *Extracting.*

  - For each of the l positions of the outer code, the merchant uses the identification procedure of the underlying code Γ_0 to identify a symbol $sym_{red,i}$ ("*red*" for "redistributed"). He decrypts it using $subst_i^{-1}$ and separates it into its halves of length κ_1 and κ_2, respectively. We call the resulting outer word $word_{red}$ and the word consisting of all the first halves $halfword\_search_{red}$.

  - The merchant searches among his purchase records for the given data item for one where $halfword\_search_T$ has at least $l/coll\_size$ (half-)symbols in common with $halfword\_search_{red}$.

  - He now tries to extract the value emb_T from the second halfsymbols of $word_{red}$. First he excludes all those symbols $sym_{red,i}$ that definitely do not belong to this traitor because their first halfsymbols are different from those in $halfword\_search_T$. The remaining second halfsymbols, $halfsym\_emb_{red,i}$, constitute a word with many erasures. The merchant applies the decoding procedure of $EECC$ to it and hopes that the result is emb_T.

5.3 Security of the Construction and Requirements on the Code

We now consider the security of the scheme and find out how many errors the code $EECC$ has to tolerate in addition to the erasures. The effectiveness of the scheme, i.e., that

---

[2] Using the index B is only a notational help for us to distinguish the values used with different buyers; of course it does not mean that the merchant has to know this buyer's identity.

embedding yields a reasonable data item for the buyer if nobody cheats, is clear if it holds for the underlying marking scheme. Recall from the proof sketch of Construction 1 what security requirements we made on a scheme for embedding and extracting:

- Security of the buyer. The merchant should not gain knowledge about emb_B during embedding.

- Security of the merchant. As long as there are at most $coll\_size$ traitors, extracting will recover the value emb_T used by a traitor with high probability.

The same requirements make the application in a non-anonymous fingerprinting scheme secure.

Security for the buyer. This is clear because the only output the merchant gets from the steps that involve emb_B are commitments, a zero-knowledge proof, and his view of a secure 2-party computation without output to him.

Security for the merchant, overview. First, the properties of the underlying code Γ_0 guarantee that all symbols $sym_{red,i}$, and thus all halfsymbols in $halfword\_search_{red}$, will belong to one of the traitors, with an error probability of at most $2^{-\sigma}$ overall. At least one traitor T^* must therefore have contributed at least $l/coll\_size$ halfsymbols. Thus the merchant's search in the second step of extracting succeeds.

We show in 1. below that for suitably chosen parameters, the merchant almost certainly really identifies the record of a traitor, i.e., no record of an honest buyer fulfils the search criterion.

However, it is not clear that the traitor T whom the merchant identifies contributed at least $l/coll\_size$ entire symbols, nor that all the symbols that she did not contribute will lead to erasures, because different symbols can agree on their first half. But at least we show in 2. below that in a position i where a symbol from a traitor other than T was used, the first halfsymbol is random. Intuitively, this means that the traitors cannot introduce errors instead of erasures on purpose.

Hence there are at most $2^{-\kappa_1}l$ errors on average. We show in 3. below that there are almost always at most $3 \cdot 2^{-\kappa_1}l$ errors. Moreover, the merchant's search criterion immediately implies that there are at most $l - l/coll\_size$ erasures. Hence it is sufficient to use a code $EECC$ that tolerates $e = 3 \cdot 2^{-\kappa_1}l$ errors and $r = l - l/coll\_size$ erasures.

Details. We now prove the three statements from the overview and state the necessary constraints on the parameters. As the worst case, we assume that the traitors know their own codewords completely, i.e., they know to which indices the marks they found belong and which version of the data in one mark encodes 0 and 1, respectively.

1. We have to show that almost certainly no honest buyer's $halfword\_search_B$ will have $l/coll\_size$ symbols in common with $halfword\_search_{red}$. This is a standard proof of collusion tolerance since [CFN94]: The traitors have no information about the randomly chosen $halfword\_search_B$ because the merchant is honest in this part of the proof. Hence, when selecting $halfword\_search_{red}$, the probability that they guess a particular halfsymbol of a particular buyer correctly is $p = q_1^{-1}$. Let S be the random variable denoting the number of symbols guessed correctly. By the Chernoff bound, $P(S \geq 3pl) < 2^{-pl}$, i.e., $P(S \geq 3q_1^{-1}l) < 2^{-q_1^{-1}l}$. If we want to bound the overall probability for all N_M buyers by $2^{-\sigma}$, we need $q_1 \geq 3coll\_size$ and $l \geq q_1(\sigma + log_2(N_M))$.

2. We have to show that in every position i where the traitors use a symbol $sym_{red,i} \neq sym_{T,i}$, the equality $halfsym\_search_{red,i} = halfsym\_search_{T,i}$ will independently hold with probability at most $2^{-\kappa_1}$. As the merchant has chosen both these halfsymbols randomly and independently, it suffices to show that the traitors have no information what values of $halfsym\_search$ are encrypted by any symbol sym_{red}. We can consider each position i separately because the merchant does not use any common information in different positions.

The only knowledge the traitors have about the encryption function $subst_i$ is their own symbols $sym_{T^*,i}$ and the corresponding halfsymbols $halfsym\_emb_{T^*,i}$. This is at most as much information as if they knew the precise range of the restricted substitution $subst_i(\bullet , halfsym\_emb)$ for each value of $halfsym\_emb$. These substitutions are completely independent random permutations (onto renamed domains). If the attackers select $sym_{red,i} \neq sym_{T,i}$ from the range of $subst_i(\bullet , halfsym\_emb_{T,i})$, then $halfsym\_search_{red,i} = halfsym\_search_{T,i}$ is impossible because of the one-to-one property. Otherwise, they have no information whether the first halfsymbols agree because of the independence of the permutations.

3. Finally, we show that there are almost always at most $3 \cdot 2^{-\kappa_1} l$ errors. We know from 2. that in each position, there is an error with respect to the word of a particular traitor T with probability at most $p = 2^{-\kappa_1} = q_1$. Hence we can use the Chernoff bound as in 1. This leads to the constraint $l \geq q_1(\sigma + log_2(coll\_size))$, if we want to bound the probability by $2^{-\sigma}$ for all traitors together. This constraint is weaker than that in 1.

5.4 Reed-Solomon Codes for Error-and-Erasure Decoding

We first recall the properties of Reed-Solomon codes. All the results mentioned here can be found in [B83]. Reed-Solomon codes are a class of cyclic codes. Any finite field GF(q) can serve as the alphabet; the blocklength is then $l = q - 1$. That the blocklength for a given alphabet is fixed is a certain restriction. For any $t < l/2$, there is a Reed-Solomon code of minimum distance $d = 2t + 1$ and dimension $dim = l - 2t$, and it can be constructed efficiently.[3] This is the maximum dimension possible for the given minimum distance for any linear code; reaching this bound is the main advantage of Reed-Solomon codes.

Usually, a code with minimum distance $d = 2t + 1$ is used to correct up to t errors. However, such a code can also tolerate any combination of e errors and r erasures with $2e + r + 1 \leq d$. This can easily be seen because the restriction of the code to the positions where no erasure occurred still has a minimum distance of at least $d - r$. Furthermore, all BCH codes, of which Reed-Solomon codes are a subclass, can be efficiently decoded for $2e + r + 1 \leq d^*$, where d^* is their so-called designed distance, which equals d for Reed-Solomon codes.

5.5 Setting the Parameters

If we use Reed-Solomon codes in Construction 2, the alphabet size $q_2 = 2^{\kappa_2}$ equals the blocklength l plus 1. To tolerate the up to $e = 3 \cdot 2^{-\kappa_1} l$ errors and $r = l - l/coll\_size$ erasures,

---

[3] For concreteness: If α is a primitive element of GF(q), the generator polynomial of this code is $g(x) = (x-\alpha)(x-\alpha^2)...(x-\alpha^{2t})$, i.e., the code consists of the multiples of $g(x)$ by polynomials of degree less than $l-2t$.

we need a minimum distance $d = 2t + 1 \geq 2e + r + 1$, which means $2t \geq 6 \cdot 2^{-\kappa_1} l + l - l/coll\_size$. To encode the secrets to be embedded, we need $dim = l - 2t \geq len(k)/\kappa_2 = len(k)/\log_2(l+1)$. Both inequalities for t can be fulfilled iff l and κ_1 are chosen such that (neglecting rounding errors)

$$-6 \cdot 2^{-\kappa_1} l + l/coll\_size \geq len(k)/\log_2(l).$$

Certainly, the left side must be positive; let us require $2^{\kappa_1} \geq 24 coll\_size$. Then l remains to be chosen such that $l \cdot \log_2(l) \geq 4/3 len(k) coll\_size$. Let $l^* := len(k) coll\_size$. One can easily verify that $l \geq 2l^*/\log_2(l^*)$ is a sufficient condition.

6 Conclusion

We have introduced the concept of anonymous fingerprinting, a cryptologic copyright mechanism where honest buyers need not identify themselves to merchants, but merchants can nevertheless find out the identity of traitors who redistribute data without permission. We (informally) presented a precise definition of the concept, mentioned some variants, and presented a provably secure framework construction. It can be instantiated with some known schemes for fingerprinting without much collusion tolerance and for collusion-tolerant traitor tracing. To obtain collusion-tolerant fingerprinting, too, we constructed the first collusion-tolerant asymmetric fingerprinting scheme with 2-party trials. Such trials have practical advantages. However, the complexity in the current instantiation with Reed-Solomon codes is somewhat higher than that of known schemes with 3-party trials. A code where the same amount of data could be encoded with a smaller alphabet and a longer blocklength would reduce this problem; however, we are not aware of one where the minimum distance can be very near the blocklength and an efficient procedure for error-and-erasure-decoding is known. Actually, we regard our constructions rather as constructive proofs of existence. However, the 2-party protocol used for actually fingerprinting the data can be replaced by an efficient scheme from [PS96], and so can the preceding step where the outer codeword is expanded using Γ_0. Thus no general primitives are needed on the overwhelming part of the data. We are confident that one could also improve upon the remaining ones, but shortening the codes seems more important.

Acknowledgments

We thank *Matthias Schunter* for interesting discussions and *Rudi Piotraschke* for helpful advice with coding theory.

References

[B83]   Richard E. Blahut: *Theory and Practice of Error Control Codes*; Addison-Wesley, Reading 1983.

[B94]   Stefan Brands: *Untraceable Off-line Cash in Wallet with Observers*; Crypto '93, LNCS 773, Springer-Verlag, Berlin 1994, 302-318.

[BCC88]   Gilles Brassard, David Chaum, Claude Crépeau: *Minimum Disclosure Proofs of Knowledge*; Journal of Computer and System Sciences 37 (1988) 156-189.

[BM97]   Ingrid Biehl, Bernd Meyer: *Protocols for Collusion-Secure Asymmetric Fingerprinting*; accepted for 14th Symposium on Theoretical Aspects of Computer Science (STACS) 1997.

[BMP86]   G. R. Blakley, C. Meadows, G. B. Purdy: *Fingerprinting Long Forgiving Messages*; Crypto '85, LNCS 218, Springer-Verlag, Berlin 1986, 180-189.

[BP90]   Holger Bürk, Andreas Pfitzmann: *Value Exchange Systems Enabling Security and Unobservability*; Computers & Security 9/8 (1990) 715-721.

[BRD95]   F. M. Boland, J. J. K. Ó Ruanaidh, C. Dautzenberg: *Watermarking Digital Images for Copyright Protection*; 5th IEE International Conference on Image Processing and its Applications, Edinburgh 1995, 326-330.

[BS95]   Dan Boneh, James Shaw: Collusion-Secure Fingerprinting for Digital Data; Crypto '95, LNCS 963, Springer-Verlag, Berlin 1995, 452-465.

[C81]   David Chaum: *Untraceable Electronic Mail, Return Addresses, and Digital Pseudonyms*; Communications of the ACM 24/2 (1981) 84-88.

[C85]   David Chaum: *Security without Identification: Transaction Systems to make Big Brother Obsolete*; Communications of the ACM 28/10 (1985) 1030-1044.

[CFN94]   Benny Chor, Amos Fiat, Moni Naor: *Tracing Traitors*; Crypto '94, LNCS 839, Springer-Verlag, Berlin 1994, 257-270.

[CKLS96]   Ingemar Cox, Joe Kilian, Tom Leighton, Talal Shamoon: *A Secure, Robust Watermark for Multimedia*; Information Hiding, LNCS 1174, Springer-Verlag, Berlin 1996, 185-206.

[DY91]   Yvo Desmedt, Moti Yung: *Arbitrated Unconditionally Secure Authentication can be Unconditionally Protected Against Arbiter's Attacks*; Crypto '90, LNCS 537, Springer-Verlag, Berlin 1991, 177-188.

[P96]   Birgit Pfitzmann: *Trials of Traced Traitors*; Information Hiding, LNCS 1174, Springer-Verlag, Berlin 1996, 49-64.

[PS96]   Birgit Pfitzmann, Matthias Schunter: *Asymmetric Fingerprinting*; Eurocrypt '96, LNCS 1070, Springer-Verlag, Berlin 1996, 84-95.

[PW97]   Birgit Pfitzmann, Michael Waidner: *Asymmetric Fingerprinting for Larger Collusions*; accepted for 4th ACM Conference on Computer and Communications Security, 1997.

[W83]   Neal R. Wagner: *Fingerprinting*; 1983 Symposium on Security and Privacy, IEEE, Oakland, California, 18-22.

[ZK95]   Jian Zhao, Eckhard Koch: *Embedding Robust Labels Into Images For Copyright Protection*; International Congress on Intellectual Property Rights for Specialized Information, Knowledge and New Technologies, Oldenbourg-Verlag, Vienna 1995.

A Secure and Optimally Efficient Multi-Authority Election Scheme

Ronald Cramer* Rosario Gennaro** Berry Schoenmakers***

Abstract. In this paper we present a new multi-authority secret-ballot election scheme that guarantees privacy, universal verifiability, and robustness. It is the first scheme for which the performance is optimal in the sense that time and communication complexity is minimal both for the individual voters and the authorities. An interesting property of the scheme is that the time and communication complexity for the voter is *independent* of the number of authorities. A voter simply posts a *single* encrypted message accompanied by a compact proof that it contains a valid vote. Our result is complementary to the result by Cramer, Franklin, Schoenmakers, and Yung in the sense that in their scheme the work for voters is linear in the number of authorities but can be instantiated to yield information-theoretic privacy, while in our scheme the voter's effort is independent of the number of authorities but always provides computational privacy-protection. We will also point out that the majority of proposed voting schemes provide computational privacy only (often without even considering the lack of information-theoretic privacy), and that our new scheme is by far superior to those schemes.

1 Introduction

In the cryptographic literature, electronic voting protocols are known as the prime examples of secure multi-party computations. Many papers have been written on the subject and by now an extensive list of properties and requirements is generally accepted as desirable. We will consider these properties in this paper, among which are privacy, universal verifiability, and various forms of robustness. Recent advancements have also been particularly concerned with the performance aspect. In this paper we will show under which circumstances it is possible to achieve a scheme with optimal performance for large-scale elections, while at the same time keeping the system simple and provably secure.

In considering the performance of elections it is clear that the main consideration should be the effort required of a voter. Indeed, while governments can (and do nowadays) afford a large organizational effort to hold elections, it is mandatory to make the voting protocol as simple and efficient as possible for the voter—who might be participating from home using a PC or a Web TV.

* Inst. for Theoretical Comp. Sc., ETH-Z, CH-8092 Zurich, Switzerland. cramer@inf.ethz.ch

** IBM T.J. Watson Research Center, P.O. Box 704, Yorktown Heights, NY 10598, USA. rosario@watson.ibm.com

*** DigiCash, Kruislaan 419, NL-1098 VA Amsterdam, The Netherlands. berry@digicash.com

W. Fumy (Ed.): Advances in Cryptology - EUROCRYPT '97, LNCS 1233, pp. 103-118, 1997.
© Springer-Verlag Berlin Heidelberg 1997

In this paper we present a simple multi-authority election scheme in which the task of the voter is reduced to the bare minimum. Basically, the voter posts a *single* encrypted message (ballot) accompanied with a proof that it contains a valid vote. For security parameter k, the size of the ballot as well as of its proof of validity is $O(k)$ bits. Moreover, due to the homomorphic properties of the encryption method used, the final tally is verifiable to any observer of the election, while due to the use of a matching fault-tolerant threshold decryption technique, the individual votes will remain private and the (benign or malign) failure of authorities can be tolerated.

We work in the model set forth by Benaloh *et al.* [CF85, BY86, Ben87], where the active parties are divided into l voters $V_1, \ldots, V_l$ and n tallying authorities (talliers) $A_1, \ldots, A_n$. To achieve universal verifiability all parties have access to a so-called bulletin board. A *bulletin board* is like a broadcast channel with memory to the extent that any party (including passive observers) can see the contents of it, and furthermore that each active participant can post messages by appending the message to her own designated area. No party can erase anything from the bulletin board.

In this model, voters cast their votes by posting ballots to the bulletin board. The ballot does not reveal any information on the vote itself but it is ensured by an accompanying proof that the ballot indeed contains a valid vote and nothing else. Due to a homomorphic property of the ballots, the final tally ("sum" of all votes) can be obtained and verified (by any observer) against the "product" of all submitted ballots. This ensures universal verifiability.

Although we are emphasizing the application of our scheme to large-scale elections, it is also suitable for small-scale elections such as boardroom elections. In the latter case it is even conceivable that each voter plays the role of tallying authority as well; a PC network will suffice as computing platform.

1.1 Computational versus information-theoretic privacy

By far, the majority of election protocols that support some level of verifiability (either universal or limited to voters, who can check their own vote) merely provide computational protection of the voter's privacy. For example, the schemes presented by Benaloh *et al.* [CF85, BY86, Ben87, BT94] all rely on the so-called r-th residuosity assumption. Once this assumption is broken (e.g., when the public modulus is factorized), the content of each individual ballot can be decrypted. Similarly, schemes using anonymous channels or mixes [Cha81] usually rely on computational assumptions. By recovering the private keys of the mixes, an adversary is able to "open" all ballots posted to the first mix. For example, the scheme of [SK95] relies on the difficulty of computing discrete logs, both for the secrecy of the mixes' private keys and for the contents of the ballots.

The extent to which the lack of information-theoretic privacy is harmful may be difficult to estimate. For instance, it is hard to predict what happens if fifty-year old votes of a U.S. president are published—although breaking the encryption methods for the currently widely used security parameters will probably be much more harmful.

Whither democracy, from a cryptographic standpoint it is necessary to determine the limits for computational and information-theoretic privacy. As an aside we note that the mere use of multiple authorities can be considered a condition as well. Indeed, election protocols have been proposed that try to eliminate this condition, e.g., see [PW92], but the methods used still require conditions regarding the channels connecting the participants. Since in our case the bulletin board is implemented from multiple servers anyway, and it is seen as a necessary primitive for achieving universal verifiability, we will not consider eliminating the use of a distributed tallying authority. Yet, to some extent we will take into account that authorities may be compromised over time, see below.

1.2 Our contributions

In this paper we will see how far one can go if computational privacy is the goal. For computational privacy it suffices to assume a public broadcast channel (bulletin board) as communication model. To make an election scheme information-theoretically secure, it is generally believed that private channels between voters and authorities are required. In Section 6.1 we will look into this aspect.

The main result of this paper is a fair election scheme in which the complexity of the voter's protocol is *linear* in the security parameter k--hence optimal. This comprises the computational as well as the communication complexity (in bits). The voter needs to communicate only $O(k)$ bits and to perform $O(k)$ modular multiplications.[4] Moreover, the dominating factor for the work of an authority is $O(lk)$. Compared to the scheme of [CFSY96], we thus achieve a reduction of the work for each participant by a factor of n.

In the new scheme, the voter just sends a particular ElGamal encryption of the vote plus a proof that it indeed contains a valid vote. The proof prevents the voters from casting bogus ballots, and should be such that no information whatsoever leaks about the actual vote contained in a ballot. The crux is to keep this proof $O(k)$, and here we follow the approach of [CFSY96]. We will need a novel application of the technique of [CDS94] for constructing efficient witness hiding protocols. The resulting proof of validity is a little bit more complicated than in [CFSY96], but still requires only a few modular exponentiations. A proof of knowledge similar to our proof of validity has been used by Chen and Pedersen to construct efficient group signatures [CP95].

Unlike previous schemes based on Benaloh's approach, however, we will achieve robustness w.r.t. faulty authorities without increasing the work for the voter. To this end, we will employ *fault-tolerant threshold cryptosystems* instead of (verifiable) secret sharing schemes. In our case there will be only one public key for which the matching private key is shared among the authorities using threshold cryptography techniques (see [Des94] for a survey.) The voter posts the ballot encrypted with the public key of the authorities. The private key is never reconstructed, and only used implicitly when the authorities cooperate to decrypt the final tally. The correctness of the decryption will be assured, even in the presence of malicious authorities.

Apart from achieving a strong set of properties, three major achievements of our scheme are: (i) The work required of the voter is minimal. Compared to [CFSY96] the work is reduced by a multiplicative factor of n. Although n is

---

[4] Throughout, we will take a modular multiplication of two $O(k)$ sized numbers as our unit of work.

usually much smaller than k, this is still a substantial gain in practice. The work for the authorities and observers is reduced accordingly. (ii) The protocol for the voter remains the same even if n is variable. Usually n grows with the desired security of the scheme (the more authorities the less potential that an adversary can corrupt, say, half of them). Using our protocol this growth is "transparent" to the user. (iii) As a bonus, the new scheme can easily be extended using techniques for proactive threshold cryptosystems [HJJ+97] to leave the system (and its keys) in place for a really long time without fearing that the secret key gets compromised (see Section 6.3).

The security of the main scheme presented in the paper is related to the difficulty of the discrct log problem. In Section 5 we describe an alternative construction related to the hardness of factoring. Finally, in Section 4 we show how our approach can be extended to more general classes of elections, and in Section 6.2 we consider the issue of receipt-free or incoercible elections and discuss the relevance of our paper in this area.

2 The building blocks

2.1 Bulletin board

The communication model required for our election scheme is best viewed as a public broadcast channel with memory, which is called a bulletin board. All communication through the bulletin board is public and can be read by any party (including passive observers). No party can erase any information from the bulletin board, but each active participant can append messages to its own designatcd section.

To make the latter requirement publicly verifiable, we assume that digital signatures are used to control access to the various sections of the bulletin board. Here we may take advantage of any public-key infrastructure that is already in place. Also note that by postulating that each participant can indeed append messages to its section, it is implicitly assumed that denial-of-service attacks are excluded. This property is realized by designing the bulletin board as a set of replicated servers implementing Byzantine agreement, for instance, such that access is never denied as long as at most a third of the servers is compromised. Reiter's work on the Rampart system shows that this can be done in a secure and practical way (see, e.g., [Rei94, Rei95]).

2.2 ElGamal cryptosystem

Our election scheme relies on the ElGamal cryptosystem [DH76, ElG85]. It is well-known that the ElGamal cryptosystem works for any family of groups for which the discrete logarithm is considered intractable. Part of the security of the scheme actually relies on the Diffie-Hellman assumption, which implies the hardness of computing discrete logarithms [DH76]. Although all our constructions can easily be shown to work in this general discrete log setting, we will present our results for subgroups G_q of order q of $\mathbb{Z}_p^*$, where p and q are large primes such that $q \mid p - 1$. Other practical families can be obtained for elliptic curves over finite fields.

We will now briefly describe the ElGamal cryptosystem, where the primes p and q and at least one generator g of G_q are treated as system parameters. These parameters as well as other independent generators introduced in the sequel should be generated jointly by (a designated subset) of the participants. This can be done by letting the participants each run a copy of the same probabilistic algorithm, where the coinflips are generated mutually at random.

The key pair of a receiver in the ElGamal cryptosystem consists of a private key s (randomly chosen by the receiver) and the corresponding public key $h = g^s$, which is announced to the participants in the system.

Given a message $m \in G_q$, encryption proceeds as follows. The sender chooses a random $\alpha \in \mathbb{Z}_q$, and sends the pair $(x, y) = (g^\alpha, h^\alpha m)$ as ciphertext to the receiving party. To decrypt the ciphertext (x, y) the receiver recovers the plaintext as $m = y/x^s$, using the private key s.

2.3 Robust threshold ElGamal cryptosystem

The object of a threshold scheme for public-key encryption is to share a private key among a set of receivers such that messages can only be decrypted when a substantial set of receivers cooperate. See [Des94] for a survey. The main protocols of a threshold system are (i) a *key generation* protocol to generate the private key jointly by the receivers, and (ii) a *decryption* protocol to jointly decrypt a ciphertext without explicitly reconstructing the private key. For the ElGamal system described above, solutions for both protocols have been described by Pedersen [Ped91, Ped92], also taking robustness into account.

Key generation As part of the set-up procedure of the election scheme, the authorities will execute a key generation protocol due to Pedersen [Ped91]. The result of the key generation protocol is that each authority A_j will possess a share $s_j \in \mathbb{Z}_q$ of a secret s. The authorities are committed to these shares as the values $h_j = g^{s_j}$ are made public. Furthermore, the shares s_j are such that the secret s can be reconstructed from any set Λ of t shares using appropriate Lagrange coefficients, say:

$$s = \sum_{j \in \Lambda} s_j \lambda_{j,\Lambda}, \qquad \lambda_{j,\Lambda} = \prod_{l \in \Lambda \setminus \{j\}} \frac{l}{l - j}. \qquad (1)$$

This is exactly as in Shamir's (t, n)-threshold secret sharing scheme [Sha79]. The public key $h = g^s$ is announced to all participants in the system. Note that no single participant learns the secret s, and that the value of s is only computationally protected.[5]

Decryption To decrypt a ciphertext $(x, y) = (g^\alpha, h^\alpha m)$ without reconstructing the secret s, the authorities execute the following protocol:

1. Each authority A_j broadcasts $w_j = x^{s_j}$ and proves in zero-knowledge that

$$\log_g h_j = \log_x w_j.$$

---

[5] The private channels assumed in Pedersen's key generation protocol may be implemented using public key encryption and the bulletin board. This suffices for computational security.

$$
\begin{array}{cc}
\text{Prover} & \text{Verifier} \\
[(x,y) = (g^\alpha, h^\alpha)] & \\
\end{array}
$$

Prover   Verifier
$[(x,y) = (g^\alpha, h^\alpha)]$

$w \in_R \mathbb{Z}_q$

$(a,b) \leftarrow (g^w, h^w)$ $\xrightarrow{\quad a,b \quad}$

$\xleftarrow{\quad c \quad}$ $c \in_R \mathbb{Z}_q$

$r \leftarrow w + \alpha c$ $\xrightarrow{\quad r \quad}$ $g^r \stackrel{?}{=} ax^c$

$h^r \stackrel{?}{=} by^c$

Fig. 1. Proof of knowledge for $\log_g x = \log_h y$.

2. Let Λ denote any subset of t authorities who passed the zero-knowledge proof. By raising x to both sides of equation (1), it follows that the plaintext can be recovered as

$$
m = y / \prod_{j \in \Lambda} w_j^{\lambda_{j,\Lambda}}.
$$

Note that step 2 assures that the decryption is correct and successful even if up to $n - t$ authorities are malicious or fail to execute the protocol. The zero-knowledge proof of step 1 will be described in the next section.

2.4 Proofs of knowledge for equality of discrete logs

Using the same notation as above, we present proofs of knowledge for the relation $\log_g x = \log_h y$, whereby a prover shows possession of an $\alpha \in \mathbb{Z}_q$ satisfying $x = g^\alpha$ and $y = h^\alpha$. An efficient protocol for this problem is due to Chaum and Pedersen [CP93], see Figure 1. This protocol is not known to be zero-knowledge or witness hiding. The following result however suffices for our application (see also [CDS94] for definitions of the notions involved).

Lemma 1. *The Chaum-Pedersen protocol is a three-move, public coin proof of knowledge for the relation $\log_g x = \log_h y$. The proof satisfies special soundness, and is special honest-verifier zero-knowledge.*

Proof. The protocol inherits its properties from the underlying Schnorr protocol [Sch91]. Special soundness holds because from two accepting conversations with the same first move (a,b,c,r) and (a,b,c',r'), $c \neq c'$, a witness $w = \frac{r-r'}{c-c'}$ can be extracted satisfying $x = g^w$ and $y = h^w$. Honest-verifier zero-knowledge holds because, for random c and r we have that $(g^r x^{-c}, h^r y^{-c}, c, r)$ is an accepting conversation with the right distribution. Since the challenge c can be chosen freely, we also have special honest-verifier zero-knowledge.

Notice that the above protocol is zero-knowledge only against the honest verifier, but this suffices for our purpose (see, e.g., [Cha91] for an efficient zero-knowledge protocol). Indeed, jumping ahead a little, in order to make our protocols non-interactive, the verifier will be implemented using either a trusted source of random bits (a beacon as in [Rab83, Ben87]) or using the Fiat-Shamir heuristic [FS87] which requires a hash function. In the latter case security is obtained for the random oracle model.

2.5 Homomorphic encryption

Homomorphic encryption schemes form an important tool for achieving universally verifiable election schemes. A general definition of the notion is as follows. Let $\mathcal{E}$ denote a probabilistic encryption scheme. Let M be the message space and C the ciphertext space such that M is a group under operation $\oplus$ and C is a group under operation $\otimes$. We say that $\mathcal{E}$ is a $(\oplus, \otimes)$-homomorphic encryption scheme if for any instance E of the encryption scheme, given $c_1 = E_{r_1}(m_1)$ and $c_2 = E_{r_2}(m_2)$, there exists an r such that

$$c_1 \otimes c_2 = E_r(m_1 \oplus m_2).$$

Homomorphic encryption schemes are important to the construction of election protocols. If one has a $(+, \otimes)$ scheme, then if c_i are the encryptions of the single votes, by decrypting $c = c_1 \otimes \ldots \otimes c_m$ one obtains the tally of the election, without decrypting single votes.

The ElGamal cryptosystem as presented above already satisfies this definition, where the message space is G_q with multiplication modulo p as group operation, and the ciphertext space is $G_q \times G_q$ with componentwise multiplication modulo p as group operation. Namely, given an ElGamal encryption (x_1, y_1) of m_1 and an ElGamal encryption (x_2, y_2) of m_2, we see that $(x_1 x_2, y_1 y_2)$ is an ElGamal encryption of $m_1 m_2$.

For the reasons sketched above however, we need to take this one step further to a homomorphic scheme with addition as group operation for the message space. That is, instead of G_q, our message space will be $\mathbb{Z}_q$ with addition modulo q as group operation. Given a fixed generator $G \in G_q$, the encryption of a message $m \in \mathbb{Z}_q$ will be the ElGamal encryption of G^m. The observation is now that, given two such encryptions of m_1 and m_2, respectively, the product is an encryption of $m_1 + m_2$ modulo q. Notice that for such a scheme decryption involves the computation of a discrete log, which is a hard task in general. Nevertheless it can be done efficiently for "small" messages, as will be the case in our election scheme (see Section 3).

2.6 Efficient proofs of validity

In our election each voter will post an ElGamal encryption of either m_0 or m_1, where m_0 and m_1 denote distinct elements of G_q. (Later we will consider suitable values for m_0 and m_1.) The encryption should be accompanied by a proof of validity that proves that the encryption indeed contains one of these values. Furthermore, the proof should not reveal any information about which one.

Consider an ElGamal encryption of the following form:

$$(x, y) = (g^\alpha, h^\alpha m), \qquad \text{with } m \in \{m_0, m_1\},$$

Voter		Verifier

$v = 1$	$v = -1$
$\alpha, w, r_1, d_1 \in_R \mathbb{Z}_q$	$\alpha, w, r_2, d_2 \in_R \mathbb{Z}_q$

$v = 1$	$v = -1$	
$x \leftarrow g^\alpha$	$x \leftarrow g^\alpha$	
$y \leftarrow h^\alpha G$	$y \leftarrow h^\alpha / G$	
$a_1 \leftarrow g^{r_1} x^{d_1}$	$a_1 \leftarrow g^w$	
$b_1 \leftarrow h^{r_1}(yG)^{d_1}$	$b_1 \leftarrow h^w$	
$a_2 \leftarrow g^w$	$a_2 \leftarrow g^{r_2} x^{d_2}$	
$b_2 \leftarrow h^w$	$b_2 \leftarrow h^{r_2}(y/G)^{d_2}$	$\xrightarrow{\ x, y, a_1, b_1, a_2, b_2\ }$
$d_2 \leftarrow c - d_1$	$d_1 \leftarrow c - d_2$	$\xleftarrow{\quad c \quad}$ $\quad c \in_R \mathbb{Z}_q$
$r_2 \leftarrow w - \alpha d_2$	$r_1 \leftarrow w - \alpha d_1$	$\xrightarrow{\ d_1, d_2, r_1, r_2\ }$ $\quad c \stackrel{?}{=} d_1 + d_2$
		$a_1 \stackrel{?}{=} g^{r_1} x^{d_1}$
		$b_1 \stackrel{?}{=} h^{r_1}(yG)^{d_1}$
		$a_2 \stackrel{?}{=} g^{r_2} x^{d_2}$
		$b_2 \stackrel{?}{=} h^{r_2}(y/G)^{d_2}$

Fig. 2. Encryption and Proof of Validity of Ballot (x, y)

where the prover knows the value of m. To show that the pair (x, y) is indeed of this form without revealing the value of m boils down to a witness indistinguishable proof of knowledge of the relation given by:

$$\log_g x = \log_h(y/m_0) \qquad \vee \qquad \log_g x = \log_h(y/m_1).$$

The prover either knows a witness for the left part or a witness for the right part (but not both at the same time), depending on the choice for m.

By the techniques of [CDS94], we can now immediately obtain a very efficient witness indistinguishable proof of knowledge for the above relation. To prove either of the two equalities we have the efficient proof of knowledge by Chaum and Pedersen, described above, for which we have prepared Lemma 1. On account of this lemma, we have that the protocol exactly satisfies the conditions for the construction of [CDS94]. See Figure 2 for a preview of the protocol, as it is used in the election scheme of the next section.

3   Multi-authority election scheme

Given the primitives of the previous section we now assemble a simple and efficient election scheme. The participants in the election protocol are n authorities $A_1, \ldots, A_n$ and l voters $V_1, \ldots, V_l$. Recall that the requirements for a ballot are that it must contain a vote in an unambiguous way such that (i) votes accumulate when ballots are aggregated, and (ii) the proof of validity shows that a ballot contains either a yes-vote or a no-vote, without revealing any information on which of the two is the case.

To show that the same masking technique as in [SK94, CFSY96] can be used, we instantiate the scheme of Section 2.6 with $m_1 = G$ and $m_0 = 1/G$, where G is a fixed generator of G_q. Thus a ballot is prepared as an ElGamal encryption of the form $(x, y) = (g^\alpha, h^\alpha G^b)$ for random $b \in_R \{1, -1\}$, and the corresponding proof of knowledge is depicted in Figure 2. To cast a ballot the voter posts an additional number $e \in \{1, -1\}$ such that $v = be$ is equal to the desired vote. Alternatively, voters may adapt the precomputed values before sending the ballot out, i.e., precompute (x, y) and then post (x^e, y^e).

In order to make vote casting non-interactive we compute the challenge c as a hash value of the first message of the proof. In this case security is retained in the random oracle model, but some care is required to prevent vote duplication. Each challenge must be made voter-specific (see [Gen95]), i.e., the challenge c is computed by voter V_i as $H(ID_i, x, y, a_1, b_1, a_2, b_2)$, where ID_i is a unique public string identifying V_i.

As part of the initialization the designated parties generate the system parameters p, q, g, G, as described in Section 2.2, where we may safely assume that $l < q/2$ for any reasonable security parameter k. Secondly, the authorities execute the robust key generation protocol as described in Section 2.3. The transcripts of these protocol should appear on the bulletin board. Note that this also shows to any observer that indeed n authorities are taken part in the scheme, which is otherwise not visible to the voters.

The main steps of the voting protocol now are, where we assume w.l.o.g. that only correct ballots are cast:

1. Voter V_i posts a ballot (x_i, y_i) to the bulletin board accompanied by a non-interactive proof of validity.
2. When the deadline is reached, the proofs of validity are checked by the authorities and the product $(X, Y) = (\prod_{i=1}^{l} x_i, \prod_{i=1}^{l} y_i)$ is formed.
3. Finally, the authorities jointly execute the decryption protocol of Section 2.3 for (X, Y) to obtain the value of $W = Y/X^s$. A non-interactive proof of knowledge is used in Step 1 of the decryption protocol.

We thus get $W = G^T$ as a result, where T is equal to the difference between the number of yes-votes and no-votes, $-l \leq T \leq l$. Hence, $T = \log_G W$ which is in general hard to compute. However, in our case we can now fully exploit the fact that the number of voters l is relatively small—certainly polynomial in the security parameter! The value of T can be determined easily using $O(l)$ modular multiplications only, by iteratively generating $G^{-l}, G^{-l+1}, G^{-l+2}, \ldots$ (each time using one multiplication) until W is found. Asymptotically, the work does therefore not increase for the authorities (at most two multiplications per voter). Note also that the computation of $\log_G W$ may be done by any party because the result is verifiable.[6]

The time and communication complexity of the scheme is as follows. The work for a voter is clearly linear in k, independent of the number of authorities.

---

[6] If this $O(l)$ search method is considered too slow for a large-scale election, Shanks' baby-step giant-step algorithm (see, e.g., [LL90, Section 3.1]) can be applied to find T in $O(\sqrt{l})$ time using $O(\sqrt{l}k)$ bits of storage.

The work for the authorities is only $O(lk+nk)$ (assuming that the zero-knowledge proof used in step 3 is $O(k)$, hence negligible). Since we may safely assume that the number of voters is larger than the number of authorities, the work for the authorities is actually $O(lk)$. Similarly, the work for an observer who wants to check the outcome of the election is $O(lk)$.

Theorem 2. *Under the Diffie-Hellman assumption, our election scheme provides universal verifiability, computational privacy, robustness, and prevents vote duplication.*

Actually, parts of this theorem also hold under the discrete log assumption, but for conciseness we are only referring to the Diffie-Hellman assumption (which is required to show that the ElGamal encryptions used do not leak information about the votes). For the non-interactive version of the scheme based on the Fiat-Shamir heuristic, the result holds in the random oracle model.

4 Extension to multi-way elections

Instead of offering a choice between two options, it is often required that a choice between several options can be made. There are numerous approaches to tackle this problem. Below, we sketch an approach fow which the size of the ballots does not increase (but the size of the proof of validity does)), which again relies on the construction of [CDS94]. To get an election for a 1-out-of-K choice, we simply take K (independently generated) generators G_i, $1 \leq i \leq K$, and accumulate the votes for each option separately. The proof of validity of a ballot (x, y) now boils down to a proof of knowledge of

$$\log_g x = \log_h(y/G_1) \quad \vee \quad \cdots \quad \vee \quad \log_g x = \log_h(y/G_K).$$

Since the voter can only generate this proof for at most one generator G_i, it is automatically guaranteed that the voter cannot vote for more than one option at a time.

The problem of computing the final tally is in general more complicated. After decryption by the authorities, a number W is obtained that represents the final tally, $W = G_1^{T_1} \cdots G_K^{T_K}$, where the T_i's form the result of the election. Note that the T_i's are uniquely determined by W in the sense that computation of a different set T_i''s satisfying $W = G_1^{T_1'} \cdots G_K^{T_K'}$ would contradict the discrete log assumption, using the fact that the generators G_i are independently generated. Since $T_i \geq 0$ and $\sum_{i=1}^{K} T_i = l$, computation of the T_i's is feasible for reasonable values of l and K.[7]

---

[7] Note that the condition $\sum_{i=1}^{K} T_i = l$ can be exploited by reducing the problem to a search for $T_1, \ldots, T_{K-1}$ satisfying

$$W/G_K^l = (G_1/G_K)^{T_1} \cdots (G_{K-1}/G_K)^{T_{K-1}},$$

where $T_i \geq 0$ and $\sum_{i=1}^{K-1} T_i \leq l$. The naive $O(l^{K-1})$ method (which checks all possible combinations) can now be improved considerably by a generalization of the baby-step giant-step algorithm of time $O(\sqrt{l}^{K-1})$.

Prover   Verifier
$[x = \alpha^q]$

$w \in_R \mathbf{Z}_N^*$
$a \leftarrow w^q$ $\xrightarrow{\quad a \quad}$
                    $\xleftarrow{\quad c \quad}$ $c \in_R \mathbf{Z}_q$

$r \leftarrow w\alpha^c$ $\xrightarrow{\quad r \quad}$ $r^q \stackrel{?}{=} ax^c$

Fig. 3. Proof that x is a q-th residue.

5   Alternative number-theoretic assumption

To show the generality of our approach we now present a scheme for which the security is related to the difficulty of factoring. Specifically, we present a scheme based on the q-th residuosity assumption (as in the original Benaloh schemes). The notion of q-th residues is an extension of quadratic residues. A number x is a q-th residue modulo N if there exists an α such that $\alpha^q = x(\bmod N)$. It is believed to be hard to distinguish between q-residues and non q-residues.

This suggests the following homomorphic encryption scheme. We present a specific implementation which is suitable to threshold cryptography techniques. The parameters of the scheme are a modulus $N = PQ$, where $P = 2P' + 1$ and $Q = 2qQ' + 1$, with P, Q, P', Q', q all large primes. As before, the prime q can thus be assumed to be larger than twice the number of voters l. Also the public key must include a fixed number $Y \in \mathbf{Z}_N^*$ which is not a q-th residue modulo N.

We will consider messages from $\mathbf{Z}_q$. The ciphertext for a message m is now $E_\alpha(m) = \alpha^q Y^m$, where $\alpha \in_R Z_N^*$. As before, decryption is hard, in general, but in our case an exhaustive search for all possible values suffices. The right m is detected when by computing $(cY^{-m})^{\phi(N)/q} \bmod N$ one gets back 1. Note that $c' = c^{\phi(N)/q} \bmod N$ and $Y' = (Y^{-1})^{\phi(N)/q} \bmod N$ can be computed first, and then test for $c'Y'^m$, where m is selected from all possible messages.

Next we discuss a robust threshold cryptosystem for this setting. Notice that the value $d = \phi(N)/q$ could be considered the secret key of the scheme, and that decryption is carried out by simply computing exponentiations (modulo N) with exponent d. As the setting is very similar to an RSA decryption, we can apply the result of [GJKR96] to obtain an efficient and robust threshold decryption procedure. The result in [GJKR96] holds for RSA moduli which are the product of safe primes (i.e., $P = 2P' + 1$ and $Q = 2Q' + 1$), but it can easily be extended to work for our specific needs.

The key generation protocol, however, relies on secure multiparty computations as there is no known efficient way to perform a distributed key generation algorithm for factoring based schemes. However, since this task is part of the set-up of the scheme, this may be acceptable as a one-time operation.

Our final task is to construct an efficient proof of validity that shows that a ballot x is correctly formed. This amounts to showing that $x = \alpha^q Y^v$, for some α, with $v \in \{1, -1\}$, hence that either x/Y or xY is a q-th residue. As before, Lemma 3 below guarantees the existence of an efficient proof of validity, based on the construction of [CDS94].

Lemma 3. *The protocol of Figure 3 is a three-move, public coin proof of knowledge for r-th residuosity. The proof satisfies special soundness, and is special honest-verifier zero-knowledge.*

Proof. Similar to proof of Lemma 1. Special soundness now holds because for any two accepting conversations (a, c, r) and (a, c', r'), $c > c'$, it follows that $(r/r')^q = x^{c-c'}$. Since $0 < c - c' < q$ we have that there exist integers k, l s.t. $(c - c')k = 1 + lq$, hence $(r/r')^{kq} = x^{lq+1}$, which yields $((r/r')^k x^{-l})^q = x$.

Theorem 4. *Under the q-th residuosity assumption, our election scheme provides universal verifiability, computational privacy, robustness, and prevents vote duplication.*

6 Discussion

6.1 Information-theoretically secure elections

The scheme of [CFSY96] in principle provides information-theoretic protection of the voter's privacy. This is due to the fact that voters post (a number of) information-theoretically hiding commitments to the bulletin board and that these commitments are opened to the authorities using private channels. A general problem with such a solution is that the use of private channels opens the possibility for disputes: on the one hand a dishonest voter may just skip sending a message to an authority, while on the other hand a dishonest authority may claim not to have received a message.

It is therefore worthwhile to limit the possibility for disputes to the set-up process for the election. During the election protocol itself no disputes on the usage of the private channel should be possible. The idea is to use a public broadcast channel (such as a bulletin board) on which the parties post commitments to mutually selected keys. Each pair of parties first agrees on a key using a secure channel. Only if both parties broadcast the same commitment, the set-up of the private channel succeeded. Otherwise, there is dispute that must be solved at this stage. It is important that (i) the commitment is information-theoretically hiding and (ii) the encryption method is information-theoretically secure (a one-time pad). More concretely, the two phases are as follows:

Set-up Both parties agree on a mutually at random selected key K and a commitment B on this key. Both parties broadcast a signed copy of the commitment. The key set-up is only succesful if both parties broadcast the same commitment. Disputes in this stage have to be resolved in a procedural way.

Communication To send a message m, the sender will broadcast the encryption $E_K(m)$ over the public channel. Only the intended receiver is able to recover the message.

Using this method, private channels can be set up from each voter V_i to each authority A_j. Once set up succeeds there can be no dispute on the use of the private channel. Anybody sees if the voter abstains from posting the required values to the bulletin board. If what the voter submits consists of incorrect shares, the respective authorities open the commitments to the key so that this fact can be verified. Note that for the scheme of [CFSY96] the use of the private channels is limited to two elements of $\mathbb{Z}_q$ per channel.

6.2 Incoercible protocols

Receipt-free or incoercible election scheme that have been proposed so far all rely on some form of physical assumption [BT94, NR94, SK95]. The minimal assumption required (as in [SK95]) is the existence of a private channel between the voters and the authorities. These schemes allow a voter to lie about the vote cast even if under coercion, but not up to the level that coercer who exactly prescribe which private random bits the voter must use can be withstood. Indeed given the execution of the protocol the voter will be able to create two different histories of his computations, both consistent with the execution but corresponding to two different votes. All these schemes also require that the authorities are incoercible, or alternatively that voters know which ones have been coerced. Moreover, as pointed out in the previous section, the use of private channels gives rise to disputes. (Another viable approach is to assume that the voters dispose of a tamper-proof encryption box such as a smartcard, but we consider this beyond the scope of this paper.)

Recently, Canetti and Gennaro in [CG96] proved that general secure multiparty computation protocols can be made incoercible without the above assumptions, in particular *without* assuming untappable channels. Their scheme is based on a new type of encryption called *deniable encryption* introduced in [CDNO96] that allows a sender to encrypt a bit b in such a way that the resulting ciphertext can be "explained" as either b or $1 - b$ to a coercer. The construction in [CG96] works for the general problem of secure multi-party computation; as such it is described in terms of a complete network of communication and the result holds as long as at most half of the players in the network are coerced. For the case of election schemes, the construction of [CG96] can be scaled down to the bulletin board model (thus not requiring communication between voters). In this model all voters can withstand coercion provided the coercer is not able to prescribe the random bits of the voters, and at most half of the authorities can be completely coerced. The complexity of the resulting scheme is high (although polynomial), but opens the door to the search for efficient incoercible schemes.

In order to make our election scheme incoercible (without physical assumptions) we would need a deniable encryption scheme which is (i) homomorphic, (ii) suitable to threshold cryptography techniques. An interesting open problem is thus to construct such a scheme.

6.3 Proactive Security

The secrecy of the votes is protected against coalitions of up to $t - 1$ authorities. In other words, an attacker must recover t shares of the private key in order to be able to decrypt single votes. This is similar to previous protocols in which the vote is (t, n)-shared among the authorities. We note that the use of threshold cryptography instead of secret sharing presents also some advantages in this area. Using proactive security techniques (see [HJKY95, HJJ+97, FGMY96]) it is possible to leave the public key of the system in place for a really long time without fearing it being compromised. Indeed, when using proactive schemes the shares of the private key are periodically "refreshed" so that an attacker is forced to recover t shares in *one* single period of time that can be as short as a day. Both schemes presented in this paper can be made proactive, the discrete-log based one using the techiniques in [HJJ+97] and the factoring one by adapting the work of [FGMY96]. The idea is that the authorities run the key generation protocol every day at midnight, say, but now sharing a zero value. The new shares are added to the old shares of the secret key s. The resulting shares still interpolate to s (since the free term of the polynomial is unchanged) but lie on an otherwise different polynomial.

7 Concluding remarks

We have shown a very efficient scheme for secure elections based on the discrete log assumption, and a somewhat more complicated scheme based on the q-th residuosity assumption. The new schemes satisfy all well-known requirements, except for receipt-freeness. An open problem is to construct efficient incoercible election protocols, preferably without relying on physical assumptions.

In our scheme the work for the voter is minimal and independent of the number of authorities. Election schemes based on the mix channel of [PIK94] also have this property but for several reasons our approach is preferable over those schemes. In mix-based schemes the final tally is computed by somehow decrypting the individual ballots, while in our approach a single decryption of the aggregate of the ballots suffices. In mix-based schemes disrupters may submit invalid ballots which are detected only after decryption has taken place; in our scheme disruption by voters is automatically prevented because of the required proof of validity for ballots. Another important difference is that due to the use of a threshold cryptosystem we achieve robustness in a stronger sense. Indeed in mix-based schemes the failure of a single authority would compromise the whole protocol. In our case we can tolerate malicious behavior of a constant fraction (half) of authorities. Finally, the security of our scheme can be proven from its construction, while some security problems with the schemes of [PIK94, SK95] exist, as shown for instance in [Pfi95, MH96].

We would like to emphasize that the work for the voter is really low. For example, for the discrete log scheme, we have for $|p| = 64$ bytes and $|q| = 20$ bytes, that the size of the ballot plus its proof plus a signature on it is only 272 bytes in total. Clearly, this is an order of magnitude better than [CFSY96], which was already two orders of magnitude better than any previous scheme. Furthermore, computation of the ballot and its proof require a few exponentiations only (see Figure 2). A direct consequence of the reduced ballot size is also that the task of verifying the final tally is much simpler.

References

[Ben87]   J. Benaloh. *Verifiable Secret-Ballot Elections*. PhD thesis, Yale University, Department of Computer Science Department, New Haven, CT, September 1987.

[BT94]   J. Benaloh and D. Tuinstra. Receipt-free secret-ballot elections. In *Proc. 26th Symposium on Theory of Computing (STOC '94)*, pages 544–553, New York, 1994. A.C.M.

[BY86]   J. Benaloh and M. Yung. Distributing the power of a government to enhance the privacy of voters. In *Proc. 5th ACM Symposium on Principles of Distributed Computing (PODC '86)*, pages 52–62, New York, 1986. A.C.M.

[CDNO96] R. Canetti, C. Dwork, M. Naor, and R. Ostrovsky. Deniable encryption, 1996. Manuscript.

[CDS94]   R. Cramer, I. Damgård, and B. Schoenmakers. Proofs of partial knowledge and simplified design of witness hiding protocols. In *Advances in Cryptology—CRYPTO '94*, volume 839 of *Lecture Notes in Computer Science*, pages 174–187, Berlin, 1994. Springer-Verlag.

[CF85]   J. Cohen and M. Fischer. A robust and verifiable cryptographically secure election scheme. In *Proc. 26th IEEE Symposium on Foundations of Computer Science (FOCS '85)*, pages 372–382. IEEE Computer Society, 1985.

[CFSY96]   R. Cramer, M. Franklin, B. Schoenmakers, and M. Yung. Multi-authority secret ballot elections with linear work. In *Advances in Cryptology— EUROCRYPT '96*, volume 1070 of *Lecture Notes in Computer Science*, pages 72–83, Berlin, 1996. Springer-Verlag.

[CG96]   R. Canetti and R. Gennaro. Incoercible multiparty computation. In *37th IEEE Symposium on Foundations of Computer Science (FOCS' 96)*, 1996. To appear.

[Cha81]   D. Chaum. Untraceable electronic mail, return addresses, and digital pseudonyms. *Communications of the ACM*, 24(2):84–88, 1981.

[Cha91]   D. Chaum. Zero-knowledge undeniable signatures. In Damgård, editor, *Advances in Cryptology—EUROCRYPT '90*, volume 473 of *Lecture Notes in Computer Science*, pages 458–464, Berlin, 1991. Springer-Verlag.

[CP93]   D. Chaum and T. P. Pedersen. Wallet databases with observers. In *Advances in Cryptology—CRYPTO '92*, volume 740 of *Lecture Notes in Computer Science*, pages 89–105, Berlin, 1993. Springer-Verlag.

[CP95]   L. Chen and T. P. Pedersen. New group signature schemes. In *Advances in Cryptology—EUROCRYPT '94*, volume 950 of *Lecture Notes in Computer Science*, pages 171–181, Berlin, 1995. Springer-Verlag.

[Des94]   Y. Desmedt. Threshold cryptography. *European Transactions on Telecommunications*, 5(4):449–457, 1994.

[DH76]   W. Diffie and M. E. Hellman. New directions in cryptography. *IEEE Transactions on Information Theory*, 22(6):644–654, 1976.

[ElG85]   T. ElGamal. A public-key cryptosystem and a signature scheme based on discrete logarithms. *IEEE Transactions on Information Theory*, IT-31(4):469–472, 1985.

[FGMY96] Y. Frankel, P. Gemmell, P. McKenzie, and M. Yung. Proactive RSA, 1996. Manuscript.

[FS87]   A. Fiat and A. Shamir. How to prove yourself: Practical solutions to identification and signature problems. In *Advances in Cryptology—CRYPTO '86*, volume 263 of *Lecture Notes in Computer Science*, pages 186–194, New York, 1987. Springer-Verlag.

[Gen95]   R. Gennaro. Achieving independence efficiently and securely. In *Proc. 14th ACM Symposium on Principles of Distributed Computing (PODC '95)*, New York, 1995. A.C.M.

[GJKR96]   R. Gennaro, S. Jarecki, H. Krawczyk, and T. Rabin. Robust and efficient sharing of RSA functions. In *Advances in Cryptology—CRYPTO '96*, volume 1109 of *Lecture Notes in Computer Science*, pages 157–172, Berlin, 1996. Springer-Verlag.

[HJJ+97]   A. Herzberg, M. Jakobsson, S. Jarecki, H. Krawczyk, and M. Yung. Proactive public-key and signature schemes. 4th Annual Conference on Computer and Communications Security, 1997. To appear.

[HJKY95]   A. Herzberg, S. Jarecki, H. Krawczyk, and M. Yung. Proactive secret sharing, or: How to cope with perpetual leakage. In *Advances in Cryptology—CRYPTO '95*, volume 963 of *Lecture Notes in Computer Science*, pages 339–352, Berlin, 1995. Springer-Verlag.

[LL90]   A. K. Lenstra and H. W. Lenstra, Jr. Algorithms in number theory. In J. van Leeuwen, editor, *Handbook of Theoretical Computer Science*, pages 673–715. Elsevier Science Publishers B.V., Amsterdam, 1990.

[MH96]   M. Michels and P. Horster. Some remarks on a receipt-free and universally verifiable mix-type voting scheme. In *Advances in Cryptology—ASIACRYPT '94*, volume 1163 of *Lecture Notes in Computer Science*, pages 125–132, Berlin, 1996. Springer-Verlag.

[NR94]   V. Niemi and A. Renvall. How to prevent buying of votes in computer elections. In *Advances in Cryptology—ASIACRYPT '94*, volume 739 of *Lecture Notes in Computer Science*, pages 141–148, Berlin, 1994. Springer-Verlag.

[Ped91]   T. Pedersen. A threshold cryptosystem without a trusted party. In *Advances in Cryptology—EUROCRYPT '91*, volume 547 of *Lecture Notes in Computer Science*, pages 522–526, Berlin, 1991. Springer-Verlag.

[Ped92]   T. P. Pedersen. *Distributed Provers and Verifiable Secret Sharing Based on the Discrete Logarithm Problem*. PhD thesis, Aarhus University, Computer Science Department, Aarhus, Denmark, March 1992.

[Pfi95]   B. Pfitzmann. Breaking an efficient anonymous channel. In *Advances in Cryptology—EUROCRYPT '94*, volume 950 of *Lecture Notes in Computer Science*, pages 332–340, Berlin, 1995. Springer-Verlag.

[PIK94]   C. Park, K. Itoh, and K. Kurosawa. Efficient anonymous channel and all/nothing election scheme. In *Advances in Cryptology—EUROCRYPT '93*, volume 765 of *Lecture Notes in Computer Science*, pages 248–259, Berlin, 1994. Springer-Verlag.

[PW92]   B. Pfitzmann and M. Waidner. Unconditionally untraceable and fault-tolerant broadcast and secret ballot election. Hildesheimer informatikberichte, Institut für Informatik, May 1992.

[Rab83]   M. Rabin. Transaction protection by beacons. *Journal of Computer and System Sciences*, 27(2):256–267, 1983.

[Rei94]   M. Reiter. Secure agreement protocols: Reliable and atomic group multicast in Rampart. 2nd ACM Conference on Computer and Communications Security, Fairfax, November 1994.

[Rei95]   M. Reiter. The Rampart toolkit for building high-integrity services. In *Theory and Practice in Distributed Systems*, volume 938 of *Lecture Notes in Computer Science*, pages 99–110, Berlin, 1995. Springer-Verlag.

[Sch91]   C. P. Schnorr. Efficient signature generation by smart cards. *Journal of Cryptology*, 4(3):161–174, 1991.

[Sha79]   A. Shamir. How to share a secret. *Communications of the ACM*, 22(11):612–613, 1979.

[SK94]   K. Sako and J. Kilian. Secure voting using partially compatible homomorphisms. In *Advances in Cryptology—CRYPTO '94*, volume 839 of *Lecture Notes in Computer Science*, pages 411–424, Berlin, 1994. Springer-Verlag.

[SK95]   K. Sako and J. Kilian. Receipt-free mix-type voting scheme—a practical solution to the implementation of a voting booth. In *Advances in Cryptology—EUROCRYPT '95*, volume 921 of *Lecture Notes in Computer Science*, pages 393–403, Berlin, 1995. Springer-Verlag.

Binding ElGamal: A Fraud-Detectable Alternative to Key-Escrow Proposals

Eric R. Verheul[*1] and Henk C.A. van Tilborg[2]

[1] Ministry of the Interior, P.O. Box 20010, 2500 EA, The Hague, The Netherlands. Eric.Verheul@pobox.com
[2] Department of Math. and Comp. Sc., P.O. Box 513, Eindhoven University of Technology, 5600 MB, Eindhoven, The Netherlands. henkvt@win.tue.nl

Abstract. We propose a concept for a worldwide information security infrastructure that protects law-abiding citizens, but not criminals, even if the latter use it fraudulently (i.e. when not complying with the agreed rules). It can be seen as a middle course between the inflexible but fraud-resistant KMI-proposal [8] and the flexible but non-fraud-resistant concept used in TIS-CKE [2]. Our concept consists of adding *binding data* to the latter concept, which will not *prevent* fraud by criminals but makes it at least *detectable* by third parties without the need of any secret information. In [19], we depict a worldwide framework in which this concept could present a security tool that is flexible enough to be incorporated in any national cryptography policy, on both the domestic and foreign use of cryptography. Here, we present a construction for binding data for ElGamal type public key encryption schemes. As a side result we show that a particular simplification in a multiuser version of ElGamal does not affect its security.

Key words ElGamal, Traceable ElGamal, Key Escrow, Key Recovery

1 Introduction

We'll briefly summarize the *technical* position taken in [19]. A robust, worldwide information security infrastructure (ISI) must be set up which includes a Key Management Infrastructure which will (likely) be based on public key cryptography. Proper certification of public keys will be a crucial (and elaborate) service within this ISI. However, the unconditional use of encryption by criminals poses a threat to law enforcement, a problem that is hard to solve. Consequently, most governments feel that they have to realize two tasks. The first is to stimulate the establishment of an ISI which protects the legitimate interests of all relevant parties (businesses, governments, citizens), but which does not aid criminals. The second task is to cope with the use of other encryption techniques by criminals. How to achieve the second goal is outside the scope of this contribution, but it is our feeling that an ISI, that is widely accepted and trusted, will make it

---

* Views expressed here are personal and not necessarily shared by my employer.

W. Fumy (Ed.): Advances in Cryptology - EUROCRYPT '97, LNCS 1233, pp. 119-133, 1997.
© Springer-Verlag Berlin Heidelberg 1997

easier to achieve the second task. We also feel that without strong cooperation of governments such a widely accepted and trusted ISI will never be established at all. In this paper we address a construction of a reliable ISI, which does not aid criminals.

In public key encryption (pke) encrypted messages - ideally - consist of two components:

C1. The (actual) message M encrypted with a symmetric system, using a random session key S.

C2. The session key S encrypted using the public key(s) of the addressee(s).

A straightforward method to prevent facilitation of criminals is outlined in the U.S.-government (draft) Key Management Infrastructure (KMI) proposal [8]. Here, participating users have to deposit their private keys with a private-sector Trusted Recovery Party (TRP).[3] When a law-enforcement agency (LEA), that has obtained legal authority to access a user's communication, strikes upon data encrypted within this scheme, the TRP will "relinquish information sufficient to access" these data. One of the problems mentioned in [19], is that the scheme is inflexible in an international context: in order to let the principle work for *any* country, *every* participating country - irrespective of its national policy on cryptography - has to escrow the private keys of its users also. Also, international cooperation of a TRP with a LEA outside the country of the TRP might be difficult and time-consuming. Although the latter problem is resolved in the "Royal Holloway" variant [11] of this scheme, it can be argued that the resulting flexibility here is not better than that of the KMI-proposal. Compare [1].

A more flexible method to prevent facilitation of criminals consists of *virtual addressing* session keys to Trusted Recovery Parties (see, for instance, the TIS Commercial Key Escrow [2]). In this scheme, participating users agree to add a third component to an encrypted message:

C3. The same session key S encrypted using the public key(s) of one or more Trusted Recovery Parties.

In effect, any TRP is treated as a virtual addressee, although the message is not sent to it. When a LEA is conducting a lawful intercept and strikes upon an enciphered message, they take the information component of one of the TRP's to that TRP. If shown an appropriate warrant, the TRP decrypts ("recovers") the information component and (only) hands over the session key S, so that the LEA agency can access the message.

This concept has been the base of several escrow products (Translucent Cryptography, AT&T Crypto Backup, RSA secure). Observe that users do not have to deposit secret key information to TRP's *beforehand*. This makes this approach

---

[3] We use the notion "Trusted Recovery Party" as it forms a combination of the (recent) U.S. notion "recovery" (replacing "key-escrow") and the European notion "Trusted Third Party".

more feasible (and acceptable to users) than the KMI-proposal; an important advantage as - also pointed out in the study of the National Research Council (NRC) [14, p.329] - feasibility of key-recovery solutions is a significant issue. We remark that one could incorporate information in the session key identifying the sender (as is done in TIS-CKE). However, as this, in principle, makes possible a (partially) known-plaintext attack (cf. [4]) one should be careful with this.

Although this concept is very flexible (see below), its main drawback is that it offers no possibility, at least for others than the TRP, to check whether the third component actually contains the (right) session key; moreover the TRP can only discover "fraud" (i.e. not complying with the agreement) after a lawful wiretap. Hence, by sending noise instead of a third component unilateral abuse (i.e. without help of the addressee) is easily possible. This can be prevented in the software of the addressee by a recalculation and validation of C3 prior to decryption. However, abuse by colluding of a sender and receiver - through a one-time manipulation of this validation in software - is still easily possible. So the solution is almost entirely unenforceable. According to the NRC-study [14, p.214] U.S. senior Administration officials have said that this matter is the reason for the limitation to (only) 64 bits in the (draft) 1995 U.S. Key Escrow Export Criteria for cryptographic applications in software:"the limitation to 64 bits is a way of hedging against the possibility of finding easily proliferated ways to break the escrow binding built into software, with that result that U.S. software products without effective key escrow would become available worldwide". On the other hand, it is noted in the NRC-study [14, p.211] that a recovery encryption product does not have to be perfectly resistant to breaking the recovery binding: it should only be more difficult to bypass the recovery features than to build a system without recovery.

In [19] we looked for a middle course between the inflexible but fraud-resistant KMI-proposal and the flexible but non-fraud-resistant virtual addressing. We found one by not *preventing* colluding of sender and receiver, but by making it at least *detectable* by third parties without having access to secret (key) information. More specifically, we proposed the *binding alternative*, which adds a fourth component to the encrypted message:

C4. Binding data.

The idea is that any (third party) monitor, e.g., a network or (Internet) service provider, who has access to components C2, C3, and C4 (but not to any additional secret information) can determine that the session keys encrypted in components C2 and C3 coincide but it can not determine any information on the actual session key S. In this way, fraud is easily detectable (and punishable). Metaphorically speaking, binding data consists of equipping public-key encryption schemes used for confidentiality with a metal detector, as used at boarding gates on airports.

The binding concept supports the virtual addressing of session keys to several TRP's (or none for that matter), for instance, one to a TRP in the country of the sender S and one in the country of the addressee A. Note that this can be easily implemented: S's software can (once) be adjusted to the public key of S's TRP; the public key of A's TRP can be part of A's (certified) public key. The solution therefore offers the same advantage for worldwide usability as [11]. We also remark that the binding concept also supports the functionality of controllable key splitting in the sense of Micali [13], even in several fashions. For instance, the private TRP key can be splitted in several parts and be deposited at several sub-TRP's. It turns out that the ElGamal system very conveniently supports the splitting and the reconstruction of private keys (see the end of Section 2). Finally, we remark that the time-boundedness condition (cf. [12, p.199]), i.e. the condition that time-limits on warrants can be enforced, can be fulfilled by additionally demanding that encrypted information (or all components) be time-stamped and signed by the sender. These can be easily verified by any third party monitor as well. A much simpler solution is to let the time be an (unencrypted) part of the message and to incorporate it in the binding data (as indicated in Section 4).

An additional feature could prevent the threat of the "tempted policemen" This tempted policemen might conspire with a criminal and have the criminal resent (or "receive") an unrelated, highly confidential business message intercepted by the policemen. The TRP, thinking the message originated from the (wiretapped) criminal, would assist the policemen in decrypting. In the binding scheme, this can be prevented by additionally requiring senders to virtually address the session key to themselves as well. The TRP could check this component before assisting a law-enforcement agency, and monitors could check on compliance. Incidentally, this feature can also solve similar problems in TIS-CKE and in the U.S. KMI-proposal. In the latter, it also overcomes the problem of international communications: the TRP has got the private key of the sender and can therefore recover the session key. Thus, binding cryptography can also benefit other proposals.

In [19], we depict a general framework in which the binding concept (as general notion) could present a security tool that is flexible enough to be incorporated in any national cryptography policy, for both the domestic and foreign use of cryptography, and that offers a flexible choice of trust for users. Here, we present a construction for binding data for the ElGamal type of pke schemes; this is particularly interesting as on 29 April 1997, ElGamal will no longer be encumbered by patents in the U.S..

A difficulty one faces in the construction of binding data for a pke scheme, apart from the binding data itself, is finding a suitable multiuser extension of it, allowing the secure (!) encryption of exactly the same session-key (i.e. including "padding" data) with different public keys. For the RSA scheme, for instance, this presents a problem (cf. [10]). In Section 2 we will introduce a secure multiuser extension of ElGamal. Section 3 deals with proving knowledge of equality of certain logarithmic values. Section 4 presents the construction of binding data techniques for ElGamal's protocol. Finally, many of the constructions for the ElGamal scheme can be extended to Desmedt's traceable variant of ElGamal ([6]). We will sketch some of these extensions in Appendix B.

2 The Multiuser ElGamal Encryption Scheme

The ElGamal [7] pke system makes use of a subgroup G of a multiplicative, cyclic group H in which the discrete logarithm problem is intractable. Let q be the order of G and let g be a generator of G. The elements g, G, and H are given to all participants by an Issuing Party (IP). We will not further specify G, H, but in a typical example H is the multiplicative group of Z/pZ for a (large) prime p and $G = H$.

To participate in the system, each participant P chooses his own secret key x_P (a random number less than q) and publishes his (certified) public key $y_P = g^{x_P} \in G$. If a person, say Ann, wants to encrypt a message $S \in H$ meant for participant Bob, she chooses a random number k less than q and sends the pair $(t, u) = (g^k, y_{Bob}^k \cdot S)$ to Bob. When Bob receives (t, u) he just calculates $u/t^{x_{Bob}}$ to find S back.

We focus on the following multiuser extension of ElGamal,

Definition 2.1 *In the* Multi-ElGamal *protocol, participant P, when going to encrypt message $S \in H$ for n participants with public ElGamal keys $y_1, y_2, \ldots, y_n$, will generate a random number k less than q and send pair $(g^k, y_i^k \cdot S)$ to the i-th participant, $1 \le i \le n$.*

The question that arises of course is whether Multi-ElGamal is less secure than choosing a different k for each participant (which is less efficient). We shall show it is not.

The following terminology is convenient. Let g be an element of G, y an element of the cyclic group $< g >$ generated by g, $S \in H$ and $k \in Z/qZ$. Then the 4-tuple $(g, y, g^k, y^k \cdot S)$ is called an *encryption* of g, y, k, S and will be denoted by $[g, y, k, S]$. The elements $k, S, \log_g y$ will be called the *secret* (or *unknown*) components of the encryption.

Lemma 2.2 *Let $[g, y_P, k_i, S_i]$, $1 \le i \le h$, be a sequence ("history") of encryptions for user P. Then anyone can construct a second sequence of encryptions $[g, \hat{y}, k_i, S_i]$, $1 \le i \le h$, with $\hat{y}$ random in G (but with the same k_i's and S_i's) such that the computation of $\log_g(y_P)$ is as difficult as that of $\log_g \hat{y}$.*

Proof: For $i = 1, 2, \ldots, h$, denote $(g^{k_i}, y_P^{k_i} \cdot S_i)$ by (A_i, B_i). Let i be one of $1, 2, \ldots, h$. Choose j randomly in Z/qZ, and compute $C = g^j$, $D_i = (A_i)^j$ and $\hat{y} = y_P \cdot C$. First of all, we observe that $\hat{y} = g^{x_P + j}$. So $\hat{y}$ is a random element in $< g > = G$.

Now $(g, \hat{y}, A_i, B_i \cdot D_i)$ can be computed. We shall prove that it is indeed an encryption $[g, \hat{y}, k_i, S_i]$. To this end the only condition that needs to be verified is $B_i \cdot D_i = \hat{y}^{k_i} \cdot S_i$. This follows from:

$$B_i \cdot D_i = y_P^{k_i} \cdot S_i \cdot g^{j \cdot k_i} = g^{x_P \cdot k_i} \cdot g^{j \cdot k_i} \cdot S_i = g^{(x_P + j)k_i} \cdot S_i = \hat{y}^{k_i} \cdot S.$$

Finally, we observe that $\log_g \hat{y} = \log_g y_P + j$, so $\log_g(y_P)$ can be determined directly from $\log_g \hat{y}$ and vice versa. $\square$

Theorem 2.3 *Let n be a natural number. Then breaking Multi-ElGamal for n addressees is as difficult as breaking ElGamal.*

Proof: Clearly, any algorithm that breaks ElGamal also breaks the Multi-version of it. So, only the implication the other way around needs to be shown. Suppose there exists an efficient algorithm $\mathcal{A}$ that on input of n sequences of h Multi-ElGamal encryptions (in the i-th encryption, $1 \leq i \leq h$, the same message S_i has been sent to all n users - with random public keys - using the same random number k_i) has a non-negligible chance of outputting (all) secret information. Now let a sequence of ElGamal encryptions for a participant P be given, say $[g, y, k_i, S_i]$ for $i = 1, 2, \ldots, h$. Then by the first part of Lemma 2.2 we can construct a sequence of outputs of a Multi-ElGamal encryption with n participants using the same k_i and S_i: the public keys of the participants will be random and the secret key of P follows from any of the secret keys of the participants. Combining this output with $\mathcal{A}$ we obtain an algorithm $\mathcal{B}$, as efficient as algorithm $\mathcal{A}$, which breaks the ElGamal encryptions for participant P with the same non-negligible chance. $\square$

Using the ideas of [13], the ElGamal scheme can very conveniently support the construction of public keys in which the secret key is secretly shared among n share-holders (TRP's in our situation) in an n out of n secret sharing scheme. Suppose all share-holders have chosen a secret key x_i less than q and have publicized the resulting ElGamal public key $y_i = g^{x_i}$. Then, their product denoted by y, will be the shared public key. Observe that the associated secret key x is given by $\log_g y = \sum_{i=1}^{n} x_i$. The ElGamal encryption $(g^k, y^k \cdot S) = (A, B)$ of a message S with respect to the public key y, can be decrypted by a third party (a LEA in our situation) by first asking the i-th share-holder to return $A_i = A^{x_i}$ and then to calculate S by $B / \prod_{i=1}^{n} A_i$. Observe that the share-holders do not have to come together and explicitly reconstruct the secret. If, in our situation, many TRP's have publicized their public key, then users *themselves* can choose the share-holders (they trust) and form the resulting public key.

By following Pedersen [15], [16] one can, for any $1 \leq k \leq n$, construct an ElGamal public key $y = g^x$ in which the secret key x is shared in a k out of n secret sharing scheme as the constant term of a polynomial f of degree $k - 1$. Also, shareholders can verify the validity of their share. In [15] a (trusted) dealer is required to construct f. In [16] f is interactively and securely constructed by the share-holders themselves (in our situation, for instance on request of a user). As a dealer forms a single point of failure, the latter construction is preferred in our situation. As above, one can construct a protocol (also used in [5]) in which a third party (a LEA in our situation) can decrypt an ElGamal encryption $(g^k, y^k \cdot S) = (A, B)$ of a message S without the share-holders need to come together and explicitly reconstruct their secret. More precisely, consider k share-holders in the scheme with public computable $a_1, \ldots, a_k$ and shares $s_1, \ldots, s_k$ (see [15, p.223]). Then the party first asks the i-th share-holder to return $A_i = A^{s_i}$ and subsequently determines S by calculating $B / \prod_{i=1}^{m} A_i^{a_i}$. We note that for $k = n$, the earlier mentioned scheme is more efficient.

3   A proof of knowledge on the equality of logarithms

The following result seems to be part of the mathematical "folklore", but for the sake of completeness a proof is given in Appendix A. The result is an extension of the Chinese Remainder Theorem in the situation that not necessarily all moduli are relatively prime in pairs.

Proposition 3.1 *Let a_i, b_i for $i = 1, 2, \ldots, n$, be integers and let C_i denote the cosets $a_i + (b_i)$ in Z, where (b_i) stands for $b_i Z$. Then the following assertions are equivalent:*

1. *The intersection of all C_i's is non-empty and can be written as $y + (\mathrm{lcm}(b_1, b_2, \ldots, b_n))$ for some integer y.*
2. *Every pair of C_i's has a non-empty intersection.*
3. *$\gcd(b_i, b_j)$ divides $a_i - a_j$ for all $1 \leq i \neq j \leq n$.*

Now consider elements $g_1, g_2, \ldots, g_n$ (not necessarily distinct) in G. Suppose that person P (for prover) gives $h_1, h_2, \ldots, h_n \in G$ to person V (for verifier) and states:

S. There exists a number $0 < k < q$, such that for all $1 \leq i \leq n$

$$g_i^k = h_i, \tag{1}$$

or equivalently, there exists a number $0 < k < q$, simultaneously satisfying:

$$k \equiv \log_{g_i} h_i \pmod{\mathrm{ord}(g_i)}. \tag{2}$$

where the "ord" of a group element stands for its multiplicative order. Note that if all g_i are generators then all $\log_{g_i} h_i$ will coincide.

The following protocol lets P prove statement **S** without revealing anything about k; it is inspired by the authentication schemes of Schnorr [17] and Guillou-Quisquater [9]. Moreover, it is an extension of a signature scheme introduced by Chaum and Pedersen in [3] (an anonymous referee is thanked for this reference). In this protocol a positive integer v occurs, that will be called the *confidence level* of the protocol. We will demand that this number satisfies:

$$v \leq \min\{v' \mid v' > 1 \text{ and, for some } i \neq j,$$
$$v' \text{ divides both } \mathrm{ord}(g_i) \text{ and } \mathrm{ord}(g_j) \}. \tag{3}$$

Note that the smallest prime factor of $q = |G|$ is a lowerbound for v; equality holds if all g_i are generators of G. As a large v is desired, q should not have small prime factors.

Protocol 3.2

1. *P generates a random number l less than q, calculates $a_i = g_i^l$ for $1 \leq i \leq n$ and hands the a_i's over to V.*

2. *V generates a random $0 < w \leq v$ and presents w as a challenge to P.*
3. *P calculates $z = w \cdot k + l$ (mod q) and hands z over to V.*
4. *V verifies for all $1 \leq i \leq n$ that $g_i^z = h_i^w \cdot a_i$. If so, V will accept S, otherwise he rejects it.*

We will now show that this protocol satisfies the following properties:

Completeness If statement S is true, then V will accept it.

Soundness If S is not true, then with a probability less than $1/v$ (so small) it will still be accepted by V.

Security If S is true, then V can not learn secret information on k by following the protocol.

The verification of the first property is straightforward. For the verification of Soundness, suppose that equality (2) does not hold, so there is no common solution to the n congruences in (2). Then, by Proposition 3.1, there exist $1 \leq i \neq j \leq n$ such that $\gcd(\text{ord}(g_i), \text{ord}(g_j))$ does not divide $\log_{g_i} h_i - \log_{g_j} h_j$. Let D denote the greatest common divisor of the latter two numbers, and let $v' = \gcd(\text{ord}(g_i), \text{ord}(g_j))/D$. Now, although P has (some) freedom in choosing $\log_{g_i} h_i$ prior to the protocol, and $\log_{g_i} a_i$ in the first step of the protocol, he has to come up with a number z in the third step satisfying for all i, $1 \leq i \leq n$, and for all (or at least sufficiently many) w, $0 < w < v$:

$$z \equiv w \cdot \log_{g_i} h_i + \log_{g_i} a_i \pmod{\text{ord}(g_i)}.$$

The i-th and j-th congruences above (resp. modulo $\text{ord}(g_i)$ and $\text{ord}(g_j)$) will also hold modulo the common factor $\gcd(\text{ord}(g_i), \text{ord}(g_j))$, yielding:

$$w \cdot \log_{g_i} h_i + \log_{g_i} a_i \equiv w \cdot \log_{g_j} h_j + \log_{g_j} a_j \pmod{\gcd(\text{ord}(g_i), \text{ord}(g_j))}.$$

As $(\log_{g_i} h_i - \log_{g_j} h_j)/D$ is relatively prime with v', w is uniquely determined modulo v'. Hence the probability that V chooses the "right" w (in V's opinion) is equal to $1/v'$ which is less than or equal to $1/v$.

Finally, as an argument for Security, we assume that both P and V really choose l resp. w randomly. Observe that it is in P's best interest to do so: more uncertainty on l will give more uncertainty on k to V in the third step of the protocol. Now we will proceed with the standard zero-knowledge argument: we will show that V can generate a typical transcript $(a_1, \ldots, a_n; w; z)$ of the protocol himself, i.e. without communicating with P. To this end, V can choose w and z at random and evaluate a_i, $1 \leq i \leq n$, such that they satisfy $g_i^z = h_i^w \cdot a_i$. Then it easily follows - provided P's statement is correct - that $a_i = g_i^l$ for $l = z - k \cdot w$.

Note that for Security it is required that the verifier follows the protocol, i.e. the verifier must choose his challenges w in a random way. Although intuitively clear, we can not prove that V learns no secret information by deviating from the protocol by choosing his challenges in a non-random way (cf. [3]). In the terminology of [18, Ch. 13] the above proof system for equality of logarithms is perfect zero-knowledge for an honest verifier, but we do not know whether it is perfect zero-knowledge without qualification, i.e. for any (dishonest) verifier. In our application of it in Section 4 we will enforce the verifier to be honest, i.e. to choose his challenges in a random way, thereby ensuring security.

We remark that the verification in the fourth step of the protocol can be rewritten as $g_i^z \cdot h_i^{-w} = a_i$. The use of data in the protocol can be reduced if P hands over the hash values $H_i = H(a_i)$ of the a_i - for some secure hash function $H(.)$ - instead of the a_i themselves. The verification step in the fourth step of the protocol then becomes:

$$H(g_i^z \cdot h_i^{-w}) = H_i. \tag{4}$$

A similar technique is employed in the U.S. Digital Signature Algorithm. To achieve the same level of security the number of bits in the output of the hash should not be less than $\log_2(v)$.

4 Binding the ElGamal Encryption Scheme

In this section we will present a construction for binding the ElGamal schemes using the multiuser extension discussed in Section 2. We shall do this with a (detailed) illustration, in which we will use the notation of Section 2. We will also make use of a conventional symmetric cipher $E(.)$ and of a public one-way (hash) function $H(.)$.

Suppose that Ronald from America wants to send a confidential document D to Margaret in Britain using a (government supported) Public Key Infrastructure (PKI) that incorporates binding ElGamal. Part of the PKI-policy is the choice of a confidence parameter v: the probability that binding data are accepted while the values of S sent to B and the TRP differ should be less than $1/v$. We assume that the parameters of the ElGamal system are chosen such that inequality (3) holds, that is q has no prime factors less than v. Now suppose that the national PKI-policy of America (resp. Britain) states that Ronald has to virtually address his messages to an American TRP (resp. a British TRP). Also suppose that the American PKI-policy allows the use of "splitted" public keys as explained at the end of Section 2. Let TRP_{A_1}, TRP_{A_2} respectively TRP_B be Trusted Recovery Parties from respectively America and Britain that Ronald trusts and chooses; TRP_{A_1}, TRP_{A_2} together form TRP_A. Let the splitted secret keys and public keys of TRP_{A_1}, TRP_{A_2} be respectively denoted by $x_{A_1}, x_{A_2}, y_{A_1}, y_{A_2}$, the shared secret key and public key (of TRP_A) will be denoted by $x_A(= x_{A_1} + x_{A_2})$ and $y_A(= y_{A_1} \cdot y_{A_2})$. Also, the secret key and public key of TRP_B will be denoted respectively by x_B and y_B. Finally, the secret and public key of Margaret will be simply denoted by x and y.

Ronald chooses a random $k < q$ and a session key $S \in H$ and sends the following data-block to Margaret: $(E, C, R_M, R_A, R_B, bind)$ where:

C1. $E = E_S(D)$: the document encrypted by E under session key S.
C2. $(C, R_M) = (g^k, y^k \cdot S)$: the session key S enciphered with Margaret's public key;

C3. $(C, R_A) = (g^k, y_A^k \cdot S)$, $(C, R_B) = (g^k, y_B^k \cdot S)$: the session key S enciphered with the public keys of resp. TRP$_A$ and TRP$_B$.

C4. *bind.*

First observe that if Ronald uses the scheme correctly, then Margaret can determine S by calculating R_M/C^x; TRP$_B$ can offer S to a British LEA by calculating R_B/C^{x_B}. An American LEA can ask TRP$_{A_1}$ (resp. TRP$_{A_2}$) to calculate $C^{x_{A_1}}$ (resp. $C^{x_{A_2}}$), and then calculate S by $R_A/(C^{x_{A_1}} \cdot C^{x_{A_2}})$. This is just an application of the multiuser ElGamal scheme which we showed to be as secure as the original ElGamal scheme.

Now we come to the construction of the binding data *bind*. Observe that the three numbers $C, R_A/R_M, R_B/R_M$ are respectively equal to $g^k, (y_A/y)^k$, and $(y_B/y)^k$, that is, they are equal to the group elements $g, y_A/y, y_B/y$ raised to the same power k. Hence, k can be viewed as the solution of the equality:

$$g^\kappa = C , \quad (y_A/y)^\kappa = R_A/R_M , \quad (y_B/y)^\kappa = R_B/R_M. \qquad (5)$$

Now suppose we know that equality (5) has a solution k'. Given that the C and R_M are formed correctly (they are meant for Margaret to decrypt the message using ElGamal). It follows that $R_A = (y_A/y)^{k'} \cdot R_M = (y_A/y)^{k'} \cdot y^{k'} \cdot S = (y_A)^{k'} \cdot S$. That is, (C, R_A) is a well-formed ElGamal encryption of the same S for TRP$_A$. A similar conclusion holds for TRP$_B$.

We conclude that to construct binding data for the ElGamal scheme one only has to construct data which shows that (5) has a solution. For this one would like to use a non-interactive version of Protocol 3.2. To this end, Ronald generates a random $j < q$ and forms $bind = (D, F, I, z)$, where $D = g^j$, $F = (y_A/y)^j$, $I = (y_B/y)^j$ and $z = w \cdot k + j \pmod{q}$, where $w < v$ is the result of letting the one-way function $H(.)$ work - in a fixed, public way - on $E, C, R_M, R_A, R_B, D, F, I$ and possibly other public data such as Margaret's full identity and the date/time. In effect, w can not be predicted by Ronald beforehand and behaves like the random challenge in Protocol 3.2, Step 2.

Now by Protocol 3.2 anybody who has access to $R_M, R_A, R_B, bind$ and the public keys of Margaret, TRP$_A$, and TRP$_B$ can determine that (5) has a solution by first calculating w and then by verifying that

$$g^z = C^w \cdot D \; ; \; (y_A/y)^z = (R_A/R_M)^w \cdot F \; ; \; (y_B/y)^z = (R_B/R_M)^w \cdot I. \quad (6)$$

The probability that this verification gives the wrong answer is less than $1/v$.

As explained at the end of Section 3, one can use hashes of D, F, I in *bind* instead. The involved binding data can then be reduced to approximately the length of q. Observe that this technique can be generalized to the situation where more than two TRP's are used. For each extra TRP the binding data increases with the length of the used hash, which is rather unfortunate.

However, reducing the binding data can be done more effectively by using a standard trick of the trade (as pointed out to us by Berry Schoenmakers). Observe that from (6) it follows that one can deduce (D, F, I) if one knows (w, z).

Now we let (in the above notation) the binding data consist of (w, z) (instead of (D, F, I, z)). Verification of the binding data now consists of three steps. First one calculates (D, F, I) as indicated in (6), that is:

$$D = g^z \cdot C^{-w} \quad ; \quad F = (y_A/y)^z \cdot (R_A/R_M)^{-w} \quad ; \quad I = (y_B/y)^z \cdot (R_B/R_M)^{-w}.$$

Second (as before), let the one-way function $H(.)$ work - in a fixed, public way - on $E, C, R_M, R_A, R_B, D, F, I$ and possibly other public data such as Margaret's full identity and the date/time resulting in a $w' < q$. Third (and finally), check if w' equals w. If so accept the binding data (and conclude that (5) has a solution), otherwise reject it (and conclude that (5) has no solution). Note that one can easily convert the "new" (w, z) type of binding data to the "old" (D, F, I, z) type (and vice versa). Hence it follows that the probability that this verification gives the wrong answer is less than $1/v$.

Note that these "new" binding data are of fixed (small) length, namely the length of q plus the length of the output of $H(.)$ which is approximately equal to the length of q. Also, one can easily generalize this technique to the situation where more than two TRP's are used. The length of the binding data is independent of the number of TRP's which is very fortunate. As this technique is also more easily and securely implemented than the one using hashes of D, F, I we prefer it.

5 Conclusion

We have introduced a new concept for the establishment of an Information Security Infrastructure that does not hamper law-enforcement, using *binding data*. More in particular, we have presented a construction for binding data for the ElGamal type of public key encryption schemes using well-understood cryptographic techniques and primitives. As a side result we show that a particular simplification in a multiuser version of ElGamal does not affect its security. We expect that many more public key encryption schemes can be equipped with binding data.

A special property of the binding concept is that abuse of the system is not only difficult but also detectable by any third party (e.g. network or service provider) without harming the privacy of law-abiding users. Other properties of the binding alternative include giving users in principle a flexible choice on who to trust with their confidential communication; moreover, there need be no vulnerable parties holding (master) keys in deposit.

In our opinion, the properties of the binding alternative are flexible enough to allow cooperating countries to implement different cryptography policies on the domestic and international use of encryption in a coherent framework, which will be acceptable to many (most?) citizens in the information society. We emphasize that the binding alternative does not solve criminal encryption outside of this framework or even *within* using super-encryption - it is not meant to. Criminals can use encryption anyhow; our sole aim is that they should only be kept from effectively gaining advantage in using the (government supported) framework for this.

6 Acknowledgments

We are very grateful to Berry Schoenmakers for his valuable comments, references to existing literature and his suggestion to improve the size of the binding data at the end of Section 4.

A Proof of Proposition 3.1

We shall only show implication 2) $\Rightarrow$ 1) as the other implications are rather straightforward. To this end, we first claim that the following equality holds for all natural numbers x:

$$\gcd(x, \operatorname{lcm}(b_1, \ldots, b_n)) = \operatorname{lcm}(\gcd(x, b_1), \ldots, \gcd(x, b_n)). \tag{7}$$

This equality simply expresses that the lattice $(Z, \gcd, \operatorname{lcm})$ is distributive. For a direct verification express the integers above in terms of prime powers and use $\min\{\chi, \max\{\beta_1, \ldots, \beta_n\}\} = \max\{\min\{\chi, \beta_1\}, \ldots, \min\{\chi, \beta_n\}\}$.

The implication 2) $\Rightarrow$ 1) is trivial for $n = 2$. We shall now use induction to n. For the step $n \to n + 1$ we may assume (by the induction hypothesis) the existence of y such that $\bigcap_{i=1}^n C_i = y + (\operatorname{lcm}(b_1, b_2, \ldots, b_n))$. Hence:

$$\bigcap_{i=1}^{n+1} C_i = (y + (\operatorname{lcm}(b_1, \ldots, b_n))) \bigcap (a_{n+1} + (b_{n+1})). \tag{8}$$

According to the last assertion of the proposition this intersection is non-empty and of the appropriate form iff $y - a_{n+1}$ is a multiple of $\gcd(b_{n+1}, \operatorname{lcm}(b_1, \ldots, b_n))$. By equality (7) this latter equals $\operatorname{lcm}(\gcd(b_{n+1}, b_1), \ldots, \gcd(b_{n+1}, b_n))$. Hence the lefthand side of equality (8) is non-empty iff $y - a_{n+1}$ is a multiple of $\gcd(b_{n+1}, b_i)$ for $i = 1, \ldots, n$.

Now, fix i in $\{1, \ldots, n\}$ and write $y - a_{n+1} = (y - a_i) + (a_i - a_{n+1})$. Then the first term in the right hand side is a multiple of b_i and hence of $\gcd(b_{n+1}, b_i)$. The second term is a multiple of $\gcd(b_{n+1}, b_i)$ as the cosets C_i and C_{n+1} meet. So $y - a_{n+1}$ is a multiple of $\gcd(b_{n+1}, b_i)$ for each $1 \le i \le n$.

We conclude that the lefthand side of (8) is non-empty. That $\bigcap_{i=1}^{n+1} C_i$ is of the form $\hat{y} + (\operatorname{lcm}(b_1, b_2, \ldots, b_{n+1})$ now easily follows from the $n = 2$ case.

B An Extension for Desmedt's traceable variant of ElGamal

We use the notation of Section 2, in particular we recall that g denotes a generator of a group G. In [6], Desmedt proposes a variant of ElGamal in which all participants are given different generators by the Issuing Party (IP). Here q is a number of the form $\prod_{i=1}^m q_i$ where all q_i are different prime numbers. For each participant P a unique divisor $d_P \neq 1$, called P's *order*, of q is chosen (linked to P and stored). P is also given the (base-)generator $g_P = g^{q/d_P}$, the order of which equals d_P. This generator is part of his public key of P, which also (as in the standard ElGamal) includes a $y_P \in\ <g_P>$ of the form $y_P = g_P^{x_P}$ where x_P (a random number less than q) is P's secret key. A message $S \in H$ encrypted by Ann using P's public key takes the form $(g_P^k, y_P^k \cdot S)$ where k is a number less than q randomly chosen by Ann. It is shown in [6] that addressees can be identified from the (orders of the) encrypted messages sent to them. We shall refer to Desmedt's variant of ElGamal as *D-ElGamal*.

In principle, there is no need for the IP to reveal d_P to participant P. However, as can be easily seen (cf.[6]), knowledge of d_P enables the Issuing Party IP to determine S^{d_P} from the encrypted message with P's public key. So, IP can use the knowledge of the d_P to determine secret information. It can be argued (cf.[6]) that breaking the system for the IP should not be significantly easier than for an outsider. Hence, we come to the following:

Assumption B.1 *With respect to the (encryption) security of D-ElGamal we assume that the orders d_P's of participants and the factorization of q, are publicly known.*

Extending D-ElGamal to a multiuser version in a similar way as in Definition 2.1 is insecure. Indeed, suppose that a participant P wants to encrypt a message $S \in H$ meant for n participants with public keys $(g_1, y_1), \ldots, (g_n, y_n)$ in the D-ElGamal scheme; the order of i-th participant will be denoted by d_i. It seems natural, as in the conventional ElGamal scheme, that P generates one random number k and sends to the i-th participant $(g_i^k, y_i^k \cdot S)$. However, by Assumption B.1 an eavesdropper Eve can determine S^{d_i} for $i = 1, \ldots n$. So, if d is the greatest common divisor of the d_i's then Eve can also determine S^d. In other words if these d_i are relatively prime (which is likely) then Eve can determine S. Although this might be an interesting feature for some countries (sending a message to a "wrong" group of people will expose the message), it is an unacceptable security risk. Also observe that generating different k_i's for each participant doesn't help to resolve this insecurity. So, even in general, the multiuser extension of D-ElGamal is insecure.

To remedy this, we will demand in the above extension of D-ElGamal that all d_i's except for d_1 are equal to q; the resulting scheme will be called *Multi-D-ElGamal*. It should be understood that later d_1 will be used for P, the addressee. The other d_i's are for the TRP's. Of course, all k_i's are still equal to each other. Below we shall show that Multi-D-ElGamal is as secure as ElGamal with respect to g. So if the orders of all TRP's are equal to q, then session keys can be virtually addressed (as explained in the introduction) to them in a secure way. Moreover, the construction of binding data for the Multi-D-ElGamal scheme is similar to that for the Multi-ElGamal scheme, as is the splitting of private keys of TRP's. However, for reasons explained above, users should have confidence that the orders of their TRP's are in fact equal to q. A fact that is difficult to check without the factorization of q.

Let (g_P, y_P) be participant P's public key in the D-ElGamal scheme, that is $g_P = g^{q/d_P}$. For technical reasons only we introduce the *alternative D-ElGamal scheme*, in which the encryption of $S \in H$ takes the form $(g^k, g_P{}^k, y_P{}^k \cdot S)$, i.e. the (superfluous) element g^k is added. The *alternative Multi-D-ElGamal scheme* is formed from the Multi-D-ElGamal scheme by sending the first participant (whose order may differ from q) the alternative D-ElGamal encryption.

Lemma B.2 *If d_P is known by an attacker Ada, then breaking the alternative D-ElGamal scheme w.r.t. (g_P, y_P) is as difficult as breaking the ElGamal scheme w.r.t. g.*

Proof [sketch]: Suppose there exists an efficient algorithm $\mathcal{A}$ that after analyzing a history of encrypted messages $(g^{k_i}, g_P{}^{k_i}, y_P{}^{k_i} \cdot S_i)$, $i = 1, \ldots h$, has a non-negligible change of outputting S on input of an encrypted message $(g^k, g_P{}^k, y_P{}^k \cdot S)$.
Now suppose that participant Q has as public key y in the ElGamal scheme w.r.t. g. From this an attacker can form two public keys for two (imaginary) participants V_1 and V_2 in the D-ElGamal scheme, namely (g^d, y^d) and $(g^{q/d}, y^{q/d})$. Moreover an encryption $(A, B) = (g^k, y^k \cdot S)$ of a message $S \in H$ with Q's public key can be transformed in an encryption of $S^d \in H$ with V_1's public key, by forming (A^d, B^d). Hence, after some time, by using $\mathcal{A}$, Ada, has a non-negligible change of outputting S^d. Similarly, Ada has a non-negligible change of outputting $S^{q/d}$. As q and q/d are relatively prime (q is square-free), Ada has a non-negligible change of outputting S. □

Theorem B.3 *Let n be a natural number. Then breaking Multi-D-ElGamal for n addressees is as least as difficult as breaking ElGamal with respect to g.*

Proof [sketch]: Breaking the Multi-D-ElGamal scheme is as least as difficult as breaking the alternative Multi-D-ElGamal scheme. Now consider a sequence ("history") of h encryptions of messages S_i ($i = 1, \ldots, h$) in the alternative D-ElGamal scheme: $(g^{k_i}, g_P{}^{k_i}, y_P{}^{k_i} \cdot S_i)$.
Observe that y_P can be seen as public key with respect to g. In fact, as $g_P = g^{q/d_P}$ and as d_P can be considered publicly known by Assumption B.1 the computation of $log_g y_P$ is as difficult as that of $log_{g_P} y_P$.
By Lemma 2.2, from a sequence of encryptions $(g^{k_i}, y_P{}^{k_i} \cdot S_i)$ anyone can construct a second sequence of encryptions of type $(g^{k_i}, \hat{y}^{k_i} \cdot S_i)$ with $\hat{y}$ random in G such that the computation of $log_g \hat{y}$ is as difficult as that of $log_g(y_P)$.
Anyone that chooses a random number j less than, relatively prime with q, can calculate the generator $\hat{g} = g^j$ and construct a third sequence of encryptions of type $(\hat{g}^{k_i}, \hat{y}^{k_i} \cdot S_i)$ with $\hat{g}$ a random generator in in G. It also follows that the computation of $log_{\hat{g}} \hat{y}$ is as difficult as that of $log_g \hat{y}$, which is as difficult as the computation of $log_{g_P} y_P$.
Hence - like in the proof of Theorem 2.3 - from the history of encryptions of messages in the alternative D-ElGamal scheme, anyone can construct a typical history of encryption of messages in the alternative Multi-D-ElGamal scheme. By a similar argument as used in Theorem 2.3, breaking the latter, means breaking the alternative D-ElGamal scheme which by Lemma B.2 and Assumption B.1 means breaking ElGamal with respect to g. □

References

1. R. Anderson, M. Roe, *The GCHQ Protocol and its Problems*, these proceedings.
2. D.M. Balenson, C.M. Ellison, S.B. Lipner, S.T. Walker (TIS Inc.), *A New Approach to Software Key Escrow Encryption*, in: L.J. Hoffman (ed.), Building in Big Brother (Springer, New York, 1996), pp. 180-207. See also http://www.tis.com.
3. D. Chaum, T.P. Pedersen, *Wallet Databases with Observers* Advances in Cryptology - CRYPTO '92 Proceedings, Springer-Verlag, 1993, pp. 89-105.
4. D. Coppersmith, *Finding a Small Root of a Univariate Modular Equation*, Advances in Cryptology - EUROCRYPT '96 Proceedings, Springer-Verlag, 1995, pp. 155-165.
5. R. Cramer, R. Gennaro, B. Schoenmakers *A Secure and Optimally Efficient Multi-Authority Election Scheme*, these proceedings.
6. Y. Desmedt, *Securing Traceability of Ciphertexts - Towards a Secure Key Escrow System*, Advances in Cryptology - EUROCRYPT '95 Proceedings, Springer-Verlag, 1995, pp. 147-157.
7. T. ElGamal, *A Public Key Cryptosystem and a Signature scheme Based on Discrete Logarithms*, IEEE Transactions on Information Theory 31(4), 1985, pp. 469-472.
8. Interagency Working Group on Cryptography Policy, *Enabling Privacy, Commerce, Security and Public Safety in the Global Information Infrastructure*, 17 May 1996, see http://www.cdt.org/crypto/clipper_III.
9. L.C. Guillou, J.-J. Quisquater *A Practical Zero-Knowledge Protocol Fitted to Security Microprocessor Minimizing Both Transmission and Memory*, Advances in Cryptology - EUROCRYPT '86 Proceedings, Springer-Verlag, 1986, pp. 123-128.
10. J. Hastad, *On Using RSA with Low Exponent in a Public Key Network*, Advances in Cryptology - CRYPTO '85 Proceedings, Springer-Verlag, 1993, pp. 403-405.
11. N. Jefferies, C. Mitchell, M. Walker, *A Proposed Architecture for Trusted Third Party Services*, Cryptography: Policy and Algorithms, Proceedings of the conference, Springer-Verlag (LNCS 1029), 1996, pp. 98-104.
12. A.K. Lenstra, P. Winkler, Y. Yacobi *A Key-Escrow System with Warrants Bounds*, Advances in Cryptology - CRYPTO '95 Proceedings, Springer-Verlag, 1995, pp. 197-207.
13. S. Micali, *Fair Public-key Cryptosystems*, Advances in Cryptology - CRYPTO '92 Proceedings, Springer-Verlag, 1993, pp. 113-138.
14. National Research Council, *Cryptography's Role in Securing the Information Society*, K.W. Dam, H.S. Lin (Editors), National Academy Press Washington, D.C. 1996, pp.720.
15. T.P. Petersen, *Distributed Provers with Applications to Undeniable Signatures*, Advances in Cryptology - EUROCRYPT '91, Springer-Verlag, 1991, pp. 221-242.
16. T.P. Petersen, *A Treshold Cryptosystem Without a Trusted Party*, Advances in Cryptology - EUROCRYPT '91, Springer-Verlag, 1991, pp. 522-526.
17. C.P. Schnorr, *Efficient Signature Generation for Smart Cards*, Advances in Cryptology - CRYPTO '89 Proceedings, Springer-Verlag, 1990, pp. 225-232.
18. D.R. Stinson *Cryptography: theory and practice*, CRC press, 1995, pp.434.
19. E.R. Verheul, B.-J. Koops, H.C.A. van Tilborg, *Binding Cryptography. A fraud-detectible alternative to key-escrow solutions*, Computer Law and Security Report, January-February 1997, pp. 3-14.

The GCHQ Protocol and Its Problems

Ross Anderson, Michael Roe

Cambridge University Computer Laboratory
Pembroke Street, Cambridge CB2 3QG
Email: (rja14,mrr)@cl.cam.ac.uk

Abstract. The UK government is fielding an architecture for secure electronic mail based on the NSA's Message Security Protocol, with a key escrow scheme inspired by Diffie-Hellman. Attempts have been made to have this protocol adopted by other governments and in various domestic applications. The declared policy goal is to entrench commercial key escrow while simultaneously creating a large enough market that software houses will support the protocol as a standard feature rather than charging extra for it.

We describe this protocol and show that, like the 'Clipper' proposal of a few years ago, it has a number of problems. It provides the worst of both secret and public key systems, without delivering the advantages of either; it does not support nonrepudiation; and there are serious problems with the replacement of compromised keys, the protection of security labels, and the support of complex or dynamic administrative structures.

1 Introduction

Over the last two years, the British government's crypto policy has changed completely. Whereas in 1994 the Prime Minister assured the House of Commons that no further restrictions on encryption were envisaged, we now find the government proposing to introduce a licensing scheme for 'trusted third parties', and licenses will only be granted to operators that escrow their customers' confidentiality keys to the government's satisfaction [11, 21].

In March 1996, a document describing the cryptographic protocols to be used in government electronic mail systems was issued by CESG, the department of GCHQ concerned with the protection of government information; it has since been made available on the worldwide web [4]. According to this document, policy goals include *'attempting to facilitate future inter-operability with commercial users, maximising the use of commercial technology in a controlled manner, while allowing access to keys for data recovery or law enforcement purposes if required*[1].

---

[1] A UK official who chairs the EU's Senior Officials' Group - - Information Security (SOGIS) has since admitted that 'law enforcement' in this context actually refers to national intelligence [10].

W. Fumy (Ed.): Advances in Cryptology - EUROCRYPT '97, LNCS 1233, pp. 134-148, 1997.

A document on encryption in the National Health Service, issued in April, had already recommended that medical traffic should be encrypted, and keys should be managed, using mechanisms compatible with the future 'National Public Key Infrastructure' [26]; part of the claimed advantages for the health service were that the same mechanisms would be used to protect electronically filed tax returns and applications from industry for government grants. Furthermore, attempts are being made to persuade other European countries to standardise on this protocol suite.

So the soundness and efficiency of the GCHQ protocol proposals could be extremely important. If an unsound protocol were to be adopted across Europe, then this could adversely affect not just the secrecy of national classified data, the safety and privacy of medical systems, and the confidentiality of tax returns and government grant applications. It could also affect a wide range of commercial systems too, and make Europe significantly more vulnerable to information warfare. If the protocols were sound but inefficient, then they might not be widely adopted; or if they were, the costs imposed on the economy could place European products and services at a competitive disadvantage.

In this paper, we present an initial analysis of the security and efficiency of the GCHQ protocol.

2　The GCHQ Protocol

The precursor of the government protocol was first published by Jefferies, Mitchell and Walker at a conference in July 1995 [13]. A flaw was pointed out there[2] and a revised version was published in the final proceedings of that conference; this version also appeared at the Public Key Infrastructure Invitational Workshop at MITRE, Virginia, USA, in September 1995 and at PKS '96 in Zürich on 1st October 1996 [14]. The final, government approved, version of the protocol fixes some minor problems and adds some new features.

The document [4] is not complete in itself, as the protocol is presented as a series of extensions to the NSA's Message Security Protocol [18]. In the next section we will attempt for the first time to present the whole system in a complete and concise way, suitable for analysis by the cryptologic and computer security communities. We will then discuss some of its more obvious faults.

The GCHQ architecture assumes administrative domains 'corresponding approximately to individual departments', although there may be smaller domains where a department is scattered over a large geographical area. Each will have a 'Certificate Management Authority', under the control of the departmental security officer, which will be responsible for registering users and supplying them with keys. Key management will initially be under the control of GCHQ but might, in time, be devolved.

---

[2] The original protocol allowed the same base and modulus to be used in different domains and was thus vulnerable to Burmester's attack [2]

The basic idea is that if Alice wants to send email to Bob, she must go to her certificate management authority, whom we will call *TA*, and obtain from him secret information that enables her to calculate a key for communicating with Bob. She also receives a certificate of this secret information, and sends this to Bob along with the encrypted message. On receipt of the message Bob contacts his certificate management authority *TB* and obtains the secret information that he needs to decrypt the message. Thus two individuals can communicate only if both their departmental security officers decide to permit this.

The communication flow can be visualised in the following diagram:

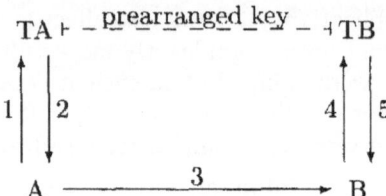

We will now describe the content of these messages. The protocol is a derivative of Diffie Hellman [5] and the basic idea is that, in order to communicate with Bob, Alice must obtain a 'public receive key' for him from *TA* and operate on this with a 'secret send key' that *TA* also issues her, along with a certificate on the corresponding 'public send key'. At the other end, Bob will obtain a 'secret receive key' for her from *TB* and will use this to operate on her 'public send key' whose certificate he will check.

The secret receive keys are known to both users' authorities, and are calculated from their names using a shared secret master key. Each pair of domains *TX*, *TY* has a 'top level interoperability key', which we will call K_{TXY} for managing communication. The relevant key here is K_{TAB} which is shared between *TA* and *TB*. The mechanisms used to establish these keys are not described.

We will simplify the GCHQ notation by following [3] and writing $\{X\}_Y$ for the block X encrypted under the key Y using a conventional block cipher. Then the long term seed key that governs Bob's reception of traffic from all users in the domain of *TA* is:

$$\text{rseed}_{B,A} = \{B\}_{K_{TAB}} \tag{1}$$

A secret receive key of the day is then derived by using this seed key to encrypt a datestamp:

$$\mathrm{SRK}_{B,A,D} = \{D\}_{\mathrm{rseed}_{B,A}} \tag{2}$$

and Bob's public key of the day, for receiving messages from users in the domain *TA*, is

$$\mathrm{PRK}_{B,A,D} = g_A^{\mathrm{SRK}_{B,A,D}} \pmod{N_A} \tag{3}$$

where the 'base' g_A and the modulus N_A are those of *TA*'s domain (the document does not specify whether N_A should be prime or composite, or the properties that the group generated by g_A should possess).

Finally, *TA* certifies Bob's public key of the day as $S_{TA}(B,D,\mathrm{PRK}_{B,A,D})$.

Only receive keys are generated using secrets shared between authorities. Send keys are unilaterally generated by the sender's authority from an internal master key, which we will call K_{TA} for *TA*, and the user's name. Thus Alice's seed key for sending messages is $\mathrm{sseed}_A = \{A\}_{K_{TA}}$; her secret send key of the day is derived as $\mathrm{SSK}_{A,D} = \{D\}_{\mathrm{sseed}_A}$ and her public send key is $\mathrm{PSK}_{A,D} = g_A^{\mathrm{SSK}_{A,D}} \pmod{N_A}$. *TA* sends her the secret send key, plus a certificate $Cert(A,D,\mathrm{PSK}_{A,D})$ on her public send key. Send seed keys may be refreshed on demand.

Now Alice can finally generate a shared key of the day with Bob as

$$k_{A,B,D} = (\mathrm{PRK}_{B,A,D})^{\mathrm{SSK}_{A,D}} \pmod{N_A} \tag{4}$$

This key is not used directly to encipher data, but as a 'token key' to encipher a token containing a session key. Thus, when sending the same message to more than one person, it need only be encrypted once, and its session key can be sent in a number of tokens to its authorised recipients.

Anyway, Alice can now send Bob an encrypted version of the message M. According to the GCHQ protocol specification, certificates are sent with the object 'to simplify processing', so the packet that she sends to Bob (in message 3 of the diagram overleaf) is actually

$$\{M\}_{k_{sess}}, \{k_{sess}\}_{k_{A,B,D}}, Cert(B,D,\mathrm{PRK}_{B,A,D}), Cert(A,D,\mathrm{PSK}_{A,D}) \tag{5}$$

This protocol is rather complex. But what does it actually achieve?

2.1 Problem 1 — why not just use Kerberos?

The obvious question to ask about the GCHQ protocol is why public key techniques are used at all. After all, if *TA* and *TB* share a secret key, and Alice and Bob have to interact with them to obtain a session key, then one might just as well use the kind of protocol invented by Needham and Schroder [19] and since deployed in products like Kerberos [20]. Where Alice shares the key K_A with *TA* and Bob shares K_B with *TB*, a suitable protocol might look like

$$A \rightarrow TA : A, B$$
$$TA \rightarrow A \quad : \{A, B, K_{AB}, d, \{K_{AB}, A, B, d\}_{K_{TAB}}\}_{K_A}$$
$$A \rightarrow B \quad : A, \{K_{AB}, A, B, d\}_{K_{TAB}}, \{k\}_{K_{AB}}, \{M\}_k$$
$$B \rightarrow TB : A, B, \{K_{AB}, A, B, d\}_{K_{TAB}}$$
$$TB \rightarrow B \quad : \{K_{AB}, d, A, B\}_{K_B}$$

This protocol uses significantly less computing than the GCHQ offering, and no more messages. It can be implemented in cheap commercial off-the-shelf tokens such as smartcards, and with only minor modification of the widely available code for Kerberos. This would bring the further advantage that the implications of 'Kerberising' existing applications have been widely studied and are fairly well understood in a number of sectors (see, e.g. [12]). On the other hand, the integration of a completely new suite of authentication and encryption software would mean redoing this work. Given that the great majority of actual attacks on cryptosystems exploit blunders at the level of implementation detail [1], this will mean less secure systems.

The GCHQ response to this criticism is [15]:

> This is not so much an attack on the recommendations as an objection to the Trusted Third Party concept and the need for key recovery. The recommendations offer a realistic architectural solution to a complex problem and, as with any system, will require professional implementation.

This completely misses the point. Given that the UK government has decided to adopt key escrow in its own internal systems, exactly the same functionality could have been provided by a simple adaptation of Kerberos at much less cost and risk. The only extra feature that appears to be provided by the GCHQ protocol is that users who receive mail from only a small number of other departments, and who operate under security rules that permit seed keys to persist for substantial periods of time, may save some communications with their TTPs by storing receive seed keys locally. This leads us to consider the issue of scalability.

2.2 Problem 2 — where are the keys administered?

How well the GCHQ protocol (or for that matter Kerberos) will scale will depend on how many key management authorities there are. With a large number of them — say, one per business enterprise — the problem of inter-enterprise key management would dominate and the above protocol would have solved nothing.

The British government may be aware of this problem, as they propose to minimise the number of authorities. Under the legislation currently proposed, large companies would be permitted to manage their own keys — the rationale being that having significant assets they would be responsive to warrants — while small to medium enterprises and individuals would have to use the services of licensed TTPs — organisations such as banks that would undertake the dual role of certificate management authority and escrow agent.

We do not believe that this will work. One of us has experience of a bank with 25,000 employees, managed through seven regional personnel offices, trying to administer mainframe passwords at a central site. With thirty staff and much message passing to and from the regions, the task was just about feasible, but compelling a million small businesses to conduct a transaction with the 'Trusted Third Party' every time a staff member was hired, fired or moved, would do little for national economic competitiveness.

Medicine is another application to consider, as the issue of encryption and signature of medical records is the subject of debate in a number of European and other countries. There is relevant experience from New Zealand, where a proposal to have doctors' keys managed by officials in the local district hospitals turned out to be impractical. It is now proposed that keys there should be managed at the practice level [9]. In the UK, with some 12,000 general practices, hospitals and community care facilities, centralised key management is even less likely to be workable.

The GCHQ response to this criticism is [15]:

> It has also been suggested that a TTP network could become large and that some users would have to keep a large number of public keys. This problem is overcome in the Royal Holloway architecture since any user can obtain all the necessary key material from its local TTP. This is inherently more scalable than other approaches.

This again misses the point. If the UK health service, with 12,000 providers, has 12,000 TTPs, then the inter-TTP communications would be the bottleneck.

There is also the issue of trust. In the UK, the medical profession perceived the recommendation in [26] that key management should be centralised in a government body as an attempt to undermine the independence of the institutions currently responsible for professional registration — the General Medical Council (for doctors), the UK Central Council (for nurses), and so on. Retaining these organisations as top level CAs is essential for creating professional trust without which a security system would deliver little value.

But with the GCHQ protocol, this would appear to mean that a doctor who wished to send an encrypted email to a nurse working in the same practice would have to send a message to the GMC to get a key to encrypt the message, and the nurse would have to contact the UKCC to get a key to decrypt it. This is clearly ludicrous.

In short, the GCHQ protocol may work for a strictly hierarchical organisation like government may be thought to be (though if that were the case, a Kerberos like system would almost certainly work better). But it is not flexible enough to accommodate real world applications such as small business and professional practice. This raises the question of whether it will even work in government. We suspect it would work at best badly — and impose a structural rigidity which could frustrate attempts to make government more efficient and accountable.

The GCHQ response to this criticism is [15]:

> The frameworks for confidentiality and authentication have been de-signed to cater for a wide range of environments. A hierarchy is defined only for the authentication framework and this is necessary because good security requires tight control.

This claim is inconsistent with the protocol document according to which *'As the Certificate Management Authority is responsible for generating the confiden-tiality keys, it should also take on the role of a certification authority in order to authenticate them'*. Thus the confidentiality and authentication hierarchies are clearly intended to be identical.

Rossnagel made the point that trust structures in the electronic world should mirror those in existing practice [23]; a point which all security engineers should consider carefully.

2.3   Problem 3 - should signing keys be escrowed?

The next problem is the plan to set up an escrowed trust structure of confiden-tiality keys first, and then bootstrap authentication keys from this [4] [26].

The GCHQ protocol defines a structure called a token to transfer private keys in an encrypted form. What is also required is a mechanism to convey public signature verification keys to the authority for certification, as well as a means to revoke signature keys (which should be independent of the 'key of the day' system that provides implicit revocation of encryption keys). Such mechanisms are not provided.

Similar considerations apply to MACs. The original US MSP has a mode of operation which provides confidentiality and integrity but not non-repudiation. In this mode, the message is not signed, and instead the confidentiality key (or a key derived from it) is used to generate a MAC on the message. As the GCHQ protocol is specified by citing the US MSP specification and explaining the differences, it would appear that this mode will also be a part of it; but when combined with the GCHQ key management, the effect is that an escrowed confidentiality key is used to authenticate the message.

Even if confidentiality keys are eventually required by law to be escrowed, the keys used for authentication must be treated differently, and there is a risk that programmers and managers responsible for implementing the GCHQ protocol might overlook this distinction and produce a flawed system. So it is worth explaining explicitly.

The stated purpose of key escrow is to enable law enforcement and other government employees to monitor the contents of encrypted traffic (and, in some escrow schemes, to facilitate data recovery if users lose or forget their keys). Its stated purpose does not include allowing government employees to create forged legal documents (such as contracts or purchase orders). It would be highly undesirable if people with access to the escrow system were able to use this access to forge other people's digital signatures. The scope for insider fraud and conspiracy to pervert the course of justice would be immense.

Any police officer will appreciate that if he can get copies of my bank statements, then perhaps he can use them in evidence against me; but if he can tracelessly forge my cheques, then there is no evidence at all any more. So if there is any possibility that a digital signature might be needed as evidence, then the private key used to create it must not be escrowed.

In fact, we would go further than this: keys which are used only for authentication (and not non-repudiation) should not be escrowed either. For example, suppose that some piece of equipment (e.g. a power station, or a telephone exchange) is controlled remotely, and digital signatures or MACs are used to authenticate the control messages. Even if these messages are not retained for the purposes of evidence, it is clearly important to distinguish between authorising a law enforcement officer to monitor what is going on and authorising him to operate the equipment. If authentication keys are escrowed, then the ability to monitor and the ability to create seemingly authentic control messages become inseparable: this is almost certainly a bad thing. Returning to the medical context, it is unlikely that either doctors or patients would be happy with a system that allowed the police to forge prescriptions, or the intelligence services to assume control of life support equipment. We doubt that a well informed minister would wish to expose himself and his officers in such a way.

In such applications, we need an infrastructure of signature keys that is as trustworthy as we can make it. Bootstrapping the trust structure from a system of escrowed confidentiality keys is unacceptable.

The GCHQ response to this criticism is [15]:

> This confuses the authentication and confidentiality frameworks. There is no intention to bootstrap signature keys required for non-repudiation purposes within the authentication framework.

The protocol document states (2.2.1) that *'to provide a non-repudiation service users would generate their own secret and public authentication key pairs, then pass the public part to a certification authority'*. But no mechanism for this is provided; in the rest of the document, it is assumed that all secret keys are generated by the certification authority, and both the secret and public parts passed to the user. Given GCHQ's response, we conclude that their protocol is not intended to provide a non-repudiation service at all.

Furthermore, both authentication and confidentiality key material is under the control of the Departmental Security Officer. This leads to an interesting 'plausible deniability' property. If there is a failure of security, and an embarrassing message is leaked, then it is always possible to claim that the message was forged (perhaps by the very security officer whose negligence permitted the leak in the first place).

For these reasons, if non-governmental use of the GCHQ protocol is contemplated — or compelled by legislation — then signing keys should be managed by some other means (and not escrowed). We also recommend that normal policy should prohibit the sending of MAC-only messages; if a MAC-only message is received, the purported sender should be asked to resend a properly signed version (there are some special purpose uses in which the MAC-only mode is useful, but we won't describe them here).

2.4 Problem 4 — clear security labels

In the original NSA Message Security Protocol, the label describing the security classification of the contents of an encrypted message is also encrypted. The GCHQ version adds an extension which contains the label in clear (we will refer to this as the 'cleartext' security label, while the actual classification is the 'plaintext' security label).

There is a problem with doing this. An attacker can often derive valuable information from the cleartext label, taken together with the identity of the sender and recipient and the message volume. Indeed, with some labels, the attacker learns all she wants to know from the label itself, and cryptanalysis of the message body is unnecessary. This is why the US does not use cleartext security labels.

The GCHQ response to this criticism is [15]:

> CESG's modifications have been made after careful consideration of government requirements and in consultation with departments; they are sensible responses to these requirements.

We understand that these 'requirements' concern the national rules concerning the forms of protection which are deemed appropriate for various types of information.

Under the UK rules, it is possible for a combination of physical and cryptographic mechanism taken together to be deemed adequate, whereas either mechanism on its own is deemed inadequate. For example, a message classified SECRET can be enciphered with RAMBUTAN and then transmitted over a link which lies entirely within the UK. The protection provided by RAMBUTAN is deemed insufficient if the same message is being transmitted across the Atlantic.

So British enciphered messages need to divided into two or more types: those that require various forms of additional physical protection, and those that don't. The message transfer system needs to be able to tell which messages are which, so it can use physically protected communications lines for some messages but not for others. The easiest way to achieve this is to mark the ciphertext with the classification of the plaintext.

However, if an opponent can get past the physical protection (which is often quite easy), then she can carry out the attacks described above. It would clearly be desirable for UK to follow the American lead and encrypt all security labels.

It may be argued that the rules are so entrenched that this is infeasible. A technical alternative is to reduce the cleartext security label to a single bit indicating only the handling requirements. In this way, routers have the information they need, and attackers get no more information than this (which they could arguably derive in any case by observing the route that the message takes). Using a completely incompatible (and information-losing) syntax for cleartext labels would also prevent lazy or careless implementers using them as plaintext labels. Such robustness would have been prudent design practice.

In fact, the GCHQ protocol does not protect the integrity of the cleartext security label either, and so the attacker can manipulate it. If it is ever used to determine the sensitivity of the decrypted plaintext, then the recipient could be tricked into believing that the message had a different classification, which might lead to its compromise.

2.5   Problem 5 — identity based keys

The GCHQ protocol gives users seed keys which they hash with timestamps to get user keys. But it is quite likely that some seed keys will be compromised (e.g. by Trojan horses previously inserted by attackers; via theft of computers; if smart cards holding them are lost etc). In that case, the user's certificates can be revoked, but the user cannot be issued with a new seed key, as it is a deterministic function of her name. All the CA can do is reissue the same (compromised) key.

To recover from this situation, either the user has to change her name, or the CA has to change the interoperability key and reissue new seed keys for every user in the domain. Both of these alternatives are unacceptable, and this is a serious flaw in the GCHQ protocol. It might be remedied by making the seed key also depend on an initial timestamp (which would also have to be added at several other places in the protocol).

2.6   Problem 6 — scope of master key compromise

The compromise of the interoperability key between two domains would be catastrophic, as all traffic between users in those domains could now be read. In our experience, the likelihood of master key compromise is persistently underestimated. We know of cases in both the banking and satellite TV industries where organisations have had to reissue millions of customer cards as a result of a key compromise that they had considered impossible and for which they therefore had no disaster recovery plan. Introducing such a vulnerability on purpose is imprudent.

The GCHQ response to this criticism is [15]:

> CESG is fully aware of the need adequately to secure such high level exchanges and there are a number of ways this could be done.

Indeed, and comparison with other escrow systems such as Clipper shows that it is possible to provide some degree of protection against accidental disclosure and rogue insiders, by using two escrow agents in different departments rather than the single crypto custodians proposed by GCHQ. Clipper is not perfect in this regard, but it at least shows that it is possible to do better. At the very least, it would be prudent to change the interoperability keys frequently; this would remove the need for seed keys (and thus strengthen the argument for using Kerberos instead).

Of course, if corrupt law enforcement officers are allowed to abuse the system indefinitely, then no cryptographic or dual control protocol can put things right. Any discussion of insider attacks must assume that there exist procedures for dealing with misbehaving insiders, and indeed for detecting misbehaviour in the first place. This can be done with non-escrowed key management protocols (see for example [16]) but appears more difficult when escrow is a requirement.

2.7   Problem 7 — MOAC

The GCHQ protocol defines an extension which provides a "simple message origin authentication check". This is a digital signature computed on the contents of the message, and nothing else. By way of contrast, the original US MSP provided message origin authentication by computing a digital signature on a hash of the message and some additional control information. This additional control information can contain the data type of the message (e.g. whether it is an interpersonal text message or an EDI transaction).

The GCHQ proposal is an extension, rather than a replacement. That is, messages will contain two forms of digital signature: the old US form and the new UK form. As a result, this extension has not made the protocol simpler; it has made it more complex.

In nearly all circumstances, it would best to use the original US form of signature rather than the new UK one. The US form is very nearly as quick to compute, and it protects against some attacks to which the UK version is vulnerable. (It is possible for a bit string to have two different interpretations, depending on which data type the receiver believes it to be. The UK signature does not protect the content type, so an attacker could change this field and trick the receiver into misinterpreting the message.)

One situation in which there might be a use for this UK extension is in implementing gateways between the GCHQ protocol and other security protocols. For example, it would be possible for a gateway to convert between Internet Privacy Enhanced Mail and UK MSP by replacing the PEM header with an MSP header and copying the PEM signature into the UK signature extension field. The important point to note about this is that such a gateway does not need access to any cryptographic keys, as it does not need to re-compute the signature. By way of contrast, a gateway between US MSP and PEM would need access to the sender's signature key: this is a very bad idea for obvious reasons.

2.8 Problem 8 — choice of encryption algorithm

GCHQ wants people to use an unpublished block cipher with 64 bit block and key size called Red Pike. According to a report on the algorithm prepared in an attempt to sell it to the Health Service, it is similar to RC5 [22] but with a different key schedule. It will apparently be the standard for government traffic marked up to 'Restricted', and it is claimed that systems containing it may be less subject to export controls: health service officials have claimed that US companies operating in the UK may be allowed by the US government to use Red Pike in products in which the use of DES would be discountenanced by the US State Department.

More significantly, Red Pike will shortly be fielded in mass market software, and will thus inevitably be reverse engineered and published, as RC2 and RC4 were. So it is hard to understand why the UK government refuses to publish it, or why anyone should trust it, at least until it has been exposed to the attention of the cryptanalytic community for a number of years. If GCHQ scientists have found a weakness in RC5 and a fix for it — or even a change that speeds it up without weakening it — then surely the best way to gain acceptance for such an innovation would be to publish it.

The GCHQ response to this criticism is [15]:

> Another common misconception is that the CESG Red Pike algorithm is being recommended for use in the public arena. No confidentiality algorithm is mandated in the recommendations; for HMG use, however, approved algorithms will be required; Red Pike was designed for a broad range of HMG applications.

Vigorous efforts are still being made to promote the use of Red Pike in the health service, and as noted above, it is supposed to be used in a wide range of citizens' interactions with government such as filing tax returns and grant applications. Thus the accuracy of the above response is a matter of how one interprets the phrase 'public arena'.

3   Conclusion

The GCHQ protocol is very poorly engineered.

1. The key management scheme gives us all the disadvantages of public key crypto (high computational complexity, long key management messages, difficult to implement on cheap devices such as smartcards), and all the disadvantages of secret key crypto (single point of failure, little forward security, little evidential force, difficulty of 'plug and play' with shrink-wrapped software). It does not provide any of the advantages that one could get from either of these technologies; and its complexity is likely to lead to the subtle and unexpected implementation bugs which are the cause of most real world security failures.

2. It is designed for tightly hierarchical organisations, and cannot economically cope with the more complex trust structures in modern commerce, industry and professional practice. Its main effect in government may to perpetuate rigid hierarchies and frustrate the efficiency improvements that modern management techniques might make possible.

3. It goes about establishing trust in the wrong way. To plan to bootstrap signature keys from a 'national public key infrastructure' of escrowed confidentiality keys shows a cavalier disregard of the realities of evidence and of safety-critical systems.

4. There are a number of serious technical problems with the modifications that have been made to the US Message Security Protocol, which underlies the UK government's offering. Quite independently of the key management scheme and trust hierarchy that are eventually adopted, these modifications are unsound and should not be used.

The above four conclusions appeared in an earlier draft of this paper. The GCHQ response that that draft, which we have cited here, has not persuaded us to change a single word of their text.

We call on the cryptologic and computer security communities to subject this protocol to further study. If adopted as widely as the British government clearly hopes it to be, it would be a single point of failure of a large number of applications on which the security, health[3], privacy and economic wellbeing of Europe's citizens would come to depend.

Acknowledgement: We are grateful to the security group at Cambridge for discussions, and to Paul van Oorschot for pointing out that the second version

---

[3] GCHQ has since claimed that the NHS proposals and its are 'similar but distinct' [15]. They are indeed similar, with many of the undesirable features described below being incorporated into NHS crypto pilots. The only respect in which they are clearly distinct is that the DH/DSA mechanisms were replaced by RSA after RSA was adopted as a European standard for healthcare. However, many of the undesirable features which we discuss above, such as the central generation of signature keys, have been retained in the NHS pilots.

of this protocol was presented at two other conferences as well as appearing in the Queensland conference proceedings [14].

References

1. RJ Anderson, "Why Cryptosystems Fail", in *Communications of the ACM* v 37 no 11 (Nov 94) pp 32–40
2. M Burmester, "On the Risk of Opening Distributed Keys", in *Advances in Cryptology — CRYPTO '94*, Springer LNCS v 839 pp 308–317
3. M Burrows, M Abadi, RM Needham, "A Logic of Authentication", in *Proceedings of the Royal Society of London A* v 426 (1989) pp 233–271
4. CESG, "Securing Electronic Mail within HMG — part 1: Infrastructure and Protocol" 21 March 1996, document T/3113TL/2776/11; available at URL http://www.rdg.opengroup.org/public/tech/security/pki/casm/casm.htm
5. W Diffie, ME Hellman, "New Directions in Cryptography", in *IEEE Transactions on Information Theory*, IT-22 no 6 (November 1976) p 644–654
6. Electronic Privacy Information Center, *1996 EPIC Cryptography and Privacy Sourcebook*, Washington, DC
7. US Department of Commerce, *'Escrowed Encryption Standard'*, FIPS PUB 185, February 1994
8. Y Frankel, M Yung, Escrow Encryption Systems Visited: Attacks, Analysis and Designs", in *Advances in Cryptology — CRYPTO 95*, Springer LNCS v 963 pp 222–235
9. P Gutman, *personal communication*, July 96
10. D Herson, in *interview with Kurt Westh Nielsen and Jérôme Thorel*, 25 September 1996; Ingeniøren/Engineering Weekly 10/04/1996; available at http://www.ingenioeren.dk/redaktion/herson.htm
11. N Hickson, Department of Trade and Industry, speaking at *'Information Security — Is IT Safe?'*, IEE, Savoy Place, London, 27th June 1996
12. I Hollander, P Rajaram, C Tanno, "Kerberos on Wall Street", in *Usenix Security 96* pp 105–112
13. N Jefferies, C Mitchell, M Walker, "A Proposed Architecture for Trusted Third Party Services", in proceedings of *Cryptography Policy and Algorithms Conference*, 3–5 July 1995, pp 67–81; published by Queensland University of Technology
14. N Jefferies, C Mitchell, M Walker, "A Proposed Architecture for Trusted Third Party Services", in *Cryptography: Policy and Algorithms*, Springer LNCS v 1029 pp 98–104; also appeared at the Public Key Infrastructure Invitational Workshop at MITRE, Virginia, USA, in September 1995 and PKS '96 in Zürich on 1st October 1996
15. ID Jones, *letter to R Anderson on behalf of GCHQ's Communications Electronics Security Group*; available at http://www.cs.berkeley.edu/~daw/GCHQ/
16. TMA Lomas, B Crispo, "A New Certification Scheme", in *Proceedings of the Fourth Cambridge Workshop on Cryptographic Protocols* (1996), Springer LNCS series pp 19-32
17. TMA Lomas, MR Roe, "Forging a Clipper Message", in *Communications of the ACM* v 37 no 12 (Dec 94) p 12
18. U.S. National Security Agency, *'Secure Data Network System : Message Security Protocol (MSP)'*, SDN.701, revision 4.0 (January 1996)

19. RM Needham, MD Schroder, "Using Encryption for Authentication in Large Networks of Computers", in *Communications of the ACM* vol 21 no 12 (Dec 78) pp 993–999

20. BC Neuman, T Ts'o, "Kerberos: An Authentication Service for Computer Networks", in *IEEE Communications Magazine* v 32 no 9 (Sep 94) pp 33–38

21. Press Association, "Move to Strengthen Information Security", 06/10 1808

22. RL Rivest, "The RC5 Encryption Algorithm", in *Fast Software Encryption* (1994), Springer LNCS v 1008 pp 86–96

23. Roßnagel A, "Institutionell-organisatorische Gestaltung informationstechnischer Sicherungsinfrostrukturen", in *Datenschutz und Datensicherung (5/95) pp 259-269*

24. RL Rivest, B Lampson, "A Simple Distributed Security Infrastructure", at `http://theory.lcs.mit.edu/~rivest/publications.html`

25. B Schneier, *'Applied Cryptography — Protocols, Algorithms, and Source Code in C'* (second edition), John Wiley & Sons, New York, 1996

26. Zergo Ltd., '*The use of encryption and related services with the NHSnet*', published by the NHS Executive Information Management Group 12/4/96, reference number E5254; available from the Department of Health, PO Box 410, Wetherby LS23 7LN; Fax +44 1937 845381

Bucket Hashing with a Small Key Size

Thomas Johansson

Department of Information Technology, Lund University,
PO Box 118, S-221 00 Lund, Sweden
Email:thomas@it.lth.se

Abstract. In this paper we consider very fast evaluation of strongly universal hash functions, or equivalently, authentication codes. We show how it is possible to modify some known families of hash functions into a form such that the evaluation is similar to "bucket hashing", a technique for very fast hashing introduced by Rogaway. Rogaway's bucket hash family has a huge key size, which for common parameter choices can be more than a hundred thousand bits. The proposed hash families have a key size that is close to the key size of the theoretically best known constructions, typically a few hundred bits, and the evaluation has a time complexity that is similar to bucket hashing.

Keywords. Universal hash functions, message authentication, authentication codes, bucket hashing, software implementations.

1 Introduction

Universal hashing is a concept that was introduced by Carter and Wegman [8] in 1979. Since then, many results in theoretical computer science use different kinds of universal hashing. One of the main topics in universal hashing is called *strongly universal hashing*, and has a large amount of applications in computer science. The most widely known application in cryptography is the construction of unconditionally secure authentication codes. The model for unconditionally secure authentication codes was originally developed by Simmons [25, 26], see also [10]. One of the most important aspect of strongly universal hash functions is that the constructions should be simple to implement in software and/or hardware. Such implementation aspects have recently been in focus, and there are several papers addressing this topic [13, 16, 17, 22, 24, 11, 1].

Message authentication is one of the most common cryptographic settings today. In this setting a transmitter and a receiver share a secret key e. When the transmitter wants to send the receiver a message s, he computes a socalled *message authentication code*[1] (MAC), $\text{MAC} = f_e(s)$, and sends the pair (s, MAC). Here $f_e()$ denotes the function producing the MAC using key e. Receiving a pair (s', MAC') the receiver checks that $\text{MAC}' = f_e(s')$. If this is the case, the message is accepted as authentic, otherwise it is rejected.

---

[1] In the theory of universal hashing, this is usually referred to as a *tag* (or an *authenticator*).

W. Fumy (Ed.): Advances in Cryptology - EUROCRYPT '97, LNCS 1233, pp. 149-162, 1997.
© Springer-Verlag Berlin Heidelberg 1997

The fastest software MACs in common use today are based on software efficient cryptographic hash functions, such as MD5 [21, 7]. We refer to such an approach as the *MAC scheme approach*. For an overview, see [18, 19, 20]. Since we are computing one of the fastest types of cryptographic primitives[2] on a string essentially identical to the message, one might think that it is not possible to do much better. However, as was shown by Wegman and Carter already in 1981 [28], this is not the case. It was noted that one does not need to work with a "cryptographically strong primitive". A "cryptographically strong primitive" needs some complexity to resist attacks (e.g. many rounds), and this complexity is also time consuming. Through Wegman and Carter's universal hashing, one can instead work with a very simple function (the universal hash function) to produce a MAC. We refer to this approach as the *universal hash approach*. The details of such an approach are given in the last section of this paper. We review some advantages of using the universal hash approach instead of the usual MAC scheme approach.

- *Speed:* The universal hash function can be very simple to implement, and experimental implementations (e.g. [11]) indicate that producing the MAC using universal hash functions is faster than for example MD5 based techniques.
- *Parallelizable:* For this self-explaining property to hold, it is sufficient that (a part of) the universal hash function is a linear function, which is usually the case.
- *Incremental:* If a small part of the message is modified or a part is added to the message, we do not need to perform the new MAC calculation over the whole message but only over the small part that was modified/added. This is again a consequence of the linearity of (a part of) the universal hash function.
- *Unconditional security/Provable security:* Universal hashing is "unconditionally secure", i.e., the probability of success in an attack is independent of computational resources. The universal hash approach sometimes includes usage of some cryptographic primitive to provide multiple use. This usage can be done in the form of provable security, i.e., an adversary who can break the scheme can also break the underlying cryptographic primitive [22].

Note that MAC schemes are highly nonlinear, hence usually neither parallelizable nor incremental[3]. Also, reductions for MAC schemes to show provable security are not at all as tight as for the universal hash approach, for details see [4, 22, 2].

This paper studies very fast software implementations of strongly universal hash functions. One of the most important steps in this direction was taken by Rogaway when he introduced a technique for hashing called "bucket hashing" [22]. It is a very efficient way of producing a MAC, ideally requiring only 6 - 10 simple instruction per word to be authenticated. The drawback of this approach

---

[2] MD5 can probably not be considered to be a "cryptographically strong primitive", due to an attack by Dobbertin [9].

[3] In [3], a MAC scheme (XOR-MAC) was presented, which is incremental.

is the huge key size that is included, which for common parameter choices can be more than a hundred thousand bits. This requires the key to be generated through a pseudo-random number generator.

As mentioned before, there have been some previous work on software efficiency of universal hash functions, [17, 24, 11, 1]. The recent paper [11] considers evaluation of universal hash functions on processors supporting very fast integer multiplication. On such processors, they get an extremely high speed. Another recent paper [1] is more in the line of our work, focusing on evaluation in hash families with a small key size.

Our contribution is to show how it is possible to modify some known families of hash functions into a form such that the evaluation is similar to "bucket hashing". The proposed hash functions have a key size that is close to the key size of the theoretically best known constructions, which for common parameter choices can be around a hundred bits for a single use. Furthermore, the evaluation has a time complexity that is similar to bucket hashing and use the same simple instructions.

The paper is organized as follows. In Section 2 the basic definitions in universal hashing and authentication theory are given, as well as connections between them. Section 3 reviews bucket hashing, and in Section 4 we introduce our new approach to bucket hashing. In Section 5 we discuss implementation and parameter choices and finally, in Section 6, we review how the proposed hash families are used to produce a MAC.

2 Universal hash functions and authentication codes

In universal hashing, we consider a hash family $\mathcal{G}$, which is a set $\mathcal{G}$ of $|\mathcal{G}|$ functions such that $g : X \to Y$ for each $g \in \mathcal{G}$. Interesting cardinality parameters for a hash family are $|\mathcal{G}|$, $|X|$, and $|Y|$. Two relevant definitions are the following.

Definition 1. A hash family $\mathcal{G}$ is called ϵ-*almost universal₂* if for any two distinct elements $x_1, x_2 \in X$, there are at most $\epsilon|\mathcal{G}|$ functions $g \in \mathcal{G}$ such that $g(x_1) = g(x_2)$. We use the abbreviation ϵ-AU₂ for the family.

Definition 2. A hash family $\mathcal{G}$ is called ϵ-*almost strongly universal₂* if

i) for any $x \in X$ and any $y \in Y$, there are exactly $|\mathcal{G}|/|Y|$ functions $g \in \mathcal{G}$ such that $g(x) = y$.
ii) for any two distinct elements $x_1, x_2 \in X$, and for any two elements $y_1, y_2 \in Y$, there are at most $\epsilon|\mathcal{G}|/|Y|$ functions $g \in \mathcal{G}$ such that $g(x_1) = y_1$, and $g(x_2) = y_2$.

We here use the abbreviation ϵ-ASU₂.

For a more thorough treatment of universal hashing, we refer to [27], where these concepts are derived further. We will instead consider the known equivalences between strongly universal hashing and authentication codes.

Authentication theory as originally described by Simmons [25], [26], see also [10], considers the problem of two trusting parties, who want to send information from the transmitter to the receiver in the presence of an adversary. The adversary may introduce false messages to the receiver or replace a legal message with a false one. To protect against these threats, the sender and the receiver share a secret key. The key is then used in an authentication code (A-code).

A *systematic* (or Cartesian) A-code is a code where the information to be transmitted appears in plaintext in the transmitted message. Such a code is a triple $(\mathcal{S}, \mathcal{E}, \mathcal{Z})$ of finite sets and a map $f : \mathcal{S} \times \mathcal{E} \to \mathcal{Z}$. Here $\mathcal{S}$ is the set of source states, i.e., the information that is to be transmitted, $\mathcal{E}$ is the set of keys, and $\mathcal{Z}$ is the tag alphabet. When the transmitter wants to send the information $s \in \mathcal{S}$ using his secret key $e \in \mathcal{E}$, he transmits the message $m = (s, z)$, where $z = f(s, e)$, and $m \in \mathcal{M} = \mathcal{S} \times \mathcal{Z}$. When the receiver receives a message $m' = (s', z')$, he checks the authenticity by calculating whether $z' = f(s', e)$ or not. If equality holds, the message m is called valid. The adversary has two different attacks to choose between. He might introduce a false message $m = (s, z)$, and hence impersonating the transmitter, called the *impersonation attack*. He can also choose to observe a transmitted message $m = (s, z)$, and then replace this message with another message $m' = (s', z')$, where $s' \neq s$. This is called the *substitution attack*. The probability of success for the adversary when trying either of the two attacks, denoted by P_I and P_S respectively, are formally defined by $P_I = \max_{s,z} P(m = (s, z) \text{ valid})$ and $P_S = \max_{s,z} \max_{s' \neq s, z'} P(m' = (s', z') \text{ valid}|m = (s, z) \text{ observed})$. We assume that the keys are uniformly distributed. Then these probabilities can be written as

$$P_I = \max_{s,z} \frac{|\{e \in \mathcal{E} : z = f(s, e)\}|}{|\{e \in \mathcal{E}\}|}, \tag{1}$$

$$P_S = \max_{s,z} \max_{s' \neq s, z'} \frac{|\{e \in \mathcal{E} : z = f(s, e), z' = f(s', e)\}|}{|\{e \in \mathcal{E} : z = f(s, e)\}|}. \tag{2}$$

For a review of different bounds and constructions of A-codes, we refer to [15]. The main result on the equivalence between strongly universal hashing and authentication/coding theory is the following.

Theorem 3 [5, 28, 27].

i) *If there exists a q-ary code with codeword length n, cardinality M, and minimum Hamming distance d, then there exists an ϵ-AU_2 family of hash functions where $\epsilon = 1 - d/n$, $|\mathcal{G}| = n$, $|X| = M$, and $|Y| = q$. Conversely, if there exists an ϵ-AU_2 family of hash functions, then there exists a code with parameters as above.*

ii) *If there exists an A-code with parameters $|\mathcal{S}|$, $|\mathcal{E}|$, $P_I = 1/|\mathcal{Z}|$, and P_S, then there exists an ϵ-ASU_2 family of hash functions where $\epsilon = P_S$, $|\mathcal{G}| = |\mathcal{E}|$, $|X| = \mathcal{S}$, and $|Y| = |\mathcal{Z}|$. Conversely, if there exists an ϵ-ASU_2 family of hash functions, then there exists an A-code with parameters as above.*

We review the equivalence ii) above. Each key $e \in \mathcal{E}$ in the A-code corresponds to a unique function g_e in $\mathcal{G}$, and $S = X$. The tag z in the authentication code is then obtained as

$$z = g_e(s).$$

The significance of ϵ-AU_2 families in strongly universal hashing lies in the fact that they are very useful when constructing strongly universal hash families. This is due to the following result by Stinson.

Lemma 4 [27]. Let $\mathcal{G}_1$ be ϵ_1-AU_2 from X_1 to Y_1 and let $\mathcal{G}_2$ be ϵ_2-ASU_2 from Y_1 to Y_2. Then $\mathcal{G} = \{g_2(g_1(x)) : g_1 \in \mathcal{G}_1, g_2 \in \mathcal{G}_2\}$ is ϵ-ASU_2 with $\epsilon = \epsilon_1 + \epsilon_2$.

Most constructions of ϵ-ASU_2 families of hash functions for large $|X|$ use this composition construction. The constructions giving best performance in terms of key size [5] (see also [12]) uses Reed-Solomon codes as the ϵ-AU_2 family in the above composition construction. Another useful result, originally used in the Wegman-Carter construction [28], is obtained through the Cartesian product.

Lemma 5 [28, 27]. Let $\mathcal{G}$ be ϵ-AU_2 from X to Y. Let $\mathcal{G}^m = \{g^m(x_1, x_2, \ldots, x_m) = (g(x_1), g(x_2), \ldots, g(x_m)) : g \in \mathcal{G}_1\}$ be a set of hash functions from X^m to Y^m. Then $\mathcal{G}^m = \{g^m\}$ is ϵ-AU_2.

3   Bucket hashing

The bucket hashing technique was introduced by Rogaway in [22]. It gave rise to ϵ-AU_2 families that are extremely fast to compute, at the cost of a very large key. Rogaway's arguments was to produce this long key through a pseudo-random number generator. We review some details of the bucket hashing technique.

Fix a "word size" $w \geq 1$. For $n \geq N$ the hash function is defined to map from $X = \{0,1\}^{wn}$ to $Y = \{0,1\}^{wN}$. The number N is referred to as "the number of buckets". It is further required that $N(N-1)(N-2) \geq 6n$.

Let $\mathcal{H}_B[w, n, N]$ denote the hash family. Then each $h \in \mathcal{H}_B[w, n, N]$ is specified by a length n list where each entry contains 3 integer numbers in the interval $[0, N-1]$. Denote this list by $h = (h_0, h_1, \ldots, h_{n-1})$, where $h_i = (h_i^1, h_i^2, h_i^3)$. The hash family $\mathcal{H}_B[w, n, N]$ is given by the hash functions taken over the set of all possible lists h subject to the constraint that no two of the 3-element sets in the list are the same, i.e., $h_i \neq h_j, \forall i \neq j$.

With a given hash function $h = (h_0, h_1, \ldots, h_{n-1})$, the output value $h(x)$ is defined as follows. Let $x = x_0 x_1 \cdots x_{n-1}$, where each x_i is a bit vector of length w. Initialize y_j to 0^w for $0 \leq j \leq N-1$. Then, for each i, replace $y_{h_i^1}$ with $y_{h_i^1} \oplus x_i$, $y_{h_i^2}$ with $y_{h_i^2} \oplus x_i$, and $y_{h_i^3}$ with $y_{h_i^3} \oplus x_i$. Then set the output to be

$h(x) = y_0 y_1 \cdots y_{n-1}$. In pseudocode, we can write the algorithm as follows.

```
for j = 0 to N - 1 do
    y[j] = 0^w
for i = 0 to n - 1 do
    y[h_i^1] = y[h_i^1] ⊕ x_i
    y[h_i^2] = y[h_i^2] ⊕ x_i
    y[h_i^3] = y[h_i^3] ⊕ x_i
return y[0]y[1] ⋯ y[n - 1]
```

The computation of $h(x)$ gives rise to the name "bucket hashing", since it can be envisioned in the following way. We have N initially empty buckets. The first word of x is then thrown into three buckets, specified by h_0. Then the second word of x is thrown into three buckets, specified by h_1, and so on. Finally, the xor of the content in each of the buckets is computed, and the hash function output is the concatenation of the final content of the buckets. This is shown in Figure 1.

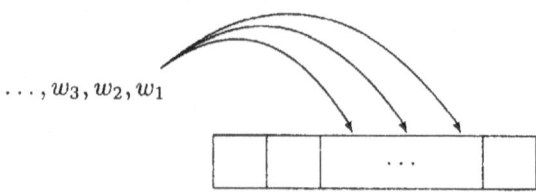

Fig. 1. A word is thrown into three buckets in Rogaway's bucket hashing.

The collision probability ϵ is given by a complicated expression [22] and instead of giving it here, we will just transfer some numerical values from [22] whenever needed. For example, for $n = 1024$ and $N = 100$, the collision probability is approximately 2^{-28}, i.e., $\mathcal{H}_B[w, n, N]$ is an ϵ-AU_2 hash family where $\epsilon = 2^{-28}$.

The bucket hashing approach gives a very fast implementation, since it only requires simple word operations as load, store and xor. Rogaway estimates that one word can be processed using only $6 - 10$ such simple instructions. Usually such simple instructions require only one clock cycle each, and can even be executed in parallel on many processors.

The drawback of the bucket hashing approach is the long key that is used. The key size is approximately $3n \log_2 N$, which is huge. For $n = 1024$ and $N = 100$, this is already more than 20000 bits, whereas a theoretically good construction [6, 14] for the same ϵ would require 76 key bits. Hence, the key bits in the bucket hashing construction must be generated by a pseudo-random number generator. This might be time consuming and the hash families are no longer unconditionally secure.

4 Bucket hashing with a small key size

The purpose of this section is to slightly modify some existing constructions of ϵ-AU$_2$ families of hash functions and then show that they can be implemented in a way that resembles the bucket hashing technique. The approach taken here is based on evaluation of polynomials similar to [6, 14]. The difference is essentially that we only consider polynomials over $GF(2)$, whereas the previous approaches consider polynomials over a larger field.

The following is a description of an ϵ-AU$_2$ family of hash functions. Let $\mathcal{P}_D$ be the set of all polynomials over $GF(2)$ without constant term and with degree at most D, i.e.,

$$\mathcal{P}_D = \{p(x) : p(x) = p_1 x + p_2 x^2 + \cdots + p_D x^D, p_i \in GF(2), 1 \le i \le D\}.$$

The hash family $\mathcal{G}_1$ is defined as follows. Let the functions in $\mathcal{G}_1$ map from $X = \mathcal{P}_D$ to $Y = GF(2^m)$, let $p \in \mathcal{P}_D = X$, $\alpha \in GF(2^m)$, and define

$$g_\alpha(p) = p(\alpha).$$

Theorem 6. *The family*

$$\mathcal{G}_1 = \{g_\alpha(p) : \alpha \in GF(2^m)\},$$

is an ϵ-AU$_2$ family of hash functions where

$$|\mathcal{G}_1| = 2^m, \quad |X| = 2^D, \quad |Y| = 2^m, \quad \epsilon = \frac{D}{2^m}.$$

Proof.

$$
\begin{aligned}
\epsilon &= \max_{x_1 \ne x_2} \frac{|\{g \in \mathcal{G}_1 : g(x_1) = g(x_2)\}|}{|\mathcal{G}_1|} \\
&= \max_{x_1 \ne x_2} \frac{|\{\alpha \in GF(2^m) : p_{x_1}(\alpha) = p_{x_2}(\alpha)\}|}{2^m} \\
&= \max_{x_1 \ne x_2} \frac{|\{\alpha \in GF(2^m) : p_{x_1 - x_2}(\alpha) = 0\}|}{2^m} \\
&\le \frac{D}{2^m},
\end{aligned}
$$

since any nonzero polynomial of degree D has at most D zeros. $\qquad\square$

Note that this is a slightly weaker result than in [6], where the polynomials have coefficients from $GF(2^m)$ and this does not change ϵ. However, as we will see, our approach will give a very efficient evaluation.

A generalization of the above construction is the hash family $\mathcal{G}_2$, constructed as follows. Let the functions in $\mathcal{G}_2$ map from $X = \mathcal{P}_D^n$ to $Y = GF(2^m)$, let $p = (p_1, p_2, \ldots, p_n) \in \mathcal{P}_D^n = X$ and define

$$g_{\alpha_1, \ldots, \alpha_n}(p) = p_1(\alpha_1) + \cdots + p_n(\alpha_n).$$

Theorem 7. *The family*

$$\mathcal{G}_2 = \{g_{\alpha_1,\ldots,\alpha_n}(p) : \alpha_1, \ldots, \alpha_n \in GF(2^m),\}$$

is an ϵ-AU_2 family of hash functions where

$$|\mathcal{G}_2| = 2^{nm}, \ |X| = 2^{nD}, \ |Y| = 2^m, \ \epsilon = \frac{D}{2^m}.$$

Proof. Similar to Theorem 6. $\qquad\qquad\qquad\qquad\qquad\qquad\qquad\qquad\qquad$ □

The central topic is to have a fast evaluation. We will now describe a hash family, denoted $\mathcal{G}_B[w, n, N]$, which has a fast evaluation. Then we show that this hash family is an implementation of $\mathcal{G}_1$ or $\mathcal{G}_2$, depending on a parameter choice in $\mathcal{G}_B[w, n, N]$.

Description of $\mathcal{G}_B[w, n, N]$: Fix w as the "word size" and let $N = 2^{m/L}$. For $n \geq N$ the hash function is defined to map from $X = \{0,1\}^{wn}$ to $Y = \{0,1\}^{wm}$. In the implementation there is an intermediate level using L arrays with $N = 2^{m/L}$ words in each, so the hash function can be described to map

$$X = \{0,1\}^{wn} \to \{0,1\}^{wNL} \to \{0,1\}^{wm} = Y.$$

The number N can be interpreted as "the number of buckets" and the number L can be interpreted as "the number of rows of buckets".

Each $h \in \mathcal{G}_B[w, n, N]$ is specified by a length n list where each entry contains L integer numbers in $[0, N-1]$. Denote this list by $h = (h_0, h_1, \ldots, h_{n-1})$, where $h_i = (h_i^0, \ldots, h_i^{L-1})$. The hash family $\mathcal{G}_B[w, n, N]$ is given by a set of such lists, which we call the set of all allowed lists. Different choices of this set will give different hash families.

With a given hash function $h = (h_0, h_1, \ldots, h_{n-1})$, the output value $h(x)$ is defined as follows. Let $x = x_0 x_1 \cdots x_{n-1}$, where each x_i is a bit vector of length w. Introduce L arrays of length N, called y_k, $0 \leq k \leq L-1$. Initialize $y_k[j]$ to 0^w for $0 \leq k \leq L-1$ and $0 \leq j \leq N-1$. Then, for each i, replace $y_0[h_i^0]$ with $y_0[h_i^0] \oplus x_i$, $y_1[h_i^1]$ with $y_1[h_i^1] \oplus x_i$, continuing in this way, and finally replacing $y_{L-1}[h_i^{L-1}]$ with $y_{L-1}[h_i^{L-1}] \oplus x_i$. This first step has hashed the input to the intermediate level of L rows of buckets, each containing N words. The procedure for $L = 2$ is shown in Figure 2.

Next, for each array, we compress the array in the following way. In $GF(2^{m/L})$ we have a primitive element γ which satisfies $\gamma^{m/L} = g_{m/L-1}\gamma^{m/L-1} + \cdots + g_1\gamma + g_0$, where $g_i \in GF(2)$. From $j = N-2$ down to m/L we add (xor) $y_k[j]$ to $y_k[j-i]$ for all i such that $g_i = 1$. Finally, set the output to be $h(x) = y_0 \cdots y_{L-1}$, where y_i denotes the content of the array $(y_i[0] \cdots y_i[m/L - 1])$.

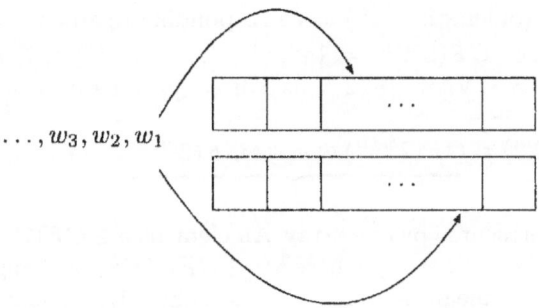

Fig. 2. A word is thrown into one bucket in each "row of buckets", here $L = 2$.

Assuming a generated list h, we can give a pseudocode for the case $L = 2$ with $\gamma^{m/2} = \gamma^{m/2-b} + 1$, for some integer b with $1 \leq b \leq m/2 - 1$, as follows.

```
for j = 0 to N − 2 do
    y_0[j] = 0^w,  y_1[j] = 0^w
for i = 0 to n − 1 do
    y_0[h_i^0] = y_0[h_i^0] ⊕ x_i
    y_1[h_i^1] = y_1[h_i^1] ⊕ x_i
for j = N − 2 to m/2 do
    y_0[j − b] = y_0[j − b] ⊕ y_0[j],  y_0[j − m/2] = y_0[j − m/2] ⊕ y_0[j]
    y_1[j − b] = y_1[j − b] ⊕ y_1[j],  y_1[j − m/2] = y_1[j − m/2] ⊕ y_1[j]
return y_0[0]y_0[1] ··· y_0[m/2 − 1]y_1[0]y_1[1] ··· y_1[m/2 − 1]
```

Observe that $\mathcal{G}_B[w, n, N]$ can be evaluated efficiently using only simple instructions as load, store and xor. Next we prove equivalences between $\mathcal{G}_B[w, n, N]$ and the hash families $\mathcal{G}_1$ and $\mathcal{G}_2$. Let $[z]$ denote the vector $(z_0, \ldots, z_{L-1})$, where $\gamma^{z_i} \in GF(2^{m/L})$, and by convention $\gamma^{N-1} = 0$, such that $z = \gamma^{z_0} + \gamma^{z_1}\beta + \cdots + \gamma^{z_{L-1}}\beta^{L-1} \in GF(2^m)$. Here $\beta \in GF(2^m)$ and $h(\beta) = 0$ for some irreducible polynomial $h(x)$ of degree L over $GF(2^{m/L})$.

Theorem 8. *Let the set of allowed lists be*

$$\{([\alpha], [\alpha^2], \ldots, [\alpha^n]), \forall \alpha \in GF(2^m)\}.$$

Then the hash family $\mathcal{G}_B[w, n, N]$ is equivalent to $\mathcal{G}_1^m$, i.e., the Cartesian product of w hash families $\mathcal{G}_1$ as in Lemma 5.

Proof. The proof is in two steps.

1. The ith bit of each output word is only dependent on the ith bit of each input word x_j and independent of all the other bit positions in the input words. Hence we can view the hash family $\mathcal{G}_B[w, n, N]$ as a Cartesian product of w hash families each having an input word size of one, as in Lemma 5. So w.l.o.g we assume $w = 1$.

2. Regard each array (of length $2^{m/L}$) as corresponding to an enumeration of elements in $GF(2^{m/L})$, i.e., $GF(2^{m/L}) = \{\gamma^0, \gamma^1, \gamma^2, \ldots, \gamma^{2^{m/L}-2}, 0\}$ and entry z_i corresponds to element γ^{z_i}. View $GF(2^m)$ as a direct product of such subfields, i.e.,

$$GF(2^m) = \underbrace{GF(2^{m/L}) \otimes \cdots \otimes GF(2^{m/L})}_{L},$$

where each subfield is represented by one array. An element $z \in GF(2^m)$ is represented by the vector $z = (z_0, \ldots, z_{L-1})$, where $\gamma^{z_i} \in GF(2^{m/L})$. Putting an input word x_i ($w = 1$) in bucket z_i means adding γ^{z_i} in the subfield, and hence, putting an input word x_i in buckets represented by $[z] = (z_0, \ldots, z_{L-1})$ means adding $x_i z$ to the previous content of the buckets. Hence, the list $([\alpha], [\alpha^2], \ldots, [\alpha^n])$, for $\alpha \in GF(2^m)$ means adding $x_0\alpha + x_1\alpha^2 + \cdots + x_n\alpha^n$. The result is now represented as powers of γ in each subfield (array). In the last part, adding $y_k[j]$ to $y_k[j-i]$ for all i such that $g_i = 1$ from $j = N-2$ down to m/L simply means reducing the powers of γ to the basis $\{\gamma^{m/L-1}, \ldots, \gamma, 1\}$. Hence the output of $\mathcal{G}_B[w, n, N]$ is $x_0\alpha + x_1\alpha^2 + \cdots + x_n\alpha^n \in GF(2^m)$, where $GF(2^m) = GF(2^{m/L}) \otimes \cdots \otimes GF(2^{m/L})$ and $GF(2^{m/L})$ is represented using the basis $\{\gamma^{m/L-1}, \ldots, \gamma, 1\}$. □

A similar result can be obtained for the family $\mathcal{G}_2$. For example, let the set of allowed lists be

$$\{([\alpha_1], [\alpha_2], \ldots, [\alpha_n]), \forall \alpha_i \in GF(2^m), 1 \le i \le n\},$$

i.e., the set of all possible lists. Then $\mathcal{G}_B[w, n, N]$ is equivalent to $\mathcal{G}_2^m$ with $D = 1$, i.e., the Cartesian product of w hash families $\mathcal{G}_2$ with $D = 1$.

5 Implementation and parameter choices

Clearly, the efficiency of the evaluation will depend on the choice of parameters in the above description. Let us consider some different ways to implement $\mathcal{G}_B[w, n, N]$. Note that the situation is very similar to Rogaway's bucket hashing. We can process word by word from input, or we can process bucket by bucket. Furthermore, we can use a self-modifying code (the actual hash function is implemented in the program code), or we can read the bucket/word locations from a table in memory. The fastest choice is a self-modifying code processing bucket by bucket. Then we can keep the current bucket in a register while processing, requiring only one `load` and one `xor` instruction for each input word and each row of buckets. Hence, for L rows this requires $2L$ simple instructions. For further details we refer to [22].

Furthermore, the compression of the arrays means LNc `load`, `add` and `store` operations, where c is the number of nonzero coefficients in the primitive polynomial defining γ (this can usually be chosen to be 2). For $n \gg N$ the time to do the compression part is hence negligible compared to the first part. Initialization of the list h is done only once. Hence, when concatenating this hash function many times using Lemma 5, the time to execute this part is also negligible.

For tabulating some values, we regard $n = 8N$ as being sufficient for considering the compression to be negligible in time. Also, the generation of the list h is different depending on the actual choice of hash function. In all cases we are aware of the fact that we need to concatenate a few, say 10, hash families in order to make the time to process this part small. Alternatively, considering multiple use, we can assume that the list h is generated once and then kept fixed. We tabulate some values for different parameter values in Table 1. The input size, the output size and N are given in number of words; ϵ is the collision probability; the key size is given in bits; and the time column gives the minimal number of simple instructions per word for a self-modifying code processing bucket by bucket.

Hash function and parameters	Input size	Output size	ϵ	Key size	Time
Bucket hashing, $n = 4096$, $N = 40$	2^{12}	40	2^{-20}	94000	6
Bucket hashing, $n = 4096$, $N = 200$	2^{12}	200	2^{-34}	94000	6
$\mathcal{G}^1$, $N = 2^{10}$, $L = 3$	2^{13}	30	2^{-17}	30	6
$\mathcal{G}^1$, $N = 2^{10}$, $L = 4$	2^{13}	40	2^{-27}	40	8
$\mathcal{G}^1$, $N = 2^{10}$, $L = 5$	2^{13}	50	2^{-37}	50	10
$\mathcal{G}^1$, $N = 2^{10}$, $L = 7$	2^{13}	70	2^{-57}	70	14
$\mathcal{G}^1$, $N = 2^{20}$, $L = 4$	2^{23}	80	2^{-57}	80	8
$\mathcal{G}^2$, $N = 2^{10}$, $L = 4$, $D = 512$	2^{13}	40	2^{-31}	640	8

Table 1. A comparison for some different parameter choices.

In order to process each simple instruction in at most one clock cycle (we might execute several in parallel) on a usual processor, each reference to a memory location needs to be in the on-chip cache of the processor. Hence, for a self-modifying code processing bucket by bucket, the input to one hash function must fit the on-chip cache, giving restrictions on the input size and thus on N. Examining the sizes of the on-chip caches of todays processors, $N = 2^{10}$ is probably about the maximum size of the arrays under these circumstances.

Note the fact that some properties of Rogaway's bucket hashing and the proposed techniques are different and hence the techniques are not directly comparable. Especially, $\mathcal{G}_1$ gives a much higher compression, i.e., input size/output size is much smaller. This means that including $\mathcal{G}_1$, it is enough to concatenate two hash families using Lemma 4 to get the desired output size, whereas bucket hashing requires many concatenations to obtain the desired output size. This can be a problem for large messages since producing a large hash output that has to be written in memory and then further processed will produce cash-misses etc.

6 The universal hash approach in practice

Up to this point, we have only considered how to construct the ϵ-AU$_2$ hash family. This short section overviews how to use the ϵ-AU$_2$ hash family to produce an ϵ-ASU$_2$ hash family that gives an authentication tag (MAC) and also have the properties mentioned in Section 1.

The usage of ϵ-ASU$_2$ families of hash functions in the described way applies to the case of sending/storing one message with fixed length (variable length can easily be included [22]). Sometimes one is interested in multiple use, i.e., sending/storing many message where each message needs individual authentication. In the unconditionally secure approach, the solution is to add new random key bits for each additional messages to be hashed. If $h_{e_1}()$ is the ϵ-ASU$_2$ hash function, the MACs $(z_1, z_2, \ldots)$ for a sequence of messages $s_1, s_2, \ldots$ can be produced by

$$z_1 = h_{e_1}(s_1), \quad z_2 = h_{e_1}(s_2) + e_2, \quad z_3 = h_{e_1}(s_3) + e_3, \ldots,$$

where $e_2, e_3, \ldots$ are randomly chosen keys of same length as the MAC. It can be proved [28] that this procedure gives the same P_I and P_S as for the single message case.

In some cases, the number of messages is limited and then it is preferable to keep the unconditionally secure approach. In other cases, the set $\{e_2, e_3, \ldots\}$ of randomly chosen keys is too large to be kept secret in an unconditionally secure way. Instead, one uses a pseudo-random number generator to produce this set. In such a case, some of the motivation to consider ϵ-AU$_2$ hash families with a short key is lost, since the same pseudo-random number generator can be used to produce the hash function itself.

A complete ϵ-ASU$_2$ hash family obtained by Lemma 4 can be described as follows. Let x be the message that is to be hashed. Divide x into suitable sized substrings $x = x_1 x_2 \cdots x_n$. Apply a secretly chosen ϵ-AU$_2$ hash function h_1 and calculate $y_i = h_1(x_i)$, $1 \le i \le n$. For the obtained string $y = y_1 y_2 \cdots y_n$ (now of modest size) we have secretly selected another ϵ-ASU$_2$ hash function h_2 and calculates $w = h_2(y)$. In an unconditionally secure authentication code we would select a secret key e and form a MAC of the form MAC $= w + e$. For the next message, we use a new value of e, etc.

If we want to produce the sequence of keys using a pseudo-random number generator we can do as follows. We have a counter, call it cnt, which is initially zero. This counter is used together with a cryptographic primitive, e.g. RC5 [23], using a secret key e. The MAC for the message is given by

$$\text{MAC} = w + \text{RC5}_e(cnt),$$

together with the used value of the counter. Finally, cnt is incremented.

Example: As a particular example for $w = 32$, choose $\mathcal{G}_1$ with $N = 1024$ and $L = 7$ as the first hash family. The key e_1 to select the hash function is 70 bits. We have 8092 word input, producing a 70 word output and $\epsilon = 2^{-57}$. As the second hash family, choose the polynomial evaluation hash [6, 14] over $GF(2^{70})$. Our key e_2 for this hash family is 70 bits as well.

Divide the input x in 32Kbyte blocks $x = x_1 x_2 \cdots x_n$. Apply the methods described in Section 4 on each block x_i, receiving n 70-word blocks called y_i, by $y_i = g_{e_1}(x_i)$. Then form the string $y = y_1 y_2 \cdots y_n$ and interpret this as a polynomial over $GF(2^{70})$. This polynomial, call it $y(x)$, will then have degree $32n$. Evaluate the polynomial in e_2, obtaining $w = y(e_2)$. Then calculate the MAC as MAC $= w + e_3$, where e_3 is a third 70 bit key. Finally, we output (x, MAC). The value of ϵ will depend on n, but for input sizes smaller than 8Mbyte we have $\epsilon < 2^{-56}$.

Alternatively, using RC5 in multiple use we calculate the MAC as MAC $= w + \text{RC5}_{e_3}(cnt)$, output (x, cnt, MAC), and increment the counter.

References

1. V. Afanassiev, C. Gehrmann, B. Smeets, Fast message authentication using efficient polynomial evaluation, *Proceedings of Fast Software Encryption Conference '97*, to appear.
2. M. Bellare, R. Canetti, H. Krawczyk, Keying hash functions for message authentication, *Lecture Notes in Computer Science* **1109** (1996), 1 15 (CRYPTO '96).
3. M. Bellare, R. Guérin, P. Rogaway, XOR MACs: New methods for message authentication, *Lecture Notes in Computer Science* **963** (1995), 15–28 (CRYPTO '95).
4. M. Bellare, J. Kilian, P. Rogaway, The security of cipher block chaining, *Lecture Notes in Computer Science* **839** (1994), 341–358 (CRYPTO '94).
5. J. Bierbrauer, T. Johansson, G. Kabatianskii, and B. Smeets, On families of hash functions via geometric codes and concatenation, *Lecture Notes in Computer Science*, **773** (1994), 331–342 (CRYPTO '93).
6. B. den Boer, A simple and key-economical unconditionally authentication scheme, *Journal of Computer Security*, **2** (1993), 65–71.
7. A. Bosselaers, R. Govaerts, J. Vandewalle, Fast hashing on the Pentium, *Lecture Notes in Computer Science* **1109** (1996), 298–313 (CRYPTO '96).
8. J.L. Carter, M.N. Wegman, Universal classes of hash functions, *J. Computer and System Sciences*, **18** (1979), 143–154.
9. H. Dobbertin, Cryptoanalysis of MD5 compress, presented at the rump session of EUROCRYPT'96.
10. E.N. Gilbert, F.J. MacWilliams, and N.J.A. Sloane, Codes which detect deception, *Bell Syst. Tech. J.*, **53** (1974), 405–424.
11. S. Halevi, H. Krawczyk, Software message authentication in the Gbit/second rates, *Proceedings of Fast Software Encryption Conference '97*, to appear.
12. T. Helleseth and T. Johansson, Universal hash functions from exponential sums over finite fields and Galois rings, *Lecture Notes in Computer Science* **1109** (1996), 31–44 (CRYPTO '96).
13. T. Johansson, A shift register construction of unconditionally secure authentication codes, *Designs, Codes and Cryptography*, **4** (1994), 69–81.
14. T. Johansson, G. Kabatianskii, B. Smeets, On the relation between A-codes and codes correcting independent errors, *Lecture Notes in Computer Science*, **765** (1994), 1–11 (EUROCRYPT'93).

15. G. Kabatianskii, B. Smeets, and T. Johansson, On the cardinality of systematic authentication codes via error correcting codes, *IEEE Trans. Inform. Theory*, **42** (1996), 566–578.

16. H. Krawczyk, LFSR-based hashing and authentication, *Lecture Notes in Computer Science*, **839** (1994), 129–139 (CRYPTO '94).

17. H. Krawczyk, New hash functions for message authentication, *Lecture Notes in Computer Science*, **921** (1995), 140–149 (EUROCRYPT '95).

18. B. Preneel, Cryptographic hash functions, *European Transactions on Telecommunications*, **5** (1994), 431–448.

19. B. Preneel, P. van Oorschot, MDx-MAC and building fast MACs from hash functions, *Lecture Notes in Computer Science*, **963** (1995), 1–14 (CRYPTO '95).

20. B. Preneel, P. van Oorschot, On the security of two MAC algorithms, *Lecture Notes in Computer Science*, **1070** (1996), 19–32 (EUROCRYPT '96).

21. R.L. Rivest, The MD5 message-digest algorithm, *Request for Comments 1321*, Internet Activities Board, Internet Privacy Task Force (1992).

22. P. Rogaway, Bucket hashing and its application to fast message authentication, *Lecture Notes in Computer Science*, **963** (1995), 29–42 (CRYPTO '95).

23. B. Schneier, *Applied Cryptography*, John Wiley & Sons (1996).

24. V. Shoup, On fast and provably secure message authentication based on universal hashing, *Lecture Notes in Computer Science*, **1109** (1996), 313–328 (CRYPTO '96).

25. G.J. Simmons, A game theory model of digital message authentication, *Congr. Numer.*, **34** (1992), 413–424.

26. G.J. Simmons, Authentication theory/coding theory, in *Lecture Notes in Computer Science*, **196** (1985), 411–431 (CRYPTO '84).

27. D.R. Stinson, Universal hashing and authentication codes, *Codes, Designs and Cryptography*, 4 (1994), 337–346.

28. M.N. Wegman and J.L. Carter, New hash functions and their use in authentication and set equality, *J. Computer and System Sciences*, **22** (1981), 265–279.

A New Paradigm for Collision-Free Hashing: Incrementality at Reduced Cost

Mihir Bellare[1] and Daniele Micciancio[2]

[1] Dept. of Computer Science & Engineering, University of California at San Diego, 9500 Gilman Drive, La Jolla, California 92093, USA. E-Mail: mihir@watson.ibm.com. URL: http://www-cse.ucsd.edu/users/mihir.

[2] MIT Laboratory for Computer Science, 545 Technology Square, Cambridge, MA 02139, USA. E-Mail: miccianc@theory.lcs.mit.edu.

Abstract. We present a simple, new paradigm for the design of collision-free hash functions. Any function emanating from this paradigm is *incremental*. (This means that if a message x which I have previously hashed is modified to x' then rather than having to re-compute the hash of x' from scratch, I can quickly "update" the old hash value to the new one, in time proportional to the amount of modification made in x to get x'.) Also any function emanating from this paradigm is parallelizable, useful for hardware implementation. We derive several specific functions from our paradigm. All use a standard hash function, assumed ideal, and some algebraic operations. The first function, MuHASH, uses one modular multiplication per block of the message, making it reasonably efficient, and significantly faster than previous incremental hash functions. Its security is proven, based on the hardness of the discrete logarithm problem. A second function, AdHASH, is even faster, using additions instead of multiplications, with security proven given either that approximation of the length of shortest lattice vectors is hard or that the weighted subset sum problem is hard. A third function, LtHASH, is a practical variant of recent lattice based functions, with security proven based, again on the hardness of shortest lattice vector approximation.

1 Introduction

A collision-free hash function maps arbitrarily long inputs to outputs of a fixed length, but in such a way that it is computationally infeasible to find a collision, meaning two distinct messages x, y which hash to the same point.[3] These functions were first conceived and designed for the purpose of hashing messages before signing, the point being to apply the (expensive) signature operation only to short data. (Whence the collision-freeness requirement, which is easily seen to be a necessary condition for the security of the signature scheme.) Although this remains the most important usage for these functions, over time many other

---

[3] The formal definition in Section 2 speaks of a *family* of functions, but we dispense with the formalities for now.

W. Fumy (Ed.): Advances in Cryptology - EUROCRYPT '97, LNCS 1233, pp. 163-192, 1997.
© Springer-Verlag Berlin Heidelberg 1997

applications have arisen as well. Collision-free hash functions are now well recognized as one of the important cryptographic primitives, and are in extensive use.

We are interested in finding hash functions that have a particular efficiency feature called "incrementality" which we describe below. Motivated by this we present a new paradigm for the design of collision-free hash functions. We obtain from it some specific incremental hash functions that are significantly faster than previous ones.

It turns out that even putting incrementality aside, functions resulting from our paradigm have attractive features, such as parallelizability.

1.1 Incremental Hashing

THE IDEA. The notion of incrementality was advanced by Bellare, Goldreich and Goldwasser [BGG1]. They point out that when we cryptographically process documents in bulk, these documents may be related to each other, something we could take advantage of to speed up the computation of the cryptographic transformations. Specifically, a message x' which I want to hash may be a simple modification of a message x which I previously hashed. If I have already computed the hash $f(x)$ of x then, rather than re-computing $f(x')$ from scratch, I would like to just quickly "update" the old hash value $f(x)$ to the new value $f(x')$. An incremental hash function is one that permits this.

For example, suppose I want to maintain a hash value of all the files on my hard disk. When one file is modified, I do not want to re-hash the entire disk contents to get the hash value. Instead, I can apply a simple update operation that takes the old hash value and some description of the changes to produce the new hash value, in time proportional to the amount of change.

In summary, what we want is a collision-free hash function f for which the following is true. Let $x = x_1 \ldots x_n$ be some input, viewed as a sequence of blocks, and say block i is modified to x'_i. Let x' be the new message. Then given $f(x), i, x_i, x'_i$ it should be easy to compute $f(x')$.

STANDARD CONSTRUCTIONS FAIL. Incrementality does not seem easy to achieve. Standard methods of hash function construction fail to achieve it because they involve some sort of iteration. This is true for constructions based on block ciphers. (For description of these constructions see for example the survey [PGV].) It is also true for the compression function based constructions that use the Merkle-Damgård meta-method [Me, Da2]. The last includes popular functions like MD5 [Ri], SHA-1 [SHA] and RIPEMD-160 [DBP]. The modular arithmetic based hash functions are in fact also iterative, and so are the bulk of number-theory based ones, eg. [Da1].

A thought that comes to mind is to use a tree structure for hashing, as described in [Me, Da2]. (Adjacent blocks are first hashed together, yielding a text half the length of the original one, and then the process is repeated until a final hash value is obtained.) One is tempted to think this is incremental because if a message block is modified, work proportional only to the tree depth needs to be done to update. The problem is you need to *store the entire tree*, meaning all the intermediate hash values. What we want is to store only the final hash value and be able to increment given only this.

PAST WORK. To date the only incremental hash function was proposed by [BGG1], based on work of [CHP]. This function is based on discrete exponentiation in a group of prime order. It uses one modular exponentiation per message block to hash the message. This is very expensive, especially compared with standard hash functions. An increment operation takes time independent of the message size, but also involves exponentiation, so again is expensive. We want to do better, on both counts.

1.2 The Randomize-then-combine Paradigm

We introduce a new paradigm for the construction of collision-free hash functions. The high level structure is quite simple. View the message x as a sequence of blocks, $x = x_1 \ldots x_n$, each block being b bits long, where b is some parameter to choose at will. First, each block x_i is processed, via a function h, to yield an outcome y_i. (Specifically, $y_i = h(\langle i \rangle . x_i)$ where $\langle i \rangle$ is a binary representation of the block index i and "." denotes concatenation). These outcomes are then "combined" in some way to yield the final hash value $y = y_1 \odot y_2 \odot \ldots \odot y_n$, where $\odot$ denotes the "combining operation."

Here h, the "randomizing" function, is derived in practice from some standard hash function like SHA-1, and treated in the analysis as an "ideal" hash function or random oracle [BR]. The combining operation $\odot$ is typically a group operation, meaning that we interpret $y_1, \ldots, y_n$ as members of some commutative group G whose operation is denoted $\odot$.

We call this the *randomize-then-combine* paradigm. It is described fully in Section 3. The security of this method depends of course on the choice of group, and we will see several choices that work. The key benefit we can observe straight away is that the resulting hash function is incremental. Indeed, if x_i changes to x_i', one can re-compute the new hash value as $y \odot h(x_i)^{-1} \odot h(x_i')$ where y is the old hash value and the inverse operation is in the group. Also it is easy to see the computation of the hash function is parallelizable.

By choosing different groups we get various specific, incremental, collision-free hash functions, as we now describe.

Notice that h needs itself to be collision-free, but applies only to fixed length inputs. Thus, it can be viewed as a "compression function." Like [Me, Da2], our paradigm can thus be viewed as constructing variable input length hash functions from compression functions. However, our construction is "parallel" rather than iterative. It is important to note, though, that even though our constructions seem secure when h is a good compression function (meaning one that is not only collision-free but also has some randomness properties) the proofs of security require something much stronger, namely that h is a random oracle.

1.3 MuHASH and its Features

MUHASH. Our first function, called MuHASH for "multiplicative hash," sets the combining operation to multiplication in a group G where the discrete logarithm problem is hard. (For concreteness, think $G = Z_p^*$ for a suitable prime p. In this case, hashing consists of "randomizing" the blocks via h to get elements of Z_p^* and then multiplying all these modulo p).

EFFICIENCY. How fast is MuHASH? The cost is essentially one modular multiplication per b-bit block. Notice that one computation of h per b-bit block is also required. However, the cost of computing h will usually be comparatively small. This is especially true if the block length is chosen appropriately. For example, if h is implemented via SHA, chosing b as a multiple of 512, the expensive padding step in computing SHA can be avoided and the total cost of computing h for every block is about the same as a single application of SHA on the whole message. The cost of h will be neglected in the rest of the paper.

At first glance the presence of modular operations may make one pessimistic, but there are two things to note. First, it is multiplications, not exponentiations. Second, we can make the block size b large, making the amortized per-bit cost of the multiplications small. Thus, MuHASH is much faster than the previous incremental hash function. In fact it is faster than any number-theory based hash function we know. Note if hardware for modular multiplication is present, not unlikely these days, then MuHASH becomes even more efficient to compute.

The increment operation on a block takes one multiplication and one division, again much better than the previous construction.

SECURITY. We show that as long as the discrete logarithm problem in G is hard and h is ideal, MuHASH is collision-free. (This may seem surprising at first glance since there does not seem to be any relation between discrete logarithms and MuHASH. In the latter we are just multiplying group elements, and no group generator is even present!) That is, we show that if there is *any* attack that finds collisions in MuHASH then there is a way to efficiently compute discrete logarithms in G. The strength of this statement is that it makes no assumptions about the cryptanalytic techniques used by the MuHASH attacker: no matter what these techniques may be, the attacker will fail as long as the discrete logarithm problem in G is hard. This proven security means we are obviated from the need to consider the effects of any specific attacks. That is, it is not necessary to have an exhaustive analysis of a list of possible attacks.

The proven security provides a strong qualitative guarantee of the strength of the hash function. However, we have in addition a strong quantitative guarantee. Namely, we have reductions that are *tight*. To obtain these we have to use the group structure more carefully. We present separate reductions, with slightly different characteristics, for groups of prime order and for the multiplicative group modulo a prime. These are Theorem 4 and Theorem 5 respectively. In practice this is important because it means we can work with a smaller value of the security parameter making the scheme more efficient.

An interesting feature of MuHASH is that its "strength in practice" may greatly exceed its proven strength. MuHASH is proven secure if the discrete logarithm problem is hard, but it might be secure even if the discrete logarithm problem is easy, because we know of no attack that finds collisions *even if it is easy to compute discrete logarithms.* And in practice, collision-freeness of h seems to suffice.

1.4 AdHASH and its Features

AdHASH (for "additive hash") uses addition modulo a large enough integer M as the combining operation in the randomize-then-combine paradigm. In other words, to hash we first randomize the blocks of the message using h and then add all the results modulo M.

Replacing multiplication by addition results in a significant improvement in efficiency. Hashing now only involves n modular additions, and the increment operation is just two modular additions. In fact AdHASH is competitive with standard hash functions in speed, with the added advantages of incrementality and parallelizability.

AdHASH also has strong security guarantees. We show that it is collision-free as long as the "weighted knapsack problem" (which we define) is hard and h is ideal. But Ajtai [Aj] has given strong evidence that the weighted subset sum problem is hard: he has shown that this is true as long as there is no polynomial time approximation algorithm for the shortest vector problem in a lattice, in the worst case. But even if this approximation turns out to be feasible (which we don't expect) the weighted subset sum problem may still be hard, so that AdHASH may still be secure.

We also prove that AdHASH is a universal one-way hash function in the sense of Naor and Yung [NY], assuming the subset sum function of [IN1, IN2] is one-way and h is ideal. (Thus, under a weaker assumption, we can show that a weaker form but still useful form of collision-freeness holds. We note our reductions here are tight, unlike those of [IN1, IN2]. These results are omitted form this abstract but can be found in [BM].)

In summary AdHASH is quite attractive both on the efficiency and on the security fronts.

1.5 Hashing from Lattice Problems

Ajtai introduced a linear function which is provably one-way if the problem of approximating the (Euclidean) length of the shortest vector in a lattice is hard [Aj]. (The function is matrix-vector multiplication, with particular parameters). Goldreich, Goldwasser and Halevi [GGH] observed that Ajtai's main lemma can be applied to show that the function is actually collision-free, not just one-way. We observe that this hash function is incremental. But we also point out some impracticalities.

We then use our randomize-then-combine paradigm to derive a more practical version of this function. (Our function is more efficient and has smaller key size). It is called LtHASH (for "lattice hash"). The group is $G = Z_p^k$ for some integers p, k, meaning we interpret the randomized blocks as k-vectors over Z_p and add them component-wise. Assuming h is ideal the security of this hash function can be directly related to the problem underlying the security of Ajtai's one-way function [Aj, GGH] so that it is collision-free as long as the shortest lattice vector approximation problem is hard.

Note that the same assumption that guarantees the security of LtHASH (namely hardness of approximation of length of the shortest vector in a lattice) also guarantees the security of AdHASH, and the efficiency is essentially the same, so we may just stick with AdHASH. However it is possible that LtHASH has some features of additional interest, and is more directly tied to the lattice hardness results, so it is worth mentioning.

1.6 Attack on XHASH

Ideally, we would like to hash using only "conventional" cryptography (ie. no number theory.) A natural thought is thus to set the combining operation to bitwise XOR. But we show in Appendix A that this choice is insecure. We present an attack on the resulting function XHASH, which uses Gaussian elimination and pairwise independence. It may be useful in other contexts.

We are loth to abandon the paradigm based on this: it is hard to imagine any other paradigm that yields incrementality. But we conclude that it may be hard to get security using only conventional cryptography to implement the combining operation. So we turned to arithmetic operations and found the above.

1.7 The balance problem

We identify a computational problem that can be defined in an arbitrary group. We call it the balance problem. It turns out that consideration of the balance problem unifies and simplifies the treatment of hash functions, not only in this paper but beyond. Problems underlying algebraic or combinatorial collision-free hash functions are often balance problems. We will see how the hardness of the balance problem follows from the hardness of discrete logs; how in additive groups it is just the weighted subset sum problem; and that it captures the matrix kernel problem presented in [Aj] which is the basis of lattice based hash functions [GGH].

The problem is simply that given random group elements $a_1, \ldots, a_n$, find disjoint subsets $I, J \subseteq \{1, \ldots, n\}$, not both empty, such that $\bigodot_{i \in I} a_i = \bigodot_{j \in J} a_j$, where $\odot$ is the group operation. Having reduced the security of our hash function to this problem in Lemma 2, our main technical effort will be in relating the balance problem in a group to other problems in the group.

1.8 Related Work

For a comprehensive survey of hashing see [MVV, Chapter 9].

DISCRETE LOGARITHM OR FACTORING BASED FUNCTIONS. To the best of our knowledge, all previous discrete logarithm or factoring based hash functions which have a security that can be provably related to that of the underlying number theoretic problem use at least one multiplication per *bit* of the message, and sometimes more. (For example this is true of the functions of [Da1], which are based on claw-free permutations [GMR].) In contrast, MuHASH uses one multiplication per b-bit block and can make b large to mitigate the cost of the multiplication. (But MuHASH uses a random oracle assumption which the previous constructions do not. And of course the previous functions, barring those of [BGG1], are non-incremental.)

COLLISION-FREE VERSUS UNIVERSAL ONE-WAY. Collision-freeness is a stronger property than the property of universal one-wayness defined by Naor and Yung [NY]. Functions meeting their conditions are not necessarily collision-free. (But they do suffice for many applications.)

SUBSET-SUM BASED HASHING. Impagliazzo and Naor [IN1, IN2] define a hash function and prove that it is a universal one-way function (which is weaker than collision-free) as long as the subset-sum function is one-way. The same function is defined in [Da2, Section 4.3]. There it is conjectured to be collision-free as well, but no proof is provided. These functions have a key length as long as the input to be hashed (very impractical) and use one addition per bit of the message. In contrast, AdHASH has short key length and uses one addition per b-bit block of the message, and b can be made large.

HASHING BY MULTIPLYING IN A GROUP. Independently of our work, Impagliazzo and Naor have also considered hashing by multiplying in a group. These results have been included in [IN2], the recent journal version of their earlier [IN1]. In their setup, a list of random numbers $a_1, \ldots, a_n$ is published, and the hash of message x is $\prod_{i=1}^{n} x_i a_i$ where x_i is the i-th bit of x and the product is taken in the group. Thus there is one group operation per bit of the message, and also the key size is proportional to the input to be hashed. Functions resulting from our paradigm use one group operation per b-bit block, which is faster, and have fixed key size. On the security side, [IN2] show that their hash function is universal one-way as long as any homomorphism with image the given group is one-way. (In particular, if the discrete logarithm problem in the group is hard.) In contrast we show that our functions have the stronger property of being collision-free. But the techniques are related and it is also important to note that we use a random oracle assumption and they do not. On the other hand our reductions are tight and theirs are not.

The general security assumption of [IN2] and their results provide insight into why MuHASH may be secure even if the discrete logarithm problem is easy.

MODULAR ARITHMETIC HASH FUNCTIONS. Several iterative modular arithmetic based hash functions have been proposed in the past. (These do not try to provably relate the ability to find collisions to any underlying hard arithmetic problems.) See Girault [Gi] for a list and some attacks. More recent in this vein are MASH-1 and MASH-2, designed by GMD (Gesellschaft fur Mathematik im Dataverarbeitung) and being proposed as ISO standards. However, attacks have been found by Coppersmith and Preneel [CP].

XOR MACs. Our paradigm for hashing is somewhat inspired by, and related to, the XOR MACs of [BGR]. There, XOR worked as a combining operation. But the goal and assumptions were different. Those schemes were for message authentication, which is a private key based primitive. In particular, the function playing the role of h was *secret*, computable only by the legitimate parties and not the adversary. (So in particular, the attack of Appendix A does not apply to the schemes of [BGR].) However, hash functions have to have a *public* description, and what we see is that in such a case the security vanishes if the combining operation is XOR.

INCREMENTALITY. Other work on incremental cryptography includes [BGG2, Mi]. The former consider primitives other than hashing, and also more general incremental operations than block replacement, such as block insertion and deletion. (Finding collision-free hash functions supporting these operations is an open problem.) The latter explores issues like privacy in the presence of incremental operations.

2 Definitions

2.1 Collision-free Hash Functions

FAMILIES OF HASH FUNCTIONS. A *family of hash functions* F has a *key space* $Keys(F)$. Each key $K \in Keys(F)$ specifies a particular function mapping $Dom(F)$ to $Range(F)$, where $Dom(F)$ is a domain common to all functions in the family, and $Range(F)$ is a range also common to all functions in the family. Formally, we view the family F as a function $F: Keys(F) \times Dom(F) \to Range(F)$, where the function specified by K is $F(K, \cdot)$.

The key space $Keys(F)$ has an associated probability distribution. When we want to pick a particular hash function from the family F we pick K at random from this distribution, thereby specifying $F(K, \cdot)$. The key K then becomes public, available to all parties including the adversary: these hash functions involve no hidden randomness.

In our constructions an "ideal hash function" h is also present. We follow the paradigm of [BR]: In practice, h is derived from a standard cryptographic hash function like SHA, while formally it is modeled as a "random oracle." The latter means h is initially drawn at random from some family of functions, and then made public. Parties have oracle access to h, meaning they are provided with a box which, being queried with a point x, replies with $h(x)$. This is the only way h can be accessed. We stress the oracle is public: the adversary too can access h.

Formally, h will be viewed as part of the key defining a hash function, and the random choice of a key includes the choice of h. Typically a key will have two parts, one being some short string σ and the other being h, so that formally $K = (\sigma, h)$. (For example, σ may be a prime p, to specify that we are working over Z_p^*). We treat them differently in the notation, writing F_σ^h for the function $F(K, \cdot)$. This is to indicate that although both σ and h are public, they are accessed differently: everyone has the complete string σ, but to h only oracle access is provided. It is to be understood in what follows that the families we discuss might involve a random oracle treated in this way, and when the key is chosen at random the oracle is specified too. For more information about random oracles the reader is referred to [BR].

We want hash functions that compress their data. A typical desired choice is that $Dom(F) = \{0,1\}^*$ and $Range(F)$ is some finite set, for example $\{0,1\}^k$ for some integer k. But other choices are possible too.

COLLISION-RESISTANCE. A *collision* for $F(K, \cdot)$ is a pair of strings $x, y \in Dom(F)$ such that $x \neq y$ but $F(K, x) = F(K, y)$. When $Dom(F)$ is larger than $Range(F)$, each $F(K, \cdot)$ will have many collisions. What we want, however, is that these are difficult to find. To formalize this, say a *collision-finder* is an algorithm C that given a key $K \in Keys(F)$ tries to output a collision for $F(K, \cdot)$. (When K includes a random oracle, this of course means the collision-finder gets oracle access to this same random oracle). We are interested in the probability that it is successful. This probability depends on the time t that is allowed C. (For convenience the "time" is the actual running time, on some fixed RAM model of computation, plus the size of the description of the algorithm C. In general we would also measure the amount of memory used, but for simplicity we only measure time. The model of computation is that used in any standard text on algorithms, for example [CLR], and we analyze the running time of algorithms in the same way as in any algorithms course). If a random oracle h is present, we consider the number of h-computations (formally, the number of oracle queries) as a separate resource of the collision-finder, and denote it by q. In this case we have the following.

Definition 1. We say that collision-finder C (t, q, ϵ)-breaks a hash family F if given a key K it runs in time t, makes at most q oracle queries, and finds a collision in $F(K, \cdot)$ with probability at least ϵ. We say that F is (t, q, ϵ)-collision-free if there is no collision-finder which (t, q, ϵ)-breaks F.

The probability above is over the choice of the key K from $Keys(F)$ (which includes the choice of the random oracle h) and the coins of C. If the random oracle is not present, we simply drop the "q", and have (t, ϵ)-breaking and (t, ϵ)-security.

2.2 Incrementality

We follow [BGG1]. Suppose we have computed the hash value $y = F(K, x)$ of a message $x = x_1 \ldots x_n$. Now x is modified: block i is replaced by a new block x_i'. We want to update y to $y' = F(K, x')$, where x' is the message resulting from replacing block i of x by x_i'. We want to do it in some way faster than re-computing $F(K, x')$ from scratch. The job will be done by an *incremental algorithm*. It takes as input K, x, y, i, x_i' and outputs y'. Ideally it runs in time that is independent of the number of blocks in the messages.

2.3 Classes of groups

We will consider groups in which computational problem (example, computing discrete logarithms or solving weighted knapsacks) is hard. Formally, we must treat families (classes) of groups.

CLASSES OF GROUPS. Formally, a *class of groups* is some finite collection of groups such that given a description $\langle G \rangle$ of a group from the class, one can compute all the group operations. Also, there is some distribution on $\mathcal{G}$ according to which we can draw a (description of a) group. Finally we assume a representation of group elements under which any group element of any group is a L-bit string for some L, meaning $G \subseteq \{0,1\}^L$ for all $G \in \mathcal{G}$. This L is called the output length. For example $\mathcal{G} = \{ Z_p^* : p \text{ is prime and } |p| = k \}$, for some large enough k, is a class of groups. Here $\langle G \rangle = p$ is the prime describing a particular group, and it is drawn at random from all k-bit primes. The output length is $L = k$.

TIMING. In the security analyses we need to estimate running times of the algorithms in the reductions. The timing estimates depend on the groups. Accordingly given a class of groups $\mathcal{G}$ we let $T_{\text{rand}}(\mathcal{G}), T_{\text{mult}}(\mathcal{G}), T_{\text{exp}}(\mathcal{G})$ denote, respectively, the time to pick a random element of G, the time to multiply two elements in G and the time to do an exponentiation in G, for $G \in \mathcal{G}$.

2.4 The balance problem in a group

For the purpose of analyzing the security of our hash functions we introduce a new computational problem, called the balance problem in a group. Lemma 2 will relate the security of our hash function to the assumed hardness of this problem. (Our task will then be reduced to finding groups with a hard balance problem. Typically we will do this by further reducing the balance problem to a conventional hard problem like discrete log finding or (weighted) subset sum.) Here we define the balance problem.

Let $\mathcal{G}$ be some family of groups and n an integer. In the $(\mathcal{G}, n)$-balance problem we are given (the description $\langle G \rangle$ of) a group $G \in \mathcal{G}$ and a sequence $a_1, \ldots, a_n$ of elements of G. We must find weights $w_1, \ldots, w_n \in \{-1, 0, +1\}$ not all zero, such that

$$a_1^{w_1} \odot \cdots \odot a_n^{w_n} = e$$

where $\odot$ is the group operation and e is the identity element in the group.[4] In other words we are asked to find two disjoint subsets $I, J \subseteq \{1, \ldots, n\}$, not both empty, such that $\bigodot_{i \in I} a_i = \bigodot_{j \in J} a_j$. We say that the $(\mathcal{G}, n)$-balance problem is (t, ϵ)-hard if no algorithm, limited to run in time t, can find a solution to an istance $G, a_1, \ldots, a_n$ of the problem with probability more than ϵ, the probability computed over a random choice of G from $\mathcal{G}$, a choice of $a_1, \ldots, a_n$ selected uniformly and independently at random in G, and the coins of the algorithm.

3 The Paradigm

We suggest a new paradigm for the construction of collision-free hash functions.

---

[4] For a multiplicative group, this means $\prod_{i=1}^n a_i^{w_i} = 1$. For an additive group it would mean $\sum_{i=1}^n w_i a_i = 0$.

3.1 The Construction

The construction is depicted in Figure 1. We fix a block size b and let $B = \{0,1\}^b$. Think of the input $x = x_1 \ldots x_n$ as a sequence of blocks, meaning $x_i \in B$ for each $i = 1, \ldots, n$. Let N be larger than the number of blocks in any message we plan to hash, and let $l = \lg(N) + b$. We are given a set G on which some operation, which we call the *combining operation* and denote by $\odot$, has been defined. (The operation is at the very least associative, but, as we will see later, we prefer it be a full-fledged group operation.) We are also given a function $h: \{0,1\}^l \to G$ which we call the *randomizer* or *compression function*. Now what we do is:

1. For each block $i = 1, \ldots, n$, concatenate a $\lg(N)$-bit binary encoding $\langle i \rangle$ of the block index i to the block content x_i to get an *augmented block* $x_i' = \langle i \rangle . x_i$

2. For each $i = 1, \ldots, n$, apply h to x_i' to get a hash value $y_i = h(x_i')$

3. Combine $y_1, \ldots, y_n$ via the combining operation to get the final hash value $y = y_1 \odot y_2 \odot \cdots \odot y_n$.

More succinctly we can write the function as

$$\mathrm{HASH}_{\langle G \rangle}^h(x_1 \ldots x_n) = \bigodot_{i=1}^n h(\langle i \rangle . x_i) , \tag{1}$$

where $\langle G \rangle$ denotes some indication of the group G which enables computation of the group operation. (For example if $G = Z_p^*$ then $\langle G \rangle = p$). We call this the *randomize then combine* construction.

If the output of our hash function (which is an element of G) is too long then optionally we can hash it to a shorter length by applying a standard collision-free hash function such as SHA-1.

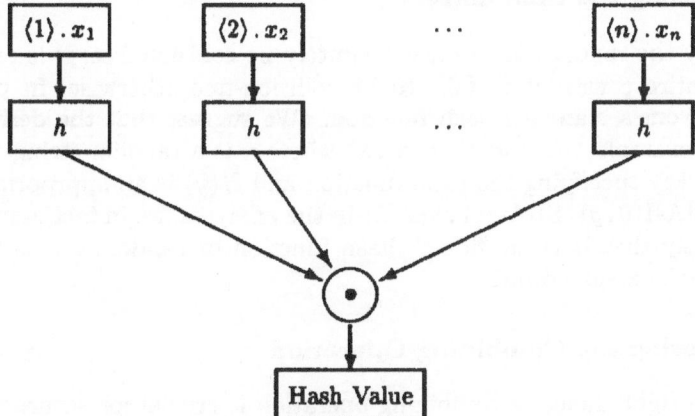

Fig. 1. Our paradigm for hashing message $x = x_1 \ldots x_n$: *Process individual blocks via a function h and then combine the results via some operation $\odot$.*

Notice that padding the blocks with (a representation of) their indexes before applying h is important for security. Without this, re-ordering of the blocks in a message would leave the hash value unchanged, leading to collisions.

THE HASH FAMILY. Equation (1) specifies an individual function, depending on the group G. Formally, we actually have a family of hash functions, because we will want to draw G from some class of groups for which some computational problem (example, computing discrete logarithms or solving weighted knapsacks) is hard.

Let $\mathcal{G}$ be a class of groups, as defined in Section 2.3. The associated family of hash functions is denoted HASH($\mathcal{G}, b$). An individual function $\text{HASH}^h_{\langle G \rangle}$ of this family, as defined in Equation (1), is specified by a random oracle h: $\{0,1\}^l \to G$ and a description $\langle G \rangle$ of a group $G \in \mathcal{G}$. Here $l = b + \lg(N)$ as above. We can set N to a constant like 2^{80}. (We will never need to hash a message with more than 2^{80} blocks!). Thus $l = b + O(1)$. So think of l as $O(b)$. This is assumed in estimates. The key defining $\text{HASH}^h_{\langle G \rangle}$ consists, formally, of $\langle G \rangle$ and h. (See Section 2.1). The domain of this hash family is $B^{\leq N} = B \cup B^2 \cup \ldots \cup B^N$ where $B = \{0,1\}^b$, namely all strings over B of length at most N. The range of this family is $\{0,1\}^L$ where L is the output length of $\mathcal{G}$.

3.2 Incrementality and parallelizability

Since the combining operation is associative, the computation is parallelizable. In order to get an incremental hash function we will work in a commutative group, so that $\odot$ is also commutative and invertible. In such a case, increments are done as follows. If block x_i changes to x'_i then the new hash is $y \odot h(\langle i \rangle . x_i)^{-1} \odot h(\langle i \rangle . x'_i)$ where $(\cdot)^{-1}$ denotes the inverse operation in the group and y is the old hash, namely the hash of x.

3.3 Choosing the randomizer

For security the randomizer h must definitely be collision-free: it is easy to see that the entire construction fails to be collision-free otherwise. In practice h is derived from a standard hash function. (We suggest that the derivation be keyed. For example, $h(x') = H(\kappa . x' . \kappa)$ where κ is a random string viewed as part of the key specifying the hash function and $H(y)$ is an apprporiate length prefix of SHA-1($0 . y$) . SHA-1($1 . y$) . . . .) In the analyses, we in fact assume much more, namely that it is an "ideal" hash function or random oracle [BR].) Its computation is assumed fast.

3.4 Choosing the Combining Operation

Making the right choice of combining operation is crucial for security and efficiency.

COMBINING BY XORING DOESN'T WORK. Ideally, we would like to hash using only "conventional" cryptography. (Ie. no number theory.) A natural thought towards this end is to set the combining operation to bitwise XOR. But this choice is insecure. Let us look at this a bit more closely.

Let $G = \{0,1\}^k$ for some fixed length k, like $k = 128$. If we set the combining operation to bitwise XOR, denoted $\oplus$, the resulting function is

$$\text{XHASH}^h(x_1 \ldots x_n) \;=\; \bigoplus_{i=1}^n h(\langle i \rangle \cdot x_i)$$

The incrementality is particularly efficient in this case since it takes just a couple of XORs. The question is whether XHASH^h is collision-free. At first glance, it may seem so. However XHASH is in fact not collision-free. Indeed, it is not even one-way. (One-wayness is necessary, but not sufficient, for collision-resistance). The attack is interesting, and may be useful in other contexts, so we present it in Appendix A. Given a string $z \in \{0,1\}^k$ we show there how to find a message $x = x_1 \ldots x_n$ such that $\text{XHASH}^h(x) = z$. (The attack succeeds with probability at least $1/2$, the probability being over the choice of h, and works for $n \geq k+1$.) The attack makes $2n$ h-computations, sets up a certain linear system, and then uses Gaussian elimination to solve it. The proof that it works exploits pairwise independence arguments.

OTHER COMBINING OPERATIONS. Thus we see that the choice of combining operation is important, and the most tempting choice, XOR, doesn't work. We are loth to abandon the paradigm based on this: it is hard to imagine any other paradigm that yields incrementality. But we conclude that it may be hard to get security using only conventional cryptography to implement the combining operation. So we turn to arithmetic operations.

We consider two: multiplication in a group where the discrete logarithm problem is hard, and addition modulo an integer of appropriate size. It turns out they work. But we need to be careful about security given the experience with XOR.

To this end, we begin below by relating the security of the hash function to the balance problem in the underlying group. A reader interested more in the constructions should skip to Section 4.

3.5 The balance lemma

The security of the hash functions obtained from our paradigm can be related to the balance problem in the underlying class of groups, as defined in Section 2.4. Specifically, in order to prove the security of a particular hash function family $\text{HASH}(\mathcal{G}, b)$, it will be sufficient to show that the balance problem associated with the corresponding group family is hard. To understand the theorem below, it may be helpful to refer to the definitions in Section 2. Recall that q refers to the number of computations of h and the theorem assumes h is ideal, ie. a random function of $\{0,1\}^l$ to G. The theorem says that if the balance problem is hard over $\mathcal{G}$ then the corresponding family of hash functions is collision-free. Moreover it tells us precisely how the parameters describing the security in the two cases relate to each other. Below $c > 1$ is a small constant, depending on the model of computation, which can be derived from the proof.

Lemma 2. *Let $\mathcal{G}$ and q be such that the $(\mathcal{G}, q)$-balance problem is (t', ϵ')-hard. Then $\mathrm{HASH}(\mathcal{G}, b)$ is a (t, q, ϵ)-collision-free family of hash functions where $\epsilon = \epsilon'$ and $t = t'/c - q \cdot b$.*

Proof. We are given a collision-finder C, which takes $\langle G \rangle$ and an oracle for h, and eventually outputs a pair of distinct strings, $x = x_1 \ldots x_n$ and $y = y_1 \ldots y_m$, such that $\mathrm{HASH}^h_{\langle G \rangle}(x) = \mathrm{HASH}^h_{\langle G \rangle}(y)$. We want to construct an algorithm K that solves the $(\mathcal{G}, q)$-balance problem. It takes as input $\langle G \rangle$ and a list of values $a_1, \ldots, a_q$ selected uniformly at random in G. K runs C on input $\langle G \rangle$, answering its oracle queries with the values $a_1, a_2, \ldots, a_q$ in order. (We assume oracle queries are not repeated.) Notice the answers to oracle queries are uniformly and independently distributed over G, as they would be if $h \colon \{0, 1\}^l \to G$ were a random function. We will let Q_i denote the i-th oracle query of C, namely the one answered by a_i, so that $h(Q_i) = a_i$, and we let $Q = \{Q_1, \ldots, Q_q\}$.

Finally, C outputs two strings $x = x_1 \ldots x_n$ and $y = y_1 \ldots y_m$, such that $x \neq y$ but $\mathrm{HASH}^h_{\langle G \rangle}(x) = \mathrm{HASH}^h_{\langle G \rangle}(y)$. We know this means

$$ h(\langle 1 \rangle . x_1) \odot \ldots \odot h(\langle n \rangle . x_n) = h(\langle 1 \rangle . y_1) \odot \ldots \odot h(\langle m \rangle . y_m) , \qquad (2) $$

the operations being in G. (Note that the strings x and y are not necessarily of the same size; that is, m may not be equal to n.) We will construct a solution to the balance problem from x and y. Let $x'_i = \langle i \rangle . x_i$ for $i = 1, \ldots, n$ and $y'_i = \langle i \rangle . y_i$ for $i = 1, \ldots, m$. We can assume wlog that $x'_1, \ldots, x'_n, y'_1, \ldots, y'_m \in Q$. We let $f_x(i)$ be the (unique) value $j \in [q]$ such that $x'_i = q_j$ and we let $f_y(i)$ be the (unique) $j \in [q]$ such that $y'_i = q_j$. We then let $I = \{ f_x(i) : i = 1, \ldots, n \}$ and $J = \{ f_y(i) : i = 1, \ldots, m \}$ be, respectively, the indices of queries corresponding to x and y. Equation (2) can be rewritten as

$$ \bigodot_{i \in I} a_i = \bigodot_{j \in J} a_j . \qquad (3) $$

We know that $x \neq y$, and so $I \neq J$. Now for $i = 1, \ldots, q$ let us define

$$ w_i = \begin{cases} -1 & \text{if } i \in J - I \\ 0 & \text{if } i \in I \cap J \\ +1 & \text{if } i \in I - J. \end{cases} $$

Then the fact that $I \neq J$ means that not all $w_1, \ldots, w_q$ are 0, and Equation (3) implies $a_1^{w_1} \odot \cdots \odot a_q^{w_q} = e$. The probability that we find a solution to the balance problem is exactly that with which C outputs a collision, and the time estimates can be checked. ∎

4 MuHASH: The Multiplicative Hash

Here we present our first concrete construction, the multiplicative hash function (MuHASH), and analyze its efficiency and security.

4.1 Construction and efficiency

We set the combining operation in our paradigm to multiplication in a group where the discrete logarithm problem is hard. (For example $G = Z_p^*$ for an appropriate prime p, or some subgroup theoreof.) To emphasize multiplication, we call the function MuHASH rather than the general HASH of Section 3. So the function is

$$\text{MuHASH}_{\langle G \rangle}^h(x_1 \ldots x_n) = \prod_{i=1}^{n} h(\langle i \rangle \cdot x_i) \, . \tag{4}$$

The product is taken in the group G over which we are working. (Thus if we are working in Z_p^*, it is just multiplication modulo p. In this case $\langle G \rangle = p$ describes G.) Here all the notation and conventions are as in Section 3.1. A class of groups $\mathcal{G}$ gives rise to a family MuHASH$(\mathcal{G}, b)$ of hash functions as described in Section 2.3.

If $G = Z_p^*$ then for security $k = |p|$ should be at least 512 or even 1024, making the final hash value of the same length. A hash of this size may be directly useful, for example for signatures, where the message is hashed before signing. (For RSA we want a string in Z_N^* where N is the modulus, and this may be 1024 bits.) In other cases, we may want a smaller hash value, say 160 bits. In such cases, we allow a final application of a standard collision-free hash function to the above output. For example, apply SHA-1 to MuHASH$_{\langle G \rangle}^h(x)$ and get a 160 bit string.

Computing our hash function takes one multiplication per block, ie. one multiplication per b bits of input. (This is in contrast to previous methods which required one multiplication per bit.) To minimize the cost, one can increase the block size. The increment operation is performed as per Section 3.2, and takes one inverse and two multiplication operations in the group, plus two applications of h. Thus it is cheap compared to re-computing the hash function.

Note that the computation of MuHASH$_{\langle G \rangle}^h$ is entirely parallelizable. The applications of h on the augmented blocks can be done in parallel, and the multiplications can also be done in parallel, for example via a tree. This is useful when we have hardware for the group operation, as well might be the case.

4.2 The discrete logarithm problem

The security of MuHASH depends on the discrete logarithm problem in the underlying group. Let us begin by defining it.

Let $\mathcal{G}$ be a class of groups, for example $\mathcal{G} = \{ Z_p^* : p \text{ is a prime with } |p| = k \}$. Let $G \in \mathcal{G}$, g a generator of G, and $y \in G$. A *discrete log finder* is an algorithm I that takes $g, y, \langle G \rangle$ and tries to output $\log_g(y)$. Its success probability is taken over a random choice of G from $\mathcal{G}$ (for the example $\mathcal{G}$ above, this means we choose a random k-bit prime p) and a random choice of $y \in G$. We say that the discrete logarithm problem in $\mathcal{G}$ is (t', ϵ')-hard if any discrete logarithm finder that runs in time t' has success probability at most ϵ'.

4.3 Security of MuHASH

The attack on XHASH we saw above indicates that we should be careful about security. Moving from XOR to multiplication as the "combining" operation kills that attack in the case of MuHASH. Are there other attacks?

We indicate there are not in a very strong way. We show that as long as the discrete logarithm problem in G is hard and h is ideal, MuHASH$^h_{\langle G \rangle}$ is collision-free. That is, we show that if there is *any* attack that finds collisions in MuHASH$^h_{\langle G \rangle}$ then there is a way to efficiently compute discrete logarithms in G. This proven security obviates us from the need to consider the effects of any specific attacks.

At first glance this relation of the security of MuHASH to the discrete logarithm problem in G may seem surprising. Indeed, the description of MuHASH$^h_{\langle G \rangle}$ makes no mention of a generator g, nor is there even any exponentiation: we are just multiplying group elements. Our proofs illustrate how the relationship is made.

We look first at general groups, then, to get better quantitative results (ie. better reductions) we look at special classes of groups.

APPROACH. All our proofs have the same structure. First it is shown that if the discrete log problem is hard in G then also the balance problem is hard in G. The security of the hash function is then derived from Lemma 2. The main technical question is thus relating the balance and discrete logarithm problems in groups.

Notice this is a question just about computational problems in groups: it has nothing to do with our hash functions. Accordingly, we have separated the materiel on this subject, putting it in Appendix B. There we prove a sequence of lemmas, showing how the quality of the reduction changes with the group. These lemmas could be of independent interest. We now proceed to apply these lemmas to derive the security of MuHASH for various groups.

SECURITY IN GENERAL GROUPS. The following theorem says that the only way to find collisions in MuHASH (assuming h is ideal) is to solve the discrete logarithm problem in the underlying group. The result holds for any class of groups with hard discrete logarithm problem. Refer to Sections 4.1, 4.2 and 2.3 for notation. Below $c > 1$ is a small constant, depending on the model of computation, which can be derived from the proof.

Theorem 3. *Let G be a class of groups with output length L. Assume the discrete logarithm problem in G is (t', ϵ')-hard. Then for any q, MuHASH(G, b) is a (t, q, ϵ)-collision-free family of hash functions, where $\epsilon = q\epsilon'$ and $t = t'/c - q \cdot [T_{rand}(G) + T_{exp}(G) + L + b]$.*

Proof. Follows from Lemma 2 and Lemma 9. ∎

In the above reduction, if the probability one can compute discrete logarithms is ϵ' then the probability of breaking the hash function may be as high as $\epsilon = q\epsilon'$. A typical choice of q is about 2^{50}. This means the discrete logarithm problem in G must be very hard in order to make finding collisions in the hash function quite hard. To make ϵ appreciably small, we must make ϵ' very small, meaning we must use a larger value of the security parameter, meaning it takes longer to do multiplications and the hash function is less efficient. It is preferable to have a stronger reduction in which ϵ is closer to ϵ'. (And we want to do this while maintaining the running time t' of the discrete logarithm finder to be within an additive amount of the running time t of the collision-finder, as it is above. Reducing the error by repetition does not solve our problem.)

We now present better reductions. They exploit the group structure to some extent. We look first at groups of prime order (where we have an essentially optimal redution), then at multiplicative groups modulo a prime (where we do a little worse, but still very well, and much better than the naive reduction above).

SECURITY IN GROUPS OF PRIME ORDER. The recommended group G in which to implement MuHASH$_{(G)}^h$ is a group of prime order. (For example, pick a large prime p of the form $p = 2p' + 1$ where p' is also prime, and let G be a subgroup of order p' in Z_p^*. The order of Z_p^* is $p - 1$ which is not prime, but the order of G is p' which is prime.) The reason is that the reduction is tight here. As usual $c > 1$ is a small constant, depending on the model of computation, which can be derived from the proof.

Theorem 4. *Let $\mathcal{G}$ be a class of groups of prime order with output length L. Assume the discrete logarithm problem in $\mathcal{G}$ is (t', ϵ')-hard. Then for any q, MuHASH$(\mathcal{G}, b)$ is a (t, q, ϵ)-collision-free family of hash functions, where $\epsilon = 2\epsilon'$ and $t = t'/c - q \cdot [T_{\mathrm{rand}}(\mathcal{G}) + T_{\mathrm{mult}}(\mathcal{G}) + T_{\exp}(\mathcal{G}) + L + b] - L^2$.*

Proof. Follows from Lemma 2 and Lemma 10. ∎

The form of the theorem statement here is the same as in Theorem 3, but this time the probability ϵ of breaking the hash function is no more than twice the probability ϵ' of computing discrete logarithms, for an attacker who runs in time which is comparable in the two cases.

SECURITY IN Z_p^*. The most popular group in which to work is probably Z_p^* for a prime p. Since its order is $p - 1$ which is not prime, the above theorem does not apply. What we can show is that an analogous statement holds. The probability ϵ of breaking the hash function may now be a little more than the probability ϵ' of computing discrete logarithms, but only by a small factor which is logarithmic in the size k of the prime p. As usual $c > 1$ is a small constant, depending on the model of computation, which can be derived from the proof.

Theorem 5. *Let $k \geq 6$ and let $\mathcal{G} = \{ Z_p^* : p$ is a prime with $|p| = k \}$. Suppose the discrete logarithm problem in $\mathcal{G}$ is (t', ϵ')-hard. Then for any q, MuHASH$(\mathcal{G}, b)$ is a (t, q, ϵ)-collision-free family of hash functions, where $\epsilon = 4\ln(0.694k) \cdot \epsilon'$ and $t = t'/c - qk^3 - qb$.*

Proof. Follows from Lemma 2 and Lemma 11. ∎

The factor multiplying ϵ' will not be too large: for example if $k = 512$ it is about 24.

SECURITY IN PRACTICE. We have shown that computation of discrete logarithms is necessary to break MuHASH as long as h is ideal. Yet, it could be that MuHASH is even stronger. The reason is that even computation of discrete logarithms does not seem *sufficient* to find collisions in MuHASH. That is, we suspect that finding collisions in MuHASH$^h_{(G)}$ remains hard even if we can compute discrete logarithms. In particular, we know of no attacks that find collisions in MuHASH even if discrete logarithm computation is easy. In this light it may be worth noting that the natural attempt at a discrete logarithm computation based attack is to try to "reduce" the problem to finding additive collisions in the exponents and then apply the techniques of Section A. But this does not work. The underlying problem is a kind of knapsack problem which is probably hard. In fact this suggests that the hash function obtained by setting the combining operation in our paradigm to addition might be already collision-free. This function and its security are discussed in Section 5.

Some evidence that breaking MuHASH is harder than computing discrete logarithms comes from the results of [IN2] who indicate that multiplication in G is a one-way hash as long as *any* homomorphism with image G is hard. We can extend their proofs, with added conditions, to our setting. This indicates that unless all such homomorphisms are invertible via discrete logarithm computation, MuHASH will be collision-free.

Also, although the proofs make very strong assumptions about the function h, it would appear that in practice, the main thing is that h is collision-free. In particular if h is set to SHA-1 then given the modular arithmetic being done on top of the h applications, it is hard to see how to attack the function.

5 AdHASH: Hashing by Adding

AdHASH is the function obtained by setting the combining operation in our paradigm to addition modulo a sufficiently large integer. Let us give the definition more precisely and then go on to look at security.

5.1 Construction and Efficiency

We let M be a k-bit integer. As usual let $x = x_1 \ldots x_n$ be the data to be hashed, let b denote the block size, let N be such that all messages we will hash have length at most N and let $l = b + \lg(N)$. We let $h\colon \{0,1\}^l \to Z_M$ be a hash function, assumed ideal. The function is–

$$\text{AdHASH}^h_M(x_1 \ldots x_n) = \sum_{i=1}^n h(\langle i\rangle . x_i) \bmod M .$$

Thus, the "key" of the function is the integer M. We let AdHASH(k, b) denote the corresponding family, consisting of the functions AdHASH^h_M as M ranges over all k-bit integers and h ranges over all functions of $\{0,1\}^l$ to Z_M. The distribution on the key space is uniform, meaning we draw M at random from all k-bit integers and h at random from all functions of $\{0,1\}^l$ to Z_M, in order to define a particular hash function from the family.

AdHASH is much faster than MuHASH since we are only adding, not multiplying. Furthermore, it would seem k can be quite small, like a few hundred, as compared to the sizes we need for MuHASH to make sure the discrete logarithm problem is hard, making the gain in efficiency even greater. In fact the speed of AdHASH starts approaching that of standard hash functions. And of course it is incremental, with the cost of incrementality also now reduced to just adding and subtracting. Thus it is a very tempting function to use. Next we look at security.

5.2 The Weighted Subset Sum Problem

The security of AdHASH can be related to the difficulty of a certain modular subset-sum or knapsack type problems which we now define.

WEIGHTED KNAPSACK PROBLEM. In the (k, q)-weighted-knapsack problem we are given a k-bit integer M, and q numbers $a_1, \ldots, a_q \in Z_M$. We must find weights $w_1, \ldots, w_q \in \{-1, 0, +1\}$, not all zero, such that

$$\sum_{i=1}^{q} w_i a_i \equiv 0 \pmod{M}$$

We say that the (k, q)-weighted-knapsack problem is (t', ϵ')-hard if no algorithm, limited to run in time t', can find a solution to an instance $M, a_1, \ldots, a_q$ of the (k, q)-weighted-knapsack problem with probability more than ϵ', the probability computed over a random choice of k-bit integer M, a choice of $a_1, \ldots, a_q$ selected uniformly and independently at random in Z_M, and the coins of the algorithm.

Notice this is just the $(\mathcal{G}, q)$-balance problem for the class of groups $\mathcal{G} = \{ Z_M : |M| = k \}$. But it is worth re-stating it for this case.

If we did not allow weights -1, and additionally asked that rather than be 0 the sum must hit some given target T, we would have the subset sum problem as used in [IN1, IN2].

We must be careful how we choose the parameters: it is well known that for certain values of k and q, even the standard problem is not hard. Specifically, make sure that $\Omega(\log q) < k < q$. It turns out this choice will not be a restriction for us anyway. Nice discussions of what is known are available in [Od] and [IN2, Section 1.2].

The hardness of the weighted problem is a stronger assumption than the hardness of the standard problem, but beyond that the relation between the problems is not known. However, there is important evidence about the hardness of the weighted knapsack problems that we discuss next.

RELATION TO LATTICE PROBLEMS. A well-known hard problem is to approximate the length of the shortest vector in a lattice. The best known polynomial time algorithms [LLL, SH] achieve only an exponential approximation factor. It has been suggested that there is no polynomial time algorithm which achieves a polynomial approximation factor. Under this assumption, Ajtai showed that both the standard and the weighted subset-sum problems are hard [Aj]. (Actually he allows any small integer weights, not just $-1, 0, +1$ like we do). That is, there is no polynomial time algorithm to solve these problems.

This is important evidence in favor of both the knapsack assumptions discussed above. As long as approximating the length of a shortest lattice vector is hard, even in the worst case, the knapsack problems are hard. This increases the confidence we can have in cryptosystems based on these knapsack assumptions.

Values of t', ϵ' for which the standard and weighted knapsack problems are (t', ϵ')-hard can be derived from Ajtai's proof, as a function of the concrete parameters for which one assumes shortest vector length approximation is hard. Since Ajtai's proof is quite complex we do not know exactly what the relation is.

We note however that even more is true. Even if the assumption about lattices fails (meaning an efficient approximation algorithm for the shortest lattice vector problem emerges), the knapsack problems may still be hard. Thus, we present all our results in terms of the knapsack assumptions.

5.3   Security of AdHASH

We relate the collision-freeness of AdHASH to the weighted knapsack problem. Below $c > 1$ is a small constant, depending on the model of computation, which can be derived from the proof.

Theorem 6. *Let k and q be integers such that the (k, q)-weighted-knapsack problem is (t', ϵ')-hard. Then* AdHASH(k, b) *is a (t, q, ϵ)-collision-free family of hash functions where $\epsilon = \epsilon'$ and $t = t'/c - qM$.*

Proof. Follows from Lemma 2 and the observation that weighted knapsack is a particular case of the balance problem, as mentioned in Section 5.2. ∎

6   Incremental Hashing via Lattice Problems

Ajtai introduced a function which he showed was one-way if the problem of approximating the shortest vector in a lattice to polynomial factors is hard [Aj]. Goldreich, Goldwasser and Halevi observed that Ajtai's main lemma could be applied to show that the same function is in fact collision-free [GGH]. Here we observe this hash function is incremental, and consider its practicality. We then use our paradigm to derive a more practical version of this function whose security is based on the same assumption as in [Aj, GGH] plus the assumption that our h is ideal. Let us begin by recalling the problem shown hard by Ajtai's main lemma.

6.1   The Matrix Kernel Problem

In the (k, n, s)-matrix-kernel problem we are given p, M where p is an s-bit integer and M is a k by n matrix with entries in Z_p. We must find a non-zero n-vector w with entries in $\{-1, 0, +1\}$ such that $Mw = 0 \bmod p$. (The operation here is matrix-vector multiplication, with the operations done modulo p). We say this problem is (t', ϵ')-hard if no algorithm, limited to run in time t', can find a solution to an instance p, M of the (k, n, s)-matrix-kernel problem with probability more than ϵ', the probability computed over a random choice of p, a random choice of matrix M, and the coins of the algorithm.

Suppose $ks < n < 2^s/(2k^4)$. Ajtai showed that with these parameters the matrix-kernel problem is hard under the assumption that there is no polynomial time algorithm to approximate the length of a shortest vector in a lattice within a polynomial factor. (Ajtai's result was actually stronger, since he allowed entries in w to be any integers of "small" absolute value. However [GGH] observed that weights of $-1, 0, +1$ are what is important in the context of hashing and we restrict our attention to these).

A close examination of Ajtai's proof will reveal specific values of t', ϵ' for which we can assume the matrix kernel problem is (t', ϵ')-hard, as a function of the assumed hardness of the shortest vector approximation problem. Since the proof is quite complex we don't know what exactly these values are.

Notice that the matrix kernel problem is just an instance of our general balance problem: it is the $(\mathcal{G}, n)$-balance problem for $\mathcal{G} = \{ Z_p^k : |p| = s \}$. This shows how the balance problem unifies so many hash functions.

6.2 The Ajtai-GGH Function

THE FUNCTION. Let M be a random k by n matrix with entries in Z_p and let x be an n vector with entries in $\{0, 1\}$. The function of [Aj, GGH] is–

$$H_{M,p}(x) = Mx \bmod p.$$

Note $Mx \bmod p$ is a k-vector over Z_p, meaning it is $k \lg(p)$ bits long. Since the parameters must obey the restriction $k \lg(p) < n < p/(2k^4)$, the function is compressing: the length n of the input x is more than the length $k \log(p)$ of the output $Mx \bmod p$. Thus it is a hash function. Now, if the matrix kernel problem is hard this function is one-way [Aj]. Moreover, under the same assumption it is collision-free [GGH]. It follows from [Aj] that the function is collision-free as long as shortest vector approximation is hard.

INCREMENTALITY. We observe the above function is incremental. Let M_i denote the i-th column of M, for $i = 1, \ldots, n$. This is a k-vector over Z_p. Let $x = x_1 \ldots x_n$ with $x_i \in \{0, 1\}$ for $i = 1, \ldots, n$. Now we can write the function as–

$$H_{M,p}(x) = \sum_{i=1}^{n} x_i M_i \bmod p.$$

In other words, we are summing a subset of the columns, namely those corresponding to bits of x that are 1. Now suppose bit x_i changes to x_i'. If y (a k-vector over Z_p) is the old hash value then the new hash value is $y + (x_i' - x_i)M_i \bmod p$. Computing this takes k additions modulo p, or $O(k \log(p))$ time, a time which does not depend on the length n of x.

DRAWBACKS OF THIS FUNCTION. A serious drawback of H is that the description of the function is very large: $(nk + 1) \lg(p)$ bits. In particular, the description size of the function grows with the number of bits to be hashed. This means we must set an a priori limit on the number of bits to be hashed and use a function of size proportional to this. This is not feasible in practice.

One way to partially overcome this problem is to specify the matrix entries via an ideal hash function. For example if $h: [k] \times [n] \to Z_p$ is such a function, set $M[i,j] = h(i,j)$. But we can do better. The function we describe next not only has small key size and no limit on input length, but is also more efficient.[5]

6.3 LtHASH

Our function is called LtHASH for "lattice based hash."

THE CONSTRUCTION. We apply the randomize-then-combine paradigm with the group G set to Z_p^k. That is, as usual let $x = x_1 \ldots x_n$ be the data to be hashed, let b denote the block size, let N be such that all messages we will hash have length at most N and let $l = b + \lg(N)$. We let $h: \{0,1\}^l \to Z_p^k$ be a hash function, assumed ideal. Think of its output as a k-entry column vector over Z_p. Our hash function is—

$$\text{LtHASH}_p^h(x_1 \ldots x_n) = \sum_{i=1}^n h(\langle i \rangle . x_i) \bmod p .$$

Namely, each application of h yields a column vector, and these are added, componentwise modulo p, to get a final column vector which is the hash value.

Notice that there is no longer any matrix M in the function description. This is why the key size is small: the key is just the s-bit integer p. Also LtHASH_p^h is more efficient than the function described above because it does one vector addition per b-bit input block rather than per input bit, and b can be made large.

We let $\text{LtHASH}(k,s,b)$ denote the corresponding family, consisting of the functions LtHASH_p^h as p ranges over s-bit integers and h ranges over all functions of $\{0,1\}^l$ to Z_p^k. The key defining any particular function is the integer p, and the distribution on the key space is uniform, meaning we draw p at random from all s-bit integers in order to define a particular hash function from the family.

Notice that AdHASH is the special case of LtHASH in which $k = 1$ and $p = M$.

SECURITY. We relate the collision-freeness of LtHASH to the hardness of the matrix-kernel problem. The relation may not be evident a priori because LtHASH does not explicitly involve any matrix. But, intuitively, there is an "implicit" k by q matrix M being defined, where q is the number of oracle queries allowed to the collision-finder. This matrix is not "fixed:" it depends on the input. But finding collisions in LtHASH_p^h relates to solving the matrix kernel problem for this matrix. Below $c > 1$ is a small constant, depending on the model of computation, which can be derived from the proof.

---

[5] Another way to reduce the key size is define $H_{M,P}$ only on relatively short data, and then, viewing it as a compression function, apply Damgård's iteration method [Da2]. But then incrementality is lost. Also, the key sizes, although no longer proportional to the data length, are still larger than for the construction we will describe.

Theorem 7. *Let k, q, s be integers such that the (k, q, s)-matrix-kernel problem is (t', ϵ')-hard. Then $\mathrm{LtHASH}(k, s, b)$ is a (t, q, ϵ)-collision-free family of hash functions where $\epsilon = \epsilon'$ and $t = t'/c - qks$.*

Proof. Follows from Lemma 2 and the observation, made in Section 6.1, that the matrix kernel problem is a particular case of the balance problem when the group is Z_p^k. ∎

We will choose the parameters so that $ks < q < 2^s/(2k^4)$. (Recall $s = |p|$). In this case, we know that the required matrix kernel problem is hard as long as shortest lattice vector approximation is hard.

To actually implement the function we must have some idea of what values to assign to the various security parameters. Opinions as to the concrete complexity of the shortest lattice vector approximation problem vary across the community: it is not clear how high must be the dimension of the lattice to get a specific desired security level. (Although the best known algorithm for shortest vector approximation is only proven to achieve an exponential factor [LLL], its in practice performance is often much better. And Schnorr and Hörner [SH] have found heuristics that do better still). In particular, it does not seem clear how big we need take k (which corresponds to the dimension of the lattice) before we can be sure of security. One must also take into account the exact security of the reductions, which are far from tight. (Some discussion is in [GGH, Section 3]).

Keeping all this in mind let us look at our case. It seems safe to set $k = 500$. (Less will probably suffice). We want to allow q, the number of oracle queries, to be quite large, say $q = 2^{70}$. To ensure $q < 2^s/(2k^4)$ we must take s about 110. Namely p is 110 bits long. This is longer than what the function of [Aj, GGH] needs, making operations modulo p slower for LtHASH, but this is compensated for by having much fewer such operations to do, since we can make the block size b large.

Of course LtHASH is still incremental. Incrementing takes one addition and one subtraction over Z_p^k.

COMPARISON WITH OUR OTHER PROPOSALS. LtHASH is very similar to Ad-HASH. In fact it is just AdHASH implemented over a different domain, and the security can be proven based on the same underlying problem of hardness of shortest lattice vector approximation. Notice also that AdHASH can be considered a special case of LtHASH, namely, the case $k = 1$. However the proof of security of LtHASH does not immediatly carry over to AdHASH because the shortest lattice vector problem in dimension $k = 1$ is easily solved by the Euclidean algorithm. So, the concrete security of LtHASH might be better because the relation to shortest lattice vector approximation is more direct.

Comparison with MuHASH is difficult, depending much on how parameters are set in both functions, but AdHASH and LtHASH are likely to be more efficient, especially because we can make the block size b large.

Acknowledgments

We thank Russell Impagliazzo for telling us about the relations between subset-sum and lattices, and for bringing [IN2] to our attention. We thank the (anonymous) referees of Eurocrypt 97 for comments which improved the presentation of this paper.

Mihir Bellare is supported in part by NSF CAREER Award CCR-9624439 and a Packard Foundation Fellowship in Science and Engineering. Daniele Micciancio is supported in part by DARPA contract DABT63-96-C-0018.

References

[Aj]   M. AJTAI, "Generating hard instances of lattice problems," *Proceedings of the* 28th *Annual Symposium on Theory of Computing*, ACM, 1996.

[BGG1]   M. BELLARE, O. GOLDREICH AND S. GOLDWASSER, "Incremental cryptography: The case of hashing and signing," *Advances in Cryptology – Crypto 94 Proceedings*, Lecture Notes in Computer Science Vol. 839, Y. Desmedt ed., Springer-Verlag, 1994.

[BGG2]   M. BELLARE, O. GOLDREICH AND S. GOLDWASSER, "Incremental cryptography with application to virus protection," *Proceedings of the* 27th *Annual Symposium on Theory of Computing*, ACM, 1995.

[BM]   M. BELLARE AND D. MICCIANCIO, "A new paradigm for collision-free hashing: Incrementality at reduced cost," full version of this paper, available at http://www-cse.ucsd.edu/users/mihir.

[BGR]   M. BELLARE, R. GUÉRIN AND P. ROGAWAY, "XOR MACs: New methods for message authentication using finite pseudorandom functions," *Advances in Cryptology – Crypto 95 Proceedings*, Lecture Notes in Computer Science Vol. 963, D. Coppersmith ed., Springer-Verlag, 1995.

[BR]   M. BELLARE AND P. ROGAWAY, "Random oracles are practical: A paradigm for designing efficient protocols," *Proceedings of the First Annual Conference on Computer and Communications Security*, ACM, 1993.

[Co]   D. COPPERSMITH, "Two Broken Hash Functions," IBM Research Report RC-18397, IBM Research Center, Yorktown Heights, NY, October 1992.

[CP]   D. COPPERSMITH AND B. PRENEEL, "Comments on MASH-1 and MASH-1," Manuscript, February 1995.

[CHP]   D. CHAUM, E. HEIJST AND B. PFITZMANN, "Cryptographically strong undeniable signatures, unconditionally secure for the signer," *Advances in Cryptology – Crypto 91 Proceedings*, Lecture Notes in Computer Science Vol. 576, J. Feigenbaum ed., Springer-Verlag, 1991.

[CLR]   T. CORMEN, C. LEISERSON AND R. RIVEST, "Introduction to Algorithms," McGraw-Hill, 1992.

[Da1]   I. DAMGARD "Collision Free Hash Functions and Public Key Signature Schemes," *Advances in Cryptology – Eurocrypt 87 Proceedings*, Lecture Notes in Computer Science Vol. 304, D. Chaum ed., Springer-Verlag, 1987.

[Da2]   I. DAMGARD "A Design Principle for Hash Functions," *Advances in Cryptology – Crypto 89 Proceedings*, Lecture Notes in Computer Science Vol. 435, G. Brassard ed., Springer-Verlag, 1989.

[DBP]   H. DOBBERTIN, A. BOSSELAERS AND B. PRENEEL, "RIPEMD-160: A strengthened version of RIPEMD," *Fast Software Encryption*, Lecture Notes in Computer Science 1039, D. Gollmann, ed., Springer-Verlag, 1996.

[Gi]   M. GIRAULT, "Hash functions using modulo-N operations," *Advances in Cryptology - Eurocrypt 87 Proceedings*, Lecture Notes in Computer Science Vol. 304, D. Chaum ed., Springer-Verlag, 1987.

[GGH]   O. GOLDREICH, S. GOLDWASSER AND S. HALEVI, "Collision-Free Hashing from Lattice Problems," *Theory of Cryptography Library* (http://theory.lcs.mit.edu/~tcryptol/) 96-09, July 1996.

[GMR]   S. GOLDWASSER, S. MICALI AND R. RIVEST, "A digital signature scheme secure against adaptive chosen-message attacks," *SIAM Journal of Computing*, Vol. 17, No. 2, pp. 281–308, April 1988.

[IN1]   R. IMPAGLIAZZO AND M. NAOR, "Efficient cryptographic schemes provably as secure as subset sum," *Proceedings of the 30th Symposium on Foundations of Computer Science*, IEEE, 1989.

[IN2]   R. IMPAGLIAZZO AND M. NAOR, "Efficient cryptographic schemes provably as secure as subset sum," *Journal of Cryptology*, Vol. 9, No. 4, Autumn 1996.

[LLL]   A. LENSTRA, H. LENSTRA AND L. LOVÁSZ, "Factoring polynomials with rational coefficients," *Mathematische Annalen* Vol. 261, pp. 515–534, 1982.

[MVV]   A. MENEZES, P. VAN OORSCHOT AND S. VANSTONE, "Handbook of Applied Cryptography," CRC Press, 1996.

[Me]   R. MERKLE "One Way Hash Functions and DES," *Advances in Cryptology - Crypto 89 Proceedings*, Lecture Notes in Computer Science Vol. 435, G. Brassard ed., Springer-Verlag, 1989.

[Mi]   D. MICCIANCIO, "Oblivious data structures: applications to cryptography," *Proceedings of the 29th Annual Symposium on Theory of Computing*, ACM, 1997.

[NY]   M. NAOR AND M. YUNG, "Universal one-way hash functions and their cryptographic applications," *Proceedings of the 21st Annual Symposium on Theory of Computing*, ACM, 1989.

[Od]   A. ODLYZKO, "The rise and fall of knapsack cryptosystems," Advances in computational number theory, C. Pomerance ed., *Proc. Symp. Applied Math* No. 42, pp. 75–88, AMS, 1990.

[PGV]   B. PRENEEL, R. GOVAERTS AND J. VANDEWALLE, "Hash functions based on block ciphers: a synthetic approach," *Advances in Cryptology - Crypto 93 Proceedings*, Lecture Notes in Computer Science Vol. 773, D. Stinson ed., Springer-Verlag, 1993.

[Ri]   R. RIVEST, "The MD5 Message-Digest Algorithm," IETF RFC 1321, April 1992.

[RS]   J. ROSSER AND L. SCHOENFELD, "Approximate formulas for some functions of prime numbers," *Illinois Journal of Math* Vol. 6, 1962.

[SH]   C. SCHNORR AND H. HÖRNER, "Attacking the Chor-Rivest cryptosystem with improved lattice reduction," *Advances in Cryptology - Eurocrypt 95 Proceedings*, Lecture Notes in Computer Science Vol. 921, L. Guillou and J. Quisquater ed., Springer-Verlag, 1995.

[SHA]   FIPS 180-1. "Secure Hash Standard," Federal Information Processing Standard (FIPS), Publication 180-1, National Institute of Standards and Technology, US Department of Commerce, Washington D.C., April 1995.

A   Attack on XHASH

In Section 3 we presented XHASH as a plausible candidate for a incremental collision-free hash function but indicated that it was in fact insecure. Here we present the attack showing this. Recall that the function is $\text{XHASH}^h(x_1 \ldots x_n) = h(\langle 1 \rangle . x_1) \oplus \cdots \oplus h(\langle n \rangle . x_n)$. Here each x_i is a b-bit block, and $l = b + \lg(N)$ is large enough to accommodate the block plus an encoding of its index, by dint of making N larger than the number of blocks in any message to be hashed. Our assumption is that $h \colon \{0,1\}^l \to \{0,1\}^k$ is ideal, ie. a random function of $\{0,1\}^l$ to $\{0,1\}^k$.

Our claim is that there is an attack that easily finds collisions in XHASH^h. We will in fact show something stronger, namely that XHASH^h is not even a one-way function. Given any k bit string z, we can efficiently compute a string x such that $\text{XHASH}^h(x) = z$. (To see that this means XHASH^h is not collision-free, let $z = \text{XHASH}^h(y)$ for some random y and then apply the algorithm to produce x. With high probability $x \neq y$ so we have a collision).

We reduce the problem to solving linear equations. See [Co] for other attacks that exploit linear equations.

THE ATTACK. Given $z \in \{0,1\}^k$ we now show how to find x so that $\text{XHASH}^h(x) = z$. Fix two messages $x^0 = x_1^0 \ldots x_n^0$ and $x^1 = x_1^1 \ldots x_n^1$ with the property that $x_i^0 \neq x_i^1$ for all $i = 1, \ldots, n$. (We will see later how to set n. In fact $n = k+1$ will suffice.) For any n-bit string $y = y[1] \ldots y[n]$ we let $x^y = x_1^{y[1]} \ldots x_n^{y[n]}$. We claim that we can find a y such that $\text{XHASH}^h(x^y) = z$. Let us first say how to find such a y, then see why the method works.

We compute the $2n$ values $\alpha_i^j = h(\langle i \rangle . x_i^j)$ for $j = 0, 1$ and $i = 1, \ldots, n$. We want to find $y[1], \ldots, y[n] \in \text{GF}(2)$ such that

$$\alpha_1^{y[1]} \oplus \alpha_2^{y[2]} \ldots \oplus \ldots \alpha_n^{y[n]} = z .$$

Let us now regard $y[1], \ldots, y[n]$ as variables. We want to solve the equation

$$\bigoplus_{i=1}^n \alpha_i^0 y[i] \oplus \alpha_i^1 (1 - y[i]) = z .$$

To solve this, we turn it into a system of equations over $\text{GF}(2)$. We first introduce new variables $\bar{y}[1], \ldots, \bar{y}[n]$. We will force $\bar{y}[i] = 1 - y[i]$. Then we turn the above into k equations, one for each bit. The resulting system is:

$$
\begin{aligned}
y[i] \oplus \bar{y}[i] &= 1 & (i = 1, \ldots, n) \\
\bigoplus_{i=1}^n \alpha_i^0[j] y[i] \oplus \alpha_i^1[j] \bar{y}[i] &= z[j] & (j = 1, \ldots, k)
\end{aligned}
$$

Here we have $n + k$ equations in $2n$ unknowns, over the field $\text{GF}(2)$. Below we show that if $n = k+1$ then there exists a solution with probability $1/2$. We now set $n = k+1$ and solve the set of equations, for example via Gaussian elimination, to get values for $y[1], \ldots, y[n] \in \text{GF}(2)$. (The system is slightly under-determined in that there are $n + k = 2k + 1$ equations in $2n = 2k + 2$ unknowns. It can be solved by setting one unknown arbitrarily.) This completes the description of the attack. Now we have to see why it works.

ANALYSIS. There are two main claims. The first is that a solution y to the above does exist (with reasonable probability as long as n is sufficiently large). The second is that given that some y exists, the algorithm finds such a y. The latter is clear from the procedure, so we concentrate on the first. The following lemma implies that with $n = k+1$ a solution exists with probability at least one-half.

Lemma 8. *Fix* $z \in \{0,1\}^k$. *Fix two messages* $x^0 = x_1^0 \dots x_n^0$ *and* $x^1 = x_1^1 \dots x_n^1$ *with the property that* $x_i^0 \neq x_i^1$ *for all* $i = 1, \dots, n$. *For any n-bit string* $y = y[1] \dots y[n]$ *let* $x^y = x_1^{y[1]} \dots x_n^{y[n]}$. *Then*

$$\Pr[\exists y \in \{0,1\}^n : \mathrm{XHASH}^h(x^y) = z] \geq 1 - \frac{2^k}{2^n}.$$

The probability here is over a random choice of h from the set of all functions mapping $\{0,1\}^l \to \{0,1\}^k$.

Proof. See [BM]. ∎

B The balance problem and discrete logs

In this section we show how the intractability of the discrete logarithm in a group implies the intractability of the balance problem in the same group. These are the technical lemmas underlying the theorems on the security of MuHASH presented in Section 4.3.

We stress that the question here is purely about computational problems in groups, having nothing to do with our hash functions. We first prove a very general, but quantitatively weak result for arbitrary groups. Then we prove strong results for groups of prime order and the group of integers modulo a prime. Refer to Section 2.4 for a definition of the balance problem and Section 4.2 for a definition of the discrete logarithm problem.

GENERAL GROUPS. The following says that if computing discrete logs in some class of groups is hard, then so is the balance problem. As usual $c > 1$ is a small constant, depending on the model of computation, which can be derived from the proof.

Lemma 9. *Let* $\mathcal{G}$ *be a class of groups with output length L. Assume the discrete logarithm problem in* $\mathcal{G}$ *is* (t', ϵ')-*hard. Then for any q, the* $(\mathcal{G}, q)$-*balance problem is* (t, ϵ)-*hard, where* $\epsilon = q\epsilon'$ *and* $t = t'/c - q \cdot [T_{\mathrm{rand}}(\mathcal{G}) + T_{\exp}(\mathcal{G}) + L]$.

Proof. We are given an algorithm A, which takes $\langle G \rangle$ and a sequence of elements $a_1, \dots, a_q$ in G and outputs weights $w_1, \dots, w_q \in \{-1, 0, +1\}$, not all zero, such that $\Pi_{i=1}^q a_i^{w_i} = 1$. Let g be a generator of the group G. We want to construct a discrete logarithm finding algorithm I. It takes as input $\langle G \rangle$, g, and $y \in G$, the last randomly chosen, and returns $\log_g(y)$.

We let $\rho = |G|$ be the order of G. We will use A to build I. I first picks at random an integer q^* in the range $1, \dots, q$. I then computes elements a_i ($i = 1, \dots, q$) as follows. If $i = q^*$ then $a_i = y$. Otherwise it chooses at random $r_i \in Z_\rho$ and sets $a_i = g^{r_i}$. (Notice that since y is random and g is a generator, all a_i are uniformly distributed over G.) Finally, I runs A on input $\langle G \rangle, a_1, \dots, a_q$ and gets a sequence of weights $w_1, \dots, w_q$, not all zero, such that $a_1^{w_1} \cdots a_q^{w_q} = 1$. Let i^* be such that $w_{i^*} \neq 0$. Since the choice of q^* was random and unknown to A, with probability at least $1/q$ it will be the case that the $q^* = i^*$. For notational convenience, assume $q^* = i^* = 1$. Now, substituting, we have

$$y^{w_1} \cdot g^{w_2 r_2} \cdots g^{w_q r_q} = 1 .$$

Re-arranging the temrs and noticing that $w_1^{-1} = w_1$ (in Z_ρ) gives us

$$y = g^{-w_1(w_2 r_2 + \cdots + w_q r_q) \bmod \rho}$$

Thus, $r = -w_1(w_2 r_2 + \cdots + w_q r_q) \bmod \rho$ is the discrete logarithm of y and I can output it and halt. The probability that I is successful is ϵ times the probability that $w_{q^*} \neq 0$, and we saw the latter was at least $1/q$. That is, $\epsilon' = \epsilon/q$.

Since I runs A it incurs time t. Computing each a_i takes one random choice and one exponentiation (except for a_{q^*} which only needs to be copied), meaning $T_{\mathrm{rand}}(\mathcal{G}) + T_{\exp}(\mathcal{G})$ steps per element. The output of C may be up to t bits long so reading it is another investment of time upto t. The final modular additions take $O(qL)$ time. The total time for the algorithm is thus $t' = t + q \cdot [T_{\mathrm{rand}}(\mathcal{G}) + T_{\exp}(\mathcal{G}) + L]$. ∎

This is a very general result, but quantitatively not the best. We now tighten the relationship between the parameters for special classes of groups.

GROUPS OF PRIME ORDER. Let $\mathcal{G}$ be some class of groups of prime order for which the discrete logarithm problem is hard, as discussed in Section 4.3. Below we see that $\epsilon = 2\epsilon'$ rather than $\epsilon = q\epsilon'$ as before, which is quite an improvement. As usual $c > 1$ is a small constant, depending on the model of computation, which can be derived from the proof.

Lemma 10. *Let $\mathcal{G}$ be a class of groups of prime order with output length L. Assume the discrete logarithm problem in $\mathcal{G}$ is (t', ϵ')-hard. Then for any q, the $(\mathcal{G}, q)$-balance problem is (t, ϵ)-hard, where $\epsilon = 2\epsilon'$ and $t = t'/c - q \cdot [T_{\mathrm{rand}}(\mathcal{G}) + T_{\mathrm{mult}}(\mathcal{G}) + T_{\exp}(\mathcal{G}) + L] - L^2$.*

Proof. We follow and modify the proof of Lemma 9. By assumption G has prime order. We let $\rho = |G|$ be this order. So $G = \{g^i : i \in Z_\rho\}$. Note that computation in the exponents is modulo ρ and takes place in a field, namely Z_ρ. We will make use of this.

Given A we are constructing I. I takes as input $\langle G \rangle$, g, and $y \in G$, the last randomly chosen. If $y = 1$ (the "1" here standing for the identity element of G), then I can immediately answer $\log_g(y) = 0$. So, we can assume that $y \neq 1$. The key point where we differ from the previous proof is in how the input to A is computed. For each $i = 1, \ldots, q$, algorithm I chooses at random $r_i \in Z_\rho$ and also chooses at random $d_i \in \{0, 1\}$ and sets $a_i = g^{d_i} y^{r_i}$. (Notice that g^{d_i} is either 1 or g and we don't need to perform a modular exponentiation to compute it. Notice also that since G has prime order every element of G except 1 is a generator. In particular y is a generator and hence a_i is uniformly distributed over G.) Now we continue to follow the proof of Theorem 3. We run A on input $\langle G \rangle, a_1, \ldots, a_q$

and get weights $w_1, \ldots, w_q$, not all zero, such that $a_1^{w_1} \cdots a_q^{w_q} = 1$. Substituting the values for a_i we have

$$y^{w_1 r_1} g^{w_1 d_1} \ldots y^{w_q r_q} g^{w_q d_q} = 1 .$$

Re-arranging terms gives us

$$y^{w_1 r_1 + \cdots + w_q r_q \bmod \rho} = g^{-w_1 d_1 - \cdots - w_q d_q \bmod \rho} ,$$

Now let

$$r = w_1 r_1 + \cdots + w_q r_q \bmod \rho$$
$$d = -w_1 d_1 - \cdots - w_q d_q \bmod \rho ,$$

so that our equation is $y^r = g^d$. Now, observe that $r \neq 0$ with probability at least $1/2$. (This is because the value of d_1 remains equi-probably 0 or 1 from the point of view of A, and is independent of other d_i values. At most one of the two possible values of d_1 can make $d = 0$ and hence $r = 0$.) If it is the case that $r \neq 0$ then, since ρ is a prime, r has an inverse modulo ρ. I computes the inverse of r modulo ρ and denotes it by r^{-1}. I outputs $r^{-1} d \bmod \rho$. We have $g^{dr^{-1}} = y^{rr^{-1}} = y$ so the output is indeed $\log_g(y)$.

To show the algorithm outputs $\log_g(y)$ with the claimed probability ϵ', we just need to observe that the input distribution to A is that required by the balance problem. A solves this problem with probability ϵ and we get $\log_g(y)$ with probability at least one half of that. ∎

THE GROUP Z_p^*. Finally we look at the group Z_p^* where p is prime. This group has order $p - 1$, which is not prime, so Lemma 10 does not apply, but we can still do much better than Lemma 9. As usual $c > 1$ is a small constant, depending on the model of computation, which can be derived from the proof.

Lemma 11. *Let $k \geq 6$ and let $\mathcal{G} = \{ Z_p^* : p$ is a prime with $|p| = k \}$. Suppose the discrete logarithm problem in $\mathcal{G}$ is (t', ϵ')-hard. Then for any q, the $(\mathcal{G}, q)$-balance problem is (t, ϵ)-hard, where $\epsilon = 4 \ln(0.694k) \cdot \epsilon'$ and $t = t'/c - qk^3 - b$.*

The following, which we will use in the proof, can be derived from inequalities in Rosser and Schoenfeld [RS].

Lemma 12. *For any integer $N \geq 23$ it is the case that*

$$\frac{\varphi(N)}{N} \geq \frac{1}{4 \cdot \ln \ln N} .$$

We will have $N = p - 1$, and it is to guarantee $N \geq 23$ that we let the length k of p be at least 6 in Lemma 11.

Proof of Lemma 11. We let $G = Z_p^*$ and let $\rho = |G| = p - 1$. Thus $\langle G \rangle = p$. We now follow and modify the proofs of Lemma 9 and Lemma 10. Given A we are constructing I.

The key point where we differ from the previous proof is in how the input to A is computed. For each $i = 1, \ldots, q$, algorithm I chooses at random $r_i \in Z_\rho$ and also chooses at random $d_i \in Z_\rho$. It sets $a_i = g^{d_i} y^{r_i}$. (Notice that a_i is uniformly distributed in G because d_i is random and g is a generator.)

Finally, we run A on input $\langle G \rangle, a_1, \ldots, a_q$. We define r and d as in the previous proof and get to the equation $y^r = g^d$. We would like to compute $r^{-1} \bmod \rho$. The problem is that since ρ is no longer prime, this inverse may not exist. However, we claim (to be justified later) that r is uniformly distributed in Z_ρ. This means that $\gcd(r, \rho) = 1$ with probability

$$\frac{\varphi(\rho)}{\rho} \geq \frac{1}{4 \ln \ln(\rho)} \geq \frac{1}{4 \ln \ln(2^k)} \geq \frac{1}{4 \ln(k \ln(2))} \geq \frac{1}{4 \ln(0.694k)} \, ,$$

having used Lemma 12 and the fact that $\rho = p - 1 \leq 2^k$. We can compute $\gcd(r, \rho)$, and, if it is one, compute $r^{-1} \bmod \rho$, in which case we can output $\log_g(y)$ as before, and the probability we succeed is the above.

Now we must justify the claim that r is uniformly distributed in Z_{p-1}. Note A has no information on the r_i values, since the a_i values are uniformly and independently distributed of the r_i values, thanks to the d_i values. So we are adding a non-zero number of uniformly distributed values. So the result is uniformly distributed. ∎

Smooth Entropy and Rényi Entropy

Christian Cachin*

Department of Computer Science
ETH Zürich
CH-8092 Zürich, Switzerland
cachin@acm.org

Abstract. The notion of smooth entropy allows a unifying, generalized formulation of privacy amplification and entropy smoothing. Smooth entropy is a measure for the number of almost uniform random bits that can be extracted from a random source by probabilistic algorithms. It is known that the Rényi entropy of order at least 2 of a random variable is a lower bound for its smooth entropy. On the other hand, an assumption about Shannon entropy (which is Rényi entropy of order 1) is too weak to guarantee any non-trivial amount of smooth entropy. In this work we close the gap between Rényi entropy of order 1 and 2. In particular, we show that Rényi entropy of order α for any $1 < \alpha < 2$ is a lower bound for smooth entropy, up to a small parameter depending on α, the alphabet size and the failure probability. The results have applications in cryptography for unconditionally secure protocols such as quantum key agreement, key agreement from correlated information, oblivious transfer, and bit commitment.

1 Introduction

Entropy smoothing is the process of converting an arbitrary random source into a source with smaller alphabet and almost uniform distribution. *Smooth entropy* is an information measure that has been proposed recently [7] to quantify the number of almost uniform bits that can be extracted by a probabilistic algorithm from any member of a set of random variables. It unifies previous work on privacy amplification in cryptography and on entropy smoothing in theoretical computer science and enables a systematic investigation of entropy smoothing and its efficiency.

The main question of entropy smoothing is: Given an arbitrary random source, how many uniformly random bits can be extracted? The formalization of smooth entropy allows for an arbitrarily small deviation of the output bits from perfectly uniform random bits that may include a small correlation with the random bits used for smoothing. The inclusion of randomized extraction functions is the main difference between entropy smoothing and "pure" random number generation in information theory [19], where no additional random sources are

---

* Supported by the Swiss National Science Foundation, grant no. 20-42105.94.

W. Fumy (Ed.): Advances in Cryptology - EUROCRYPT '97, LNCS 1233, pp. 193-208, 1997.

available. However, entropy smoothing does not consider the auxiliary random bits as a resource, unlike extractors used in theoretical computer science [17].

In cryptography, entropy smoothing is known as privacy amplification. Introduced in 1985 [3,4] and later generalized [2], it has become a key component of unconditionally secure cryptographic protocols with such various purposes as key agreement from correlated information [16], key agreement over quantum channels [1,5], oblivious transfer [6], and bit commitment [10].

Privacy amplification, for short, is a process that allows two parties to distill a secret key from common information about which an adversary has partial knowledge. The two parties do not know anything about the adversary's knowledge except that it satisfies a general bound. By using a publicly chosen compression function, they are nevertheless able to extract a short key from their common information such that the total knowledge of the adversary about the key is arbitrarily small.

Apart from the applications in cryptography, entropy smoothing is also at the core of many constructions in complexity theory. Examples are pseudorandom generation [11,14], derandomization of algorithms [15], hardness results in computational learning theory [13], and computing with degenerate, weak random sources [20]. A survey of these applications is given by Nisan [17].

Bennett et al. [4,2] and Impagliazzo et al. [12] independently analyzed entropy smoothing by universal hash functions [8] and showed that the length of the almost uniform output depends on the Rényi entropy of order 2 of the input. Privacy amplification can therefore be applied if the two parties assume a lower bound on the Rényi entropy of order 2 of the adversary's knowledge about their information. By the properties of Rényi entropy, it is straightforward to extend this result to Rényi entropy of any order $\alpha > 2$.

On the other hand, it is known that a lower bound in terms of Rényi entropy of order 1 (which is equivalent to entropy in the sense of Shannon) is not sufficient to extract a non-trivial amount of uniform bits [2].

In this work, we close this gap and prove a lower bound on smooth entropy in terms of Rényi entropy of order α for any α between 1 and 2. Our result shows that the number of almost uniform bits that can be extracted with high probability from a random variable is given by its Rényi entropy order α, for any $\alpha > 1$, up to a correcting term depending on α, the alphabet size and the failure probability. The correcting term becomes dominating for $\alpha \to 1$.

In a second part, we show that tighter lower bounds for smooth entropy can be obtained if one makes additional assumptions about the distribution. In particular, we show how an assumption about the so-called profile of the random variable leads to a lower bound on its smooth entropy that can be much tighter than the one given by Rényi entropy.

The results can be applied immediately to any of the above-mentioned scenarios using entropy smoothing and, in particular, to all applications of privacy amplification in cryptography. Our analysis shows that entropy smoothing by universal hashing is, in general, much more efficient than what was guaranteed by previous results using Rényi entropy of order 2. This has important consequences for the efficiency of these protocols.

The paper is organized as follows. Entropy and Rényi entropy are introduced in Section 2 and a review of smooth entropy is provided in Section 3. Our results are based on the spoiling knowledge proof technique, which is introduced in Section 4. The main result is proved in Section 5, and Section 6 contains the derivation of the tighter bound in terms of the profile.

2 Preliminaries

We assume that the reader is familiar with the notion of entropy and the basic concepts of information theory [9]. We repeat some fundamental definitions in this section and introduce the notation. All logarithms in this paper are to the base 2. The cardinality of a set S is denoted by $|S|$.

A random variable X induces a probability distribution P_X over an alphabet $\mathcal{X}$. Random variables are denoted by capital letters. If not stated otherwise, the alphabet of a random variable is denoted by the corresponding script letter. Families of random variables are denoted by $\mathbb{X}$.

The expected value of a real-valued random variable X is denoted by $E[X]$. The *k-th moment inequality* for any real-valued random variable X, any integer $k > 0$, and $t \in \mathbb{R}^+$ is

$$P[|X| \geq t] \leq \frac{E[|X|^k]}{t^k}. \tag{1}$$

Another useful bound for any real-valued random variable X, any $t \in \mathbb{R}^+$, and any $r \in \mathbb{R}$ is [14]

$$P[X \geq r] \leq E[e^{(X-r)t}]. \tag{2}$$

The *(Shannon) entropy* of a random variable X with probability distribution P_X and alphabet $\mathcal{X}$ is defined as

$$H(X) = -\sum_{x \in \mathcal{X}} P_X(x) \log P_X(x).$$

The *conditional entropy* of X conditioned on a random variable Y is

$$H(X|Y) = \sum_{y \in \mathcal{Y}} P_Y(y) H(X|Y = y)$$

where $H(X|Y = y)$ denotes the entropy of the conditional probability distribution $P_{X|Y=y}$. The *binary entropy function* is

$$h(p) = -p \log p - (1 - p) \log(1 - p).$$

The *relative entropy* or *discrimination* between two probability distributions P_X and P_Y with the same alphabet $\mathcal{X}$ is defined as (using $0 \log \frac{0}{q} = 0$ and $p \log \frac{p}{0} = \infty$)

$$D(P_X \| P_Y) = \sum_{x \in \mathcal{X}} P_X(x) \log \frac{P_X(x)}{P_Y(x)}. \tag{3}$$

The *Rényi entropy of order* α of a random variable X with alphabet $\mathcal{X}$ is

$$H_\alpha(X) = \frac{1}{1-\alpha} \log \sum_{x \in \mathcal{X}} P_X(x)^\alpha$$

for $\alpha \geq 0$ and $\alpha \neq 1$ [18]. Because the limiting case of Rényi entropy for $\alpha \to 1$ is Shannon entropy, we can extend the definition to $H_1(X) = H(X)$. In the other limiting case $\alpha \to \infty$, we obtain the *min-entropy*, defined as

$$H_\infty(X) = -\log \max_{x \in \mathcal{X}} P_X(x).$$

For a fixed random variable X, Rényi entropy is a continuous positive decreasing function of α. For $0 < \alpha < \beta$,

$$H_\alpha(X) \geq H_\beta(X) \tag{4}$$

with equality if and only if X is uniformly distributed over some subset of $\mathcal{X}$. In particular, $\log |\mathcal{X}| \geq H_\alpha(X) \geq 0$ for $\alpha \geq 0$ and $H(X) \geq H_\alpha(X)$ for $\alpha > 1$.

3   Review of Smooth Entropy and Privacy Amplification

Smooth entropy [7] is an abstraction and a generalized formulation of privacy amplification [2] and entropy smoothing [12,14]. As an information measure, smooth entropy is defined operationally with respect to an application scenario (similar to channel capacity [9]). Its value cannot be computed immediately for a given probability distribution. This contrasts with other entropy measures such as Shannon or Rényi entropy that are defined formally in terms of a probability distribution.

Consider a random variable X. We want to apply a *smoothing function* $f : \mathcal{X} \to \mathcal{Y}$ to X such that $Y = f(X)$ is uniformly distributed over its range $\mathcal{Y}$. The size of the largest $\mathcal{Y}$ such that Y is still sufficiently uniform is a measure for the amount of *smooth entropy* inherent in X, relative to the allowed deviation from perfect uniformity. To quantify this deviation we use a nonuniformity measure M that associates with every random variable X a positive number $M(X)$ that is 0 if and only if P_X is the uniform distribution P_U over $\mathcal{X}$. Examples for M are relative entropy $D(P_X \| P_U) = \log |\mathcal{X}| - H(X)$ or L_1 distance $\|P_X - P_U\|_1 = \sum_{x \in \mathcal{X}} |P_X(x) - \frac{1}{|\mathcal{X}|}|$.

The smoothing algorithm should be able to produce outputs that achieve some desired uniformity. More uniform outputs can usually be obtained by reducing the output size. We introduce the parameter s to control the trade-off between the uniformity of the output and the amount of entropy lost in the smoothing process.

Probabilistic smoothing functions are formalized by extending the input of f with an additional random variable T that models the random choices of f. However, T must be independent of X and its value must be known to ensure that no randomness from T is inserted into Y. The size of T is explicitly ignored.

It can be tolerated that the uniformity bound for an extraction process fails if an error event $\mathcal{E}$ occurs. $\mathcal{E}$ should have small probability, denoted by ϵ, and may depend on X. The uniformity is calculated only in the case that the complementary event $\overline{\mathcal{E}}$ occurs.

In many applications it is only known that the random variable X has some property that is shared by many others. Therefore, smooth entropy is defined for a family of random variables $\mathbb{X}$ with the same alphabet. The same smoothing algorithm is required to work for all probability distributions in the family.

Definition 1 ([7]). Let M be a nonuniformity measure and let $\Delta : \mathbb{R} \to \mathbb{R}$ be a decreasing non-negative function. A family $\mathbb{X}$ of random variables with alphabet $\mathcal{X}$ has *smooth entropy* $\Psi(\mathbb{X})$ *within* $\Delta(s)$ *[in terms of M]* with probability $1 - \epsilon$ if $\Psi(\mathbb{X})$ is the maximum of all ψ such that for any security parameter $s \geq 0$, a random variable T and a function $f : \mathcal{X} \times \mathcal{T} \to \mathcal{Y}$ exist with $|\mathcal{Y}| = \lfloor 2^{\psi-s} \rfloor$ such that for all $X \in \mathbb{X}$ there is a failure event $\mathcal{E}$ that has probability at most ϵ, and the expected value over T of the nonuniformity M of $Y = f(X, T)$, given T and $\overline{\mathcal{E}}$, is at most $\Delta(s)$. Formally,

$$\Psi(\mathbb{X}) = \max_{\psi} \Big\{ \psi \Big| \forall s \geq 0 : \exists T, f : \mathcal{X} \times \mathcal{T} \to \mathcal{Y}, |\mathcal{Y}| = \lfloor 2^{\psi-s} \rfloor :$$

$$\forall X \in \mathbb{X} : Y = f(X, T), \exists \mathcal{E} : P[\mathcal{E}] \leq \epsilon, M(Y|T\overline{\mathcal{E}}) \leq \Delta(s) \Big\}.$$

$$\triangle$$

For singleton sets $\{X\}$, we also use $\Psi(X)$ instead of $\Psi(\{X\})$. The failure probability ϵ can be integrated into the uniformity parameter $\Delta(s)$ for certain nonuniformity measures such as L_1 distance.

The principal method for extracting smooth entropy is based on universal hashing. A universal hash function [8] is a set $\mathcal{G}$ of functions $\mathcal{X} \to \mathcal{Y}$ such that for all distinct $x_1, x_2 \in \mathcal{X}$, there are at most $|\mathcal{G}|/|\mathcal{Y}|$ functions g in $\mathcal{G}$ such that $g(x_1) = g(x_2)$.

Privacy amplification is fundamental for many unconditionally secure cryptographic protocols [2]. Assume Alice and Bob share a random variable W, while an eavesdropper Eve knows a correlated random variable V that summarizes her knowledge about W. The details of the distribution P_{WV}, and thus of Eve's information V about W, are unknown to Alice and Bob, except that they assume a lower bound on the Rényi entropy of order 2 of $P_{W|V=v}$ for the particular value v that Eve observes.

Using an authentic public channel, which is susceptible to eavesdropping but immune to tampering, Alice and Bob wish to agree on a function g such that

Eve knows nearly nothing about $g(W)$. The following theorem by Bennett et al. [2] shows that if Alice and Bob choose g at random from a universal hash function $\mathcal{G} : \mathcal{W} \to \mathcal{Y}$ for suitable $\mathcal{Y}$, then Eve's information about $Y = g(W)$ is negligible.

Theorem 1 (Privacy Amplification Theorem [2]). *Let X be a random variable over the alphabet $\mathcal{X}$ with Rényi entropy $H_2(X)$, let G be the random variable corresponding to the random choice (with uniform distribution) of a member of a universal hash function $\mathcal{G} : \mathcal{X} \to \mathcal{Y}$, and let $Y = G(X)$. Then*

$$H(Y|G) \geq \log|\mathcal{Y}| - \frac{2^{\log|\mathcal{Y}| - H_2(X)}}{\ln 2}. \tag{5}$$

The theorem can be applied in the described scenario by replacing P_X with the conditional probability distribution $P_{W|V=v}$. The Privacy Amplification Theorem implies that $H_2(X)$ is a lower bound for smooth entropy. It is crucial that the same smoothing algorithm can be applied to any X from a family $\mathbb{X}$ of random variables and produce an output of the desired size and uniformity.

Corollary 2 ([7]). *The smooth entropy of a family $\mathbb{X}$ of random variables within $2^{-s}/\ln 2$ in terms of relative entropy with probability 1 is at least the minimum Rényi entropy of order 2 of any $X \in \mathbb{X}$.*

Note that Shannon entropy cannot be used as a lower bound for smooth entropy. This was observed by Bennett et al. [2] and is illustrated in the following example.

Example 1. Suppose that everything we know about a random variable X is $H(X) \geq t$. Then P_X could be such that $P_X(x_0) = p$ for some $x_0 \in \mathcal{X}$ with $p = 1 - t/\log(|\mathcal{X}| - 1)$ and $P_X(x) = (1 - p)/(|\mathcal{X}| - 1)$ for all $x \neq x_0$. X satisfies $H(X) = h(p) + (1 - p)\log(|\mathcal{X}| - 1) \geq t$. But $X = x_0$ occurs with probability p, and no matter how small a Y is extracted from X, its value can be predicted with probability p. Thus, with knowledge of a lower bound on $H(X)$ alone, the probability that X is guessed correctly cannot be reduced and only a small part of the randomness in X can be converted to uniform bits. Therefore, the entropy of a random variable is not an adequate measure of its smooth entropy. In other words, there are random variables with arbitrarily large entropy and almost no smooth entropy. $\bigcirc$

4 Spoiling Knowledge Proofs

As noted above, Rényi entropy of order 2 is a lower bound for smooth entropy. A counter-intuitive property of conditional Rényi entropy of order $\alpha > 1$ is that it can increase even on the average when conditioned on a random variable that provides side information. Suppose side information that increases the Rényi entropy is made available by an imaginary oracle. This increase can be exploited to prove lower bounds on smooth entropy that are much tighter than Rényi entropy of order 2. Side information of this kind was introduced by Bennett et al. [2] and is called *spoiling knowledge* because it leads to less information about the output of the smoothing process.

We examine side information that induces an event $\mathcal{A}$ such that $P[\mathcal{A}]$ is at least $1 - \epsilon$ and $H_2(X|\mathcal{A})$ is large. This can then be transformed into a lower bound on smooth entropy with probability $1 - \epsilon$ of X. A formal statement of this is given in the next theorem, where the binary random variable V models side information such that $\mathcal{A}$ corresponds to $V = 0$.

Theorem 3. *The smooth entropy $\Psi(X)$ within $2^{-s}/\ln 2$ with probability $1 - \epsilon$ of a random variable X is lower bounded by the maximum of the conditional Rényi entropy $H_2(X|V = 0)$, where the maximization ranges over all random variables V with alphabet $\{0, 1\}$ such that the joint distribution P_{XV} is consistent with P_X and satisfies $P_V(0) \geq 1 - \epsilon$:*

$$\Psi(X) \geq \max_{P_V \,:\, P_V(0) \geq 1-\epsilon} H_2(X|V = 0) \tag{6}$$

Note that the oracle knows the particular distribution of the random variable that is to be smoothed (e.g. the adversary's knowledge in privacy amplification) and can prepare the side information depending on that distribution.

For the construction of the lower bounds, we introduce special side information U with alphabet $\{0, \ldots, m\}$. Let $U = f(X)$ be the deterministic function of X given by

$$f(x) = \begin{cases} m & \text{if } P_X(x) \leq 2^{-m} \\ \lfloor -\log P_X(x) \rfloor & \text{otherwise.} \end{cases}$$

We call side information U of this type *log-partition spoiling knowledge* because U partitions the values of X into sets of approximately equal probability and because it is most useful with $m \approx \log |\mathcal{X}|$. For such m, the values of the probability distributions $P_{X|U=u}$ differ at most by a factor of two for all u except for $u = m$.

In the following, let

$$p_{\min} = \min_{x \in \mathcal{X}} P_X(x) \qquad \text{and} \qquad p_{\max} = \max_{x \in \mathcal{X}} P_X(x).$$

The following two lemmas show that Rényi entropy of order 2 and Shannon entropy cannot differ arbitrarily for probability distributions where $p_{\min}$ and $p_{\max}$ are a constant factor apart.

Lemma 4. *Let X be a random variable with alphabet $\mathcal{X}$ such that $p_{\max} \leq c \cdot p_{\min}$ for some $c > 1$. Then*

$$\frac{1}{|\mathcal{X}| - 1 + c} \leq p_{\min} \leq \frac{1}{|\mathcal{X}|}$$

$$\frac{1}{|\mathcal{X}|} \leq p_{\max} \leq \frac{c}{|\mathcal{X}| - 1 + c}.$$

Proof. It is easy to see that maximum of $p_{max} - p_{min}$ is reached when $P_X(x) = p_{min}$ for all x except for the one that has maximal probability $p_{max} = c \cdot p_{min}$. The lemma follows directly. □

If the minimum and maximum probability in a distribution P_X do not differ by more than a constant factor, then the Rényi entropy of order 2 of X is at most a constant below the Shannon entropy.

Lemma 5. *Let X be a random variable with alphabet $\mathcal{X}$ such that $p_{max} \leq c \cdot p_{min}$ for some $c > 1$. Then*

$$H_2(X) > H(X) - 2 \log c.$$

Proof. Lemma 4 is used in the second inequality of the following derivation:

$$
\begin{aligned}
H(X) - H_2(X) &= H(X) + \log \sum_{x \in \mathcal{X}} P_X(x)^2 \\
&\leq \log |\mathcal{X}| + \log(|\mathcal{X}| \, p_{max}^2) \\
&= 2 \log(|\mathcal{X}| \, p_{max}) \\
&\leq 2 \log\Big(|\mathcal{X}| \frac{c}{|\mathcal{X}| - 1 + c}\Big) \\
&= 2\Big(\log c + \log\Big(\frac{|\mathcal{X}|}{|\mathcal{X}| - 1 + c}\Big)\Big) \\
&< 2 \log c
\end{aligned}
$$

□

5 A Bound Using Rényi Entropy of Order $\alpha > 1$

The connection between entropy smoothing and Rényi entropy was established independently by Bennett et al. [2] and Impagliazzo et al. [12]. The Privacy Amplification Theorem shows that Rényi entropy of order 2 is a lower bound for smooth entropy. That is, for any random variable X by assuming only a lower bound t on $H_2(X)$, approximately t almost uniform random bits can be extracted from X and the deviation from a uniform distribution decreases exponentially when fewer bits are extracted.

In some applications, only the stronger bound $H_\infty(X) \geq t$ in terms of min-entropy is assumed, equivalent to bounding the maximum probability of any value of X. Indeed, Theorem 1 holds if an assumption about $H_\alpha(X)$ for any $\alpha \geq 2$ is made because $H_2(X) \geq H_\alpha(X)$ for $\alpha \geq 2$ by (4).

On the other hand, it is known from Example 1 that a lower bound on $H_1(X) = H(X)$ is not sufficient to guarantee a non-trivial amount of smooth entropy. Rather, the smooth entropy could be arbitrarily small if no further assumptions are made. In this section we examine the remaining range for $1 < \alpha < 2$. We show that, with high probability, the smooth entropy of X is lower bounded by $H_\alpha(X)$, up to the logarithm of the alphabet size and some security parameters depending on α and on the error probability.

Our approach uses a spoiling knowledge argument. We will use side information U such that for any distribution of X, with high probability, U takes on a value u for which $H_2(X|U = u)$ is not far below $H_\alpha(X)$. A simple and very weak bound that always holds follows from the next lemma.

Lemma 6. *For any random variable X and for any $\alpha > 1$,*

$$\frac{\alpha}{\alpha - 1} H_\infty(X) \geq H_\alpha(X) \geq H_\infty(X).$$

Proof. Because $\alpha > 1$,

$$\begin{aligned}
\frac{\alpha}{\alpha - 1} H_\infty(X) &= \frac{1}{1 - \alpha} \log \max_{x \in \mathcal{X}} P_X(x)^\alpha \\
&\geq \frac{1}{1 - \alpha} \log \sum_{x \in \mathcal{X}} P_X(x)^\alpha \\
&= H_\alpha(X).
\end{aligned}$$

The lower bound follows from (4). □

We conclude that

$$H_2(X) \geq H_\infty(X) \geq \frac{\alpha - 1}{\alpha} H_\alpha(X)$$

for any $\alpha > 1$. However, this bound is multiplicative in $\alpha - 1$ which limits its usefulness for $\alpha \to 1$. The tighter bound derived below is only additive in $(\alpha - 1)^{-1}$. It is based on the following theorem that provides the connection between the Rényi entropy of order $\alpha > 1$ conditioned on side information and the Rényi entropy of the joint distribution.

Theorem 7. *Let $\alpha > 1$ and let $r, t > 0$. For arbitrary random variables X and Y, the probability that Y takes on a value y for which*

$$H_\alpha(X|Y = y) \geq H_\alpha(XY) - \log |\mathcal{Y}| - \frac{r}{\alpha - 1} - t$$

is at least $1 - 2^{-r} - 2^{-t}$.

Proof. It is straightforward to expand the Rényi entropy of XY as

$$\begin{aligned}
H_\alpha(XY) &= \frac{1}{1 - \alpha} \log \sum_{x \in \mathcal{X}, y \in \mathcal{Y}} P_{XY}(x, y)^\alpha \\
&= \frac{1}{1 - \alpha} \log \sum_{y \in \mathcal{Y}} P_Y(y) \cdot P_Y(y)^{\alpha - 1} \sum_{x \in \mathcal{X}} P_{X|Y=y}(x)^\alpha \\
&= \frac{1}{1 - \alpha} \log \sum_{y \in \mathcal{Y}} P_Y(y) \, 2^{(\alpha - 1) \log P_Y(y) + (1 - \alpha) H_\alpha(X|Y=y)}.
\end{aligned}$$

We introduce the function $\beta(y) = H_\alpha(X|Y = y)$ to interpret $H_\alpha(X|Y = y)$ as a function of y and consider the random variables $P_Y(Y)$ and $\beta(Y)$. The equation above is equivalent to

$$\mathrm{E}_Y\left[2^{(1-\alpha)\beta(Y)+(\alpha-1)\log P_Y(Y)}\right] = 2^{(1-\alpha)H_\alpha(XY)}$$

or

$$\mathrm{E}_Y\left[2^{(1-\alpha)\beta(Y)+(\alpha-1)\log P_Y(Y)-(1-\alpha)H_\alpha(XY)-r}\right] = 2^{-r}.$$

Inserting this into the right-hand side of inequality (2) yields

$$\mathrm{P}_Y\left[(1-\alpha)\beta(Y)+(\alpha-1)\log P_Y(Y)-(1-\alpha)H_\alpha(XY) \geq r\right] \leq 2^{-r}$$

form which we see after dividing by $1-\alpha$ that with probability at least $1-2^{-r}$, Y takes on a value y for which

$$H_\alpha(X|Y = y) \geq H_\alpha(XY) + \log P_Y(y) - \frac{r}{\alpha-1}. \tag{7}$$

The only thing missing is a bound for the term $\log P_Y(y)$. However, large values of $|\log P_Y(Y)|$ occur only with small probability. For any $t > 0$,

$$\mathrm{P}\left[P_Y(Y) < 2^{-t}/|\mathcal{Y}|\right] = \sum_{y\,:\,P_Y(y) < 2^{-t}/|\mathcal{Y}|} P_Y(y) < 2^{-t}$$

because there are only $|\mathcal{Y}|$ terms in the summation. Therefore, with probability at least $1-2^{-t}$, Y takes on a value y for which

$$\log P_Y(y) \geq -t - \log|\mathcal{Y}| \tag{8}$$

and the theorem follows from (7) and (8) by the union bound. $\qquad\square$

Applying this bound for log-partition side information gives the main result of this paper and shows how smooth entropy is lower bounded by Rényi entropy of order α for any $\alpha > 1$.

Theorem 8. *Fix $r, t > 0$, let m be an integer such that $m - \log(m + 1) > \log|\mathcal{X}| + t$, and let s be the security parameter for smooth entropy. For any $\alpha > 1$, the smooth entropy of a random variable X within $2^{-s}/\ln 2$ in terms of relative entropy with probability $1 - 2^{-r} - 2^{-t}$ is lower bounded by Rényi entropy of order α in the sense that*

$$\Psi(X) \geq H_\alpha(X) - \log(m + 1) - \frac{r}{\alpha-1} - t - 2.$$

Proof. We again use log-partition spoiling-knowledge $U = f(X)$ with alphabet $\{0, \ldots, m\}$ as defined above. Because f is a deterministic function of X, we have $H_\alpha(XU) = H_\alpha(X)$ and Theorem 7 shows that U takes on a value u for which

$$H_\alpha(X|U = u) \geq H_\alpha(X) - \log|\mathcal{U}| - \frac{r}{\alpha-1} - t$$

with probability at least $1 - 2^{-r} - 2^{-t}$. Because $m > \log|\mathcal{X}|$, Lemma 5 can be applied with $c \leq 2$ and by (4) it follows for all $u \neq m$ that

$$H_2(X|U = u) > H(X|U = u) - 2 \geq H_\alpha(X|U = u) - 2.$$

Combining these results shows that the probability that U takes on a value $u \neq m$ for which

$$H_2(X|U = u) \geq H_\alpha(X) - \log(m + 1) - \frac{r}{\alpha - 1} - t - 2 \tag{9}$$

is at least $1 - 2^{-r} - 2^{-t}$.

Remember that in (8) in the proof of Theorem 7, values of U with probability less than $2^{-t - \log|\mathcal{U}|}$ have been excluded. Therefore, if m is chosen such that

$$\mathrm{P}[U = m] = \sum_{x\,:\,P_X(x) < 2^{-m}} P_X(x) \leq |\mathcal{X}| \cdot 2^{-m} < 2^{-t - \log|\mathcal{U}|}$$

then $U = m$ does not occur in (9). Choosing m such that $m - \log(m + 1) > \log|\mathcal{X}| + t$ achieves this and applying Theorem 3 completes the proof. $\quad\square$

Corollary 9. *Let* $\mathbb{X}$ *be a family of random variables and let* r, t, m, *and* s *be defined as in the theorem above. For any* $\alpha > 1$, *the smooth entropy of* $\mathbb{X}$ *within* $2^{-s}/\ln 2$ *in terms of relative entropy with probability* $1 - 2^{-r} - 2^{-t}$ *satisfies*

$$\Psi(\mathbb{X}) \geq \min_{X \in \mathbb{X}} H_\alpha(X) - \log(m + 1) - \frac{r}{\alpha - 1} - t - 2.$$

The corollary follows from the fact that the oracle knows the distribution of the random variable $X \in \mathbb{X}$ to be smoothed and can prepare the side information accordingly. Especially for large alphabets, these results can yield much better bounds on smooth entropy than Rényi entropy of order 2. The logarithmic term vanishes asymptotically with the alphabet size: For any $\alpha > 1$, the ratio between smooth entropy and the logarithm of the alphabet size is asymptotically lower bounded by the ratio between Rényi entropy of order α and the logarithm of the alphabet size.

Example 2. Consider the random variables X_β with alphabet $\{0, 1\}^n$ and distribution

$$P_{X_\beta}(x) = \begin{cases} 2^{-n/(2\beta)} & \text{for } x = 0^n \\ \frac{1 - 2^{-n/(2\beta)}}{2^n - 1} & \text{otherwise} \end{cases}$$

for $\beta \ll n$. (With $\beta = 2$ this is the example from [2].) The lower bound on $\Psi(X)$ by Rényi entropy of order 2 is weak because $H_2(X) < n/\beta$. However, $H(X_\beta)$ is very close to n bits. Figure 1 displays the Rényi entropy $H_\alpha(X_\beta)$ for $1 \leq \alpha \leq 2$. For α close to 1, it is almost equal to $H(X_\beta) \approx n$.

Using Rényi entropy of order 2, Corollary 2 shows that $\Psi(X_8)$ within $2^{-s}/\ln 2$ with probability 1 is at least $H_2(X_8) \approx n/8$. Allowing failure of the bound with

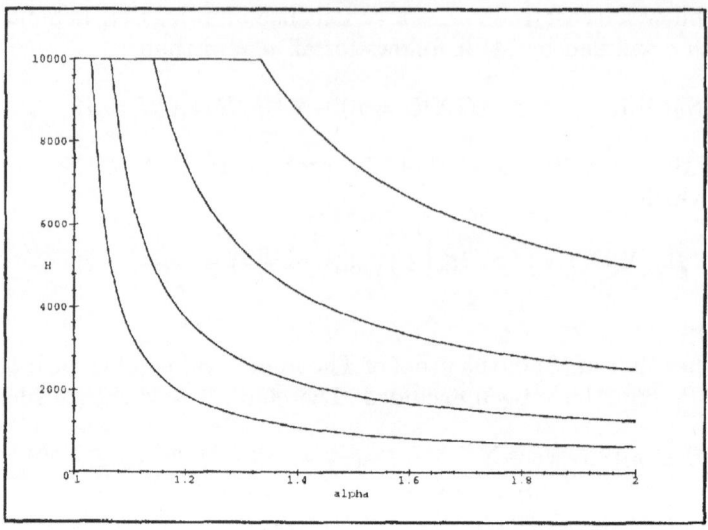

Fig. 1. Rényi entropy $H_\alpha(X_\beta)$ as function of α between 1 and 2. The random variables X_β for $\beta = 16, 8, 4, 2$ (from below) are defined as in Example 2 with $n = 10000$. The graph shows that, together with Theorem 8, Rényi entropy of order α close to 1 can yield much better bounds on smooth entropy than Rényi entropy of order 2.

probability 2^{-19}, the lower bound by Theorem 8 on $\Psi(X_8)$ with probability $1 - 2^{-19}$ is about $n - \log n - 222$ (using Rényi entropy of order $\alpha = 1.1, r = t = 20$, and simplifying the choice of m such that $m = \log |\mathcal{X}| = n$). With $n = 10000$ (as in Figure 1), $\Psi(X_8) \geq 9764$ with probability $1 - 2^{-19}$, compared to Rényi entropy of order 2 from which we can conclude only $\Psi(X_8) \geq 1250$. ◯

For $\alpha \to 1$, the bound of Theorem 8 is reduced to the Shannon entropy. But as shown in Example 1, $H(X)$ yields a weak lower bound for $\Psi(X)$. The next example shows this transition for $\alpha \to 1$.

Example 3. Let X be a random variable with alphabet $\{0, 1\}^{10000}$. We now examine the lower bounds on $\Psi(X)$ when $H_\alpha(X) \geq 9000$ is assumed for various α (see Figure 2). For $\alpha \geq 2$, $\Psi(X) \geq H_2(X) \geq 9000$ is guaranteed by Corollary 2. Theorem 8 shows that $\Psi(X)$ with probability $1 - 2^{-19}$ is close to 9000 for α between 2 and about 1.05. The bound decreases sharply with $\alpha \to 1$. For $\alpha = 1$, if only $H(X) \geq 9000$ is assumed, the random variable constructed in Example 1 has $H_2(X) = 6.64$ and has almost no smooth entropy. ◯

6 A Tighter Bound Using the Profile of the Distribution

The last section shows how smooth entropy can be lower bounded by Rényi entropy of order α for any $\alpha > 1$. This bound, however, is not tight for small

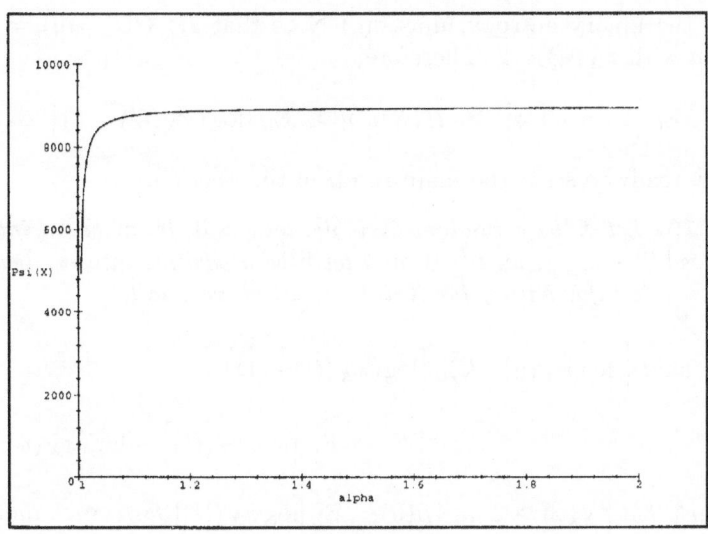

Fig. 2. The dependence of the lower bound for $\Psi(X)$ on the order α of Rényi entropy. The graph shows the lower bound of Theorem 8 on the smooth entropy $\Psi(X)$ within $2^{-s}/\ln 2$ with probability $1-2^{19}$ that can be deduced from $H_\alpha(X) \geq 9000$ as a function of α. Note the sharp decrease with $\alpha \to 1$. (See also Example 3.)

alphabet sizes. We derive a tighter bound in this section that depends on an assumption about the profile of the probability distribution (defined below). The bound is tighter than the one of Theorem 8, especially for smaller alphabets.

We use again log-partition spoiling knowledge $U \in \mathcal{U} = \{0, \dots, m\}$ as defined above. For a fixed value m, define the *profile* π_X of the random variable X as the function $\pi_X : \mathcal{U} \to \mathbb{N}$ such that for $u < m$

$$\pi_X(u) = \left| \{x \in \mathcal{X} \,|\, 2^{-u-1} < P_X(x) \leq 2^{-u} \} \right|$$

and

$$\pi_X(m) = \left| \{x \in \mathcal{X} \,|\, P_X(x) \leq 2^{-m} \} \right|.$$

The expected difference (over U) between the logarithm of the profile $\pi_X(u)$ and the conditional entropy of X given U, $H(X|U=u)$, can be used to obtain a lower bound on smooth entropy. Examining the structure of the probability distributions $P_{X|U=u}$ for all u such that $\pi_X(u) \geq 2$, we see that the logarithm of the profile, $\pi_X(u)$, is close to the conditional entropy, $H(X|U=u)$, in the sense that

$$\log \pi_X(u) \geq H(X|U=u) \geq h\left(\frac{2}{\pi_X(u)+1}\right) + \log(\pi_X(u) - 1). \tag{10}$$

(h denotes the binary entropy function.) Note that $H(X|U = u) = 0$ for the remaining u with $\pi_X(u) < 2$. Therefore,

$$\mathrm{E}_U\Big[\log \pi_X(U)\Big] \geq H(X|U) \geq \mathrm{E}_U\Big[\log(\pi_X(U) - 1)\Big]. \tag{11}$$

We are now ready to state the main result of this section.

Theorem 10. *Let X be a random variable, let $\epsilon > 0$, let m be an integer such that $m \geq \log|\mathcal{X}| + \log\frac{1}{\epsilon}$, let $t > 0$, and let k be a positive integer. Let U be the log-partition side information for X introduced above and let*

$$\mu(u) = \max\Big\{ \log\pi_X(u) - \mathrm{E}_U\Big[\log(\pi_X(U) - 1)\Big],$$
$$\mathrm{E}_U\Big[\log\pi_X(U)\Big] - \log(\pi_X(u) - 1) \Big\}.$$

for all u such that $\pi_X(u) \geq 2$ and $\mu(u) = \mathrm{E}_U[\log\pi_X(U)]$ for u such that $\pi_X(u) < 2$. If

$$\mathrm{E}_U[\mu(U)^k] \leq \epsilon \cdot t^k,$$

the following lower bound on the smooth entropy of X within $2^{-s}/\ln 2$ in terms of relative entropy holds with probability at least $1 - 2\epsilon$:

$$\Psi(X) \geq H(X|U) - t - 2 \geq H(X) - \log(m + 1) - t - 2.$$

Proof. Let $\gamma(u) = H(X|U = u)$ be a function of $u \in \mathcal{U}$ that denotes the entropy of X given $U = u$ and consider the random variable $C = \gamma(U)$. The expectation $\mathrm{E}[C]$ is equal to $H(X|U) \geq H(X) - \log(m + 1)$. Applying the k-th moment inequality (1), we see that

$$\mathrm{P}\big[|C - \mathrm{E}[C]| \geq t\big] \leq \frac{\mathrm{E}[|C - \mathrm{E}[C]|^k]}{t^k}. \tag{12}$$

If this probability is small, then $H(X|U = u) \geq H(X|U) - t$ with high probability. Using (10) and (11), we can bound the probability in (12):

$$\mathrm{E}\Big[|C - \mathrm{E}[C]|^k\Big]$$
$$= \sum_{u \in \mathcal{U}} P_U(u)\big|H(X|U = u) - H(X|U)\big|^k$$
$$= \sum_{u\,:\,\pi_X(u)<2} P_U(u)H(X|U)^k +$$
$$\sum_{u\,:\,H(X|U=u)>H(X|U)} P_U(u)\Big(H(X|U = u) - H(X|U)\Big)^k +$$
$$\sum_{u\,:\,H(X|U=u)<H(X|U)} P_U(u)\Big(H(X|U) - H(X|U = u)\Big)^k$$

$$\leq \sum_{u\,:\,\pi_X(u)<2} P_U(u)H(X|U)^k +$$

$$\sum_{u\,:\,H(X|U=u)>H(X|U)} P_U(u)\Big(\log \pi_X(u) - \mathrm{E}_U\left[\log(\pi_X(U)-1)\right]\Big)^k +$$

$$\sum_{u\,:\,H(X|U=u)<H(X|U)} P_U(u)\Big(\mathrm{E}_U\left[\log \pi_X(U)\right] - \log\big(\pi_X(u)-1\big)\Big)^k$$

$$= \sum_{u\in\mathcal{U}} P_U(u)\mu(u)^k$$

where the last step follows form the definition of $\mu(u)$. We conclude from (12) and from the assumption of the theorem that $H(X|U=u) \geq H(X|U)-t$ occurs with probability at least $1-\epsilon$. It follows from Lemma 5 that for $u \neq m$

$$H_2(X|U=u) \geq H(X|U) - t - 2. \tag{13}$$

But the event $U=m$ has small probability because the choice of m guarantees that

$$\mathrm{P}[U=m] = \sum_{x\,:\,P_X(x)<2^{-m}} P_X(x) \leq |\mathcal{X}|\cdot 2^{-m} \leq \epsilon.$$

By the union bound, the total probability that (13) fails is 2ϵ and the proof is completed by applying Theorem 3. $\qquad\square$

Example 4. Consider again the random variable X_8 from Example 2. For $n = 100$ and desired total failure probability 2^{-19}, the bound of Theorem 8 cannot be applied and we have to resort to Rényi entropy of order 2 that shows $\Psi(X_8) \geq 12.5$ (within $2^{-s}/\ln 2$ in terms of relative entropy).

Applying Theorem 10 with $\epsilon = 2^{-20}$, $t = 12$, and $k = 6$, however, shows that $\Psi(X_8) \geq 84.6$. Therefore, a 60-bit string Y can be extracted from X_8 by a randomly chosen universal hash function such that $H(Y|T) \geq 60 - 2^{-24}/\ln 2$. $\bigcirc$

As the example shows, the bound on smooth entropy by Theorem 10 can be much tighter than Rényi entropy of order 2 and also tighter than the bound of Theorem 8. However, this comes at the cost of the stronger assumption that must be made in terms of the profile of the distribution to be smoothed.

Acknowledgment

It is a pleasure to thank Ueli Maurer for his motivation and support and Jan Camenisch for helpful remarks.

References

1. C. H. Bennett, F. Bessette, G. Brassard, L. Salvail, and J. Smolin, "Experimental quantum cryptography," *Journal of Cryptology*, vol. 5, no. 1, pp. 3–28, 1992.

2. C. H. Bennett, G. Brassard, C. Crépeau, and U. M. Maurer, "Generalized privacy amplification," *IEEE Transactions on Information Theory*, vol. 41, pp. 1915–1923, Nov. 1995.

3. C. H. Bennett, G. Brassard, and J.-M. Robert, "How to reduce your enemy's information," in *Advances in Cryptology — CRYPTO '85* (H. C. Williams, ed.), vol. 218 of *Lecture Notes in Computer Science*, pp. 468–476, Springer-Verlag, 1986.

4. C. H. Bennett, G. Brassard, and J.-M. Robert, "Privacy amplification by public discussion," *SIAM Journal on Computing*, vol. 17, pp. 210–229, Apr. 1988.

5. G. Brassard and C. Crépeau, "25 years of quantum cryptography," *SIGACT News*, vol. 27, no. 3, pp. 13–24, 1996.

6. G. Brassard and C. Crépeau, "Oblivious transfers and privacy amplification." Proceedings of EUROCRYPT '97, 1997.

7. C. Cachin and U. Maurer, "Smoothing probability distributions and smooth entropy." Preprint (abstract to appear in Proceedings of International Symposium on Information Theory, ISIT 97), 1997.

8. J. L. Carter and M. N. Wegman, "Universal classes of hash functions," *Journal of Computer and System Sciences*, vol. 18, pp. 143–154, 1979.

9. T. M. Cover and J. A. Thomas, *Elements of Information Theory*. New York: Wiley, 1991.

10. C. Crépeau, "Efficient cryptographic protocols based on noisy channels." Proceedings of EUROCRYPT '97, 1997.

11. J. Håstad, R. Impagliazzo, L. A. Levin, and M. Luby, "Construction of a pseudorandom generator from any one-way function," Tech. Rep. 91-068, International Computer Science Institute (ICSI), Berkeley, 1991.

12. R. Impagliazzo, L. A. Levin, and M. Luby, "Pseudo-random generation from one-way functions," in *Proc. 21st Annual ACM Symposium on Theory of Computing (STOC)*, pp. 12–24, 1989.

13. M. Kharitonov, "Cryptographic hardness of distribution-specific learning," in *Proc. 25th Annual ACM Symposium on Theory of Computing (STOC)*, pp. 372–381, 1993.

14. M. Luby, *Pseudorandomness and Cryptographic Applications*. Princeton University Press, 1996.

15. M. Luby and A. Wigderson, "Pairwise independence and derandomization," Tech. Rep. 95-035, International Computer Science Institute (ICSI), Berkeley, 1995.

16. U. M. Maurer, "Secret key agreement by public discussion from common information," *IEEE Transactions on Information Theory*, vol. 39, pp. 733–742, May 1993.

17. N. Nisan, "Extracting randomness: How and why — a survey," in *Proc. 11th Annual IEEE Conference on Computational Complexity*, 1996.

18. A. Rényi, "On measures of entropy and information," in *Proc. 4th Berkeley Symposium on Mathematical Statistics and Probability*, vol. 1, (Berkeley), pp. 547–561, Univ. of Calif. Press, 1961.

19. S. Vembu and S. Verdú, "Generating random bits from an arbitrary source: Fundamental limits," *IEEE Transactions on Information Theory*, vol. 41, pp. 1322–1332, Sept. 1995.

20. D. Zuckerman, "Simulating BPP using a general weak random source," *Algorithmica*, vol. 16, pp. 367–391, 1996. Preliminary version presented at 32nd FOCS (1991).

Information-Theoretically Secure Secret-Key Agreement by NOT Authenticated Public Discussion[1]

Ueli Maurer

Department of Computer Science
ETH Zurich
CH-8092 Zurich, Switzerland
maurer@inf.ethz.ch

Abstract. All information-theoretically secure key agreement protocols (e.g. based on quantum cryptography or on noisy channels) described in the literature are secure only against passive adversaries in the sense that they assume the existence of an authenticated public channel. The goal of this paper is to investigate information-theoretic security even against active adversaries with complete control over the communication channel connecting the two parties who want to agree on a secret key. Several impossibility results are proved and some scenarios are characterized in which secret-key agreement secure against active adversaries is possible. In particular, when each of the parties, including the adversary, can observe a sequence of random variables that are correlated between the parties, the rate at which key agreement against active adversaries is possible is characterized completely: it is either 0 or equal to the rate achievable against passive adversaries, and the condition for distinguishing between the two cases is given.

1 Introduction

One of the fundamental problems in cryptography is the generation of a shared secret key by two parties, Alice and Bob, not sharing a secret key initially, in the presence of an adversary Eve who has access to the communication channel connecting Alice and Bob. Several scenarios, which differ in their assumptions about Eve's capabilities and possibly about the intractability of certain computational problems, have been considered in the literature.

Public-key cryptography introduced by Diffie and Hellman [9] (see also [20]) solves this problem under the two assumptions that

(1) Eve is unable to solve a certain computational problem (such as factoring integers or computing discrete logarithms in a certain finite group) in feasible time, and

---

[1] This work is supported in part by the Swiss National Science Foundation, grant no. 20-42105.94.

W. Fumy (Ed.): Advances in Cryptology - EUROCRYPT '97, LNCS 1233, pp. 209-225, 1997.
© Springer-Verlag Berlin Heidelberg 1997

(2) that Eve has only passive (read) access to the communication channel between Alice and Bob, i.e., that the communication between Alice and Bob is authenticated.

The purpose of this paper is to investigate the described key distribution problem when neither of these assumptions is made: We consider adversaries with infinite computing power and complete control over the communication channel connecting Alice and Bob. Several impossibility results are proved and some scenarios in which secret-key agreement secure against active adversaries is possible are characterized. Secret-key agreement can be possible in this scenario only if Alice and Bob (but possibly also Eve) have correlated information. More formally, while Alice and Bob share no secret key initially, they know some random variables X and Y, respectively, jointly distributed with a random variable Z known to Eve. The joint probability distribution is denoted P_{XYZ}.

One can have different opinions about whether it is reasonable to assume that a specific computational problem is difficult. Furthermore, since quantum computation has been invented as a (at least for now) theoretical model of computation, it is not completely clear whether intractability assumptions in the Turing machine model of computation are still adequate. There also exist different opinions about whether certain methods of authentication, like speaker identification on a voice channel, are strong enough to support the second assumption above. It is not a goal of this paper to discuss these issues, but we believe that avoiding both assumptions is an interesting research topic.

There exists a substantial body of results on secret-key agreement by public discussion secure against adversaries with infinite computing power (see Section 2.3 for a brief summary), but they all depend in a crucial manner on the assumption that eavesdroppers are *passive* and hence the communication between Alice and Bob can be assumed to be authenticated. Of course, as is pointed out in these papers, the authenticity can be guaranteed, even when the channel is completely insecure, when Alice and Bob initially share a secret key that is used for authentication purposes (see Section 2.2). Hence these results can be interpreted as providing information-theoretically secure protocols for expanding a short initially shared secret key to an arbitrarily long secret key.

This paper characterizes scenarios in which secret-key agreement against active adversaries is possible and shows that for an important class of scenarios of correlated random variables available to Alice, Bob and Eve, active adversaries are not more powerful than passive ones.

2 Key-agreement protocols

2.1 Scenarios and definitions

We now formalize key-agreement protocols; the security of such protocols will be defined later.

Definition 1. A *key-agreement protocol* consists of three phases:

- a (possibly missing) initialization phase[1] in which Alice, Bob and an adversary Eve receive random variables X, Y and Z, respectively, which are jointly distributed according to some probability distribution P_{XYZ}.
- During the communication phase Alice and Bob alternate sending each other messages $C_1, C_2, \ldots$ where we assume that Alice sends messages $C_1, C_3, C_5, \ldots$ and Bob sends messages $C_2, C_4, C_6, \ldots$ Each message depends possibly on the sender's entire view of the protocol at the time it is sent and possibly on privately generated random bits. Let t be the total number of messages and let $C^t = [C_1, \ldots, C_t]$ denote the set of exchanged messages.
- Finally, Alice and Bob each either accepts or rejects the protocol execution, depending on whether they believe to be able to generate a secret key. If Alice accepts, she generates a key S depending on her view of the protocol. Similarly, if Bob accepts, he generates a key S' depending on his view of the protocol.

In general, the channel connecting Alice and Bob is completely insecure, i.e. Eve can see every message C_i and replace it by an arbitrary message $\tilde{C}_i$ of her choice. She need not keep Alice and Bob synchronized and she can impersonate either party by fraudulently initiating a protocol execution.

For stating impossibility results in the strongest possible form, we also consider protocols in which certain messages can be sent in a secret or authenticated manner (by appropriate means not specified by the protocol).

Definition 2. If a message C_i is *secret* (by the protocol specification), Eve learns nothing about it except that it exists[2]. However, she may replace such a message by a different message. If a message C_i is *authenticated* (by the protocol specification), then the receiver will always (with probability 1) detect any modification to the message due to Eve, but Eve sees the message.

Considering a passive adversary is equivalent to assuming the entire communication to be authenticated. The above definition can be made information-theoretically precise.

If two parties share a secret key, they can use the one-time pad encryption to transmit a message in perfect secrecy over a completely insecure channel. They can also use part of the secret key for authenticating messages (see Section 2.2).

---

[1] The initialization phase summarizes the parties' entire initial information, for instance the history of previous executions of protocols, the information resulting from quantum transmissions (like in quantum cryptography [2]), or information received from other sources like a satellite broadcasting random bits (see Section 4.3) or the signal of a deep-space radio source. When the initialization phase is missing, this means that Alice's and Bob's complete knowledge at the beginning of the protocol is assumed to be statistically independent.

[2] It is possible that she later obtains information about C_i because subsequent messages depend on C_i, but Eve never learns anything about C_i not provided by subsequent messages. This will be formalized in the full paper.

However, in contrast to perfect secrecy, perfect authenticity cannot be achieved even if a secret key of arbitrary fixed size is used because an adversary can always guess the key with non-zero probability of success. Authenticity and confidentiality are dual security properties, and the duality can be shown in various ways (e.g., see [16]).

All the protocol steps proposed in this paper are polynomial-time computable, but there may generally be steps in subprotocols taken from the literature that are not known to be computable in polynomial time. However, for every protocol resulting in Alice and Bob sharing a secret key mentioned here, there also exist efficient protocols for generating a secret key (which may be somewhat shorter).

In general, the distribution P_{XYZ} may be under Eve's partial control and may only partly be known to Alice and Bob. Two examples are the privacy amplification scenario [3] mentioned in Section 2.3, and quantum cryptography, where both Bob's and Eve's distributions depend on the type of measurement performed by Eve on the photons sent by Alice. In this paper we assume that P_{XYZ} is known to all parties.

In the sequel we assume without loss of generality that S and S' are binary strings of length $|S| = |S'| = k$. Clearly, the goal of a protocol is that S and S' agree with very high probability and that Eve has very little information about S. An adversary can of course block the communication between Alice and Bob completely by replacing all messages by empty messages, thus preventing any secret-key agreement. The goal of the design of a protocol can thus only be to generate a (hopefully large amount of) secret key when Eve is passive, but to detect any tampering with very high probability. However, even when Eve's strategy is active, it is allowed that she goes undetected if the secret key shared by Alice and Bob at the end of the protocol nevertheless is secret. In other words, Alice and Bob should not primarily be interested in catching an active cheater but in making sure that whenever they believe (or at least one of them believes) to have agreed on a secret key, then this is indeed the case with very high probability.

Definition 3. A key-agreement protocol with $|S| = k$ is (ϵ, δ)-*secure* if, for every passive eavesdropping strategy,

$$P[S \neq S'] \leq \epsilon,$$
$$I(S; C^t Z) \leq \epsilon,$$
$$\text{and} \qquad H(S) \geq k - \epsilon,$$

and if for every active adverse strategy, with probability at least $1 - \delta$, either Eve is caught by at least Alice or Bob (i.e. they do not both accept) or they successfully generate a secret key S (and S') satisfying the above conditions.

Note that one cannot require both Alice and Bob to reject. Eve could delete the last message from Alice to Bob (or vice versa) that would make Bob accept after Alice has accepted. (Byzantine agreement is impossible between two players in the presence of an active adversary.)

Here $H(S)$ denotes the entropy[3] of S and $I(S; C^t Z) = H(S) - H(S|C^t Z)$ denotes the information about S given by Eve's total observation (consisting of C^t and Z). The condition $H(S) \geq k - \epsilon$ implies that S is virtually uniformly distributed and together with the condition $I(S; C^t Z) \leq \epsilon$ it implies $H(S|C^t Z) \geq k - 2\epsilon$ and hence that S is also virtually uniformly distributed from Eve's point of view, i.e., given Eve's total information. Such a uniformity constraint could alternatively be defined in terms of any reasonable constraint on the deviation of a distribution from the uniform distribution, without changing the results of this paper.

2.2 Unconditionally secure message authentication

Adversaries with complete control over the communication channel have previously been considered in message authentication scenarios where, unlike in this paper, a secret key is shared initially by Alice and Bob about which Eve is assumed to have no information *a priori*.

Unconditionally secure message authentication based on a shared secret key was first considered in [11] and later in a large number of papers (e.g. [22], [23]). One of the most recent papers on this topic is by Gemmell and Naor [10] who proved the surprising result that interactive protocols for authenticating an n-bit message are more efficient in terms of the length of the secret key required to restrict an adversary's cheating probability to at most p. In particular, they proposed a one-round protocol using only $\log n - 2\log p$ bits of secret key and showed that this can be reduced to $\log^{(k)} n - 5\log p$ in a k-round protocol. We will make use of these results.

2.3 Review of the literature

In this section some of the results on secret-key agreement by perfectly authenticated public discussion are reviewed. Shannon's [21] famous result on perfect secrecy, stating that a cipher can achieve perfect secrecy only if the entropy of the secret key is at least as large as the entropy of the plaintext, can be considered as a special case (for 1-round protocols) of Theorem 1 below. Although Wyner's wire-tap channel scenario [25] and Csiszár and Körner's generalization [8] thereof do not include a public channel between Alice and Bob, they should nevertheless be mentioned here. In those scenarios, Alice can send information over a so-called broadcast channel where Bob and Eve can receive different outputs of the channel. Secret information transmission (and hence secret-key agreement) was shown to be possible if and only if Eve's channel is noisier than Bob's channel [8], an assumption that is generally unrealistic.

In the scenario considered in quantum cryptography (see [2] and references therein), Alice can send polarized light pulses of very low intensity to Bob over

---

[3] $H(S) = -\sum_{s: P_S(s) > 0} P_S(s) \log_2 P_S(s)$. See [6] for an introduction to the basic concepts of information theory.

some channel (e.g. an optical fiber) controlled by Eve. The use of this quantum communication results in Alice, Bob, and Eve possessing correlated strings. By subsequent discussion over the authenticated public channel, Alice and Bob manage to generate a secret key about which Eve has arbitrarily little information.

Another special case of key agreement protocols secure against passive adversaries is privacy amplification introduced in [4] and generalized in [3]. Privacy amplification is a protocol step that would typically be used as the last step in a practical key agreement protocol, but it can itself be described in the framework of key agreement protocols. Here Alice and Bob are assumed to know a string W (i.e. $X = Y = W$) about which Eve has some partial information. The protocol of [3] is secure even when Eve specifies an arbitrary probability distribution P_{ZW} unknown to Alice and Bob, subject to the only constraint that a bound on the second order Rényi entropy of W, given Eve particular value z of Z, is known to Alice and Bob. In the privacy amplification literature only passive adversaries have been considered. It is proved in [19] that privacy amplification secure against active adversaries is possible when the adversary's min-entropy about the string is more than half its length.

3 The case of no common initial information

In this section we characterize to what extent secret and/or authenticated communication between Alice and Bob can help them to agree on a secret key. These results demonstrate an interesting difference between computational and information-theoretic cryptography. In both models a secret channel from Alice to Bob can be transformed into an authenticated channel from Bob to Alice. This is achieved by Alice sending a secret key to Bob and Bob using the key in a message authentication techniques (see Section 2.2) for authenticating a message to be sent to Alice.

In sharp contrast, only the computational model allows to transform an authenticated channel from Alice to Bob into a secret channel from Bob to Alice. This is achieved by Alice sending her public key for a public-key cryptosystem to Bob who uses it to encrypt the message to be sent secretly to Alice. The security of public-key cryptosystems is inherently bound to be computational rather than information-theoretic. (Actually, this follows from Theorem 1 below.) See also [16] for a discussion of the described and other security transformations. It is hence not surprising that in the information-theoretic model, when Alice and Bob have no common information initially, authenticated channels are of no use, in contrast to secret channels.

Theorem 1. *Consider key agreement protocols without initialization phase which allow some of the exchanged messages to be either secret or authenticated. For $\epsilon \leq 1 - 3/(|S| + 2)$ there exists no such protocol that is (ϵ, δ)-secure, even when all messages are authenticated (or, equivalently, when Eve is passive.) Moreover, even if all messages from Alice to Bob are secret and all messages from Bob to*

Alice are authenticated, there exists no such protocol that is (ϵ, δ)-secure against active adversaries for any $\delta < 1$.

Proof. To prove the first part we make use of Theorem 1 of [14] which implies that

$$H(S) \leq H(S|S') + I(S; C^t) \tag{1}$$

for all such protocols. Note that the random variables X, Y do not exist in our context and hence $I(X; Y) = 0$ in Theorem 1 of [14] . Fano's Lemma (see [6]) states that the error probability p of guessing a random variable U when given a correlated random variable U' satisfies

$$H(U|U') \leq h(p) + p \log_2(|\mathcal{U}| - 1),$$

where $\mathcal{U}$ is the set of possible values that U can take on[4]. Therefore the condition $P[S \neq S'] \leq \epsilon$ implies

$$H(S|S') < h(\epsilon) + \epsilon k$$

which together with inequality (1) and the second and third conditions of Definition 3 gives

$$k - \epsilon \leq H(S) < h(\epsilon) + \epsilon k + \epsilon.$$

Using $h(\epsilon) \leq 1$, this implies $k - 1 \leq \epsilon(k + 2)$ and hence $\epsilon > 1 - 3/(k + 2)$.

To prove the second part, notice that from Bob's point of view, Alice has no advantage compared to Eve. When Eve performs the same protocol as Alice would, pretenting to be Alice, Bob accepts with the same probability as he would accept a protocol execution with Alice which according to the definition is 1. □

Note again that the first statement of the theorem is in sharp contrast to the public-key cryptographic scenario where, under a suitable intractability assumption, secret-key agreement secure against computationally bounded adversaries is possible when a single authenticated message in each direction can be sent. A public-key cryptosystem can be interpreted [16] as a means for transforming an authenticated channel into a secret channel in the other direction. The following well-known result is an observation following from Theorem 1.

Corollary 2. *A public-key cryptosystem can be computationally secure but not information-theoretically (i.e. unconditionally) secure.*

Theorem 3. *Assume that one secret (but not necessarily authenticated) message can be sent from Alice to Bob. Then, for any $\delta > 0$, key agreement $(0, \delta)$-secure against active adversaries is possible if, in addition, either an authenticated message can be sent from Alice to Bob or a secret message can be sent from Bob to Alice.*

---

[4] $h(p) = -p \log_2 p - (1 - p) \log_2(1 - p)$ denotes the binary entropy function which measures the entropy of a binary random variable that takes on the two values with probabilities p and $1 - p$.

Proof. Note that when the same message from Alice to Bob is both secret and authenticated, then Alice can simply send a secret key as the message. When two messages can be sent from Alice to Bob, one secret and one authenticated, then Alice can send a random n-bit string R to Bob ($n \geq -2\log_2 \delta$) over the secret channel and the description of a function f in a universal class hash functions from $\{0,1\}^n$ to $\{0,1\}^n$ [7] over the authenticated channel, together with the first $n/2$ bits of $f(R)$. The other half of $f(R)$ is kept by Alice and Bob as their secret key. If Eve's capability to interfere with the secret channel is limited to sending fraudulent messages (but she is assumed to be unable to modify a message sent from Alice to Bob), then no universal hash function is needed; it could instead be replaced by the identity function.

The proof for the case of a secret channel from Bob to Alice is based on the following protocol. Bob (secretly) sends Alice a random string U of sufficient length ($\Omega(\log \delta)$). Then they use the above protocol where the authenticated channel is obtained by Alice by using an authentication scheme [10] using R as the secret key. $\qquad\square$

Theorem 1 is pessimistic: it demonstrates that information-theoretically secure secret-key agreement against active or passive adversaries is impossible to achieve when the channel between Alice and Bob is completely insecure. However, if Alice and Bob have correlated information initially (not necessarily a secret key, but possibly only two bitstrings that are somehow correlated), about which also Eve has partial knowledge, then secret-key agreement can be possible.

In the following we consider such scenarios. One of our general goals is to achieve secret-key agreement under mild conditions on such an initialization phase, for instance conditions that can be argued to occur (or can be made to occur) in a realistic communications scenario.

4 Protocols with initialization phase

4.1 Impossibility results

The following theorem on authenticated public discussion follows from Corollary 1 in [14]. Recall from Section 2 that X, Y, and Z are the random variables obtained by Alice, Bob, and Eve, respectively, during the initialization phase.

Theorem 4. *For every probability distribution P_{XYZ}, a key agreement protocol that is (ϵ, δ)-secure against passive (or active) adversaries satisfies*

$$H(S) \leq \min[I(X;Y), I(X;Y|Z)] + h(\epsilon) + \epsilon(k+1).$$

In particular, for $\epsilon = 0$, we have $H(S) \leq \min[I(X;Y), I(X;Y|Z)]$.

Note that by definition, $I(X;Y) = H(X) - H(X|Y)$ and $I(X;Y|Z) = H(X|Z) - H(X|YZ)$ and that $I(X;Y|Z) \geq I(X;Y)$ is possible. It will be demonstrated in the following section that this theorem is not as pessimistic as it looks at first sight.

Theorem 4 states that secret-key agreement is possible and only if Y gives a substantial amount of information about X, both when Z is given or when it is not. In other words, X and Y must be correlated, and this correlation must to some extent be independent of Z. The bound $\min[I(X;Y), I(X;Y|Z)]$ can be replaced by the stronger bound derived in [18], called the intrinsic mutual information between X and Y given Z. It is the minimum of $I(X;Y|Z')$ over conditional probability distributions $P_{Z'|Z}$.

Definition 4. We call the distribution P_{XYZ} *X-simulatable by Eve* if Eve can generate from Z a random variable $\tilde{X}$ such that the pairs $[X,Y]$ and $[\tilde{X},Y]$ have the same distribution, i.e. if there exists a conditional probability distribution $P_{\tilde{X}|Z}$ such that

$$P_{\tilde{X}Y}(x,y) = P_{XY}(x,y)$$

for all x and y, where $P_{\tilde{X}Y}$ is the marginal distribution of $P_{X\tilde{X}YZ} = P_{XYZ} \cdot P_{\tilde{X}|Z}$, i.e.,

$$P_{\tilde{X}Y}(x,y) = \sum_{x'} \sum_{z} P_{XYZ}(x',y,z) \cdot P_{\tilde{X}|Z}(x,z).$$

Similarly, the distribution P_{XYZ} is called *Y-simulatable by Eve* if the symmetric condition with respect to Bob, with X replaced by Y and $\tilde{X}$ replaced by $\tilde{Y}$, is satisfied.

More intuitively, P_{XYZ} is X-simulatable by Eve if she can send Z through a (simulated) channel (characterized by $P_{\tilde{X}|Z}$) whose output $\tilde{X}$ has the same *joint* distribution with Y as X. (An example of such a distribution is given in Section 4.3.) Therefore, when P_{XYZ} is X-simulatable by Eve, then there is no way Bob can distinguish between a correct message sent by Alice and an appropriately generated fraudulent message sent by Eve. Similarly, when P_{XYZ} is Y-simulatable by Eve, then there is no way Alice can distinguish between a correct message sent by Bob or a fraudulent message sent by Eve. We obtain the following generalization of Theorem 1.

Theorem 5. *When P_{XYZ} is X-simulatable (or Y-simulatable) by Eve, then no key agreement protocol can be (ϵ, δ)-secure against active adversaries for any ϵ and $\delta < 1$, even if all messages from Alice to Bob (Bob to Alice) are perfectly secret and all messages from Bob to Alice (Alice to Bob) are authenticated.*

4.2 Independent repetition of a random experiment

In order to be able to derive interesting results on secret-key agreement against active or passive adversaries, we must consider specific types of probability distributions of the random variables given to Alice, Bob, and Eve.

One natural assumption is that the random experiment generating the triple $[X, Y, Z]$ is repeated many times independently. Hence we assume that Alice, Bob and Eve receive strings $X^n = [X_1, \ldots, X_n]$, $Y^n = [Y_1, \ldots, Y_n]$, and $Z^n =$

$[Z_1, \ldots, Z_n]$, respectively, where

$$P_{X^n Y^n Z^n}(x_1, \ldots, x_n, y_1, \ldots, y_n, z_1, \ldots, z_n) \;=\; \prod_{i=1}^{n} P_{XYZ}(x_i, y_i, z_i).$$

Note that we have changed the notation here and for the rest of the paper: P_{XYZ} now denotes the distribution of one of several random experiments while it previously denoted the distribution of the overall experiment.

This particular scenario is motivated by the well-known models for discrete memoryless sources and channels of communication theory. Many concrete practical scenarios can be modeled in this way, for instance the one discussed below in which Alice, Bob, and Eve receive noisy versions of a random string broadcast by a satellite or of the signal emitted by a deep space radio source.

For such a scenario of independent repetitions of a random experiment, the quantity that is of most interest is the maximal rate at which Alice and Bob can generate secret key bits, where rate is to be understood per execution of the random experiment generating a triple $[X, Y, Z]$.

Definition 5. The *secret key rate of P_{XYZ} for passive adversaries*, denoted $S(P_{XYZ})$, is the maximum rate at which Alice and Bob can agree on a secret key S while keeping a passive adversary's information about S arbitrarily small. More formally, it is the maximal R such that for all $\epsilon > 0$, for all $R' < R$, and for all sufficiently large n there exists a protocol with $|S| = \lfloor R'n \rfloor$ that is $(\epsilon, 0)$-secure against passive adversaries[5]. The *secret key rate of P_{XYZ} for active adversaries*, denoted $S^*(P_{XYZ})$, is defined in the same way, except that the adversary is allowed to be active, and for any given $\delta > 0$, (ϵ, δ)-security is required instead of $(\epsilon, 0)$-security.

The first part of this definition is given in [15] as a considerably strengthened definition of that given in [14], and the second part is new. In particular, in [14] it was only required that the *rate* at which Eve obtains information, $I(S; C^t Z^n)/n$ be arbitrarily small for large n, and proving results for the much stronger definition involves some technical steps, including privacy amplification [3]. The following result was proved in [15] (and in [14] using the weaker definition).

Theorem 6. $S(P_{XYZ})$ *is lower and upper bounded by*

$$\max[0, \, I(Y;X) - I(Z;X), \, I(X;Y) - I(Z;Y)] \;\leq\; S(P_{XYZ})$$

and

$$S(P_{XYZ}) \;\leq\; \min[I(X;Y), \, I(X;Y|Z)].$$

The lower bound is not tight in general. In particular, for the binary scenario discussed in Section 4.3, if Eve's channels is less noisy than both Alice's and Bob's channel, the lower bound vanishes while the secret-key rate is actually strictly positive.

---

[5] For the case of passive adversaries, $\delta = 0$ can trivially be achieved.

We are primarily interested in investigating the relation between $S(P_{XYZ})$ and $S^*(P_{XYZ})$, i.e., the power of authenticated versus non-authenticated communication. Quite surprisingly, it turns out that $S^*(P_{XYZ}) = 0$ or $S^*(P_{XYZ}) = S(P_{XYZ})$. However, before treating the general case, we consider the case of binary symmetric random variables which is of particular interest.

4.3 The binary case

In this section we consider the natural special case where the random variables known to Alice, Bob and Eve are noisy versions of a random string (e.g. broadcast by a satellite) received over binary symmetric channels C_A, C_B and C_E with bit error probabilities ϵ_A, ϵ_B and ϵ_E, respectively (see Figure 1). Without loss of generality we assume that these channels are independent because any scenario of dependent channels can be transformed [14] into an equivalent scenario of independent channels (with different bit error probabilities). In other words, when U denotes the random bit generated by the source ($P_U(0) = P_U(1) = 1/2$), we have

$$P_{XYZ|U} = P_{X|U} \cdot P_{Y|U} \cdot P_{Z|U}$$

where $P_{X|U}(x,r) = 1 - \epsilon_A$ if $x = u$ and ϵ_A else, $P_{Y|U}(y,r) = 1 - \epsilon_B$ if $y = u$ and ϵ_B else and $P_{Z|U}(z,r) = 1 - \epsilon_E$ if $z = u$ and ϵ_E else.

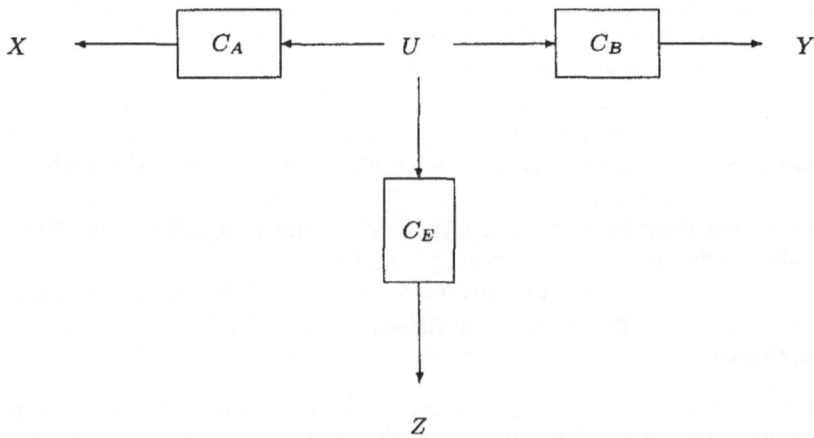

Fig. 1. The scenario of three independent channels

It is easy to verify that P_{XYZ} is X-simulatable by Eve if and only if $\epsilon_E \leq \epsilon_A$ and it is Y-simulatable by Eve if and only if $\epsilon_E \leq \epsilon_B$. Such a simulation can be achieved by Eve by sending Z through an additional (simulated) binary symmetric channel of appropriate bit error probability. Therefore, when either $\epsilon_E \leq \epsilon_B$ or $\epsilon_E \leq \epsilon_A$ in the described scenario, then $S^*(P_{XYZ}) = 0$ by Theorem 5.

Let

$$\epsilon_{AB} = \epsilon_A + \epsilon_B - \epsilon_A\epsilon_B$$

be the bit error probability between corresponding bits of Alice's and Bob's strings, and let similarly

$$\epsilon_{AE} = \epsilon_A + \epsilon_E - \epsilon_A\epsilon_E$$

and

$$\epsilon_{BE} = \epsilon_B + \epsilon_E - \epsilon_B\epsilon_E$$

be the bit error probabilities between corresponding bits of Alice's and Eve's and between Bob's and Eve's strings, respectively.

Assuming that Alice and Bob share no secret key initially, authentication for messages transmitted from Alice to Bob can nevertheless be achieved when Eve's channel is noisier than Alice's channel ($\epsilon_E > \epsilon_A$). This implies that $\epsilon_{BE} > \epsilon_{AB}$, i.e. that Alice's bits agree with Bob's bits with higher probability than Eve's bits agree with Bob's bits.

To demonstrate this fact, consider the following (very wasteful) authentication method.[6] A more efficient scheme will be considered below. In order to authenticate a single bit ($k = 1$) sent from Alice to Bob, Alice appends a substring of X^n of length l. The two substrings of X^n appended to authenticate a 0 or a 1 are disjoint. For instance, a 0 or a 1 is authenticated by appending (for some q) the string $[X_q, \ldots, X_{q+l-1}]$ or $[X_{q+m}, \ldots, X_{q+2l-1}]$, respectively, as the authenticator, and these $m = 2l$ bits of X^n are never used again for any other purpose. Bob expects to receive as an authenticator either a version of $[Y_q, \ldots, Y_{q+l-1}]$ or of $[Y_{q+m}, \ldots, Y_{q+2l-1}]$ with a fraction of close to ϵ_{AB} bit errors. Informally, Bob hence accepts the received bit if and only if the fraction of bits in the authenticator that agree with his noisy version of the authenticator ($[Y_q, \ldots, Y_{q+l-1}]$ or $[Y_{q+m}, \ldots, Y_{q+2l-1}]$) is not much smaller than $1 - \epsilon_{AB}$. It is easy to see that for any fixed $\epsilon_{BE} > \epsilon_{AB}$, the probability that Eve can successfully deceive Bob is exponentially small in l.

The described scheme is quite inefficient in terms of the number of bits used from the sequence. A much better approach is described in the proof of the following theorem.

Theorem 7. *When $\epsilon_{BE} > \epsilon_{AB}$ in the described binary scenario, a k-bit message sent from Alice to Bob can be authenticated by an l-bit authenticator with $l = 2k$ using $m = 4k$ bits of the random string X^n and achieving an arbitrarily small deception probability for sufficiently large k.*

Proof sketch. A scheme for authenticating a k-bit message sent from Alice to Bob using m bits of X^n (e.g. $[X_q, \ldots, X_{q+m-1}]$ for some q) can be derived as follows. Every message is authenticated by appending a particular subset of bits in $[X_q, \ldots, X_{q+m-1}]$. These subsets should be sufficiently disjoint to avoid that

---

[6] In the following we consider schemes for authenticating a k-bit message by an l-bit authenticator using $m > l$ bits of the common sequence.

such an authenticator can be guessed by Eve from an observed one. Bob checks whether his version of the authenticator (i.e. his subset of $[Y_q, \ldots, Y_{q+m-1}]$) agrees with the received authenticator on a fraction roughly $1 - \epsilon_{AB}$ of the bits, as expected when Alice sends the authenticator. Security requires that given one of these sets, it should be impossible for Eve to approximate a different authenticator of Alice with a bit error fraction close to ϵ_{AB}.

When Eve has intercepted a message together with its authenticator, her best strategy for creating an authenticator for a different message (hoping that it will be accepted by Bob) is to copy those bits from the received authenticator that are also contained in the new authenticator and to take as guesses for the remaining bits her copies of the bits (in $[Z_q, \ldots, Z_{q+m-1}]$), introducing bit errors in those bits with probability ϵ_{BE}. The maximal probability of successful deception is hence determined by the number d of bits that Eve must guess and the total number l of bits in the forged authenticator.

The expected value and the standard deviation of the number of bits in the correct autenticator that agree with Bob's corresponding bits are

$$\mu = l(1 - \epsilon_{AB})$$

and

$$\sigma = \sqrt{l\epsilon_{AB}(1 - \epsilon_{AB})},$$

respectively. When Eve tries to deceive Bob, the expected value and the standard deviation of the fraction of bits in the forged autenticator that agree with Bob's corresponding bits are

$$\mu' = (l - d)\epsilon_{AB} + d\epsilon_{BE}$$

and

$$\sigma' = \sqrt{(l - d)\epsilon_{AB}(1 - \epsilon_{AB}) + d\epsilon_{BE}(1 - \epsilon_{BE})},$$

respectively. Bob accepts an authenticator if and only if the number of his bits that agree with the corresponding authenticator bits is within q standard deviations of μ, where q is a security parameter that grows with l. The difference between the two expected values is $d\epsilon_{BE}$ and the standard deviation is $\sigma = \Omega(\sqrt{l})$. When d grows substantially faster than $\sqrt{l}$ one can let $q = \Omega(d/\sqrt{l})$. The law of large numbers implies that Eve's cheating probability decreases exponentially in q.

We now investigate how this can be achieved. An appropriate set of such subsets of bit positions (i.e., subsets of $\{1, \ldots, m\}$) can be interpreted as a code: each subset corresponds to a codeword of length m, where a 1 (or a 0) indicates that the bit at the corresponding position is (is not) contained in the subset. The weight of a codeword is equal to the length of the corresponding authenticator.

The desired distance property of the code differs from the Hamming distance considered in the theory of error-correcting codes. Instead, we define the *0-1 distance* from a codeword c_1 to a codeword c_2, denoted $d(c_1 \rightarrow c_2)$, as the number of bits that Eve must guess when trying to convert the authenticator corresponding to c_1 into the authenticator corresponding to c_2. The distance $d(c_1 \rightarrow c_2)$ is hence defined as the number of transitions from 0 to 1 when going from c_1 to c_2, hence not counting the transitions from 1 to 0. Note that this distance is not symmetric, i.e. $d(c_1 \to c_2) \neq d(c_2 \to c_1)$ in general. It is required that the $0-1$ distance from any codeword to any other codeword be large, say at least d. A conventional linear code cannot be used because the $0-1$ distance from any codeword to the zero-codeword is zero.

We now give a simple construction of codes that are good with respect to this distance measure. One can convert any code of length l and minimum distance d into a (non-linear) code of length $m = 2l$ and minimum $0-1$ distance d, where each codeword has weight l. This is achieved by replacing every bit in the original code by pair of bits, namely by replacing 0 by 01 and 1 by 10.

In the context of this proof, a possible code to be used for the construction is an extended Reed-Solomon code over a finite field $GF(2^r)$ [5]. For any K there exists such a code encoding K information digits into codewords of length $N = 2^r$ and with minimum distance $N - K + 1$. By interpreting elements of $GF(2^r)$ as binary substrings of length r, we obtain a binary code with 2^{rK} codewords of length $2rN$ and with minimum $0-1$ distance at least d.

By taking r as a security parameter and letting $N = 2^r$, $K = N/2$ and $k = rK$ we obtain $l = 2k = rN$ and $m = 2l = 2rN$. This is sufficient to complete the proof. □

By symmetry, the same technique can be used to authenticate messages sent from Bob to Alice, provided that $\epsilon_E > \epsilon_B$. This theorem shows that the rate at which random bits are needed for authentication is a constant factor times the bit rate at which Alice sends messages to Bob. Therefore, the secret key rate of P_{XYZ} for active adversaries is a constant (≤ 1) times the secret key rate of P_{XYZ} for passive adversaries. In the proof of the following theorem we need to show that the number of bits needed for authentication is asymptotically negligible compared to the number of bits needed for secret-key agreement (in the passive case).

Theorem 8. *When both $\epsilon_E > \epsilon_B$ and $\epsilon_E > \epsilon_A$ in the described scenario, then $S^*(P_{XYZ}) = S(P_{XYZ})$, i.e., an active adversary is not more powerful than a passive adversary. Otherwise, if either $\epsilon_E > \epsilon_B$ or $\epsilon_E > \epsilon_A$, then $S^*(P_{XYZ}) = 0$.*

Proof. The fact that $S^*(P_{XYZ}) = 0$ when either $\epsilon_E < \epsilon_B$ or $\epsilon_E < \epsilon_A$ follows from Theorem 5 because P_{XYZ} is either X-simulatable or Y-simulatable by Eve. The fact that $S^*(P_{XYZ}) = S(P_{XYZ})$ when $\epsilon_E > \epsilon_B$ and $\epsilon_E > \epsilon_A$ can be proved as follows. A suboptimal protocol based on the authentication method of Theorem 7 can be used to generate a relatively small t-bit secret key K, using $O(t)$ bits of the random string. This key can then be used, similar to a bootstrapping process, for instance based on the protocols of [10], to authenticate the messages exchanged in an optimal passive-adversary protocol $\mathcal{P}$ achieving $S(P_{XYZ})$. The size of K must only be logarithmic in the maximal size of a message exchanged in $\mathcal{P}$ [10] and linear in the number of rounds of $\mathcal{P}$. No matter what amount of secret key must be generated by $\mathcal{P}$, this can be achieved by using messages of size proportional to the key size in a constant number of rounds. Therefore, the ratio of size of K and the size of the generated key vanishes asymptotically. □

It is known from [14] that

$$\min[h(\epsilon_{AE}), h(\epsilon_{BE})] - h(\epsilon_{AB}) \leq S(P_{XYZ}) \leq 1 - h(\epsilon_{AB}).$$

It was recently proved that $S(P_{XYZ}) > 0$ unless $\epsilon_E = 0$ [17], even when both $\epsilon_E < \epsilon_B$ and $\epsilon_E < \epsilon_A$, i.e., even when the above lower bound vanishes (or is negative).

4.4 A completeness result for the general case

Let P_{XYZ} be an arbitrary probability distribution of a random experiment that is repeated many times. In general, only lower and upper bounds on $S(P_{XYZ})$ are known and $S(P_{XYZ})$ is known exactly only for special cases. The following theorem characterizes $S^*(P_{XYZ})$ completely in terms of P_{XYZ} and $S(P_{XYZ})$ and characterizes the power of active adversaries in comparison to passive ones for the described noisy-channel initialization scenario. Determining the exact power of a passive adversary remains an open problem.

Theorem 9. *When P_{XYZ} is either X-simulatable or Y-simulatable by Eve, then $S^*(P_{XYZ}) = 0$. Otherwise, $S^*(P_{XYZ}) = S(P_{XYZ})$.*

Proof sketch. The proof of this theorem relies on the theory of typical sequences[7] and is similar to the proof of Theorem 8, which is a special case of this theorem, but the technical details are omitted from this extended abstract. In order to authenticate a k-bit message by an $l = 2k$-bit authenticator using $m = 4k$ bits of X^n (or of Y^n when Bob is the sender), the described approach based on error correcting codes can be used to select the positions of a subsequence $[X_{i_1}, \ldots, X_{i_l}]$ of X^n. The receiver accepts the message if and only if the sequence of pairs $[(X_{i_1}, Y_{i_1}), \ldots, (X_{i_l}, Y_{i_l})]$ is γ-typical for the distribution P_{XY} for some suitable small γ. One can prove that for every distribution P_{XYZ} that is neither X-simulatable nor Y-simulatable by Eve, there exists a positive γ such that for sufficiently large k Eve's cheating probability is arbitrarily small. The same argument as in the proof of Theorem 8 can be used to prove that the ratio of bits needed for authentication and of bits used for secret-key agreement vanishes asymptotically. □

Acknowledgement

I would like to thank Christian Cachin and Stefan Wolf for interesting discussions and helpful comments.

---

[7] Loosely speaking, a sequence $U_1, \ldots, U_r$ of digits of an alphabet $\mathcal{U}$ is γ-typical for a given distribution P_U over $\mathcal{U}$ if for every $u \in \mathcal{U}$ the fraction of occurrences of u in $U_1, \ldots, U_r$ deviates by at most γ from $P_U(u)$ (see for instance [6]).

References

1. R. Ahlswede and I. Csiszár, Common Randomness in information theory and cryptography – part I: secret sharing, *IEEE Transactions on Information Theory*, Vol. IT–39, 1993, pp. 1121–1132.

2. C. H. Bennett, F. Bessette, G. Brassard, L. Salvail and J. Smolin, "Experimental quantum cryptography", *Journal of Cryptology*, Vol. 5, no. 1, 1992, pp. 3–28.

3. C.H. Bennett, G. Brassard, C. Crépeau, and U.M. Maurer, "Generalized privacy amplification", to appear in *IEEE Transactions on Information Theory*, Nov. 1995.

4. C. H. Bennett, G. Brassard and J.-M. Robert, "Privacy amplification by public discussion", *SIAM Journal on Computing*, Vol. 17, no. 2, April 1988, pp. 210–229.

5. R. E. Blahut, *Theory and Practice of Error Control Codes*, Reading, MA: Addison-Wesley, 1983.

6. R. E. Blahut, *Principles and Practice of Information Theory*, Reading, MA: Addison-Wesley, 1987.

7. J. L. Carter and M. N. Wegman, "Universal classes of hash functions", *Journal of Computer and System Sciences*, Vol. 18, 1979, pp. 143–154.

8. I. Csiszár and J. Körner, "Broadcast channels with confidential messages", *IEEE Transactions on Information Theory*, Vol. IT–24, no. 3, 1978, pp. 339–348.

9. W. Diffie and M. E. Hellman, "New directions in cryptography", *IEEE Transactions on Information Theory*, Vol. IT–22, 1976, pp. 644–654.

10. P. Gemmell and M. Naor, Codes for interactive authentication *Advances in Cryptology — Proceedings of Crypto '93*, Lecture Notes in Computer Science, Vol. 773, Springer–Verlag, Berlin, 1994, pp. 355–367.

11. E. N. Gilbert, F. J. MacWilliams, and N. J. A. Sloane, Codes which detect deception, *Bell Syst. Tech. J.*, Vol. 53, No. 3, 1974, pp. 405–424.

12. R. L. Graham, D. E. Knuth and O. Patashnik, *Concrete mathematics*, Reading, MA: Addison-Wesley, 1990.

13. U.M. Maurer, Protocols for secret key agreement by public discussion based on common information, *Advances in Cryptology - CRYPTO '92*, Lecture Notes in Computer Science, Berlin: Springer-Verlag, vol. 740, pp. 461–470, 1993.

14. U. M. Maurer, Secret key agreement by public discussion from common information, *IEEE Transactions on Information Theory*, vol. IT–39, 1993, pp. 733–742.

15. U. M. Maurer, The strong secret key rate of discrete random triples, *Communications and Cryptography, Two Sides of one Tapestry*, R.E. Blahut et al. (editors), Kluwer Academic Publishers, 1994, pp. 271–285.

16. U. M. Maurer and P.E. Schmid, A calculus for security bootstrapping in distributed systems, *Journal of Computer Security*, vol. 4, no. 1, pp. 55–80, 1996.

17. U. M. Maurer and S. Wolf, Towards characterizing when information-theoretic secret key agreement is possible, *Advances in Cryptology - ASIACRYPT '96*, K. Kim and T. Matsumoto (Eds.), Lecture Notes in Computer Science, Berlin: Springer-Verlag, vol. 1163, pp. 145–158, 1996.

18. U. M. Maurer and S. Wolf, The intrinsic conditional mutual information and perfect secrecy, to appear in *Proc. 1997 IEEE Symposium on Information Theory*, (Abstracts), Ulm, Germany, June 29–July 4, 1997.

19. U. M. Maurer and S. Wolf, Privacy amplification secure against active adversaries, preprint, 1997.

20. R. L. Rivest, A. Shamir, and L. Adleman, A method for obtaining digital signatures and public-key cryptosystems, *Communications of the ACM*, Vol. 21, No. 2, 1978, pp. 120–126.

21. C. E. Shannon, Communication theory of secrecy systems, *Bell System Technical Journal*, Vol. 28, October 1949, pp. 656–715.
22. G. J. Simmons, Authentication theory/coding theory, in *Advances in Cryptology – CRYPTO 84*, G.R. Blakley and D. Chaum (Eds.), Lecture Notes in Computer Science, No. 196, Berlin: Springer Verlag, 1985, pp. 411–431.
23. D. R. Stinson, Universal hashing and authentication codes, *Advances in Cryptology — Proceedings of Crypto '91*, Lecture Notes in Computer Science, Vol. 576, Springer–Verlag, Berlin, 1994, pp. 74–85.
24. M. N. Wegman and J. L. Carter, New hash functions and their use in authentication and set equality, *Journal of Computer and System Sciences*, Vol. 22, 1981, pp. 265–279.
25. A. D. Wyner, The wire-tap channel, *Bell System Technical Journal*, Vol. 54, no. 8, 1975, pp. 1355–1387.

Linear Statistical Weakness of Alleged RC4 Keystream Generator

Jovan Dj. Golić *

School of Electrical Engineering, University of Belgrade
Bulevar Revolucije 73, 11001 Beograd, Yugoslavia

Abstract. A keystream generator known as RC4 is analyzed by the linear model approach. It is shown that the second binary derivative of the least significant bit output sequence is correlated to 1 with the correlation coefficient close to $15 \cdot 2^{-3n}$ where n is the variable word size of RC4. The output sequence length required for the linear statistical weakness detection may be realistic in high speed applications if $n \leq 8$. The result can be used to distinguish RC4 from other keystream generators and to determine the unknown parameter n, as well as for the plaintext uncertainty reduction if n is small.

1 Introduction

Any keystream generator for practical stream cipher applications can generally be represented as an autonomous finite-state machine whose initial state and possibly the next-state and output functions as well are secret key dependent. A common type of keystream generators consists of a number of possibly irregularly clocked linear feedback shift registers (LFSRs) that are combined by a function with or without memory. Standard cryptographic criteria such as a large period, a high linear complexity, and good statistical properties are thus relatively easily satisfied, see [16], [17], but such a generator may in principle be vulnerable to various divide-and-conquer attacks in the known plaintext (or ciphertext-only) scenario, where the objective is to reconstruct the secret key controlled LFSR initial states from the known keystream sequence, for a survey see [17] and [6]. Most the attacks require an exhaustive search over the initial states of a subset of the LFSRs, with the exception of a small number of faster cryptanalytic attacks which may work for long LFSRs as well, such as fast correlation attacks [13] based on iterative probabilistic decoding, the conditional correlation attack [14] based on information set decoding, and the inversion attack [10], all on regularly clocked LFSRs, and a specific fast correlation attack on irregularly clocked LFSRs whose theoretical framework is developed in [8]. In practice, the initial state is for resynchronization purposes also made dependent on a

---

* This work was done while the author was with the Information Security Research Centre, Queensland University of Technology, Brisbane, Australia. This research was supported in part by the Science Fund of Serbia, grant #04M02, through the Mathematical Institute, Serbian Academy of Science and Arts.

randomizing key, which is typically sent in the clear before every new message to be encrypted. This may open new possibilities for cryptanalytic attacks, see [2].

In the open literature, there is a very small number of proposed keystream generators that are not based on shift registers. For example, an interesting design approach, which may have originated from the table-shuffling principle [12], is to use a relatively big table that slowly varies in time under the control of itself. A keystream generator [15] publicized in [18] and known as RC4 (although a public confirmation is still missing) is such an example, which is according to [18] widely used in many commercial products, including Lotus Notes, Apple Computer's AOCE, Oracle Secure SQL, and the Cellular Digital Packet Data specification [1]. Another, somewhat similar example is a keystream generator called ISAAC [11]. Of course, one may also use a set of tables controlling each other, but this may lead to some divide-and-conquer attacks. The resulting schemes are hardly analyzable, and about the only known theoretical argument [4] concerns the period of the internal state sequence, but has probabilistic rather than deterministic nature. Namely, if the internal memory size is M and if the next-state function is randomly chosen according to the uniform distribution, then the average cycle and tail lengths are both around $2^{M/2}$, whereas if the next-state function is in addition required to be invertible, then the internal state period (cycle length) is uniformly distributed between 1 and 2^M, with the average value 2^{M-1}.

The statistical properties of the keystream sequence are typically measured by standard statistical tests, and for some sequences, including the LFSR ones, theoretical results can be derived as well. For keystream generators like RC4 such theoretical results are difficult to establish. The results typically deal with the relative frequency of occurrence of blocks of successive symbols within a period, where the block size is assumed to be smaller than the internal memory size. However, it is shown in [7], [9] that for block sizes bigger than M, a linear statistical weakness or a so-called linear model always exists and can be efficiently determined by the linear sequential circuit approximation (LSCA) method [5]. The linear statistical weakness is a linear relation among the keystream bits that holds with probability different from one half. It turns out [9] that for many practical schemes, including the clock-controlled LFSRs, the keystream sequence length needed to detect the weakness is considerably shorter than the period. Although the weakness may not lead to a significant plaintext uncertainty reduction, it is structure dependent and can be used as such to distinguish between different types of keystream generators and for secret key reconstruction as well. As well, linear models of individual components of a keystream generator can be utilized in correlation attacks, whereas multiple linear models can also be used to mount fast correlation attacks [8] on clock-controlled LFSRs.

The main objective of this paper is to derive linear models for RC4 by using the LSCA method [5], [9]. The LSCA method consists in determining and solving a linear sequential circuit that approximates a given keystream generator and yields linear models with comparatively large correlation coefficient c, where the probability of the corresponding linear relation among the keystream bits is $(1 + c)/2$. It also gives an estimate of c, but sometimes, as in the case of RC4, special techniques have to be developed to obtain more accurate estimates of c.

Given a parameter n, the internal state of RC4 consists of a balanced table (permutation) of 2^n binary words of dimension n and two pointer binary words of the same dimension, n, which, at each time, define the positions of two words in the table to be swapped to produce the table at the next time. The internal memory size[1] is thus practically given as $M = n2^n + 2n$. One of the pointers is updated by using the table content at the position defined by the other, which is in turn updated in a known way by a counter. Initially, the two pointer words are set to zero and the table content is defined by the secret key in a specified way. At each time, the output of RC4 is a binary word of dimension n which is taken from an appropriate position in the table. The output word is then bitwise added to the plaintext word to give the ciphertext word.

Let $z = (z_t)_{t=1}^\infty$ denote the least significant bit output sequence of RC4 and let $\dot{z} = (\dot{z}_t = z_t + z_{t+1})_{t=1}^\infty$ and $\ddot{z} = (\ddot{z}_t = z_t + z_{t+2})_{t=1}^\infty$ denote its first and second binary derivatives, respectively. Our main results are to show that $\dot{z}$ is correlated neither to 1 nor to 0 and that $\ddot{z}$ is correlated to 1 with the correlation coefficient close to $15 \cdot 2^{-3n}$ for large 2^n. Since the output sequence length needed to detect a statistical weakness with the correlation coefficient c is $O(c^{-2})$, the required length is around $64^n/225$. For example, if $n = 8$, as recommended in most applications, the required length is close to $2^{40} \approx 10^{12}$. Experimental results agree well with the above theoretical predictions. As the resulting correlation coefficient is significantly bigger than $2^{M/2}$, $M = n2^n + 2n$, the determined linear model should be regarded as a statistical weakness, at least on a theoretical level. Moreover, the output sequence length required for the detection may even be realistic in high speed applications if $n \leq 8$. Also note that the second binary derivative weakness involves only three successive least significant output bits which is much smaller than the memory size. The weakness is a consequence of a very simple next-state function of RC4. It is also shown that similar linear relations hold for other output bits as well, but the correlation coefficients are smaller.

In Section 2, a more detailed description of the RC4 keystream generator is presented. In Section 3, some relevant correlation properties of random boolean functions are derived, while the linear models of RC4 and the corresponding correlation coefficients are determined in Section 4. A summary and conclusions are given in Section 5. Central moments of an underlying discrete probability distribution needed for estimating the correlation coefficients are evaluated in the Appendix.

---

[1] The effective internal memory size is slightly smaller and is according to Stirling's approximation given as $\log 2^n! + 2n \approx 2^n(n - \log e) + 5n/2 + \log \sqrt{2\pi}$. All the logarithms are to the base 2 throughout.

2 Description of RC4

We will follow the description given in [18]. RC4 is in fact a family of algorithms indexed by a parameter n, which is a positive integer typically recommended to be equal to 8. The internal state of RC4 at time t consists of a table $S_t = (S_t(l))_{l=0}^{2^n-1}$ of 2^n n-bit words and of two pointer n-bit words i_t and j_t. So, the internal memory size[1] is $M = n2^n + 2n$. Let the output n-bit word of RC4 at time t be denoted by Z_t. As usual, we keep the same notation for the binary and integer representations of n-bit words, where, for example, the least significant bit is the leftmost one. Let initially $i_0 = j_0 = 0$. Then the next-state and output functions of RC4 are for every $t \geq 1$ defined by

$$i_t = i_{t-1} + 1 \tag{1}$$

$$j_t = j_{t-1} + S_{t-1}(i_t) \tag{2}$$

$$S_t(i_t) = S_{t-1}(j_t), \quad S_t(j_t) = S_{t-1}(i_t) \tag{3}$$

$$Z_t = S_t(S_t(i_t) + S_t(j_t)) \tag{4}$$

where all the additions are modulo 2^n. It is assumed that all the words except for the swapped ones remain the same (swapping itself is effective only if $i_t \neq j_t$). The output n-bit word sequence is $Z = (Z_t)_{t=1}^{\infty}$.

The initial table S_0 is defined in terms of the key string $K = (K_l)_{l=0}^{2^n-1}$ by using the same next-state function starting from the table (identity permutation) $(l)_{l=0}^{2^n-1}$. More precisely, set $j_0 = 0$ and for every $1 \leq t \leq 2^n$, compute $j_t = (j_{t-1} + S_{t-1}(t-1) + K_{t-1}) \bmod 2^n$ and then swap $S_{t-1}(t-1)$ with $S_{t-1}(j_t)$. The last produced table represents S_0. The key string K is composed of the secret key, possibly repeated, and of the randomizing key which is sent in the clear for resynchronization purposes.

There are no published results regarding RC4. The known pointer sequence $\{i_t\}_{t=0}^{\infty}$ ensures that every element in the table is affected by swapping at least once in any 2^n successive times and, also, that the next-state function is invertible (one-to-one). Accordingly, the state diagram consists of cycles only, which, according to [4], can be expected to have average length close to 2^{M-1} and are very unlikely to be short if $n \geq 5$. Of course, since the next-state function of RC4 is not randomly chosen, this remains to be proved, if possible at all.

3 Correlation Properties of Random Boolean Functions

The correlation coefficients of the linear models of RC4 to be determined in the next section are related to certain correlation properties of random boolean functions. These properties provide insight into the linear statistical weaknesses of RC4 and are as such pointed out in this section. Note that the correlation properties of boolean functions for cryptographic applications are first introduced in [19]. Let f denote an arbitrary boolean function of n variables and let $f(X)$ denote the value of f at a point $X = (x_0, \ldots, x_{n-1}) \in \{0,1\}^n$. We will use the same notation, X, for the integer representation of X too, that is, for $\sum_{i=0}^{n-1} x_i 2^i$. A boolean function f is called balanced if it has the same number of zeros and ones in its truth table. In the probabilistic analysis to follow, we will, for simplicity, keep the same notation for random variables and their values. As usual, the correlation coefficient between any two binary random variables x and y is defined as $c = \Pr\{x = y\} - \Pr\{x \neq y\}$. The correlation coefficient of a single binary random variable x is defined as the correlation coefficient between x and the constant zero variable. Accordingly, let for any two boolean functions f and g, $c(f,g)$ denote the correlation coefficient between $f(X)$ and $g(X)$, and let $c(f)$ stand for $c(f,0)$, where X is uniformly distributed. A basic result to be used is that the correlation coefficient of a sum of independent binary random variables is equal to the product of their individual correlation coefficients, see [9] (addition of binary variables is modulo 2 throughout).

Proposition 1. *Let X and Y be two independent uniformly distributed n-dimensional binary random variables and let f be a uniformly random boolean function of n variables. Let l be an arbitrary linear boolean function of n variables (including the constant zero function). Then the correlation coefficient c between $f(X) + f(Y)$ and $l(X) + l(Y)$ is equal to $1/2^n$. (Instead of being linear, l may be any boolean function of n variables.)*

Proof. Let c_f denote the correlation coefficient between $f(X) + l(X)$ and $f(Y) + l(Y)$ for any fixed f. The correlation coefficient c is then equal to the expected value of c_f over uniformly random f. The correlation coefficient c_f is clearly equal to the correlation coefficient of $f(X) + l(X) + f(Y) + l(Y)$ which is in turn equal to the product of the correlation coefficients of $f(X) + l(X)$ and $f(Y) + l(Y)$, as X and Y are independent. Since the two are equal, we get that $c_f = c(f,l)^2$. Since l is fixed, c is then equal to the expected value $E(c(f)^2)$, where $c(f)$ is itself given as $2^{1-n}(k - 2^{n-1})$ with k being the number of zeros in the truth table of f. As k has the binomial distribution $\left\{ \binom{2^n}{k} 2^{-2^n} \right\}_{k=0}^{2^n}$, it follows that $E(c(f)^2) = 2^{2(1-n)} \text{Var}(k) = 2^{-n}$, because the variance $\text{Var}(k)$ is equal to 2^{n-2}. $\qquad\square$

Proposition 2. *Let X and Y be two independent uniformly distributed n-dimensional binary random variables and let f be a uniformly random balanced boolean function of n variables. Let l be an arbitrary nonzero linear boolean function of n variables. Then the correlation coefficient of $f(X) + f(Y)$ is equal to zero and the correlation coefficient c between $f(X) + f(Y)$ and $l(X) + l(Y)$ is equal to $1/(2^n - 1)$. (Instead of being linear, l may be any balanced boolean function of n variables.)*

Proof. First note that for any balanced f, the correlation coefficient of $f(X)$ is equal to zero. Then the correlation coefficient of $f(X) + f(Y)$ is equal to zero since, for any fixed f, it is the product of two zero correlation coefficients. Second, proceeding along similar lines as in the proof of Proposition 1, we get that $c = E(c(f, l)^2)$. Since l is balanced and fixed, $c(f, l)$ is given as $2^{2-n}(k - 2^{n-2})$ where k is the number of zeros in the half of the truth table of f where $l(X) = 0$. The probability distribution of k is $\left\{ \binom{2^{n-1}}{k}^2 / \binom{2^n}{2^{n-1}} \right\}_{k=0}^{2^{n-1}}$ with the variance $\mathrm{Var}(k) = 2^{2(n-2)}/(2^n - 1)$. Hence $E(c(f, l)^2) = 2^{2(2-n)} \mathrm{Var}(k) = 1/(2^n - 1)$. $\square$

Proposition 3. *Let l be an arbitrary nonzero linear boolean function of n variables and let f be a uniformly random balanced boolean function of n variables such that $c(f, l) = c$ where c is a given constant. Then the correlation coefficient of $f(X) + l(X)$ is equal to c for any fixed X. (Instead of being linear, l may be any balanced boolean function of n variables.)*

Proposition 4. *Let X be a uniformly distributed n-dimensional binary random variable and let f be a uniformly random balanced boolean function of n variables. Let $X + 1$ denote the integer addition modulo 2^n of X and 1. Then the correlation coefficient of $f(X) + f(X + 1) + 1$ is equal to $1/(2^n - 1)$. Furthermore, let l be a linear function defined as $l(X) = x_0$ and let f be in addition such that $c(f, l) = c$ where c is a given constant. Then the correlation coefficient of $f(X) + f(X + 1) + 1$ is equal to c^2 for any fixed X. (Instead of $X + 1$, one may take any permutation $P(X)$ such that $P(X) \neq X$, $X \in \{0, 1\}^n$, but then a balanced function l has to be defined appropriately.)*

4 Linear Models

The essence of the linear sequential circuit approximation (LSCA) method [5], [9] applied to binary keystream generators is in finding good linear approximations to the output and the component next-state functions and in solving the resulting linear sequential circuit. Its objective is to obtain feedforward linear transforms (i.e., linear sequential transforms with finite input memory) of the output sequence that are correlated to linear transforms of the initial state variables (to be used in correlation attacks) and, in particular, to the constant zero sequence, in which case the output linear transform defines a linear relation among the output bits that holds with probability different from one half. The resulting probabilistic linear recursion is called a linear model [9]. Estimating the correlation coefficients can be a problem on its own. In the underlying probabilistic model, the initial state is assumed to be random and uniformly distributed, and if the next-state function is one-to-one, then the internal state at any time is also uniformly distributed, so that the resulting correlation coefficients are time independent, see [9].

In the case of RC4, the next-state function is one-to-one and the balanced initial table S_0 (each n-bit word appears exactly once) can be assumed to be uniformly random, but the initial pointer words i_0 and j_0 are both fixed to zero. It follows that for every $t \geq 0$, the table S_t is uniformly random and balanced, whereas i_t is deterministic and known and j_t is uniformly distributed for $t \geq 1$, but dependent on S_t. As a consequence, while the dependence between j_t and S_t is insignificant, the deterministic nature of i_t may in principle lead to linear models with time dependent correlation coefficients. A related approach is to fix the initial state and to consider the same linear relation at random times, in which case the average value of the correlation coefficient over time is relevant. If the tail and cycle lengths combined are big (as one should expect for RC4), then the obtained correlation coefficient should be close to the value corresponding to a fixed time and a random initial state.

Since RC4 has n binary outputs, one should first decide on a linear combination of these outputs to be linearly approximated. To maximize the correlation coefficients, we will consider the individual binary outputs. Let $Z_t^{(k)}$, $i_t^{(k)}$, $j_t^{(k)}$, and $S_t^{(k)}$ denote the kth components of Z_t, i_t, j_t, and S_t, respectively, $0 \leq k \leq n-1$, where $k = 0$ corresponds to the least significant bit of the corresponding n-bit words. Note that S_t defines a uniformly random balanced vectorial boolean function $\{0,1\}^n \to \{0,1\}^n$, so that $S_t^{(k)}$ is a uniformly random balanced boolean function of n variables. As the linearization of Z_t and j_t necessarily involves finding linear approximations to S_t, the problem is to find such approximations leading to the correlation coefficients that do not vanish for a random S_t. The main point of the LSCA method applied to RC4 is that S_t can be approximated by S_{t-1}, because of the slow change of the table due to swapping. Another point is that $S_{t-1}^{(k)}$ can be approximated by any linear function of its inputs, but to maximize the overall correlation coefficient, $S_{t-1}^{(k)}$ is approximated by its kth binary input. As before, all the additions of l-bit words are integer additions modulo 2^l (usually, $l = 1$ or $l = n$).

As a result, we get $Z_t^{(k)} \approx S_{t-1}^{(k)}(i_{t-1} + 1) + S_{t-1}^{(k)}(j_{t-1} + S_{t-1}(i_{t-1} + 1)) \approx j_{t-1}^{(k)}$, where $S_{t-1}^{(k)}$ is linearized exactly twice. It then follows that $Z_t^{(k)} + Z_{t+1}^{(k)} \approx j_{t-1}^{(k)} + j_t^{(k)} \approx i_t^{(k)}$, where $i_t^{(k)}$ is known for every $t \geq 1$. The total number of linear approximations needed is five. In order for the overall correlation coefficient not to vanish, the total number of linear approximations to $S_{t-1}^{(k)}$ should be even, because positive and negative correlation coefficients would otherwise cancel out. More precisely, Proposition 2 can be extended to deal with an arbitrary number of linear approximations, in which case the resulting correlation coefficient is related to the central moments of the probability distribution considered in the Appendix, and the odd central moments are necessarily equal to zero. So, the first binary derivative of any binary component of the n-dimensional output sequence does not represent a linear model with a nonzero correlation coefficient.

Further, by adding two successive bits of the first binary derivative sequence we get that $Z_t^{(k)} + Z_{t+2}^{(k)} \approx i_t^{(k)} + i_{t+1}^{(k)}$, which is further equal to 1 if $k = 0$ and can be approximated as 0 if $1 \leq k \leq n-1$. The total number of linear approximations needed for this is at most ten and will be shown be equal to six. Accordingly, the second binary derivative of any binary component of the output sequence defines a linear model with a nonzero correlation correlation coefficient, to be determined in the sequel. The most significant correlation coefficient is obtained for the least significant bit, that is, for $k = 0$. Other linear models for RC4 should have smaller or much smaller correlation coefficients.

Our objective now is to estimate the correlation coefficient between the second binary derivative $\ddot{Z}_t^{(0)} = Z_t^{(0)} + Z_{t+2}^{(0)}$ and 1, for any $t \geq 1$. Letting $F = S_t$, $F' = S_{t+1}$, $F'' = S_{t+2}$, $X = i_t$, and $Y = j_t$, we have

$$\ddot{Z}_t^{(0)} = F^{(0)}(F(X) + F(Y)) + $$
$$F''^{(0)}(F''(X+2) + F''(Y + F(X+1) + F'(X+2))) \qquad (5)$$

where Y is uniformly distributed, F is a uniformly random balanced vectorial boolean function, and F' and F'' are obtained from F by one and two random swappings of two n-bit words, respectively, whereas X is fixed for any particular t and is uniformly distributed for a random t.

The direct computation of the correlation coefficient by using (5) is not possible since the functions F, F', and F'' are random. The starting point of our approach is forming the following series of linear approximations:

$$\ddot{Z}_t^{(0)} \approx F^{(0)}(X) + F^{(0)}(Y) + $$
$$F''^{(0)}(F''(X+2) + F''(Y + F(X+1) + F'(X+2))) \qquad (6)$$
$$\approx F^{(0)}(X) + F^{(0)}(Y) + $$
$$F''^{(0)}(X+2) + F''^{(0)}(Y + F(X+1) + F'(X+2)) \qquad (7)$$
$$\approx F^{(0)}(X) + F^{(0)}(Y) + F''^{(0)}(X+2) + $$
$$Y^{(0)} + F^{(0)}(X+1) + F'^{(0)}(X+2) \qquad (8)$$
$$\approx F^{(0)}(X) + Y^{(0)} + F''^{(0)}(X+2) + $$
$$Y^{(0)} + F^{(0)}(X+1) + F'^{(0)}(X+2) \qquad (9)$$
$$\approx F^{(0)}(X) + F^{(0)}(X+1) \qquad (10)$$
$$\approx 1. \qquad (11)$$

The next point is to observe that the correlation coefficients of the individual linear approximations can be computed if conditioned on the random functions in an appropriate way. Let $c_f = c(F^{(0)}, X^{(0)})$, $c'_f = c(F'^{(0)}, X^{(0)})$, and $c''_f = c(F''^{(0)}, X^{(0)})$ be the correlation coefficients between $F^{(0)}$ and $X^{(0)}$, $F'^{(0)}$ and $X^{(0)}$, and $F''^{(0)}$ and $X^{(0)}$, respectively, where the subscript f indicates the dependence upon a particular balanced boolean function f (here $f = F^{(0)}$). Then the linear approximations (6), (7), (8), and (9) hold with the correlation coefficients c_f, c''_f, c''_f, and c_f, respectively, where $F^{(0)}$, $F'^{(0)}$, and $F''^{(0)}$ are fixed and X is either uniformly distributed or fixed. The linear approximation (10) holds for any fixed X with the correlation coefficient $\varepsilon'_{m'} = 1 - m'2^{1-n}$ (conditioned on m') if $F'^{(0)}$ is a uniformly random balanced boolean function and if $F''^{(0)}$ is produced from $F'^{(0)}$ by a random effective change, due to swapping, of m' bits, where, as before, m' takes values 0 and 2, each with probability $1/2$. The linear approximation (11) holds for any fixed X with correlation coefficient c_f^2 if $F^{(0)}$ is a uniformly random balanced boolean function with a fixed correlation coefficient c_f to $X^{(0)}$, see Proposition 4.

Now, let m denote the number of bits where $F^{(0)}$ and $F''^{(0)}$ are effectively different. Under the *independence assumption* that the individual linear approximations are independent when conditioned on c_f, m', and m, the correlation coefficient between $\ddot{Z}_t^{(0)}$ and 1 is given as $c_f^4 c_f''^2 \varepsilon_{m'}'$, where $c_f'' = c_f \varepsilon_m$, $\varepsilon_m = 1 - m2^{1-n}$, if $F^{(0)}$ is a uniformly random balanced boolean function with a fixed correlation coefficient c_f to $X^{(0)}$, where X is either uniformly distributed or fixed. The resulting correlation coefficient conditioned on c_f, m', and m is thus equal to $c_f^6 \varepsilon_m^2 \varepsilon_{m'}'$. Note that the independence assumption seems to be the only tractable way of combining the individual linear approximations.

Consequently, the overall correlation coefficient is then given as

$$c = E(c_f^6) \cdot E(\varepsilon_m^2) \cdot E(\varepsilon_{m'}') \tag{12}$$

where the expectations are over random c_f, m, and m', respectively (for simplicity, it is assumed that the random variables m' and m are independent). From the proof of Proposition 2, recall that c_f can be expressed as $2^{2-n}(k - 2^{n-2})$ where k (standing for the number of zeros in the half of the truth table of $f = F^{(0)}$ where $X^{(0)} = 0$) has the probability distribution

$$\Pr\{k\} = \frac{\binom{2^{n-1}}{k}^2}{\binom{2^n}{2^{n-1}}}, \quad 0 \le k \le 2^{n-1}. \tag{13}$$

The random variable m' takes values 0 and 2 each with probability $1/2$, so that

$$E(\varepsilon_{m'}') = \varepsilon_{E(m')}' = 1 - 2^{1-n} \tag{14}$$

which tends to 1 as 2^n increases.

The probability distribution of m is not straightforward to derive. By careful combinatorial analysis, one can prove the following result.

Lemma 5. *Let f be a uniformly random balanced boolean function of n variables and let f'' be a boolean function obtained from f first by swapping the bits defined by input variables X and Y and, then, by additional swapping the bits defined by $X + 1$ and Y', where X is fixed or random and Y and Y' are independent uniformly distributed n-dimensional binary random variables. Let m be the number of bits where f and f'' are different and let $N = 2^n$. Then m is a random variable with the following probability distribution*

$$\Pr\{m = 0\} = \frac{N^2 - N + 2}{4N(N-1)} \tag{15}$$

$$\Pr\{m = 2\} = \frac{2N^2 + N - 6}{4N(N-1)} \tag{16}$$

$$\Pr\{m = 4\} = \frac{(N-2)^2}{4N(N-1)}. \tag{17}$$

The expected value of m is given by

$$E(m) = \frac{4N^2 - 7N + 2}{2N(N-1)}. \tag{18}$$

Note that $E(m) < 2$ since effective changes in two successive swappings can cancel out, but as N increases, we have that $\Pr\{m = 0\} \sim 1/4$, $\Pr\{m = 2\} \sim 1/2$, $\Pr\{m = 4\} \sim 1/4$, and $E(m) \sim 2$, as should be expected. Accordingly, we get

$$E(\varepsilon_m^2) = \frac{N^4 - 9N^3 + 38N^2 - 64N + 40}{N^3(N-1)} \tag{19}$$

which, of course, tends to 1 as $N = 2^n$ increases.

Finally, it remains to compute the main product factor in (12), that is, $E(c_f^6)$. According to (13), we then have

$$E(c_f^6) = 2^{-6(n-2)}\mu_6 \tag{20}$$

where μ_6 is the 6th central moment of the probability distribution (13), that is,

$$\mu_6 = \sum_{k=0}^{2^{n-1}} (k - 2^{n-2})^6 \frac{\binom{2^{n-1}}{k}^2}{\binom{2^n}{2^{n-1}}} \sim 15 \cdot 2^{3(n-4)}, \tag{21}$$

see the Appendix. It is crucial to observe that the exponent, 6, is even, so that μ_6 is necessarily different from zero.

The equation (12) together with (20), (21), (19), and (14) then determines the overall correlation coefficient c which can be easily computed for any n of interest, and, as 2^n increases we have

$$c \sim 15 \cdot 2^{-3n}. \tag{22}$$

The necessary sequence length to detect with high probability the second binary derivative statistical weakness is $O(c^{-2})$ [9], that is, neglecting a small constant less than 10,

$$L \approx 2^{6n}/225 \approx 2^{6n-7.814} \approx 10^{1.8n-2.35}. \tag{23}$$

As the memory size of RC4 is $M = n2^n + 2n$, we get $L \approx (M/(2.466 \log M))^6$.

For example, for $n = 4, 6, 8$, we computed the following values for μ_6 and c: $\mu_6 \approx 16.1716$ and $c \approx 2.2 \cdot 10^{-3}$, $\mu_6 \approx 975.762$ and $c \approx 4.97 \cdot 10^{-5}$, $\mu_6 \approx 61682.916$ and $c \approx 8.67 \cdot 10^{-7}$, respectively. In fact, for $n \geq 4$, the approximation to μ_6 included in (21) is also very good. The estimates of c obtained by computer simulations for $n = 4$ and $n = 6$ are $\hat{c} = 1.34 \cdot 10^{-3}$ and $\hat{c} = 1.95 \cdot 10^{-5}$, respectively. The first estimate is an average value for 5 output sequences each of length 10^{11} and the second one is an average value for 10 output sequences each of length 10^{11}, where each sequence is produced from a randomly chosen initial state. One may observe that the estimates are roughly by 50% smaller that the values predicted by theory. This shows that the influence of the utilized linear approximations being dependent is relatively small. The difference may also be due to the fact that the correlation coefficient estimates are essentially obtained by averaging over time rather than over random initial states.

5 Conclusions

The linear model approach aiming at finding linear relations among the keystream bits that hold with probability different from one half is applied to the RC4 keystream generator. It is first shown by the linear sequential circuit approximation method that the first and the second binary derivative of the least significant bit output sequence may yield such linear relations. A specific technique involving correlation properties of random balanced boolean functions is then developed to study the corresponding correlation coefficients. It is thus proven that the correlation coefficient for the first binary derivative is equal to zero and, more importantly, that the correlation coefficient between the second binary derivative and 1 is around $15 \cdot 2^{-3n}$ where n is the word size of RC4. The theoretical result derived agrees well with the experimental results obtained by computer simulations.

The output sequence length needed to detect the corresponding linear statistical weakness is then around $64^n/225$, which is significantly smaller than 2^M, where $M = n2^n + 2n$ is the memory size, and may even be realistic in high speed applications. Although the resulting plaintext uncertainty reduction may not be practically important unless n is small, the determined linear model can be used to distinguish RC4 from other keystream generators and, also, to recover the unknown parameter n. Whether the linear model indicates that the initial state reconstruction from the known output sequence is also possible remains to be further investigated.

Appendix

Consider a discrete probability distribution $\left\{ \binom{2\nu}{k}^2 / \binom{4\nu}{2\nu} \right\}_{k=0}^{2\nu}$ where ν is a positive integer. For any positive integer r, the central moment μ_r of this probability distribution is defined as

$$\mu_r = \sum_{k=0}^{2\nu} (k - \nu)^r \frac{\binom{2\nu}{k}^2}{\binom{4\nu}{2\nu}}. \tag{24}$$

Our objective here is to study the asymptotics of μ_r as ν increases. First note that $\mu_r = 0$ if r is odd. Assume then that r is even. By using the well-known normal approximation to the binomial coefficients, obtained by Stirling's formula $n! \sim \sqrt{2\pi}\, n^{n+\frac{1}{2}} e^{-n}$, along with a uniform convergence argument regarding this approximation (e.g., see [3, pp. 179-186]), it is easy to see that

$$\mu_r \sim \frac{\nu^{r/2}}{2^r \sqrt{2\pi}} \int_{-\infty}^{\infty} x^r e^{-x^2/2} dx \tag{25}$$

as $\nu \to \infty$. For r even, this reduces to

$$\mu_r \sim \frac{\nu^{r/2}}{2^r} \sqrt{\frac{2^r}{\pi}} \, \Gamma\left(\frac{r+1}{2}\right) \tag{26}$$

where $\Gamma(z) = \int_0^\infty x^{z-1} e^{-x} dx$ is the well-known gamma function. Finally, we obtain

$$\mu_r \sim \frac{\nu^{r/2}}{2^r} (r-1)!! \tag{27}$$

where $(r-1)!! = 1 \cdot 3 \cdots (r-1)$.

Acknowledgments

The author is grateful to Lars Knudsen and Andrew Klapper for providing the correlation coefficient estimates by computer simulations. Part of this work was carried out while the author was on leave at the Isaac Newton Institute for Mathematical Sciences, Cambridge, United Kingdom.

References

1. Ameritech Mobile Communications et al., "Cellular digital packet data system specifications, part 406: airlink security," CDPD Industry Input Coordinator, Costa Mesa, Calif., July 1993.
2. J. Daemen, R. Govaerts, and J. Vandewalle, "Resynchronization weakness in synchronous stream ciphers," Advances in Cryptology - EUROCRYPT '92, *Lecture Notes in Computer Science*, vol. 765, T. Helleseth ed., Springer-Verlag, pp. 159-167, 1994.
3. W. Feller, *An Introduction to Probability Theory and its Applications*. New York: Wiley, 3. edition, vol. 1, 1968.
4. P. Flajolet and A. M. Odlyzko, "Random mapping statistics," Advances in Cryptology - EUROCRYPT '89, *Lecture Notes in Computer Science*, vol. 434, J.-J. Quisquater and J. Vandewalle eds., Springer-Verlag, pp. 329-354, 1990.
5. J. Dj. Golić, "Correlation via linear sequential circuit approximation of combiners with memory," Advances in Cryptology - EUROCRYPT '92, *Lecture Notes in Computer Science*, vol. 658, R. A. Rueppel ed., Springer-Verlag, pp. 113-123, 1993.
6. J. Dj. Golić, "On the security of shift register based keystream generators," Fast Software Encryption - Cambridge '93, *Lecture Notes in Computer Science*, vol. 809, R. J. Anderson ed., Springer-Verlag, pp. 90-100, 1994.
7. J. Dj. Golić, "Intrinsic statistical weakness of keystream generators," Advances in Cryptology - ASIACRYPT '94, *Lecture Notes in Computer Science*, vol. 917, J. Pieprzyk and R. Safavi-Naini eds., Springer-Verlag, pp. 91-103, 1995.
8. J. Dj. Golić, "Towards fast correlation attacks on irregularly clocked shift registers," Advances in Cryptology - EUROCRYPT '95, *Lecture Notes in Computer Science*, vol. 921, L. C. Guillou and J.-J. Quisquater eds., Springer-Verlag, pp. 248-262, 1995.

9. J. Dj. Golić, "Linear models for keystream generators," *IEEE Trans. Computers*, vol. C-45, pp. 41-49, Jan. 1996.

10. J. Dj. Golić, "On the security of nonlinear filter generators," Fast Software Encryption - Cambridge '96, *Lecture Notes in Computer Science*, vol. 1039, D. Gollmann ed., Springer-Verlag, pp. 173-188, 1996.

11. R. J. Jenkins Jr., "ISAAC," Fast Software Encryption - Cambridge '96, *Lecture Notes in Computer Science*, vol. 1039, D. Gollmann ed., Springer-Verlag, pp. 41-49, 1996.

12. M. D. MacLaren and G. Marsaglia, "Uniform random number generation," *J. ACM*, vol. 15, pp. 83-89, 1965.

13. W. Meier and O. Staffelbach, "Fast correlation attacks on certain stream ciphers," *Journal of Cryptology*, vol. 1(3), pp. 159-176, 1989.

14. W. Meier and O. Staffelbach, "Correlation properties of combiners with memory in stream ciphers," *Journal of Cryptology*, vol. 5(1), pp. 67-86, 1992.

15. R. L. Rivest, "The RC4 encryption algorithm," RSA Data Security, Inc., Mar. 1992.

16. R. A. Rueppel, *Analysis and Design of Stream Ciphers*. Berlin: Springer-Verlag, 1986.

17. R. A. Rueppel, "Stream ciphers," *Contemporary Cryptology: The Science of Information Integrity*, G. Simmons ed., pp. 65-134. New York: IEEE Press, 1991.

18. B. Schneier, *Applied Cryptography*. New-York: Wiley, 1996.

19. T. Siegenthaler, "Correlation immunity of nonlinear combining functions for cryptographic applications," *IEEE Trans. Inform. Theory*, vol. IT-30, pp. 776-780, Sept. 1984.

Cryptanalysis of Alleged A5 Stream Cipher

Jovan Dj. Golić *

School of Electrical Engineering, University of Belgrade
Bulevar Revolucije 73, 11001 Beograd, Yugoslavia

Abstract. A binary stream cipher, known as A5, consisting of three short LFSRs of total length 64 that are mutually clocked in the stop/go manner is cryptanalyzed. It is allegedly used in the GSM standard for digital cellular mobile telephones. Very short keystream sequences are generated from different initial states obtained by combining a 64-bit secret session key and a known 22-bit public key. A basic divide-and-conquer attack recovering the unknown initial state from a known keystream sequence is first introduced. It exploits the specific clocking rule used and has average computational complexity around 2^{40}. A time-memory trade-off attack based on the birthday paradox which yields the unknown internal state at a known time for a known keystream sequence is then pointed out. The attack is successful if $T \cdot M \geq 2^{63.32}$, where T and M are the required computational time and memory (in 128-bit words), respectively. The precomputation time is $O(M)$ and the required number of known keystream sequences generated from different public keys is about $T/102$. For example, one can choose $T \approx 2^{27.67}$ and $M \approx 2^{35.65}$. To obtain the secret session key from the determined internal state, a so-called internal state reversion attack is proposed and analyzed by the theory of critical and subcritical branching processes.

1 Introduction

A common type of keystream generators for additive stream cipher applications consists of a number of possibly irregularly clocked linear feedback shift registers (LFSRs) that are combined by a function with or without memory. Standard cryptographic criteria such as a large period, a high linear complexity, and good statistical properties are thus relatively easily satisfied, see [12]. However, such a generator may in principle be vulnerable to various divide-and-conquer attacks in the known plaintext (or ciphertext-only) scenario, where the objective is to reconstruct the secret key controlled LFSR initial states from the known keystream sequence, for a survey see [12] and [5]. In practice, for resynchronization purposes, the internal state of a keystream generator is reinitialized once in

---

* This work was done while the author was with the Information Security Research Centre, Queensland University of Technology, Brisbane, Australia. Part of this work was carried out while the author was on leave at the Isaac Newton Institute for Mathematical Sciences, Cambridge, United Kingdom. This research was supported in part by the Science Fund of Serbia, grant #04M02, through the Mathematical Institute, Serbian Academy of Science and Arts.

W. Fumy (Ed.): Advances in Cryptology - EUROCRYPT '97, LNCS 1233, pp. 239-255, 1997.
© Springer-Verlag Berlin Heidelberg 1997

a while by combining the same secret session key with different randomizing keys (typically transmitted in the clear and called here *public*) into the secret message keys defining different initial internal states. This may open new possibilities for the secret key recovery cryptanalytic attacks, see [3].

In this paper, a keystream generator consisting of three short binary LFSRs with known primitive feedback polynomials that are mutually clocked in the stop/go manner is cryptanalyzed. The LFSR lengths are 19, 22, and 23, respectively, and the total length is thus 64. Middle taps in each of the LFSRs are used to define the clock-control sequence, the clocking rule is such that at least two LFSRs are effectively clocked per each output bit, and the keystream sequence is formed as the bitwise sum of the three stop/go clocked LFSR sequences. The 64-bit long secret key is nonlinearly combined with a 22-bit long public key (frame number) to form the LFSR initial states. The first 100 output bits are discarded and the message length is only 114 bits (frequent resynchronization). However, the full-duplex communication mode makes the effective message length of 228 bits. The scheme along with the code has been made public in [1] and is allegedly used under the name A5 for stream cipher encryption in the GSM standard for digital cellular mobile telephones, see [13]. For simplicity, the name A5 is used here throughout. In a yet unpublished paper [14], it has been observed, perhaps surprisingly, that the period of the keystream sequence is only slightly bigger than the period, $\approx 2^{23}$, of the longest LFSR. A possibility for a divide-and-conquer attack of average complexity 2^{40} has been mentioned in [1] and [13]. The attack would consist in guessing the initial states of the two shorter LFSRs and, then, in computing the longest LFSR sequence from the known keystream sequence. However, this attack can not work, because the clocking depends on the unknown longest LFSR sequence as well. In addition, one has to take care of the first 100 output bits being discarded as well.

Although one may in principle imagine that edit distance or edit probability correlation attacks [4] can be adapted to deal with stop/go clocking, such attacks are not likely to be successful on A5, because of a very short available keystream sequence. Due to the bitwise summation, to achieve a divide-and-conquer effect, one or two LFSRs have to be replaced by their linear models [7], where linear models of individual LFSRs can be based on the repetition property only, while linear models of pairs of the LFSRs must involve their feedback polynomials as well. Instead of the so-called shrunk feedback polynomials [7], we now have to introduce the expanded feedback polynomials. If the whole scheme is replaced by the corresponding linear model, one may then even conceive of a fast correlation attack framework similar to the one from [6], but the required keystream sequence length would be much bigger than the one at disposal. On the other hand, the conditional correlation attack [11] based on the repetition property can not be extended to deal with A5, because of the specific clocking rule.

The objective of this paper is to develop cryptanalytic attacks on A5 that can reconstruct the 64-bit secret key in the known plaintext scenario with the computational complexity smaller than 2^{64}. In Section 2, a more detailed description of the A5 stream cipher is presented. It is shown that the known plaintext attacks are very realistic in the GSM applications. In Section 3, a basic divide-and-conquer attack on A5 with the average computational complexity $2^{40.16}$ is introduced. It essentially consists in guessing some bits of the LFSR states, in recovering the others by solving appropriate linear equations, and in the LFSR states reversion via the unknown binary clocking sequences to obtain the LFSR initial states. The last step is needed since the first 100 output sequence bits are discarded. In Section 4, a time-memory trade-off attack based on the birthday paradox probabilistic argument is pointed out. This attack is feasible due to relatively short internal state size of 64 bits. It can recover the LFSR internal states for a particular keystream sequence at a particular time and is successful if $T \cdot M \geq 2^{63.32}$, where T and M are the required computational time and memory, respectively. The precomputation time is $O(M)$ and a sample of $T/102\,228$-bit long observed keystream sequences generated from the same secret session key and different public keys is needed. To obtain the secret key, a low-complexity internal state reversion attack is then proposed in Section 5. It consists in the reversion of the LFSR internal states, first when the output sequence is known, then when the output sequence is unknown, and finally when the secret key is nonlinearly combined with the known public key. The complexity of the attack is analyzed by the theory of critical and subcritical branching processes, briefly outlined in the Appendix. Conclusions are given in Section 6.

2 Description of the Stream Cipher

The stream cipher algorithm to be defined is for simplicity called A5 according to [1], [13]. The A5 type keystream generator considered is shown in Fig. 1.

Let $f_i(z) = \sum_{l=0}^{r_i} f_{i,l}\, z^l$ denote a known binary primitive feedback polynomial of LFSR$_i$ of length r_i, $i = 1, 2, 3$, and let $r_1 = 19$, $r_2 = 22$, and $r_3 = 23$. The feedback polynomials specified in [1], [13] are sparse, but our cryptanalytic methods to be presented do not depend on their choice. Let $S_i(0) = (x_i(t))_{t=0}^{r_i-1}$ denote the initial state of LFSR$_i$ and let $x_i = (x_i(t))_{t=0}^{\infty}$ denote the corresponding maximum-length sequence of period $2^{r_i} - 1$ produced by LFSR$_i$ via the linear recursion $x_i(t) = \sum_{l=1}^{r_i} f_{i,l}\, x_i(t - l)$, $t \geq r_i$.

Let $S_i(t) = (s_{i,l}(t))_{l=1}^{r_i}$ denote the state of LFSR$_i$ at time $t \geq 0$ in a scheme with stop/go clocking to be defined below, and let τ_i denote a middle tap from LFSR$_i$ used for clock-control. The values suggested in [1] are $\tau_1 = 10$, $\tau_2 = 11$, and $\tau_3 = 12$. Then the clock-control sequence $C = (C(t))_{t=1}^{\infty}$ is defined by

$$C(t) = g(s_{1,\tau_1}(t - 1), s_{2,\tau_2}(t - 1), s_{3,\tau_3}(t - 1)) \tag{1}$$

where g is a 4-valued majority function of three binary variables such that $g(s_1, s_2, s_3) = \{i, j\}$ if $s_i = s_j \neq s_k$ for $i < j$ and $k \neq i, j$, and $g(s_1, s_2, s_3) = \{1, 2, 3\}$ if $s_1 = s_2 = s_3$. The clock-control value $C(t)$ defines which LFSRs are clocked to produce an output bit $y(t)$ as the sum

$$y(t) = s_{1,1}(t) + s_{2,1}(t) + s_{3,1}(t), \quad t \geq 1. \tag{2}$$

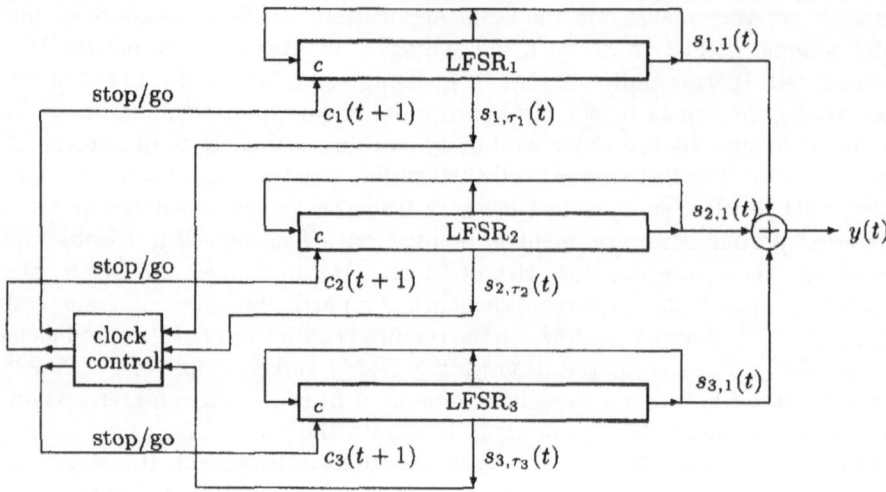

Fig. 1. Alleged A5 type keystream generator.

Let $c_i = (c_i(t))_{t=1}^\infty$ denote the binary clocking sequence for LFSR_i (it is clocked if $c_i(t) = 1$ and not clocked if $c_i(t) = 0$) which is derived from the clock-control sequence C in an obvious way. Equation (2) can formally be used to generate the initial bit $y(0)$ from $S(0)$, so that $y = (y(t))_{t=0}^\infty$ is called the output sequence. The first 100 output bits, $(y(t))_{t=1}^{100}$, are discarded, the following 114 bits are used as the keystream for one direction of communication in the full-duplex mode, then the next 100 bits are again discarded, and the following 114 bits are used as the keystream for the reverse direction of communication. The encrypted messages are thus very short and the resynchronization is frequent.

The LFSR initial states are defined in terms of the secret and public keys. The public key is a known 22-bit frame number generated by a counter and hence different for every new message. The 64-bit secret session key is first loaded into the LFSRs (the all-zero initial state is avoided by setting the output of the last stage to 1) and the 22-bit public key is then bitwise added into the feedback path of each of the LFSRs that are mutually clocked as above. More precisely, if $p = (p(t))_{t=-21}^0$ denotes the public key, then for every $-21 \le t \le 0$, the LFSRs are first stop/go clocked as before and, then, the bit $p(t)$ is added to the last stage of each of the LFSRs. The LFSR states after these 22 steps, as a secret message key, represent the initial LFSR states for the keystream generation.

The A5 stream cipher is allegedly used to encrypt the links between individual cellular mobile telephone users and the base station in the GSM system, see [13]. Therefore, if two users want to communicate to each other via their base station(s), the same messages get encrypted twice which makes the known plaintext cryptanalytic attack possible, provided a cooperative insider user can be established. Note also that the links between the base stations are not encrypted. For any user, a 64-bit secret session key is generated by another algorithm from the secret master key specific to the user and a public random 128-bit key transmitted in the clear from the base station to the user. So, a possible reconstruction of one or more session keys for a user opens a door for a cryptanalytic attack on the master key of that user.

3   Basic Attack

The objective of a divide-and-conquer attack to be presented in this section is to determine the LFSR initial states from a known keystream sequence corresponding to only one known plaintext-ciphertext pair. In fact, only about 64 known successive keystream bits are required. Let $S(t) = (S_1(t), S_2(t), S_3(t))$ denote the whole internal state of A5 at time $t \geq 0$, where $S(0)$ is the initial internal state defined by the secret message key. The known keystream sequence is in fact composed of two segments $(y(t))_{t=101}^{214}$ and $(y(t))_{t=315}^{428}$. The first goal is to reconstruct the internal state $S(101)$ and the second one is to determine $S(0) = (S_1(0), S_2(0), S_3(0))$ from $S(101)$.

Recall that $c_i = (c_i(t))_{t=1}^{\infty}$ denotes the binary clocking sequence for LFSR$_i$, which is clocked if $c_i(t) = 1$ and not clocked if $c_i(t) = 0$. If A_i denotes the state-transition matrix of regularly clocked LFSR$_i$, then

$$S_i(t) = S_i(0) \cdot A_i^{\sum_{l=1}^{t} c_i(l)} \tag{3}$$

with the integer summation in the exponent. Also, let $\tilde{x}_i = (\tilde{x}_i(t))_{t=0}^{\infty}$ denote the stop/go clocked LFSR$_i$ sequence, where $\tilde{x}_i(t) = s_{i,1}(t)$. In the probabilistic analysis to follow, a sequence of independent uniformly distributed random variables, over any finite set, is called purely random. As usual, we keep the same notation for random variables and their values.

Proposition 1. *Assume that the three regularly clocked LFSR sequences are mutually independent and purely random. Then the 4-valued clock-control sequence C is purely random and, hence, the binary clocking sequence c_i is a sequence of independent identically distributed binary random variables with the probability of zero being equal to 1/4.*

Proposition 2. *Assume that the three regularly clocked LFSR sequences are mutually independent and purely random. Then the bitwise sum of any two or more stop/go clocked sequences $\tilde{x}_i$ is purely random.*

It is shown in Section 5 that the state-transition function of A5 is not one-to-one, so that the set of all reachable internal states at time t, $t \geq 1$, is a subset of the set S_0 of all 2^{64} initial states. In particular, only $5 \cdot 2^{61} \approx 2^{63.32}$ internal states are reachable for $t = 1$. As a consequence, different initial states can give rise to the same internal state at some time in future or even to the same output sequence too. This is explained in terms of the theory of branching processes in Section 5. More precisely, the number of different initial states giving rise to the same internal state at some time in future is very likely linear in that time and, therefore, relatively small for the times of interest (internal state reversion when the output is not known, Subsection 5.1). On the other hand, the number of different initial states yielding the same internal state at some time in future and the same output sequence is very likely to be a very small integer (internal state reversion when the output is known, Subsection 5.2). In addition, since the individual LFSR sequences are maximum-length sequences with good (low) autocorrelation and crosscorrelation properties and the combining function is maximum-order correlation immune, it is highly likely that different output sequences $y = (y(t))_{t=0}^{\infty}$ are different on the first successive 64 positions, $(y(t))_{t=0}^{63}$.

Consequently, it takes about 64 successive keystream bits to check if an assumed preceding internal state is consistent with the subsequent output sequence. The expected number of solutions for $S(101)$ is with high probability a small integer, whereas the the number of solutions for $S(0)$ (equivalent initial states) is very likely to be relatively small.

3.1 Internal state reconstruction

Let $S(101)$ be the internal state to be determined in the first stage of the attack. Since the number of reachable states $S(101)$ is not bigger than $2^{63.32}$ and the unreachable ones can be simply characterized by a set of linear equations, in the average complexity analysis given below we can simply take 63.32 instead of 64. For every $i = 1, 2, 3$, first guess n bits $(s_{i,l}(101))_{l=\tau_i}^{\tau_i+n-1}$ of $S_i(101)$ if $n \leq r_i - \tau_i + 1$, and, if not, then also guess the next $n - r_i + \tau_i - 1$ bits produced by the linear recursion from $S_i(101)$. In any case, one thus obtains $3n$ linearly independent equations for unknown bits of $S(101)$, provided that $n \leq 19$. Since the assumed bits on average define $4n/3$ elements of the clock-control sequence, one can thus form $1 + 4n/3$ additional linear equations, where the first one is clearly obtained from the first keystream bit $y(101)$ without using the clock-control sequence. The additional linear equations are mutually linearly independent, provided that $n \leq 18$, because each one then contains at least two new bits that have not appeared before. They are linearly independent of the first $3n$ equations if and only if each of them contains at least one new bit that is not already guessed. This happens with high probability if

$$n < \max(\tau_1, \tau_2, \tau_3) - 1. \tag{4}$$

If not, then the last among the additional equations will necessarily involve some of the already guessed bits and will with high probability be linearly dependent on the first $3n$ equations. Suppose first that the condition (4) is satisfied. Then all the obtained linear equations are with high probability linearly independent, so that the internal state can be determined uniquely if $1 + 3n + 4n/3 \geq 63.32$, that is, if $n \geq 14.38$ (it follows that $\max(\tau_1, \tau_2, \tau_3) \geq 16$). The obtained state should then be tested for correctness on additional $3n$ keystream bits on average. The computational complexity is then $\approx 2^{43.15}$ and the total required keystream sequence length is about 64 successive bits (we keep the fractions since we deal with the average case complexity).

Suppose now that $\max(\tau_1, \tau_2, \tau_3) \leq 15$, which means that the condition (4) is not satisfied, as is the case in the particular proposal from [1], where $\max(\tau_1, \tau_2, \tau_3) = \tau_3 = 12$. In this case, the last of the additional equations are with high probability linearly dependent and as such can not be used as before, but can be used to test the linear consistency of the initial guess. If the previous analysis was extended, then one would get that n has to be bigger than 14.38 and that the average complexity would hence increase, contrary to the intuition. Indeed, one can do better than that. Let initially $n = 10$, so that (4) is satisfied. One thus obtains the total of $1 + 3n + 4n/3 \approx 44.3$ linearly independent equations on average. Now, instead of guessing the next $m \approx 19.02/3$ bits on average in each of the LFSR sequences, we will build a tree structure to sequentially store all the possibilities for the next bits that are consistent with the additional linear equations. In each node of the tree one stores the next three input bits to the majority clock-control function such that the resulting clocking is consistent with the equations. This approach is in spirit similar to the inversion attack [8] on nonlinear filter generators. The average number of branches leaving each node would have been $\frac{3}{4} \cdot 4 + \frac{1}{4} \cdot 8 = 5$ if it were not for the additional equations. They on average reduce this number to 2.5. The required depth of the tree should on average be $4m/3$ to obtain the next m guessed bits in each of the LFSR sequences. So, instead of 2^{3m} possibilities for the next m bits, we have to check only $2.5^{4m/3} \approx 2^{1.76\,m} \approx 2^{11.16}$ possibilities on average, under the reasonable independence assumption valid for the so-called supercritical branching processes, see Theorem 6 from the Appendix. The overall complexity is then $2^{30+11.16} \approx 2^{41.16}$. For comparison, suppose that the clock-control bits are used to produce the output, that is, $\tau_1 = \tau_2 = \tau_3 = 1$. Then, clearly, only the part of the process involving the tree applies and the overall complexity is minimum possible, that is, $2^{1.76 \cdot 63.32/3} \approx 2^{37.15}$.

To get the average number of trials needed to find the correct internal state $S(101)$, one should in fact divide by two the complexity figures given above, e.g., $2^{41.16}$ thus reduces to $2^{40.16}$.

3.2 Internal state reversion via clocking sequences

In the second stage, our objective is to recover the initial LFSR states from $S(101)$. In view of (3), this can be done by guessing the number of ones in individual binary clocking sequences, that is, the number of clocks needed to get $S_i(101)$ from $S_i(0)$, for each $i = 1, 2, 3$. According to Proposition 1, the underlying probability distribution is binomial with the average number of clocks $0.75 \cdot 101 \approx 76$ and the standard deviation $0.25 \cdot \sqrt{303} \approx 4.35$, for each of the LFSR sequences. If the search is organized in order of decreasing probabilities for each of the LFSR sequences independently, the number of trials required to find the correct numbers of clocks is with high probability not bigger than about 10^4 and is at worst about 10^6. For each guess, one first recovers $S_i(0)$ from $S_i(101)$ by backward linear recursion, for each $i = 1, 2, 3$, and then tests the guess by running the keystream generator forwards to obtain $S(101)$. Note that multiple solutions for $S(0)$, if they exist, are all obtained by checking all $\approx 10^6$ possibilities for the clocking sequences, for any possible $S(101)$ obtained in the first stage. This number can clearly be reduced by assuming the mutually constrained rather than independent clocking sequences for individual LFSRs. In any case, reconstructing the initial state $S(0)$ from $S(101)$ is much faster than obtaining $S(101)$ itself.

4 Time-Memory Trade-Off Attack

As was already explained in the previous section, the first 64 successive output bits of A5, $(y(t))_{t=0}^{63}$, represent a vectorial boolean function of 64 initial state bits $S(0)$ such that the number of different initial states $S(0)$ producing the same 64-bit initial output block is in most cases only 1 or a very small integer. In fact, since the initial 101 output bits are not used for the keystream, the initial state bits $S(0)$ should be confined to the $2^{63.32}$ values achievable by $S(1)$ which are easily characterized. As a consequence, for any observed 64 successive keystream bits, one can find all the preceding internal states yielding these bits either by exhaustive search over all reachable internal states requiring $2^{63.32}$ 64-bit computations and bitwise comparisons or by only one table lookup requiring $2^{63.32}$ 64-bit words of memory to store the inverse of the vectorial boolean function considered. The inverse function, with multiple preimages if they exist, is found and stored in $2^{63.32}$ precomputation time. Let the time and memory required in these two extreme cases be denoted as $T = 2^{63.32}$, $M = 1$ and $T = 1$, $M = 2^{63.32}$, respectively. Is any meaningful time-memory trade-off based on the birthday paradox possible?

Assume that the objective is to recover the preceding internal states for any observed 64 successive keystream bits in the known plaintext scenario. Each known keystream sequence of effective length 228 bits provides $102 \approx 2^{6.67}$ 64-bit blocks, and, due to the very small keystream sequence length, it is very likely that the cryptanalyst knows either all 228 bits or none of them. So, any time-memory trade-off solely based on these 102 keystream blocks is meaningless. However, we may consider a sample of all the keystream sequences corresponding to different initial states (secret message keys) derived from K (at most 2^{22}) different known public keys and a single secret session key. The reconstruction of any internal state corresponding to a particular public key is then meaningful if $K < 2^{22}$ and if it leads to the recovery of the secret session key, which can then be used to decrypt the ciphertexts obtained from the remaining public keys.

Let the cryptanalyst form a table of M possibly multiple 64-bit words defining the reachable initial states corresponding to a random sample of M different 64-bit output blocks, and let the table be then sorted out with respect to the output blocks, which are also stored. Multiple preimages are all obtained by the internal state reversion given a known output, in $O(M)$ time, see Subsection 5.2. The required precomputation time for sorting is $M \log M$ or, approximately, just M if the logarithmic factor, smaller than 64, is neglected. Altogether, the required precomputation time is thus $O(M)$. By the standard birthday paradox (used in meet-in-the-middle attacks), it then follows that with high probability at least one of the $102 \cdot K$ keystream blocks in the observed sample will coincide with one of the output blocks used to form the table if

$$102 \cdot K \cdot M \geq 2^{63.32} \tag{5}$$

where a small multiplicative constant is neglected for simplicity. The time T needed to find such a keystream block is $102 \cdot K \cdot \log M$ or simply $102 \cdot K$ neglecting the logarithmic factor. Then only one table lookup gives the desired internal state(s). So, the time-memory trade-off is possible with $T \cdot M \geq 2^{63.32}$ and $T < 102 \cdot 2^{22}$. For example, if $K = 2^{15}$, then the time and memory required are $T \approx 2^{21.67}$ and $M \approx 2^{41.65}$ (in 128-bit words), respectively, and the precomputation time is $O(M)$. In an extreme case, when $K = 2^{21}$, we get $T \approx 2^{27.67}$ and $M \approx 2^{35.65} \approx 862$ Gbytes, but the secret session key to be determined can then only be used to decrypt ciphertexts obtained from the remaining half of the public keys.

A more general approach for the cryptanalyst would be to analyze the traffic corresponding to L different sessions for each out of N users. This increases the sample size (and time) to $102 \cdot K \cdot L \cdot N$, so that further reduction in M is possible, which makes the attack quite realistic. In this case, a particular user whose secret session key is to be determined is not known in advance. This, of course, does not make a difference if the objective is cloning rather than decryption. Even more generally, one may also allow that K be maximum possible, 2^{22}, if the cryptanalyst is capable of attacking the algorithm that combines the secret master key of a user and a public random 128-bit key into the secret session key. Namely, the determined session key may be useless for decryption, but may be used for the secret master key reconstruction with devastating consequences regarding both decryption and cloning.

The time-memory trade-off attack described clearly applies to arbitrary keystream generators, and is feasible in the case of A5 because of its relatively short memory size of only 64 bits. It yields an internal state of A5 at a known time and is meaningful when coupled with a cryptanalytic attack to be introduced in the next section which gives all the candidates for the secret session key. If the internal state is determined at time $101 \leq t \leq 151$, then the attack consists in the reversion of the internal state to $S(101)$ based on known output, then to $S(0)$ when the output is not known (due to the first 100 output bits discarded), and finally to the secret session key when the known public key is incorporated. If the internal state is determined at time $315 \leq t \leq 365$, then the attack consists in the reversion of the internal state to $S(315)$ based on known output, then to $S(214)$ when the output is not known, and the rest is the same as in the first case with $S(214)$ as the internal state. Note that possible multiple solutions are all obtained. Multiple candidates for the secret session key are then easily reduced to only one, correct solution by comparing a small number of already known keystream sequences with the ones generated from the assumed candidates and known public keys.

5 Internal State Reversion via Branching

The objective of the internal state reversion attack to be described in this section is to find all the secret session keys that combined with a known public key give rise to a given internal state at a known time. All the internal states at a known time that are consistent with a known keystream sequence can be obtained either by the basic internal state reconstruction attack from Subsection 3.1 or by the time-memory trade-off attack from Section 4.

The performance of the attack is analyzed by the theory of critical and subcritical branching processes and its time and space complexities are thus shown to be both small. Extensive computer experiments on nonlinear filter generators regarding the so-called generalized inversion attack [9] (where the whole internal state is recovered starting from its finite input memory part in a way similar to the internal state reversion) show that the size of the generated search trees can be well described by the theory of branching processes.

5.1 Unknown output

Given an internal state $S(t)$ at time t, $t \geq 1$, $S(t) \in \mathcal{S}_0$, the objective of the reversion attack when the output sequence is not known is to determine all the internal states $S(t')$ at a given previous time $t' < t$ that produce $S(t)$ at time t by the state-transition function, whereas the output sequence is not considered at all. For the reversion to work, the state-transition function, $\mathcal{F}$, must be easily computable in the reverse direction. Letting $\mathcal{F}^{-1}$ denote the reverse state-transition function, $\mathcal{F}^{-1}(S(t))$ denotes the set of all $S(t-1)$ such that $\mathcal{F}(S(t-1)) = S(t)$. The reversion attack then consists in the recursive computation of the reverse state-transition function starting from $S(t)$ and up to $S(t')$. The internal states obtained can all be stored as nodes in a tree with $t - t' + 1$ levels where the initial level, $n = 0$, has one initial node representing $S(t)$, and the level n, $1 \leq n \leq t - t'$, contains the nodes representing all possible $S(t - n)$ giving rise to $S(t)$. The end nodes thus give all the desired internal states $S(t')$. The main problem here is to estimate the size of the trees arising from a random $S(t)$, that is, the number of the nodes obtained at each level n if $S(t)$ is chosen uniformly at random, and especially if n is not small.

The state-transition function of A5 is essentially determined by the clock-control sequence, see (1) and (3). Accordingly, the number of different states $S(t-1)$ in $\mathcal{F}^{-1}(S(t))$ is derived by backward clocking from all the possibilities for $C(t-1)$ and hence only depends on the following six bits: the three bits $(s_{1,\tau_1}(t), s_{2,\tau_2}(t), s_{3,\tau_3}(t))$ which define the clock-control sequence at the current time t, $C(t)$, and the three preceding bits in the regularly clocked LFSR sequences which, if $\min(\tau_1, \tau_2, \tau_3) \geq 2$, all belong to $S(t)$ and are given as $(s_{1,\tau_1-1}(t), s_{2,\tau_2-1}(t), s_{3,\tau_3-1}(t))$. Denote these bits by s_1, s_2, s_3 and s_1', s_2', s_3', respectively.

Proposition 3. *Let (i, j, k) denote a permutation of $(1, 2, 3)$. Then the following six events can occur:*

- A :   for any k, if $s_i' = s_j' \neq s_k' = s_k$, then $C(t-1) = \{i, j\}$
- B :   for any k, if $s_i' = s_j' \neq s_k' \neq s_k$, then $C(t-1)$ can take no values
- C :   if $s_1' = s_2' = s_3' = s_1 = s_2 = s_3$, then $C(t-1) = \{1, 2, 3\}$
- D :   if $s_1' = s_2' = s_3' \neq s_1 = s_2 = s_3$, then $C(t-1)$ can take every of the four values $\{1, 2\}, \{1, 3\}, \{2, 3\},$ and $\{1, 2, 3\}$
- E :   for any k, if $s_1' = s_2' = s_3' = s_i = s_j \neq s_k$, then $C(t-1)$ can take every of the two values $\{i, j\}$ and $\{1, 2, 3\}$
- F :   for any i, if $s_1' = s_2' = s_3' = s_i \neq s_j = s_k$, then $C(t-1)$ can take every of the three values $\{i, j\}, \{i, k\},$ and $\{1, 2, 3\}$.

Proposition 4. *If an internal state $S(t)$ is randomly chosen from S_0 according to uniform distribution, then the number of solutions for $S(t-1)$ is a nonnegative integer random variable Z with the probability distribution*

$$\Pr\{Z = 0\} = \frac{3}{8}, \quad \Pr\{Z = 1\} = \frac{13}{32},$$

$$\Pr\{Z = 2\} = \Pr\{Z = 3\} = \frac{3}{32}, \quad \Pr\{Z = 4\} = \frac{1}{32}. \tag{6}$$

It follows that the state-transition function of A5 is not one-to-one and that the fraction of the internal states from S_0 not reachable in one step is $3/8$ (they are simply characterized by a set of three linear equations). Let $\{S(t-n)\}$ denote the set of all the internal states/nodes at level n in the tree spanned by the reversion from a given $S(t)$, and let $Z_n = |\{S(t-n)\}|$ and $Y_n = \sum_{l=1}^{n} Z_l$. Both the time and space complexities of the reversion attack are determined by Y_n. Our objective now is to estimate how large Z_n and Y_n can be when $S(t)$ is randomly chosen. Of course, each particular $S(t)$ uniquely determines the tree (model $\mathbf{M'}$), and if we assume that regularly clocked LFSR sequences are mutually independent and purely random (model $\mathbf{M}$), then the tree is random rather than unique even when $S(t)$ is fixed. From the internal state reversion via the clocking sequences, Subsection 3.2, we know that in both the models $Z_n \leq n^3$ necessarily holds. The trees spanned in both the models are expected to be similar if the depth n is smaller than $4/3$ of the period of the shortest LFSR, $\approx \frac{4}{3}2^{19}$, which is when on average the LFSR sequences start to repeat themselves in model $\mathbf{M'}$.

Proposition 4 shows that the associated Galton-Watson *branching process*, described in the Appendix, has the branching probability distribution defined by (6), with the expected value and variance $\mu = 1$ and $\sigma^2 = 9/8$, respectively. The branching process is *critical*. The random trees produced by model $\mathbf{M}$ and by the associated branching process are not exactly the same, as random variables. The reason for this is that in the branching process the branching probability distribution for a given node is independent of the nodes at the same or the preceding levels (the history), whereas in model $\mathbf{M}$ there is a weak dependence between the nodes as a result of different internal states having some clock-control bits in common. This weak dependence affects the expected values and variances of both Z_n and Y_n, but insignificantly.

Consequently, if $S(t)$ is uniformly distributed over $\mathcal{S}_0$, then in model $\mathbf{M}$, in view of Theorem 6 from the Appendix, $E(Z_n) \approx 1$, $\text{Var}(Z_n) \approx \sigma^2 n$, and $\Pr\{Z_n > 0\} \approx 2/(\sigma^2 n)$, where $\sigma^2 = 9/8$. So, the fraction of the internal states reachable in n steps is about $2/(\sigma^2 n)$. On the other hand, both the computational time and the storage required for the reversion attack are determined by the total number of nodes Y_n. Theorem 7 from the Appendix then gives that $E(Y_n) \approx n$ and $\text{Var}(Y_n) \approx \sigma^2 n^3/3$. In view of the Chebyshev's inequality $\Pr\{|Y_n - E(Y_n)| \geq \varepsilon\} \leq \text{Var}(Y_n)/\varepsilon^2$, we then get that the total number of nodes Y_n is with high probability $O(n\sqrt{n})$ and the multiplicative constant is not big. Note that in the case of interest, $n = 101$, and the approximations are expected to be very good.

It is also interesting to see how large Z_n and Y_n can grow when conditioned on the event that the internal state $S(t)$ is reachable in n steps. We know that at least one such state results from both the basic internal state reconstruction attack and the time-memory trade-off attack. Theorem 8 from the Appendix yields that in this case $E(Z_n|Z_n > 0) \approx \sigma^2 n/2$ and $\text{Var}(Z_n|Z_n > 0) \approx \sigma^4 n^2/4$. This means that the number of solutions for $S(t - n)$ is with high probability linear in n, provided at least one such solution exists. As for the total number of nodes Y_n, we noted in the Appendix that $E(Y_n|Z_n > 0) = O(\sigma^2 n^2)$ and $\text{Var}(Y_n|Z_n > 0) = O(\sigma^2 n^4)$, so that Y_n is then with high probability $O(\sigma^2 n^2)$. So, for $n = 101$ both the time and space complexities are small, although somewhat bigger than in the case of a uniformly distributed $S(t)$.

The number, N, of starting internal states in the real reversion attack may be bigger than just one, but is still small, as will be shown in the following subsection. The time complexity clearly increases proportionally with N, whereas the space complexity, determined as the maximum tree size over all the starting states, increases only logarithmically with N, due to the exponential probability distribution (21) in Theorem 8 from the Appendix.

5.2 Known output

Given an internal state $S(t)$ at time t, $t \geq 1$, $S(t) \in \mathcal{S}_0$, the objective of the reversion attack when the output sequence is known is to determine all the internal states $S(t')$ at a given previous time $t' < t$ that produce $S(t)$ at time t by the state-transition function as well as the known output sequence $(y(t - l))_{l=1}^{t-t'}$. This reversion attack then goes along the same lines as the one when the output sequence is not known, with a difference that from each level in the tree spanned, the nodes whose internal states produce the output bits different from the one known are all removed. The size of the resulting tree is hence much smaller.

The output bit produced from $S(t - 1)$ at time $t - 1$ depends on the following six bits: $(s_{1,1}(t), s_{2,1}(t), s_{3,1}(t))$ and the preceding three bits in the regularly clocked LFSR sequences. They are denoted as z_1, z_2, z_3 and z_1', z_2', z_3', respectively. The produced output bit is then equal to $z_i' + z_j' + z_k$ if $C(t - 1) = \{i, j\}$, for any $\{i, j\}$ (as usual, (i, j, k) is a permutation of $(1, 2, 3)$), and to $z_1' + z_2' + z_3'$ if $C(t - 1) = \{1, 2, 3\}$. An analog of Proposition 3 can then be established, with a difference that in each of the given six events, $C(t - 1)$ can take every specified value for which, in addition, the produced output bit coincides with the one known, $y(t - 1)$. If $\min(\tau_1, \tau_2, \tau_3) \geq 3$, one can then derive the following analog of Proposition 4.

Proposition 5. *If an internal state $S(t)$ is randomly chosen from S_0 according to uniform distribution, then the number of solutions for $S(t-1)$ is a nonnegative integer random variable Z with the probability distribution*

$$\Pr\{Z = 0\} = \frac{315}{512}, \quad \Pr\{Z = 1\} = \frac{75}{256}, \quad \Pr\{Z = 2\} = \frac{9}{128},$$

$$\Pr\{Z = 3\} = \frac{5}{256}, \quad \Pr\{Z = 4\} = \frac{1}{512}. \tag{7}$$

The probabilistic models $\mathbf{M}$ and $\mathbf{M'}$ are defined in the same way as before, with a difference that the known output sequence is assumed to be either fixed or purely random and independent of the LFSR sequences. The dependence between the nodes in the trees produced by $\mathbf{M}$ and $\mathbf{M'}$, although still relatively weak, is stronger than before due to the six additional bits controlling the output. The associated branching process is now *subcritical* with $\mu = 1/2$ and $\sigma^2 = 17/32$. The results regarding the probability distribution, the expected values, and the variances for the random variables Z_n and Y_n are then obtained analogously, by applying the parts of Theorems 6-8 from the Appendix relating to subcritical branching processes. Consequently, we get that in model $\mathbf{M}$, $E(Z_n) \approx 2^{-n}$, $\mathrm{Var}(Z_n) \approx \sigma^2 2^{-(n-2)}$, and $\Pr\{Z_n > 0\} \approx c\,2^{-n}$, where c is a positive constant that is obtained numerically as $c = \lim_{n \to \infty} 2^n (1 - f^{(n)}(0)) \approx 0.63036$, where $f^{(n)}$ is the self-composition of the generating function f of the probability distribution defined by (7), see the Appendix. Also, $E(Y_n) \approx 1$ and $\mathrm{Var}(Y_n) \approx 8\,\sigma^2$.

Conditioning on the event that the starting internal state is reachable in n steps, we get $E(Z_n|Z_n > 0) \approx 1/c \approx 1.586$, $\mathrm{Var}(Z_n|Z_n > 0) \approx 4\sigma^2/c - 1/c^2 \approx 0.854$, $E(Y_n|Z_n > 0) = O(n)$, and $\mathrm{Var}(Y_n|Z_n > 0) = O(n^2)$. The size Y_n of the resulting tree is then $O(n)$ with high probability. In the case of interest, resulting from the time-memory trade-off attack, we have that $n \leq 50$, so that the obtained trees are very small, whereas the number of possible solutions for $S(t - n)$ ($S(101)$) is with high probability only 1 or a very small positive integer.

5.3 Secret key reconstruction

Our goal now is to obtain all possible secret session keys from all the determined initial states $S(0)$ given a known public key $p = (p(t))_{t=-21}^{0}$. Recall that the secret session key is in fact an internal state of the initialization scheme, which works in the same way as the keystream generator A5, except that the public key is bitwise added, in 22 steps, into the feedback path of each of the LFSRs. Given an initial state $S(0)$, $S(0) \in S_0$, the objective of the secret key reconstruction attack is to determine all the internal states $S(t')$ at the previous time $t' = -22$ that produce $S(0)$ by the modified state-transition function $S(t) = \mathcal{F}_0(S(t - 1), p(t))$, $-21 \leq t \leq 0$, which also depends on the known public key sequence

p. The modified reverse state-transition function $\mathcal{F}_0^{-1}(S(t), p(t))$ then consists of two stages: first, the bit $p(t)$ is added to the last stage of each of the LFSRs and, second, the LFSRs are clocked backwards according to all possible values $C(t-1)$ for the clock-control sequence.

It is readily seen that the secret key reconstruction can be achieved by the reversion attack when the output sequence is not known in which the reverse state-transition function is modified according to the public key p as explained above. Consequently, both the analysis based on the theory of critical branching processes and the conclusions derived remain valid for the secret key reconstruction attack. Since now $n = 22$ instead of $n = 101$, the trees spanned are much smaller in size. Multiple solutions for the secret session key $S(-22)$ giving rise to the same $S(0)$ are still possible, but their number is relatively small. All the resulting candidates for the secret session key are consistent with the used keystream sequence. These multiple candidates for the secret session key are then easily reduced to only one, correct solution by comparing a small number of already known keystream sequences with the ones generated from the assumed candidates and known public keys.

6 Conclusions

Several cryptanalytic attacks on a binary stream cipher known as A5 are proposed and analyzed. The objective of the attacks is to reconstruct the 64-bit secret session key from one or several known keystream sequences produced by different 22-bit (randomizing) public keys, in the known plaintext scenario which is shown to be very realistic in the GSM applications. A basic divide-and-conquer attack with the average computational complexity $2^{40.16}$ and negligible memory requirements is first introduced. It requires only about 64 known successive keystream bits and gives all possible LFSR initial states consistent with a known keystream sequence. A time-memory trade-off attack based on the birthday paradox is then pointed out. The objective of the attack is to find the LFSR internal states at a known time for a known keystream sequence corresponding to a known public key. The attack is feasible as the internal state size of A5 is only 64 bits.

To obtain the secret session key from the determined LFSR internal states, an internal state reversion attack is proposed and analyzed by the theory of critical and subcritical branching processes. It is shown that there typically exist multiple, but not numerous, candidates for the secret session key that are all consistent with the used keystream sequence. The unique, correct solution is then found by checking on a small number of additional keystream sequences. The secret session key recovered can be used to decrypt the ciphertexts obtained from the remaining public keys and, possibly, to mount a cryptanalytic attack on the secret master key of the user as well.

A simple way of increasing the security level of the A5 stream cipher with respect to the cryptanalytic attacks introduced is to make the internal memory size larger. For example, doubling the memory size, from 64 to 128 bits, is very likely to push the attacks beyond the current technological limits. Note that the secret session key size need not be increased to 128 bits. In addition, one can make the clock-control dependent on more than just a single bit in each of the shift registers by using a balanced nonlinear filter function applied to each of them individually. The inputs to the filter functions should be spread over the shift register lengths, respectively, and their outputs can be combined in the same way as in A5. This increases the complexity of the basic internal state reconstruction attack.

Appendix

Branching processes

The so-called Galton-Watson process, see [10], [2], is a Markov chain $\{Z_n\}_{n=0}^{\infty}$ on the nonnegative integers whose transition function is defined in terms of a given probability distribution $\{p_k\}_{k=0}^{\infty}$. The initial random variable Z_0 takes value 1 with probability 1, and for any $n \geq 1$, the random variable Z_n conditioned on $Z_{n-1} = i$ is the sum of i independent identically distributed random variables with the probability distribution $\{p_k\}_{k=0}^{\infty}$. The process can be regarded as a random (finite or infinite) tree with Z_n being the number of nodes at level $n \geq 0$, where the number of branches leaving any node in the tree is equal to k with probability p_k, independently of other nodes at the same or previous levels. The generating function characterizing the probability distribution of Z_n can be expressed as the self-composition of the generating function $f(s) = \sum_{k=0}^{\infty} p_k s^k$ of $\{p_k\}_{k=0}^{\infty}$, which is the probability distribution of Z_1. Precisely, if $f^{(n)}(s)$, $0 \leq s \leq 1$, denotes the generating function of the probability distribution of Z_n and if $f^{(0)} = s$, then for every $n \geq 1$, $f^{(n)}(s) = f(f^{(n-1)}(s))$.

The basic characteristic of a branching process is the expected number of branches leaving any node, that is,

$$\mu = E(Z_1) = \sum_{k=0}^{\infty} k\, p_k. \tag{8}$$

A branching process is called subcritical, critical, or supercritical if $\mu < 1$, $\mu = 1$, or $\mu > 1$, respectively. The extinction probability defined as the probability of a tree being finite is 1 for subcritical and critical (provided $p_0 > 0$) processes and smaller than 1 for supercritical processes. We are here only interested in sub-critical and critical processes, whose main properties are given by the following theorem, see [2], [10]. Let $\sigma^2 = \text{Var}(Z_1)$ be the variance of Z_1.

Theorem 6. *In the subcritical case, $\mu < 1$, for any $n \geq 1$,*

$$E(Z_n) = \mu^n \tag{9}$$

$$\text{Var}(Z_n) = \sigma^2 \mu^{n-1} \frac{1-\mu^n}{1-\mu} \tag{10}$$

and if $E(Z_1 \log Z_1) < \infty$, then as $n \to \infty$,

$$\Pr\{Z_n > 0\} \sim c\mu^n \tag{11}$$

where a constant c, $0 < c \leq 1$, depends on the probability distribution of Z_1.
In the critical case, $\mu = 1$, if $0 < \sigma^2 < \infty$, then for any $n \geq 1$,

$$E(Z_n) = 1 \tag{12}$$

$$\mathrm{Var}(Z_n) = \sigma^2 n \tag{13}$$

$$\Pr\{Z_n > 0\} \sim \frac{2}{\sigma^2 n}. \tag{14}$$

The same equations (9) and (10) hold for supercritical processes too. It is also interesting to study the total number of nodes in a random tree up to level n, not counting the initial node, that is, the random variable $Y_n = \sum_{l=1}^{n} Z_l$, for any $n \geq 1$. Another random variables to be considered are Z_n and Y_n conditioned on the event $\{Z_n > 0\}$ meaning that a random tree has depth at least n.

Theorem 7. In the subcritical case, $\mu < 1$, for any $n \geq 1$,

$$E(Y_n) = \frac{\mu}{1-\mu}(1 - \mu^n) \sim \frac{\mu}{1-\mu} \tag{15}$$

$$\mathrm{Var}(Y_n) = \frac{\sigma^2}{(1-\mu)^3}\left((1 - \mu^n)(1 + \mu^{n+1}) - 2(1 - \mu)\,n\,\mu^n\right) \sim \frac{\sigma^2}{(1-\mu)^3}. \tag{16}$$

In the critical case, $\mu = 1$, if $\sigma^2 > 0$, then for any $n \geq 1$,

$$E(Y_n) = n \tag{17}$$

$$\mathrm{Var}(Y_n) = \frac{\sigma^2}{6} n(n + 1)(2n + 1) \sim \frac{\sigma^2}{3} n^3. \tag{18}$$

Theorem 8. In the subcritical case, $\mu < 1$, as $n \to \infty$, the probability distribution of $Z_n | \{Z_n > 0\}$ converges to a limit probability distribution, and if $E(Z_1 \log Z_1) < \infty$, then

$$\lim_{n\to\infty} E(Z_n | Z_n > 0) = \frac{1}{c} \tag{19}$$

$$\lim_{n\to\infty} \mathrm{Var}(Z_n | Z_n > 0) = \frac{\sigma^2}{c\mu(1-\mu)} - \frac{1}{c^2} \tag{20}$$

where c is the same positive constant as in (11).
In the critical case, $\mu = 1$, if $0 < \sigma^2 < \infty$, then

$$\lim_{n\to\infty} \Pr\left\{\frac{Z_n}{n} > z | Z_n > 0\right\} = e^{-2z/\sigma^2}, \quad z \geq 0, \tag{21}$$

$$E(Z_n | Z_n > 0) \sim \frac{\sigma^2}{2} n \tag{22}$$

$$\mathrm{Var}(Z_n | Z_n > 0) \sim \frac{\sigma^4}{4} n^2. \tag{23}$$

The probability distribution of the conditioned random variable $Y_n | \{Z_n > 0\}$ is not treated in the standard books on branching processes like [10] and [2]. Nevertheless, the previous theorems and the results regarding the conditioned random variable $Z_n | \{Z_{n+k} > 0\}$ presented in [2] lead us to conclude that in the subcritical case, $E(Y_n | Z_n > 0) = O(n)$ and $\mathrm{Var}(Y_n | Z_n > 0) = O(n^2)$, whereas in the critical case, $E(Y_n | Z_n > 0) = O(\sigma^2 n^2)$ and $\mathrm{Var}(Y_n | Z_n > 0) = O(\sigma^4 n^4)$.

References

1. R. J. Anderson, Internet communication.
2. K. B. Athreya and P. E. Ney, *Branching Processes*. Berlin: Springer-Verlag, 1972.
3. J. Daemen, R. Govaerts, and J. Vandewalle, "Resynchronization weakness in synchronous stream ciphers," Advances in Cryptology – EUROCRYPT '92, *Lecture Notes in Computer Science*, vol. 765, T. Helleseth ed., Springer-Verlag, pp. 159-167, 1994.
4. J. Dj. Golić and M. J. Mihaljević, "A generalized correlation attack on a class of stream ciphers based on the Levenshtein distance," *Journal of Cryptology*, vol. 3(3), pp. 201-212, 1991.
5. J. Dj. Golić, "On the security of shift register based keystream generators," Fast Software Encryption – Cambridge '93, *Lecture Notes in Computer Science*, vol. 809, R. J. Anderson ed., Springer-Verlag, pp. 90-100, 1994.
6. J. Dj. Golić, "Towards fast correlation attacks on irregularly clocked shift registers," Advances in Cryptology – EUROCRYPT '95, *Lecture Notes in Computer Science*, vol. 921, L. C. Guillou and J.-J. Quisquater eds., Springer-Verlag, pp. 248-262, 1995.
7. J. Dj. Golić, "Linear models for keystream generators," *IEEE Trans. Computers*, vol. C-45, pp. 41-49, Jan. 1996.
8. J. Dj. Golić, "On the security of nonlinear filter generators," Fast Software Encryption – Cambridge '96, *Lecture Notes in Computer Science*, vol. 1039, D. Gollmann ed., Springer-Verlag, pp. 173-188, 1996.
9. J. Dj. Golić, A. Clark, and E. Dawson, "Generalized inversion attack on nonlinear filter generators," submitted.
10. T. H. Harris, *The Theory of Branching Processes*. Berlin: Springer-Verlag, 1963.
11. R. Menicocci, "Cryptanalysis of a two-stage Gollmann cascade generator," *Proceedings of SPRC '93*, Rome, Italy, pp. 62-69, 1993.
12. R. A. Rueppel, "Stream ciphers," *Contemporary Cryptology: The Science of Information Integrity*, G. Simmons ed., pp. 65-134. New York: IEEE Press, 1991.
13. B. Schneier, *Applied Cryptography*. New York: Wiley, 1996.
14. S. Shepherd and W. Chambers, private communication.

Lower Bounds for Discrete Logarithms and Related Problems

Victor Shoup

IBM Research–Zürich, Säumerstr. 4, 8803 Rüschlikon, Switzerland
sho@zurich.ibm.com

Abstract. This paper considers the computational complexity of the discrete logarithm and related problems in the context of "generic algorithms"—that is, algorithms which do not exploit any special properties of the encodings of group elements, other than the property that each group element is encoded as a unique binary string. Lower bounds on the complexity of these problems are proved that match the known upper bounds: any generic algorithm must perform $\Omega(p^{1/2})$ group operations, where p is the largest prime dividing the order of the group. Also, a new method for correcting a faulty Diffie-Hellman oracle is presented.

1 Introduction

The discrete logarithm problem plays an important role in cryptography. The problem is this: given a generator g of a cyclic group G, and an element g^x in G, determine x. A related problem is the Diffie-Hellman problem: given g^x and g^y, determine g^{xy}.

In this paper, we study the computational power of "generic algorithms"—that is, algorithms which do not exploit any special properties of the encodings of group elements, other than the property that each group element is encoded as a unique binary string. For the discrete logarithm problem, as well as several other related problems, including the Diffie-Hellman problem, we present lower bounds that match the known upper bounds for these problems. We also give a new method for correcting a faulty Diffie-Hellman oracle.

Generic Algorithms

Let $\mathbf{Z}/n$ be the additive group of integers mod n, and let S be a set of bit strings of cardinality at least n. An *encoding function* of $\mathbf{Z}/n$ on S is an injective map σ from $\mathbf{Z}/n$ into S.

A generic algorithm A for $\mathbf{Z}/n$ on S is a probabilistic algorithm that behaves as follows. It takes as input an *encoding list* $(\sigma(x_1), \ldots, \sigma(x_k))$, where each x_i is in $\mathbf{Z}/n$, and σ is an encoding function of $\mathbf{Z}/n$ on S. As the algorithm executes, it may from time to time consult an *oracle*, specifying two indices i and j into the encoding list, and a sign bit. The oracle computes $\sigma(x_i \pm x_j)$, according to the specified sign bit, and this bit string is appended to the encoding list (to which A always has access). The output of A is a bit string denoted $A(\sigma; x_1, \ldots, x_k)$.

W. Fumy (Ed.): Advances in Cryptology - EUROCRYPT '97, LNCS 1233, pp. 256-266, 1997.
© Springer-Verlag Berlin Heidelberg 1997

Note that the algorithm A depends on n and S, but not on σ; information about σ is only available to A through the oracle.

To measure the running time of such an algorithm, we count both the number of bit operations, and the number of group operations (i.e., oracle queries).

It is readily seen that the classical Pohlig-Hellman algorithm [8] is a generic algorithm. Let p denote the largest prime divisor of n. Assuming the strings in S have a length that is polynomial in $\log n$, this algorithm has a running time of $p^{1/2}(\log n)^{O(1)}$, and this bound holds uniformly for all possible encoding functions. Note that this algorithm makes essential use of the fact that group elements are uniquely encoded as bit strings, which facilitates the use of fast sorting-and-searching techniques.

Pollard's discrete logarithm algorithm [9] also falls into this generic class. This algorithm is much more space efficient than the Pohlig-Hellman algorithm, but its efficiency relies on the heuristic assumption that the encoding function behaves like a random mapping.

As an example, consider the multiplicative group $(\mathbf{Z}/q)^*$ for a prime q, together with a generator g for this group. Here, $n = q - 1$, and the relevant encoding function sends $a \in \mathbf{Z}/n$ to the binary encoding of $g^a \bmod q$.

Of course, not all algorithms for the discrete logarithm problem are generic. Index-calculus methods for $(\mathbf{Z}/q)^*$, for example, do not fall in this category, and our results have no bearing on such algorithms. For groups associated with elliptic curves, however, the only known algorithms for discrete logarithms are generic. Our results imply that for elliptic curves, one cannot substantially improve upon the Pohlig-Hellman algorithm using generic algorithms: some method must be devised to exploit the particular representation of group elements.

Summary of Results

In §2 we consider the discrete logarithm problem. Theorem 1 says that any generic algorithm that solves (with high probability) the discrete logarithm problem on $\mathbf{Z}/n$ must perform at least $\Omega(p^{1/2})$ group operations, where p is the largest prime dividing n. The theorem shows that for any algorithm, there must be an encoding function for which it makes $\Omega(p^{1/2})$ queries to the group oracle; we do this by showing that this must hold for a *random* encoding function, and a *random* input.

Theorem 2 deals with the analog for the discrete logarithm problem in non-cyclic groups, which was suggested to the author by Buchmann [3]. Suppose G is the product of r cyclic groups of prime order p. Then any generic algorithm that (with high probability) expresses a given element on a given basis for G must perform at least $\Omega(p^{r/2})$ group operations.

In §3 we consider the Diffie-Hellman problem. Theorem 3 proves the analog of Theorem 1 for the Diffie-Hellman problem.

Theorem 4 shows that if the group order is divisible by only large primes, then it is hard to simply determine which of two possible solutions is correct.

Theorem 5 deals with the problem of solving the Diffie-Hellman problem in subgroups. Suppose we are given an oracle for solving the Diffie-Hellman problem in a group G, and now want to solve the Diffie-Hellman problem in a proper subgroup H. This problem is interesting, as it plays an important role in Maurer's [5] and Boneh and Lipton's [2] reductions from the discrete logarithm problem to the Diffie-Hellman problem: they require Diffie-Hellman oracles for prime-order subgroups. Theorem 5 implies that in the context of generic algorithms, there are situations where the oracle for G does not help at all in solving the problem in H.

In §4 we consider the security of an identification scheme due to Schnorr [10] based on the discrete logarithm problem. While this scheme is known to be secure against "passive" attacks, its security against "active" attacks is not well understood. Theorem 6 shows that this scheme is indeed secure against active attacks when the adversary is a generic algorithm.

In §5 we consider a quite different problem: given a faulty oracle for the Diffie-Hellman problem, how to make it highly reliable? One reason that this problem is interesting is that the reductions of Maurer and Boneh/Lipton mentioned above require reliable oracles. That is, these reductions say that if Diffie-Hellman is "easy," then the discrete logarithm is "easy." However, in proving the security of a cryptosystem based on the Diffie-Hellman problem, one normally assumes that this problem is "hard." The above reductions do not allow one to directly weaken this to an assumption that the discrete logarithm is "hard": that a problem is not "hard" does not imply that it is "easy". For this, one must solve precisely the problem we address: making a faulty oracle reliable.

In light of our Theorem 4, standard techniques for amplifying correctness do not apply to the Diffie-Hellman problem. Theorem 7 and its corollary show how to efficiently turn an oracle that is occasionally correct into one that is almost always correct. The theorem is also useful in the application of the Goldreich-Levin theorem to hard bits of the Diffie-Hellman problem.

Related Work

Babai and Szemerédi [1] proved lower bounds in a "black box" model in which the encoding of group elements is not necessarily unique, and the group oracle must be consulted to test for equality. For a cyclic group of order n, if p is the largest prime divisor of n, their results give an $\Omega(p)$ lower bound. Note that the Pohlig-Hellman algorithm does not work in this model.

More recently, Nechaev [7] considered algorithms for the discrete logarithm problem in the following computational model: an algorithm is allowed to perform group operations and equality tests, but no other operations on group elements are allowed—the notion of encodings of elements does not enter into this model at all. While the above $\Omega(p)$ lower bound still applies to the total running time, Nechaev proves an $\Omega(p^{1/2})$ lower bound on the number of group operations alone. These bounds match a variant of the Pohlig-Hellman algorithm in which only linear searching techniques are used.

One can view our results as an extension of Nechaev's results to a broader and more natural class of algorithms, and to a wider range of problems related to the discrete logarithm problem.

For the problem of correcting a faulty Diffie-Hellman oracle, Maurer and Wolf [6] independently devised a scheme based on techniques quite different from ours. It seems that our scheme is substantially simpler and more efficient than theirs.

2 The Discrete Logarithm Problem

The main result of this section is the following.

Theorem 1 *Let n be a positive integer whose largest prime divisor is p. Let $S \subset \{0,1\}^*$ be a set of cardinality at least n. Let A be a generic algorithm for $\mathbf{Z}/n$ on S that makes at most m oracle queries. If $x \in \mathbf{Z}/n$ and an encoding function σ are chosen at random, then the probability that $A(\sigma; 1, x) = x$ is $O(m^2/p)$.*

Note that the above probability is taken over the random choices of σ and x, as well as the coin flips of A. The theorem implies that for any algorithm, there exists an encoding function σ for which it succeeds with probability $O(m^2/p)$, taking the probability over x and the coin flips of A. If we insist that A succeed with probability bounded away from 0 by a constant, this translates into a lower bound of $\Omega(p^{1/2})$ on the number of group operations.

To prove this and several other theorems, we need the following lemma.

Lemma 1 *Let p be prime and let $t \geq 1$. Let $F(X_1, \ldots, X_k) \in \mathbf{Z}/p^t[X_1, \ldots, X_k]$ be a nonzero polynomial of total degree d. Then for random $x_1, \ldots, x_k \in \mathbf{Z}/p^t$, the probability that $F(x_1, \ldots, x_k) = 0$ is at most d/p.*

Proof. For $t = 1$, this is proved in Schwartz [11]. For $t > 1$, one divides the equation $F = 0$ by the highest possible power of p, and obtains a nonzero equation of no greater degree that holds modulo p. If $x_1, \ldots, x_k$ are chosen from $\mathbf{Z}/p^t$ at random, then their images in $\mathbf{Z}/p$ are random as well, and so we can apply the result for $t = 1$. $\square$

We now sketch the proof of Theorem 1. Let $n = p^t s$, where $(p, s) = 1$. Instead of letting the algorithm interact with the actual oracle, we play the following game. Let X be an indeterminant. At any step in the game, the algorithm has computed a list $F_1, \ldots, F_k$ of linear polynomials in $\mathbf{Z}/p^t[X]$, along with a list $z_1, \ldots, z_k$ of values in $\mathbf{Z}/s$, and a list $\sigma_1, \ldots, \sigma_k$ of *distinct* values in S. At the beginning of the game, $k = 2$; $F_1 = 1$ and $F_2 = X$; $z_1 = 1$ and z_2 is chosen at random; σ_1 and σ_2 are chosen at random, subject to $\sigma_1 \neq \sigma_2$. When the oracle is given two indices i and j, we append new values $F_{k+1}, z_{k+1}, \sigma_{k+1}$ to the appropriate lists as follows. We compute $F_{k+1} = F_i \pm F_j \in \mathbf{Z}/p^t[X]$ and $z_{k+1} = z_i \pm z_j \in \mathbf{Z}/s$. If $F_{k+1} = F_l$ and $z_{k+1} = z_l$ for some l with $1 \leq l \leq k$, we set $\sigma_{k+1} = \sigma_l$; otherwise, we set σ_{k+1} to a random element in S distinct from $\sigma_1, \ldots, \sigma_k$.

When the algorithm terminates, it outputs some $y \in \mathbf{Z}/n$. Let y' be the image of y in $\mathbf{Z}/p^t$. Now we choose a random $x \in \mathbf{Z}/p^t$. We say the algorithm wins the game if $F_i(x) = F_j(x)$ for any $F_i \neq F_j$ or if $x = y'$.

Fix i, j with $F_i \neq F_j$, and set $F = F_i - F_j$. Now, since $F \neq 0$, and $\deg F \leq 1$, then by Lemma 1, the probability that $F(x) = 0$ is at most $1/p$. Likewise, the probability that $x = y'$ is at most $1/p$. It follows that the probability that the algorithm wins the above game is $O(m^2/p)$.

To finish the proof, one must only observe that the behavior of this game differs from an actual interaction between the algorithm and oracle only when the algorithm wins the above game. Therefore, the probability that the algorithm outputs the correct answer is bounded by the probability that the algorithm wins the above game.

To make the above argument completely rigorous, one can easily construct a single probability space that is shared by both the actual interaction and the above game, such that

(1) the shared probability space does not change the behavior of either the actual interaction or the above game, and
(2) in this shared space, the event that A outputs the correct answer in the actual interaction is contained in the event that A wins the above game.

The details of this are quite straightforward, and are omitted. That completes the proof of Theorem 1.

We now consider a variation of the discrete logarithm problem that applies to non-cyclic groups.

Suppose that $G = \mathbf{Z}/p \times \cdots \times \mathbf{Z}/p$ is the product of r cyclic groups of order p, where p is prime. The input consists of the encodings of the unit vectors $e_1, \ldots, e_r$, along with the encoding of an element $(x_1, \ldots, x_r) \in G$. The output should be $(x_1, \ldots, x_r)$. Here, an encoding function is an injective map σ from G into some set S of at least p^r bit strings. The following theorem establishes an $\Omega(p^{r/2})$ lower bound for this problem with respect to generic algorithms. Note that a simple generalization of the Pohlig-Hellman algorithm gives a matching upper bound.

Theorem 2 *Let A be a generic algorithm for G on S for the above problem that makes at most m oracle queries. If $(x_1, \ldots, x_r) \in G$ and an encoding function σ are chosen at random, then the probability that*

$$A(\sigma; e_1, \ldots, e_r, (x_1, \ldots, x_r)) = (x_1, \ldots, x_r)$$

is $O(m^2/p^r)$.

The proof is similar to that of Theorem 1. We sketch the differences. Let $X_1, \ldots, X_r$ be indeterminants. We play the same game as before, but instead of a list of polynomials, we maintain a list of r-tuples, each of which has the form

$$(aX_1 + b_1, aX_2 + b_2, \ldots, aX_r + b_r),$$

where $a, b_1, \ldots, b_r \in \mathbf{Z}/p$. The key observation is that when we add or subtract (component-wise) two r-tuples of this form, we get an r-tuple of the same form.

Also, by Lemma 1, the probability that a nonzero r-tuple of this form vanishes when $X_1, \ldots, X_r$ are substituted with random values is at most $1/p^r$. The rest of the proof goes as before.

One can easily extend the above theorem to an arbitrary finite abelian group $G = \mathbf{Z}/n_1 \times \cdots \times \mathbf{Z}/n_r$, obtaining a lower bound of $\Omega(p^{k/2})$, where p is a prime and k is the number of moduli n_i divisible by p.

3 The Diffie-Hellman Problem

In this section, we prove a lower bound for the Diffie-Hellman problem.

Theorem 3 *Let n be a positive integer whose largest prime divisor is p. Let $S \subset \{0,1\}^*$ be a set of cardinality at least n. Let A be a generic algorithm for $\mathbf{Z}/n$ on S that makes at most m oracle queries. If $x, y \in \mathbf{Z}/n$ and an encoding function σ are chosen at random, then the probability that $A(\sigma; 1, x, y) = \sigma(xy)$ is $O(m^2/p)$.*

The proof of this is similar to that of Theorem 1. We may assume that the output of A is one of the encodings obtained from the oracle, since otherwise the success probability is bounded by $1/(p - m)$. We play precisely the same game as there, except that we maintain a list of polynomials F_i in the variables X, Y over $\mathbf{Z}/p^t$, where each polynomial has total degree 1. When the algorithm terminates, we pick $x, y \in \mathbf{Z}/p^t$ at random, and we say that the algorithm wins the game if $F_i(x, y) = F_j(x, y)$ for some $F_i \neq F_j$, or if $F_i(x, y) = xy$ for some i. Applying Lemma 1, for fixed i, j, the probability that $F_i - F_j$ vanishes is at most $1/p$, and for fixed i, the probability that $F_i - XY$ vanishes is at most $2/p$. It follows that the probability that the algorithm wins this game is $O(m^2/p)$.

That completes the proof of Theorem 3.

When n is divisible by only small primes, just determining which of two possible answers is the correct one is hard.

Theorem 4 *Let n be a positive integer whose smallest prime divisor is p. Let $S \subset \{0,1\}^*$ be a set of cardinality at least n. Let A be a generic algorithm for $\mathbf{Z}/n$ on S that makes at most m oracle queries. Let $x, y, z \in \mathbf{Z}/n$ be chosen at random, let σ be a random encoding function, and let b be a random bit. Also, let $w_0 = xy$ and $w_1 = z$. Then the probability that $A(\sigma; 1, x, y, w_b, w_{1-b}) = b$ is $1/2 + O(m^2/p)$.*

We sketch the proof. We play a similar game as before, this time maintaining a list of polynomials $F_i(X, Y, U, V)$ over $\mathbf{Z}/n$ of total degree 1, assigning to each distinct polynomial a distinct random encoding. We say the algorithm wins the game if for any $F_i \neq F_j$, we have $F_i(x, y, xy, z) = F_j(x, y, xy, z)$, or $F_i(x, y, z, xy) = F_j(x, y, z, xy)$. For a fixed $F_i \neq F_j$, the polynomial $F_i - F_j$ must be nonzero modulo some prime power q^t that exactly divides n. Since the images x, y, and z in $\mathbf{Z}/q^t$ are also uniformly distributed, by Lemma 1, the above condition holds with probability at most $4/q \leq 4/p$. Thus, the probability that the algorithm wins the game is $O(m^2/p)$. Moreover, it is clear that in the actual interaction between the algorithm and the oracle, the probability that the algorithm determines b is bounded by $1/2$ plus the probability that the algorithm wins the above game.

We close this section with a look at the following question. Suppose we have a cyclic group G, and we have an oracle for the Diffie-Hellman problem in G. Can we use this oracle to solve the Diffie-Hellman problem efficiently in a proper subgroup H? It is not difficult to see that if $(|H|, |G|/|H|)$ is divisible only by small primes, then this problem can be solved efficiently. More specifically, if p is the largest prime dividing $(|H|, |G|/|H|)$, the problem can be solved in time $p^{1/2}(\log n)^{O(1)}$. The following theorem shows that this bound is essentially tight, and thus for large p the problem can not be solved efficiently using a generic algorithm.

To study this problem, we extend the notion of a generic algorithm so as to include a Diffie-Hellman oracle: given indices i and j, the oracle computes $\sigma(x_i \cdot x_j)$. The output of such an algorithm A is denoted by $A_{DH}(\sigma; x_1, \ldots, x_k)$.

Theorem 5 *Let n be a positive integer, and let l be a divisor of n such that for some prime p, $l = l'p^s$, $n = n'p^t$, and $t > s > 0$. Let $S \subset \{0,1\}^*$ be a set of cardinality at least n. Let A be a generic algorithm for $\mathbf{Z}/n$ on S that makes at most m oracle queries. If $x \in \mathbf{Z}/n$ and an encoding function σ are chosen at random, then the probability that $A_{DH}(\sigma; 1, lx, ly) = \sigma(lxy)$ is $O((t/s) \cdot m^2/p)$.*

We sketch the proof in the case $l' = n' = 1$. The more general case is dealt with as in Theorem 1. Let $d = \lceil t/s \rceil - 1$. We play the usual game, this time maintaining a list of polynomials $F_i(X, Y)$ in the variables X and Y over $\mathbf{Z}/p^t$, each of which has the form

$$\sum_{k=0}^{d} p^{sk} \sum_{i+j=k} a_{ij} X^i Y^j.$$

The key observation is that when we add, subtract, *or even multiply* two polynomials of this form, we get a polynomial that is also of this form. When the algorithm terminates, we select $x, y \in \mathbf{Z}/p^t$ at random, and the algorithm wins the game if for some $F_i \neq F_j$, $F_i(x, y) = F_j(x, y)$ or for some i, $F_i(x, y) = p^s xy$. By Lemma 1, this happens with probability at most $O(dm^2/p)$.

4 Analysis of an Identification Scheme

An identification scheme is an interactive protocol that allows one party P to prove its identity to another party V. To do this, P has a public key, which is known to all parties, and a private key, which is known only to himself.

Such a scheme is considered secure if an adversary can not feasibly make V believe it is conducting the protocol with P. One can allow the adversary to first interact with P, pretending to be V (but not not necessarily following V's protocol), in order to gain some information about P's secret key that will be of use in its impersonation attempt. Such an attack is called "active." An attack where no prior interaction with P is allowed is called "passive." Clearly, security against active attacks is preferable to security against passive attacks.

An identification scheme due to Schnorr [10] runs as follows. Let G be a cyclic group of order n, with a publicly known generator g. P's private key is an element $x \in \mathbf{Z}/n$, and its public key is $h = g^x$. The value x is randomly chosen. In the first step of the protocol, P generates $r \in \mathbf{Z}/n$ at random, computes $h' = g^r$, and sends h' to V. Upon receiving h', V chooses $e \in \mathbf{Z}/n$ at random, and sends e to P. Upon receiving e, P computes $y = r + xe \in \mathbf{Z}/n$ and sends y to V. Upon receiving y, V checks that $g^y = h'h^e$. If this identity holds, V accepts; otherwise, V rejects.

In his paper, Schnorr shows that this protocol is secure against passive attacks, assuming the discrete logarithm is hard. We prove that the scheme is secure against a active adversary that behaves as a generic algorithm.

Theorem 6 *Consider the above identification scheme in a generic setting; that is, there is an encoding function σ mapping elements of $\mathbf{Z}/n$ into a set S of bit strings. Suppose that the adversary makes no more than m interactions with P or queries to the group oracle, and that σ is chosen at random. Suppose also that the private key x is chosen at random. Then the probability that the adversary successfully impersonates P is $O(m^2/p)$, where p is the largest prime dividing n.*

The above probability is taken over σ, x, and the coin tosses of all of the players. In proving this theorem, we allow the adversary to interact with several instances of P in parallel—we do not require that one interaction ends before the next one begins.

We sketch the proof for $n = p$; the more general case is dealt with as in Theorem 1.

We use the same type of game argument that we used in proving the other theorems, but with a few changes. In this game, we maintain a list of degree 1 polynomials $F_i(X, R_1, R_2, \ldots, R_m)$ in $m + 1$ variables over $\mathbf{Z}/p$, corresponding to the group elements the adversary has seen so far, along with a corresponding list of random encodings.

Initially, the list of polynomials contains the two polynomials 1 and X. Whenever the adversary starts an interaction with P for the kth time, we add the polynomial R_k to the polynomial list, and a distinct random encoding to the list of encodings. Whenever the adversary consults the group oracle, we add to the polynomial list the sum of the appropriate polynomials; we either either re-use an encoding or generate a distinct random encoding, as appropriate.

Now suppose the adversary sends a challenge e to the lth instance of P. In our game, we do the following: we choose $y \in \mathbf{Z}/p$ at random, and send y to the adversary as the response from P; we also go through our list of polynomials and substitute $y - eX$ for the variable R_l wherever it appears. If upon making this substitution any two distinct polynomials in the list become equal, we quit and we say the adversary wins. Otherwise, we continue the game.

Now suppose the adversary attempts an impersonation. Without loss of generality, we may assume the adversary has completed all interactions with P that it started. So it has collected a list $F_1, \ldots, F_{m+2} \in \mathbb{Z}/p[X]$ of polynomials along with a list of encodings. In the first step of the protocol, the adversary presents the encoding of some group element corresponding to one of these polynomials, say F_l. Next V chooses $e \in \mathbb{Z}/p$ at random. If $F_l + eX$ is a constant polynomial, we quit and say the adversary wins. Otherwise, the adversary chooses $y \in \mathbb{Z}/p$. Finally, we choose $x \in \mathbb{Z}/p$ at random, and we say that the adversary wins if $y = F_l(x) + ex$ or if $F_i(x) = F_j(x)$ for any $F_i \neq F_j$.

That completes the description of the game. First observe that the behavior of this game deviates from that of the actual interaction only if the adversary wins the game. So it suffices to bound the probability that the adversary wins the game. It is relatively straightforward to show that this is $O(m^2/p)$. One observation to bear in mind is the following. When making a substitution $y - eX$ for a variable R_k, one need only count pairs of polynomials $F_i \neq F_j$ such that $F_i - F_j \in \mathbb{Z}/p[X, R_k]$ and the coefficient of R_k is nonzero. But note that if we count this pair when substituting for R_k, we will not count this pair when we later make a substitution for some other R_l. Thus, the total number of pairs we need to count is $O(m^2)$.

5   A Diffie-Hellman Self-Corrector

In this section, we consider the following problem. Let G be a cyclic group of order n with generator g. Suppose we have a "faulty" oracle for the Diffie-Hellman problem; that is, given g^a and g^b, the oracle outputs g^c, such that $c \equiv ab \pmod{n}$ with probability at least ϵ. We take this probability to be over the random choice of a and b mod n, and any coin tosses of the oracle. Here, ϵ is small, but nonnegligible. The problem is to use this oracle to build an efficient algorithm for the Diffie-Hellman problem whose output is almost certainly correct for all inputs. One motivation for this problem is again the reductions of [5] and [2] from the discrete logarithm problem to the Diffie-Hellman problem; these reductions require a nearly-perfect oracle—a faulty oracle will simply not do.

Given such an oracle, using the standard random self-reduction, we can run it $O(1/\epsilon)$ times so that with high probability one of its outputs is correct. However, as we have seen, in the generic model we have no hope of determining which output is correct.

We consider the following, more general, problem. We define a (k, δ) Diffie-Hellman oracle as follows: for all inputs g^a, g^b, it produces a list of k elements in G such that this list contains g^{ab} with probability at least δ. The problem is to use this oracle to solve the Diffie-Hellman problem.

Another situation in which this type of oracle arises is in the hard-bit construction of Goldreich and Levin [4], where a bit-predicting oracle can be turned into this type of oracle.

Theorem 7 *Given a (k, δ) Diffie-Hellman oracle with $\delta > 7/8$, we can construct a probabilistic (generic) algorithm for the Diffie-Hellman problem with the following properties. For given α, with $0 < \alpha < 1$, the algorithm makes $O(\log(1/\alpha))$ queries to the (k, δ) oracle, and performs an additional $O(\log(1/\alpha)k \log n + (\log n)^2)$ group operations. For all inputs, the output of the algorithm is correct with probability at least $1 - \alpha$.*

As an immediate corollary, we have:

Corollary 1 *Given a faulty Diffie-Hellman oracle that has a success probability of ϵ, we can construct a probabilistic algorithm for the Diffie-Hellman problem with the following properties. For given α, with $0 < \alpha < 1$, the algorithm makes $O(\epsilon^{-1} \log(1/\alpha))$ queries to the faulty oracle, and performs an additional $O(\epsilon^{-1} \log(1/\alpha) \log n + (\log n)^2)$ group operations. For all inputs, the output of the algorithm is correct with probability at least $1 - \alpha$.*

To prove Theorem 7, we assume that n is known, and that for all prime factors p of n, $k^2/p < 1/8$. If this does not hold, we can partially factor n, and apply the Pohlig-Hellman algorithm to the "smooth" part. A straightforward calculation shows that this takes $O(k \log n + (\log n)^2)$ group operations. So we can assume that n is of the desired form.

For given g^a, g^b, the following algorithm either reports failure, or outputs a single value g^c. The algorithm makes two queries to the (k, δ) oracle and performs an additional $O(k \log n)$ group operations. The probability that it reports failure is at most $3/8$. The conditional probability that $g^c \neq g^{ab}$, given that it does not report failure, is $2/7$. The Diffie-Hellman algorithm then simply runs the above algorithm $O(\log(1/\alpha))$ times, taking the majority of the non-failure outputs.

We call the (k, δ) oracle twice, first with g^a, g^b, obtaining a list $g_1, \ldots, g_k$ of group elements. Next, we choose $x, y \in \{0, \ldots, n-1\}$ at random, and send $(g^a)^x g^y, g^b$ to the (k, δ) oracle, obtaining a list $g'_1, \ldots, g'_k$ of group elements. Next, for all $1 \leq i \leq k$ and $1 \leq j \leq k$, we test if

$$g_i^x (g^b)^y = g'_j. \tag{1}$$

If (1) is satisfied for a unique pair (g_i, g'_j), we output g_i; otherwise, we report failure. Note that standard sorting-and-searching techniques can be used to make this last step efficient.

The claimed running-time bound is easily verified. We now analyze its correctness. Let $z = ax + y$. Fix i and j, and suppose $g_i = g^c$ and $g'_j = g^d$. Suppose $c \equiv ab \pmod{n}$. Then (1) is satisfied if and only if and $d \equiv zb \pmod{n}$. Now suppose $c \not\equiv ab \pmod{n}$. Then for some prime power p^t that exactly divides n, we must have $c \not\equiv ab \pmod{p^t}$. In this case, the probability that (1) holds is at most the conditional probability that for random $x, y \bmod p^t$, $cx + by \equiv d \pmod{p^t}$, given that $ax + y \equiv z \pmod{p^t}$. This is equal to the probability that for fixed z and random x, $(c - ab)x + bz - d \equiv 0 \pmod{p^t}$, which is by Lemma 1 at most $1/p$.

There are three mutually exclusive events of interest: the algorithm either (F) reports failure, (I) produces an incorrect output, or (C) produces a correct output.

$\Pr[F] + \Pr[I]$ is bounded by the probability that one of the lists does not contain a correct output, or that any extraneous relations (1) hold. This happens with probability at most $1/8 + 1/8 + k^2/p \leq 3/8$.

$\Pr[I]$ is bounded by the probability that one of the lists does not contain a correct output. This is because if both lists contain a correct output, any extraneous relations (1) that hold will cause the algorithm to report failure. This probability is thus bounded by $1/8 + 1/8 = 1/4$.

It trivially follows that $\Pr[F]$ is bounded by 3/8. Moreover, by a simple calculation, $\Pr[I]/(\Pr[I] + \Pr[C])$ is bounded by 2/7.

References

1. L. Babai and E. Szemerédi. On the complexity of matrix group problems I. In *25th Annual Symposium on Foundations of Computer Science*, pages 229–240, 1984.
2. D. Boneh and R. J. Lipton. Algorithms for black-box fields and their application to cryptography. In *Advances in Cryptology–Crypto '96*, pages 283–297, 1996.
3. J. Buchmann, 1995. Personal communication.
4. O. Goldreich and L. A. Levin. A hard-core predicate for all one-way functions. In *21st Annual ACM Symposium on Theory of Computing*, pages 25–32, 1989.
5. U. Maurer. Towards the equivalence of breaking the Diffie-Hellman protocol and computing discrete logarithms. In *Advances in Cryptology–Crypto '94*, pages 271–281, 1994.
6. U. Maurer and S. Wolf. Diffie-Hellman oracles. In *Advances in Cryptology–Crypto '96*, pages 268–282, 1996.
7. V. I. Nechaev. Complexity of a determinate algorithm for the discrete logarithm. *Mathematical Notes*, 55(2):165–172, 1994. Translated from *Matematicheskie Zametki*, 55(2):91–101, 1994.
8. S. Pohlig and M. Hellman. An improved algorithm for computing logarithms over GF(p) and its cryptographic significance. *IEEE Trans. Inf. Theory*, 24:106–110, 1978.
9. J. M. Pollard. Monte Carlo methods for index computation mod p. *Mathematics of Computation*, 32:918–924, 1978.
10. C. Schnorr. Efficient signature generation by smart cards. *J. Cryptology*, 4:161–174, 1991.
11. J. T. Schwartz. Fast probabilistic algorithms for verification of polynomial identities. *J. ACM*, 27(4):701–717, 1980.

Stronger Security Proofs for RSA and Rabin Bits

R. Fischlin and C.P. Schnorr

Fachbereich Mathematik/Informatik
Universität Frankfurt
PSF 111932
60054 Frankfurt/Main, Germany

Abstract. The RSA and Rabin encryption function are respectively defined as $E_N(x) = x^e \bmod N$ and $E_N(x) = x^2 \bmod N$, where N is a product of two large random primes p, q and e is relatively prime to $\varphi(N)$. We present a much simpler and stronger proof of the result of ALEXI, CHOR, GOLDREICH and SCHNORR [ACGS88] that the following problems are equivalent by probabilistic polynomial time reductions: (1) given $E_N(x)$ find x; (2) given $E_N(x)$ predict the least-significant bit of x with success probability $\frac{1}{2} + \frac{1}{\mathrm{poly}(n)}$, where N has n bits. The new proof consists of a more efficient algorithm for inverting the RSA/Rabin-function with the help of an oracle that predicts the least-significant bit of x. It yields provable security guarantees for RSA-message bits and for the RSA-random number generator for moduli N of practical size.

1 Introduction

Randomness is a fundamental computational resource and the efficient generation of provably secure pseudorandom bits is a basic problem. YAO [Y82] and BLUM, MICALI [BM84] have shown that perfect random number generators (RNG) exist under reasonable complexity assumptions. Some perfect RNG's are based on the RSA-function $E_N(x) = x^e \bmod N$ and the Rabin-function $E_N(x) = x^2 \bmod N$, where the integer N is a product of two large random primes p, q and e is relatively prime to $\varphi(N) = (p-1)(q-1)$. The corresponding RNG transforms a random seed $x_0 \in [1, N)$ into a bit string $b_1, \ldots, b_m$ of arbitrary polynomial length $m = n^{O(1)}$ according to the recursion $x_i := E_N(x_{i-1})$, $b_i := x_i \bmod 2$, where N has n bits. The security of these RNG's is related to a result of [ACGS88] that the RSA/Rabin-function can be inverted in polynomial time if one is given an oracle which predicts from given $E_N(x)$ the least-significant bit of x with success probability $\frac{1}{2} + \frac{1}{\mathrm{poly}(n)}$. While the ACGS-result shows that the RSA/Rabin RNG is perfect in an asymptotic sense the practicality of this result has been questionable as the transformation of attacks against these RNG's into a full inversion of the RSA/Rabin-function (resp. the factorization of N) is rather slow.

The main contribution of this paper is a much simpler and stronger proof of the ACGS-result. The new proof gives a more efficient algorithm for the inversion of the RSA/Rabin-function if one is given an oracle that predicts the

W. Fumy (Ed.): Advances in Cryptology - EUROCRYPT '97, LNCS 1233, pp. 267-279, 1997.
© Springer-Verlag Berlin Heidelberg 1997

least significant message bit. While the new method is primarily of theoretical interest, it yields a security guarantee for moduli N of practical size. We extend our results to the Rabin-function $E_N(x) = x^2 \bmod N$. The reduction from E_N-inversion, resp. factoring N, to prediction is particular efficient for the *absolute* Rabin-function $E_N^a(x) = |x^2 \bmod N|$, where $|y| = \min(y, N - y)$.

Notation. Let N be product of two large primes p, q. Let $\mathbb{Z}_N = \mathbb{Z}/N\mathbb{Z}$ be the ring of integers modulo N, and let $\mathbb{Z}_N^*$ denote the subgroup of invertible elements in $\mathbb{Z}_N$. We represent elements $x \in \mathbb{Z}_N$ by their least nonnegative residue in the interval $[0, N)$, i.e., $\mathbb{Z}_N = [0, N)$. We let $[ax]_N \in [0, N)$ denote the least non-negative residue of $ax(\bmod N)$. We use $[ax]_N$ for arithmetic expressions over $\mathbb{Z}$ while the arithmetic over $a, x \in \mathbb{Z}_N = [0, N)$ is done modulo N. Let n be the bit length of N, $2^{n-1} < N < 2^n$. For $x \in \mathbb{Z}$ we let $\ell(x) = x \bmod 2$ denote the *least-significant bit* of x. Let e be relatively prime to $\varphi(N) = (p-1)(q-1)$, $e \neq 1$. The RSA cryptosystem enciphers a message $x \in \mathbb{Z}_N$ into $E_N(x) = x^e \bmod N$. Let O_1 be an oracle running in expected time T which, given $E_N(x)$ and N, predicts the least-significant bit $\ell(x)$ of x with advantage ε: $\Pr_{x,w}[O_1(E_N(x)) = \ell(x)] \geq \frac{1}{2}+\varepsilon$, where the probability refers to random $x \in_R [0, N)$ and the internal coin tosses w of the oracle. We assume that the time T of the oracle also covers the n^2 steps for the evaluation of the function E_N. Throughout the paper we assume that ε^{-1} and n are powers of 2, $n \geq 2^9$. We let lg denote the logarithm function with base 2. For a finite set A let $b \in_R A$ denote a random element of A that is uniformly distributed. All time bounds count arithmetic steps using integers with $\lg(n\varepsilon^{-1})$ bits. We use integers of that size for counting the votes in majority decisions.

Our results. Consider the problem to compute from $E_N(x)$ and N the message $x \in \mathbb{Z}_N$ with the help of the oracle O_1 but without knowing the factorization of N. The new method inverts E_N by iteratively tightening approximations uN of random multiples $[ax]_N$ with known multiplier a via binary division. The basic idea is that $[\frac{1}{2}ax]_N = \frac{1}{2}[ax]_N$ for even $[ax]_N$, $[\frac{1}{2}ax]_N = \frac{1}{2}([ax]_N + N)$ for odd $[ax]_N$. Thus we get from a rational approximation uN to $[ax]_N$ and the least-significant bit $\ell(ax)$ a tighter approximation to $[\frac{1}{2}ax]_N$:
$$[\tfrac{1}{2}ax]_N - \tfrac{1}{2}(u + \ell(ax))N = \tfrac{1}{2}([ax]_N - uN).$$
Without knowing x we get $\frac{1}{2}(u + \ell(ax))N$ from the multiplier a, the previous approximation uN and $E_N(x)$. This in turn yields $E_N(ax) = E_N(a)E_N(x)$ and a guess $O_1(E_N(ax))$ for $\ell(ax)$. Binary division without an oracle has already been used by GOLDWASSER, MICALI, TONG [GMT82]. The method of binary division is more efficient than the gcd-method in [BCS83], [ACGS88]. In order to decipher $E_N(x)$ it guesses the least-significant bits and approximate locations of two random multiples $[ax]_N, [bx]_N$ whereas the gcd-method requires four random multiples. Most importantly, the number of oracle calls becomes nearly minimal.

In section 2 we present our basic algorithm that inverts the RSA-function E_N in expected time $O(n^2\varepsilon^{-2} T + n^2\varepsilon^{-6})$, where T is the time and ε the advantage of oracle O_1. The expectation refers to the internal coin tosses of O_1 and of the inversion algorithm. This greatly improves the [ACGS88]-time bound $O(n^3\varepsilon^{-8} T)$ for oracle RSA-inversion. The new time bound differentiates the costs induced by the oracle calls and the *additional overhead*. The oracle calls induce $O(n^2\varepsilon^{-2} T)$

steps, we call the $O(n^2\varepsilon^{-6})$ other steps the additional overhead. We generalize our security result to the j-th least-significant message bit for arbitrary j. This generalization affects only the additional overhead of E_N-inversion, the number of oracle calls remains unchanged.

In section 3 we introduce the *subsample majority rule*, a trick that improves the efficiency of majority decisions. Suppose we are given pairwise independent 0,1-valued votes that each has an advantage ε in predicting the target bit $\ell(a_t x)$. A large sample size m is necessary in order to make the error probability $\frac{1}{m\varepsilon^2}$ of the majority decision sufficiently small. To reduce the computational costs of the large sample we only use a small random subsample of it. While the random subsample induces only a small additional error probability the time for the subsample majority decision reduces to the size of the small subsample. The large sample is only mentally used for the analysis, it does not enter into the computation. Using this trick we gain a factor $\frac{n}{\lg n}$ in the number of oracle calls and in the time for the inversion of E_N. The reduced number of oracle calls is optimal up to factor $O(\lg n)$.

In section 4 we process all possible locations for $[ax]_N, [bx]_N$ much faster than trying them separately. This reduces the additional overhead in the time for RSA-inversion to $O(n^2\varepsilon^{-4}\lg(n\varepsilon^{-1}))$.

In section 5 we give conclusions for the security of RSA-message bits and of the RSA-random number generator for moduli N of practical size. These conclusions are preliminary as the additional overhead can be further reduced.

In section 6 we extend the oracle inversion algorithm to the Rabin-function E_N and we derive a security guarantee for the $x^2 \bmod N$ generator under the assumption that factoring is hard. The oracle inversion of the absolute Rabin-function is as fast as that of the RSA-function . For the centered Rabin-function the inversion runs in time $O(n\varepsilon^{-4}\lg(n\varepsilon^{-1})T)$. The latter improves the previous time bound $O(n^3\varepsilon^{-11}T)$ due to [VV84] in connection with [ACGS88].

2 RSA-inversion by binary division

We introduce a novel method for inverting the RSA-function without knowing the factorization of N if one is given an oracle O_1 that predicts the least-significant message bit with non-negligible advantage ε. The algorithm RSA-inversion is a simple version of the new method, that will be made more efficient by subsequent modifications. In order to invert $E_N(x)$ it picks two random multipliers a, b and guesses the least-significant bits and the approximate locations for the message multiples $[ax]_N, [bx]_N$. For $a_t := a\, 2^{-t} \bmod N$ it iteratively constructs rational approximations $u_t N$ so that $|[a_t x]_N - u_t N| \le \frac{\varepsilon N}{4 \cdot 2^t}$ for $t = 1, ..., n$. To this end it uses the method of binary division explained in the introduction. From the approximation $u_n N$ to $[a_n x]_N$ we get the message $x = a_n^{-1} \lfloor u_n N + \frac{1}{2} \rfloor \bmod N$. The main work is to determine the bits $\ell(a_t x)$ by majority decision using the oracle O_1.

The majority decision for $\ell(a_t x)$ uses multipliers $a_t + i a_{t-1} + b$ that are pairwise independent. Recall that the arithmetic on a, b, x, a_t is done modulo N.

The algorithm determines an integer $w_{t,i}$ that most likely satisfies the equation $(a_t + i a_{t-1} + b)x = [a_t x]_N + i[a_{t-1}x]_N + [bx]_N - w_{t,i} N$, in which case we call $w_{t,i}$ *correct*. The i-th measurement guesses $\ell(a_t x)$ by evaluating ℓ for both sides of the latter equation. We guess ℓ for the left hand side via the oracle O_1 and we use that the right hand side is linear in $\ell(a_t x)$. The majority decision performs $m = \min\{2^t, 2n\}\varepsilon^{-2}$ measurements, where m and the set A_m of integers i is chosen as to optimize the trade-off between error probability and efficiency of the majority decision. The use of pairwise independent multipliers for majority decision is a crucial contribution of [ACGS88].

RSA-inversion

1. INPUT $E_N(x)$, N
   $t := 0$ (t is the *stage*), pick random integers $a, b \in_R \mathbb{Z}_N^* \subset [0, N)$,
   guess rational integers $u \in \frac{\varepsilon^3}{4}[0, 4\varepsilon^{-3})$, $v \in \frac{\varepsilon}{4}[0, 4\varepsilon^{-1})$ satisfying
   $$|[ax]_N - uN| \leq \tfrac{\varepsilon^3}{8}N, \qquad |[bx]_N - vN| \leq \tfrac{\varepsilon}{8}N.$$
   Guess the least-significant bits $\ell(ax)$, $\ell(bx)$, $a_0 := a$, $u_0 := u$.
2. WHILE $t < n$ DO
   $t := t + 1$, $a_t := \frac{1}{2}a_{t-1} \bmod N$, $u_t := \frac{1}{2}(u_{t-1} + \ell(a_{t-1}x))$,
   $m := \min\{2^t, 2n\}\varepsilon^{-2}$.
   $A_m := \{i \mid |1 + 2i| \leq m\}$, $w_{t,i} := \lfloor u_t + i u_{t-1} + v \rfloor$ for all $i \in A_m$.
   Majority decision
   $$z := \#\left\{ i \in A_m \ \middle| \ \begin{array}{l} O_1(E_N((a_t + i a_{t-1} + b)x)) = \\ i\ell(a_{t-1}x) + \ell(bx) - w_{t,i}N \bmod 2 \end{array} \right\}$$
   $\ell(a_t x) := [\,0$ if $z \geq \frac{m}{2}$ and 1 otherwise $]$ END while
3. OUTPUT $x := a_n^{-1}\lfloor u_n N + \frac{1}{2}\rfloor \bmod N$

Correctness. If $\ell(a_t x)$ is always correctly determined the rational approximation $u_t N$ to $[a_t x]_N$ tightens from stage $t-1$ to stage t by a factor $\frac{1}{2}$. As $a_t = \frac{1}{2}a_{t-1} \bmod N$ we have $[a_t x]_N = \frac{1}{2}[a_{t-1}x]_N$ for even $[a_{t-1}x]_N$, $[a_t x]_N = \frac{1}{2}([a_{t-1}x]_N + N)$ for odd $[a_{t-1}x]_N$. Hence

$$[a_t x]_N - u_t N = [a_t x]_N - \tfrac{1}{2}(u_{t-1} + \ell(a_{t-1}x))N = \tfrac{1}{2}([a_{t-1}x]_N - u_{t-1}N). \quad (1)$$

Probability of success. We call $w_{t,i}$ *correct* if $0 \leq [a_t x]_N + i[a_{t-1}x]_N + [bx]_N - w_{t,i}N < N$. Correct $w_{t,i}$ satisfy the equation (as N is odd we have $-w_{t,i}N = w_{t,i} \bmod 2$) : $\ell((a_t + i a_{t-1} + b)x) = \ell(a_t x) + i\,\ell(a_{t-1}x) + \ell(bx) + w_{t,i} \bmod 2$. In the majority decision we replace in this equation $\ell((a_t + i a_{t-1} + b)x)$ by $O_1(E_N((a_t + i a_{t-1} + b)x))$, and we determine $\ell(a_t x)$ so that the equation holds for the majority of the $i \in A_m$. The algorithm succeeds if step 1 guesses correctly and if the majority decisions for $\ell(a_t x)$ are all correct. In this case we have $|[a_n x]_N - u_n N| \leq \frac{\varepsilon N}{4 \cdot 2^n} < \frac{1}{2}$ and thus $a_n x = \lfloor u_n N + \frac{1}{2}\rfloor \bmod N$ and the output is correct. All probabilities refer to the random pair $(a, b) \in_R (\mathbb{Z}_N^*)^2$ and to the coin tosses of the oracle. We use the conditional probability for the case that we are in the right alternative, where step 1 guesses correctly and the bits $\ell(a_t x)$ of previous stages have been correctly determined.

Error probability of $w_{t,i}$. Let us denote $w'_{t,i} = u_t + iu_{t-1} + v$ so that $w_{t,i} = \lfloor w'_{t,i} \rfloor$. In the right alternative we have by iteration of equation (1) $[a_j x]_N - u_j N = 2^{-j}([ax]_N - uN)$ for all $j \leq t$. Therefore and since $2^{-t}\varepsilon^2|1 + 2i| \leq 1$ for $i \in A_m$ we have

$$|[a_t x]_N + i[a_{t-1}x]_N + [bx]_N - w'_{t,i}N| \leq \tfrac{\varepsilon}{8}(2^{-t}\varepsilon^2|1 + 2i| + 1)N \leq \tfrac{\varepsilon}{4}N.$$

Hence $w_{t,i}$ is correct except that there exists an integer between $w'_{t,i}N$ and $[a_t x]_N + i[a_{t-1}x]_N + [bx]_N$. Therefore $w_{t,i}$ errs with probability at most $\tfrac{\varepsilon}{4}$. By using the m integers $i \in A_m$ instead of $i = 1, ..., m$ we save a factor 2 in $|i|$ and in the error probability of $w_{t,i}$.

Error probability of the majority decision. The multipliers $(\tfrac{1}{2} + i)a + b$ are pairwise independent for $|i| < \tfrac{1}{2}\min(p,q)$ since the matrix of the $\mathbb{Z}_N$-linear transformation $\begin{bmatrix} 1, & \tfrac{1}{2} + i \\ 1, & \tfrac{1}{2} + j \end{bmatrix} \begin{bmatrix} a \\ b \end{bmatrix}$ has determinant $j - i \neq 0 \bmod N$ and (a, b) is random in $(\mathbb{Z}_N^*)^2$. A similar argument shows that the errors of the $w_{t,i}$ for $i \in A_m$ are pairwise independent if we are in the right alternative. The i-th measurement is *correct* iff

$$O_1(E_N((a_t + ia_{t-1} + b)x)) = \ell(a_t x) + i\,\ell(a_{t-1}x) + \ell(bx) + w_{t,i} \bmod 2.$$

This is the case if the oracle guesses correctly and $w_{t,i}$ is correct. The error of the i-th measurement can be dominated by $0, 1$-valued random variables X_i with $\mathrm{E}[X_i] = \mathrm{E}[O_1(E_N((a_t + ia_{t-1} + b)x)) \neq \ell((a_t + ia_{t-1} + b)x)] + \mathrm{E}[w_{t,i} \text{ errs}]$ so that the X_i are pairwise independent for $i \in A_m$. Hence $\mathrm{E}[X_i] \leq \tfrac{1}{2} - \tfrac{3}{4}\varepsilon$, $\mathrm{Var}[X_i] \leq \tfrac{1}{4}$.

A majority decision is correct iff the majority of the m measurements is correct. A majority decision errs only if $\tfrac{1}{m}\sum_i X_i - \mu \geq \tfrac{3}{4}\varepsilon$, where $\mu := \tfrac{1}{m}\sum_i \mathrm{E}[X_i]$. We apply Chebyshev's inequality to the m pairwise independent error variables X_i with $i \in A_m$.

Chebyshev's inequality.
$$\Pr[|\tfrac{1}{m}\sum_i X_i - \mu| \geq \tfrac{3}{4}\varepsilon] \leq \sum_i \mathrm{Var}[X_i](m\tfrac{3}{4}\varepsilon)^{-2} \leq \tfrac{4}{9m\varepsilon^2}.$$

By $m = \min\{2^t, 2n\}\varepsilon^{-2}$ the majority decisions for $\ell(a_t x)$ errs with probability $\tfrac{4}{2^t 9}$ for $t \leq 1 + \lg n$ and with probability $\tfrac{2}{9n}$ for $t \geq 1 + \lg n$. The majority decision for $t = 1, ..., n$ have error probability $\sum_{t \geq 1} \tfrac{4}{2^t 9} + (2n - \lg n)/(9n) \leq \tfrac{4}{9} + \tfrac{2}{9} = \tfrac{2}{3}$.

Running time. We give an upper bound for the expected number of steps required to compute x when given $E_N(x)$ and N. We separately count the steps of the oracle calls and the other steps which form the *additional overhead*.

The oracle is queried about $E_N((a_t + ia_{t-1} + b)x)$ for $t = 1, ..., n$ for the $i \in A_m$. The oracle calls depend on a, b but not on $u, v, \ell(ax), \ell(bx)$. So we keep a, b fixed while we try all possibilities for $u, ..., \ell(bx)$. As the algorithm has success rate $\tfrac{1}{3}$ and calls the oracle at most $m \leq 2n\varepsilon^{-2}$ times per stage, there are in total at most $3 \cdot 2\, n^2\varepsilon^{-2}T$ oracle calls.

Each majority decision contributes to the additional overhead at most $2n\varepsilon^{-2}$ steps that are performed with all oracle replies given. The algorithm does not need the exact rational $u_t + v$ and merely computes $w_{t,i} = \lfloor u_t + v + iu_{t-1} \rfloor$ using

$\lg(n\varepsilon^{-1}) + O(1)$ precision bits from $u_t + v$ and iu_{t-1}. We see that the additional overhead is at most the product of the following factors

1. # of quadruples $(u, v, \ell(ax), \ell(bx))$ $4^2\varepsilon^{-4}\,2^2$
2. # of stages n
3. # of steps per majority decision $2n\varepsilon^{-2}$
4. the inverse of the success rate 3

Hence the additional overhead is at most $3 \cdot 2^7 n^2 \varepsilon^{-6}$, and thus the expected time for the inversion of E_N is $3n^2\varepsilon^{-2}(2T + 2^7\varepsilon^{-4})$.

Using an oracle for the j-th least-significant message bit. The j-th least-significant message bit $\ell_j(x)$ is called *secure* if E_N can be inverted in polynomial time via an oracle O_j that predicts $\ell_j(x)$ when given $E_N(x)$. Let oracle O_j predict $\ell_j(x)$ with advantage ε in expected time T. With the oracle O_j the RSA-inversion proceeds in a similar way as for $j = 1$. It guesses initially $L_j(ax), L_j(bx) \in [0, 2^j)$, the integers that consist of the j least-significant bits of $[ax]_N, [bx]_N$. A main point is that the majority decision for $\ell_j(a_t x)$ takes into account carry overs from the $j - 1$ least-significant bits. The equation

$$L_{j-1}((a_t + i\,a_{t-1} + b)x) + 2^{j-1}\ell_j((a_t + i\,a_{t-1} + b)x) = L_{j-1}(a_t x) +$$

$$iL_{j-1}(a_{t-1}x) + L_{j-1}(bx) + 2^{j-1}(\ \ell_j(a_t x) + i\ell_j(a_{t-1}x) + \ell_j(bx)\ ) - w_{t,i}N \bmod 2^j$$

holds for correct $w_{t,i}$. In order to predict $\ell_j(a_t x)$ we replace in this equation $\ell_j((a_t + i\,a_{t-1} + b)x)$ by $O_j(E_N((a_t + i\,a_{t-1} + b)x))$ and we recover $L_{j-1}((a_t + i\,a_{t-1} + b)x)$, $L_{j-1}(a_t x)$ and $L_{j-1}(a_{t-1}x)$ recursively from the initial values $L_j(ax), L_j(bx)$, the approximate locations uN, vN and N. We choose $\ell_j(a_t x)$ so that the equation holds for the majority of $i \in A_m$.

The time of the inversion algorithm does not change from the case $j = 1$ to arbitrary j, except that the factor under 1. increases to $2^{2j}4^2\varepsilon^{-4}$ as we have to guess $L_j(ax), L_j(bx) \in [0, 2^j)$. Now the time bound for RSA-inversion via O_j is $O(n^2\varepsilon^{-2}(T + 2^{2j}\varepsilon^{-4}))$ while it is $O(2^{4j}n^3\varepsilon^{-8}T)$ for the ACGS-algorithm. There is a double advantage in the new time bound. The factor 2^{4j} decreases to 2^{2j} and it only affects the additional overhead. The additional overhead can be reduced by the method in section 4 to $O(n^2\varepsilon^{-4}\lg(n\varepsilon^{-1}))$.

3   From pairwise to mutually independent votes.

We introduce the *subsample majority decision*, a trick that reduces the number of oracle calls for RSA-inversion by a factor $\lg n/n$. Suppose we have m pairwise independent 0,1-valued random variables (votes) V_i for $i \in A_m$ that have advantage ε in predicting the target bit $\ell(a_t x)$. The error probability of a majority decision is $\frac{1}{m\varepsilon^2}$, so we need a large m to make this error small. To reduce the computational costs of the large sample we only use a small random subsample consisting of $m' \ll m$ votes that are selected uniformly at random. Now the votes of the subsample are mutually independent, even though the original votes are merely pairwise independent, and their advantage ε' is close to ε. While the subsample induces only a small additional error probability $\exp(-2m'\varepsilon'^2)$ the time for the subsample majority decision is only m'. The large sample only appears in the mental error analysis, it does not enter into the computation. We can even fix a random subset $A'_{m'} \subset A_m$ for all SMAJ-calls, where $A_m := \{i \mid |1 + 2i| \leq m\}$ as in section 2. Theorem 3 in section 4 uses such a fixed subset $A_{m'}$.

Subsample Majority Decision (SMAJ). Pick $(\nu(1), ..., \nu(m')) \in_R (A_m)^{m'}$ and guess that $\ell(a_t x)$ is $[\sum_{i=1}^{m'} V_{\nu(i)} \geq \frac{m'}{2}]$.

As in section 2 let X_i be the error of the vote V_i so that $E[X_i] \leq \frac{1}{2} - \frac{3}{4}\varepsilon$. We denote $\mu = \frac{1}{m}\sum_{i \in A_m} E[X_i]$ and $\mu' = \frac{1}{m}\sum_{i \in A_m} X_i$. Consider the case that $|\mu' - \mu| < \frac{1}{4}\varepsilon$ which by Chebyshev's inequality holds except with probability $\frac{\text{Max}_i \text{Var}[X_i]}{m(\varepsilon/4)^2} \leq \frac{4}{m\varepsilon^2}$. The SMAJ-rule errs in this case only if $\frac{1}{m'}\sum_{i=1}^{m'} X_{\nu(i)} \geq \mu' + \frac{1}{2}\varepsilon$. For fixed values X_i with $i \in A_m$ the variables $X_{\nu(1)}, \ldots, X_{\nu(m')}$ are identically distributed and *mutually independent* with mean value μ'. So we use

Bernstein's law of large numbers. For random $(\nu(1), ..., \nu(m')) \in_R (A_m)^{m'}$:
$$\Pr[\tfrac{1}{m'}\textstyle\sum_{i=1}^{m'} X_{\nu(i)} \geq \mu' + \tfrac{1}{2}\varepsilon] < \exp(-2(\tfrac{1}{2}m'\varepsilon)^2).$$

Proposition 1. *If the errors X_i of the votes are pairwise independent and $E[X_i] \leq \frac{1}{2} - \frac{3}{4}\varepsilon$ then SMAJ errs with probability at most $\frac{4}{m\varepsilon^2} + \exp(-\frac{1}{2}m'\varepsilon^2)$.*

Proof. If $\frac{1}{m'}\sum_{i=1}^{m'} X_{\nu(i)} \geq \frac{1}{2}$ we either have $|\mu - \mu'| \geq \frac{1}{4}\varepsilon$ or $\frac{1}{m'}\sum_{i=1}^{m'} X_{\nu(i)} \geq \mu' + \frac{1}{2}\varepsilon$. The first event has probability $\leq \frac{4}{m\varepsilon^2}$ and the second $\leq \exp(-2m'(\frac{1}{2}\varepsilon)^2)$.

RSA-inversion using the SMAJ-rule. Let us modify the stages $t \geq 4 + \lg n$ of RSA-inversion so that at these stages the SMAJ-rule is used with $m = 2^4 \varepsilon^{-2} n$ and the multipliers $a_t + i a_{t-1} + b$ with $i \in A_m$ — at stages $t \leq 3 + \lg n$ the set A_m is too small for SMAJ. We apply Proposition 1 with this m and $m' = 2\varepsilon^{-2} \lg n$. Then $\frac{1}{2}m'\varepsilon^2 = \lg n > 1.4426 \ln n$, and thus a single SMAJ-call at stage $t \geq 4 + \lg n$ fails with probability $\frac{4}{m\varepsilon^2} + n^{-1.4426} < \frac{1}{3n}$ for $n \geq 2^9$. All SMAJ-calls together fail with probability $\frac{4}{9} + \frac{1}{3} = \frac{7}{9}$. As the number of oracle calls and the additional overhead decrease by a factor $\lg n / n$ we get

Theorem 2. *Using an oracle O_1 that, given $E_N(x)$ and N, predicts $\ell(x)$ with advantage ε in time T, the RSA-function E_N can be inverted in expected time $9n(\lg n)\varepsilon^{-2}(T + 2^6 \varepsilon^{-4})$.*

A main point is that the number of oracle calls for RSA-inversion is at most $9n\varepsilon^{-2}\lg n$, whereas the ACGS-algorithm requires $(64)^3 \frac{\pi^2}{3} n^3 \varepsilon^{-8}$ oracle calls, where $(64)^3 \frac{\pi^2}{3} \approx 2^{19.7}$. We can further reduce the factor 9 in Theorem 2 by guessing upon initiation closer approximations uN, vN — this merely increases the additional overhead. On the other hand the number of oracle calls is nearly minimal.

Oracle optimality. Goldreich [G96] observed that the number $9n\varepsilon^{-2}\lg n$ of oracle calls in Theorem 2 is minimal up to a factor $O(\lg n)$.

4 Processing all possible locations together.

We sketch a first step in reducing the additional overhead in the time for RSA-inversion. So far RSA-inversion processes all pairs of locations uN, vN separately. Together these pairs can be processed much faster. We simulate the algorithm RSA-inversion for fixed a, b and for all $u \in \frac{\varepsilon^3}{4} [0, 4\varepsilon^{-3})$, $v \in \frac{\varepsilon}{4} [0, 4\varepsilon^{-1})$ with all oracle replies $O_{i,t} := O_1(E_N((a_t + ia_{t-1} + b)x))$ given. The majority decision sets $\ell(a_t x)$ to 0 iff the equation (2) holds for the majority of the $i \in A'_{m'}$.

$$O_{i,t} = i\,\ell(a_{t-1}x) + \ell(bx) + \lfloor u_t + v + iu_{t-1} \rfloor \bmod 2. \qquad (2)$$

The main work of RSA-inversion is to compute for all $\bar{u} \in \frac{\varepsilon^3}{2^6 n} [0, 2^6 n\varepsilon^{-3})$, $u_{t-1} := 2u_t \bmod 1$, all v, all t and $l = (l_1, l_2) := (\ell(a_{t-1}x), \ell(bx)) \in \{0, 1\}^2$:

$$\Gamma(\bar{u}, v, l, t) := \#\{i \in A'_{m'} \mid \text{equation (2) holds with } \bar{u} = u_t, v, l, t\}.$$

This requires some technical algorithms and a tedious analysis that are contained in the full version of this paper. A main point is to separate in equation (2) the influence of $u_t + v$ — we only use a few precision bits of $u_t + v$ — and that of u_{t-1}. A key observation is that counting the i that satisfy equation (2) can easily be done simultaneously for u_{t-1} and $u_{t-1} + \frac{1}{2}$ if we separately count even and odd i. By exploiting and extending these ideas we can prove

Theorem 3. *If all pairs (u, v) are processed together, the additional overhead in RSA-inversion requires at most expected time $O(n^2 \varepsilon^{-4} \lg(n\varepsilon^{-1}))$.*

The additional overhead in Theorem 3 can be further reduced. We can discard all pairs (u, v) for which $\Gamma(u, v, l, t)/m'$ is not in the *correct* range of numbers that differ from $\frac{1}{2} \pm \varepsilon$ by at most $\frac{3}{4}\varepsilon$, where ε is the exact advantage of O_1. Thus we can restrict the set of pairs (u, v) to a small subset of $\frac{\varepsilon^3}{4} [0, 4\varepsilon^{-3}) \times \frac{\varepsilon}{4} [0, 4\varepsilon^{-1})$.

5 Security of RSA-message bits and of the RSA-RNG.

An important question of practical interest is how to generate efficiently many pseudorandom bits that are provably good under weak complexity assumptions. Provable security for the RSA-RNG follows from Theorems 2 and 3. Under the assumption that there is no breakthrough in algorithms for inverting the whole RSA-function Theorems 2 and 3 yield provable security for RSA-message bits and for the RSA-RNG for moduli N of practical size — $n = 1\,000$ and $n = 5\,000$.

Practical security of RSA-message bits. For given $E_N(x)$ it is impossible to predict $\ell(x)$ with advantage $\frac{1}{100}$ within one MIP-year ($3.16 \cdot 10^{13}$ instructions) or else the RSA-function E_N can be inverted faster than is possible by factoring N using the fastest known algorithm. For this we choose $T := 3.16 \cdot 10^{13}$, $n := 1\,000$, $\varepsilon := \frac{1}{100}$. As the O-constant in Theorem 3 is about 2^{10}, Theorems 2 and 3 yield a time bound $3 \cdot 10^{22}$ for factoring N that is clearly smaller than

$10^{25.5} \approx L_N[\frac{1}{3}, 1.9]$, the time for the fastest known factoring algorithm, see the next paragraph.

Each of the 10 least-significant RSA-message bits is individually secure for RSA-moduli N with 1 000 bits. This is because we can — see the end of section 2 — invert E_N in time $9n \lg n \, \varepsilon^{-2} T + O(n^2 \varepsilon^{-4} \lg(n\varepsilon^{-1}))$ using an oracle O_j that predicts the j-th message bit $\ell_j(x)$.

On the other hand the ACGS-result does not give any security-guarantee for moduli N of bit length 1 000, not even against one-step attackers with $T = 1$, as $2^{19.7} 1000^3 100^8 \approx 8.5 \cdot 10^{30} \gg 10^{25.5}$.

The fastest known factoring method. The fastest known algorithm for factoring N or for breaking the RSA cryptoscheme requires at least $L_N[\frac{1}{3}, 1.9]^{1+o(1)}$ steps, where $L_N[v, c] = \exp(c \cdot (\ln N)^v (\ln \ln N)^{1-v})$. $L_N[\frac{1}{3}, 1.9]$ is the conjectured run time of the number field sieve method with Coppersmith's modification using several number fields [BLP93]. Factoring even a non-negligible fraction of random RSA-moduli N requires $L_N[\frac{1}{3}, 1.9]$ steps by this algorithm.

Practical and provably secure random bit generation. Let $N = p \cdot q$ be a random RSA-modulus with primes p, q, e an RSA-exponent and let $x_0 \in_R [0, N)$. The RSA-RNG produces from random seeds (x_0, N) the bit string $b = (b_1, \ldots, b_m)$ as

$$x_i = x_{i-1}^e \bmod N, \quad b_i = x_i \bmod 2 \quad \text{for } i = 1, \ldots, m.$$

A statistical test A rejects b at tolerance level ε if for random $a \in_R \{0, 1\}^m$

$$| \Pr_b[A(b) = 1] - \Pr_a[A(a) = 1] | \geq \varepsilon.$$

A tolerance level $\frac{1}{100}$ is considered to be sufficient for practical purposes.

Theorem 4. *Let the RSA-RNG produce from random seeds (x_0, N) of length $2n$ an output $b = (b_1, ..., b_m)$ of length m. Every statistical test A, that rejects the output at tolerance level ε, yields an algorithm that inverts the whole RSA-function E_N in expected time $9n \lg n \, (m/\varepsilon)^2 T(A) + O(n^2(m/\varepsilon)^4 \lg(nm/\varepsilon))$ for a non-negligible fraction of N.*

Proof. Suppose the bit string $b \in \{0,1\}^m$ is rejected by some test A in time $T(A)$ and tolerance level ε. By Yao's argument, see eg. [K97, section 3.5, Lemma P1], and since the distribution of b is shift-invariant, there is an oracle O_1, which given $E_N(x)$ and N, predicts $\ell(x)$ in time $T(A) + mn^2$ with advantage ε/m for a non-negligible fraction of N. By Theorems 2 and 3, and assuming that $T(A)$ dominates mn^2, we can invert E_N in the claimed expected time. □

Corollary 5. *The RSA-random generator produces for $n = 5\,000$ from random seeds (x_0, N) of bit length 10^4 at least $m = 10^7$ pseudorandom bits that withstand all statistical tests doable with the 1995 world computing power at tolerance level $\frac{1}{100}$, or else the whole RSA-function E_N can be inverted in less than $L_N[\frac{1}{3}, 1.9]$ steps for a non-negligible fraction of N.*

Proof. ODLYZKO rates the 1995 yearly world computing power to $3 \cdot 10^8$ MIP-years, where a MIP-year corresponds to $3.16 \cdot 10^{13}$ instructions. Then $3 \cdot 10^8$ MIP-years correspond to 10^{22} instructions. By Theorem 4 with a O-constant of 2^{10} we can invert E_N using less than 10^{48} steps while $L_N[\frac{1}{3}, 1.9] > 3.7 \cdot 10^{50}$. $\square$

6 The $x^2 \bmod N$ generator and the Rabin-function.

The $x^2 \bmod N$ generator has been proved to be secure under the assumption that factoring integers is hard. Here we show that this even holds for moduli N of practical size. The $x^2 \bmod N$ generator transforms a random seed (x_0, N) into a bit string $(b_1, ..., b_m)$ as $x_i := E_N(x_{i-1})$, $b_i := \ell(x_i)$ for $i = 1, ..., m$. Here E_N is the Rabin-function, N is a random *Blum integer* — a product of two primes p, q that are congruent 3 mod 4 — and x_0 is a random number in $\mathbb{Z}_N$. We distinguish three variants of this generator, the *absolute*, the *centered* and the *uncentered* RNG, according to the following variants of the Rabin-function:

- the *absolute* Rabin-function $E_N^a(x) = |x^2 \bmod N| \in (0, N/2)$,
- the *centered* Rabin-function $E_N^c(x) = x^2 \underline{\bmod} N \in (-N/2, N/2)$,
- the *uncentered* Rabin-function $E_N^u(x) = x^2 \bmod N \in [0, N)$.

The centered function E_N^c outputs $x^2 \underline{\bmod} N$, the absolute smallest residue of x^2 modulo N in $(-N/2, N/2)$ whereas E_N^u outputs the residue in $[0, N)$. Historically the uncentered RNG has been introduced as the $x^2 \bmod N$ generator [BBS86]. However, the absolute and the centered RNG coincide and are more natural than the uncentered RNG. We note that
$$E_N^c(x) = \pm E_N^a(x), \quad E_N^a(x) = |E_N^u(x)|, \quad E_N^u(x) \in \{E_N^c(x), E_N^c(x) + N\},$$
where $|y| = \min(y, N - y)$ for $y \in \mathbb{Z}_N = [0, N)$. Thus E_N^c extends the output of E_N^a by one bit, the sign.

The absolute and the centered RNG coincide in the output. Let x_i^a, x_i^c, x_i^u denote the integer x_i in the i-th iteration with E_N^a, E_N^c, E_N^u and input $x_0 = x_0^a = x_0^c = x_0^u$. Using $E_N^c(x) = \pm E_N^a(x)$ we see by induction on i that $x_i^c = \pm x_i^a$ and $\ell(x_i^c) = x_i^c \bmod 2 = x_i^a \bmod 2 = \ell(x_i^a)$.

On the other hand the uncentered RNG is quite different. It outputs the xor of $\ell(x_i^c)$ and the sign-bit $[x_i^c > 0]$. The uncentered RNG is less natural. Consider the group $\mathbb{Z}_N^*(+1)$ of elements in $\mathbb{Z}_N^*$ with Jacobi symbol 1. $\mathbb{Z}_N^*(+1)$ is a subgroup of $\mathbb{Z}_N^*$ of index 2 that contains the group QR_N of quadratic residues modulo N. We see from $-1 \in \mathbb{Z}_N^*(+1) \setminus QR_N$ that E_N^a permutes the set $S_N = \mathbb{Z}_N^*(+1) \cap [1, N/2)$, E_N^c permutes the set $QR_N \cap (-N/2, N/2)$ and E_N^u permutes $QR_N \cap (0, N)$. The whole point is that $\mathbb{Z}_N^*(+1)$ can be decided in polynomial time whereas QR_N may be difficult to decide. So E_N^a permutes a nice set S_N whereas E_N^c, E_N^u permute complicated sets. It comes as no surprise that we get better security results for the absolute/centered RNG than for the uncentered one.

Oracle inversion of the absolute Rabin-function. The algorithm RSA-inversion can be directly extended from the RSA-function to the permutation E_N^a acting on $S_N = \mathbb{Z}_N^*(+1) \cap [1, N/2)$. This extension uses an oracle O_1 which given $E_N^a(x)$ and N predicts for random $x \in \mathbb{Z}_N^*(+1)$ the bit $\ell(x)$ with advantage ε. A main point is that the majority decisions must use multipliers $\bar{a} = a_t + i a_{t-1} + b$ in $\mathbb{Z}_N^*(+1)$ as we can only interpret the oracle for such inputs $E_N(\bar{a}x)$ with $x \in \mathbb{Z}_N^*(+1)$. On the average half of the multipliers $\bar{a}$ are in $\mathbb{Z}_N^*(+1)$, the usable multipliers are nearly uniformly distributed, see[P92]. For compensation the inversion algorithm guesses initially an approximate location uN for $[ax]_N$ of half the previous distance. This doubles the additional overhead, but does not affect the number of oracle calls. With these remarks Theorems 2 and 3 extend from the RSA-function to the absolute Rabin-function E_N^a, Theorem 4 and Corollary 5 extend from the RSA-RNG to the absolute/centered $x^2 \bmod N$ generator. The extended results prove security if factoring integers is hard, as the problems of inverting E_N^a and of factoring N are equivalent.

Theorem 6. *The assertions of Theorems 2 and 3 hold for the absolute Rabin-function E_N^a in place of the RSA-function E_N. Theorems 4 and Corollary 5 hold for the absolute/centered $x^2 \bmod N$ generator in place of the RSA-generator.*

Comparison with the muddle square method. It is interesting to compare the centered $x^2 \bmod N$ generator with the randomized $x^2 \bmod N$ generator proposed by GOLDREICH and LEVIN [GL89,L93]: iteratively square $x_i \bmod N$ and output the scalar products $b_i = \langle x_i, z \rangle \bmod 2$ for $i = 1, .., m$ with a random bit string z. Following [GL89, L93] KNUTH shows that N can be factored in expected time $O(n^2 \varepsilon^{-2} m^2 T(A) + n^4 \varepsilon^{-2} m^3)$ for a non-negligible fraction of the N if we are given a statistical test A that rejects $(b_1, ..., b_m)$ at tolerance level ε, see [K97, section 3.5, Theorem P]. This yields a security guarantee for the *muddle square method* that is similar to the one of Corollary 5.

The problem of inverting of the (un)centered Rabin-function. Consider the permutations E_N^c, E_N^u acting on the set of quadratic residues. The problems of inverting E_N^c and E_N^u are equivalent as we can easily transform one output into the other using that $E_N^u(x) - E_N^c(x) \in \{0, N\}$. We consider the oracle inversion of E_N^c. The problem we face in the oracle inversion of E_N^c is that for given $\pm y \in \mathbb{Z}_N^*(+1)$ we do not know which of $\pm y$ is in QR_N. A solution has been found by VAZIRANI and VAZIRANI [VV84]. We can determine the quadratic character of $\pm y$ using the oracle that predicts $\ell(z)$ for the inverse image $z \in QR_N$ with $E_N^c(z) = \pm y$.

Let O_1 be an oracle which, given $E_N^c(x)$ and N, predicts the least-significant bit of $x \in QR_N$ with advantage ε, $\Pr_{x,w}[O_1(E_N^c(x)) = \ell(x)] \geq \frac{1}{2} + \varepsilon$ for $x \in_R QR_N$ and the coin tosses w of O_1. The main problem in extending the RSA-inversion to the Rabin-function is that we can only use multipliers $\bar{a} = a_t + i a_{t-1} + b$ that are in QR_N as we can only interpret oracle values $O_1(E_N^c(\bar{a}x))$ with $\bar{a}x \in QR_N$. QR_N is a subgroup of $\mathbb{Z}_N^*$ with index 4.

Let us first suppose that 2 is in QR_N and that we are given $n\varepsilon^{-2}$ multipliers in QR_N of each of the two types $(\frac{1}{2}+i)a+b$ and $\frac{1}{2}(ia+b)$. Hereafter we show how to get rid of this assumption.

Inverting the centered Rabin-function. We describe how the algorithm differs from RSA-inversion if $2 \in QR_N$.

Initially pick random $a, b \in_R \mathbb{Z}_N^*$ and produce about $n\varepsilon^{-2}$ quadratic residues of either type $(\frac{1}{2}+i)a+b$, $\frac{1}{2}(ia+b)$ — with $|1+2i| \leq 4m\varepsilon^{-2}$ — in QR_N. On the average there are $n\varepsilon^{-2}$ residues in QR_N of either type. Guess the closest approximations uN, vN to $[ax]_N, [bx]_N$ with $u \in \frac{\varepsilon^3}{24}[0, 2^4\varepsilon^{-3})$, $v \in \frac{\varepsilon}{4}[0, 4\varepsilon^{-1})$.

At stage t determine $\ell(\frac{1}{2}ax)$ by majority decision using oracle O_1 and all sample points $(\frac{1}{2}+i)a+b \in QR_N$. Given $\ell(\frac{1}{2}ax)$ we can in the same way determine $\ell(\frac{1}{2}bx)$ using the sample points $\frac{1}{2}(ia+b)$ in QR_N. Then replace a, b by $\frac{1}{2}a \bmod N$, $\frac{1}{2}b \bmod N$ and go to the next stage. The new sample points $(i+\frac{1}{2})a+b$ and $\frac{1}{2}(ia+b)$ are again in QR_N since we only divide by the quadratic residue 2.

The case that 2 is a quadratic nonresidue. In this case we determine the quadratic residues $(\frac{1}{2}+i)a+b$ and $\frac{1}{2}(ia+b)$ at stages $t=1$ and $t=2$. We use the quadratic residues of stage 1 at the odd stages and the quadratic residues of stage 2 at the even stages. This is possible since we divide the residues by a power of 4 compared to stages 1 and 2.

Determining quadratic residuosity. Suppose $\bar{a} \in \mathbb{Z}_N^*$ has Jacobi symbol 1, then we have $\bar{a} \in QR_N$ iff $\text{Pr}_z[O_1 E_N^c(\bar{a}z) = \ell(\bar{a}z)] \geq \frac{1}{2}+\varepsilon$ for $z \in_R QR_N$. This yields an oracle that predicts quadratic residuosity with advantage ε.

The algorithm for inverting the Rabin-function requires $O(n\varepsilon^{-4}\lg(n\varepsilon^{-1})T)$ extra steps for the determination of the quadratic residues $(\frac{1}{2}+i)a+b$, $\frac{1}{2}(ia+b)$. There is an extra factor 4 induced by the density $\frac{1}{4}$ of QR_N in $\mathbb{Z}_N^*$. To compensate for the smaller density the inversion algorithm guesses initially an approximate location uN for $[ax]_N$ with $\frac{1}{4}$ times the previous distance. We reduce the additional overhead by the method of section 4. Assuming that T dominates n we get

Theorem 7. *The centered Rabin-function E_N^c can be inverted in expected time $O(n\varepsilon^{-4}\lg(n\varepsilon^{-1})T)$ with the help of an oracle that predicts $\ell(x)$ with advantage ε in time T when given N and $E_N^c(x)$.*

Conclusion. We have given a stronger security proof for RSA/Rabin bits. Our proof yields provable security for RSA-message bits, for the RSA-RNG and for the centered $x^2 \bmod N$ generator for moduli N of practical size, e.g. of bit length 1 000 and 5 000. For the first time this yields provably secure and practical RNG's under the assumption that factoring integers is hard. On the other hand there are more efficient and provably secure RNG's based on stronger complexity assumptions, e.g. [MS91], [FS96].

Acknowledgement. We gratefully acknowledge the comments of D.E. Knuth and that of an anonymous referee that led to a considerably improved presentation of the material.

References

[ACGS88] W. Alexi, B. Chor, O. Goldreich and C.P. Schnorr: RSA and Rabin Functions: certain parts are as hard as the whole. Siam J. Comp. 17 (1988), pp. 194–209.

[BCS83] M. Ben-Or, B. Chor and A. Shamir: On the Cryptographic Security of Single RSA-Bits. Proc. 15th ACM Symp. on Theory of Computation, April 1983, pp. 421–430.

[BBS86] L. Blum, M. Blum and M. Shub: A Simple Unpredictible Pseudo-Random Number Generator. Siam J. Comp. 15 (1986), pp. 364–383.

[BLP93] J.P. Buhler, H.W. Lenstra, Jr. and C. Pomerance: Factoring Integers with the Number Field Sieve. in: The Development of the number field sieve, (Ed. A.K. Lenstra, H.W. Lenstra, Jr.) Springer LNM 1554 (1993), pp. 50–94.

[BM84] M. Blum and S. Micali: How to Generate Cryptographically Strong Sequences of Pseudorandom Bits. Siam J. Comp., 13 (1984), pp. 850–864.

[FS96] J.B. Fischer and J. Stern: An Efficient Pseudo-Random Generator Provably as Secure as Syndrome Decoding. Proc. EUROCRYPT'96, Springer LNCS 1070 (1996) pp. 245–255.

[G96] O. Goldreich: personal information at the Oberwolfach workshop on Complexity Theory, November 10–16, 1996.

[GL89] O. Goldreich and L.A. Levin: Hard Core Bit for any One Way Function. Proc. of ACM Symp. on Theory of Computing (1989) pp. 25–32.

[GMT82] S. Goldwasser, S. Micali and P. Tong: Why and How to Establish a Private Code on a Public Network. Proc. 23rd IEEE Symp. on Foundations of Computer Science, Nov. 1982, pp. 134–144.

[HSS93] J. Håstad, A.W. Schrift and A. Shamir: The Discrete Logarithm Modulo a Composite Hides $O(n)$ bits. J. of Computing and Systems Science 47 (1993), pp. 376–404.

[K97] D.E. Knuth: Seminumerical Algorithms, 3rd edn. Addison-Wesley, Reading, MA (1997). Also Amendments to Volume 2. January 1997. http://www-cs-staff.Stanford.EDU/~uno/taocp.html

[L93] L.A. Levin: Randomness and Nondeterminism. J. Symbolic Logic 58 (1993), pp. 1102–1103.

[MS91] S. Micali and C.P. Schnorr: Efficient, Perfect Polynomial Random Number Generators. J. Cryptology 3 (1991), pp. 157–172.

[O95] A.M. Odlyzko: The Future of Integer Factorization. CryptoBytes, RSA Laboratories, 1 (1995), pp. 5–12.

[P92] R. Peralta: On the Distribution of Quadratic Residues and Non-residues Modulo a Prime Number. Math. Comp., 58

[R79] M.O. Rabin: Digital signatures and public key functions as intractable as factorization. TM-212, Laboratory of Computer Science, MIT, 1979.

[RSA78] R.L. Rivest. A. Shamir and L. Adleman: A Method for Obtaining Digital Signatures and Public Key Cryptosystems. Comm. ACM, 21 (1978), pp. 120–126.

[VV84] U.V. Vazirani and V.V. Vazirani: Efficient and Secure Pseudo-Random Number Generation. In Proc. 25th Symp. on Foundations of Computing Science (1984) IEEE, pp. 458–463.

[Y82] A.C. Yao: Theory and Application of Trapdoor Functions. Proc. of IEEE Symp. on Foundations of Computer Science (1982), pp. 80–91.

Round-Optimal Zero-Knowledge Arguments Based on Any One-Way Function

Mihir Bellare[1] and Markus Jakobsson[2] and Moti Yung[3]

[1] Department of Computer Science & Engineering, Mail Code 0114, University of California at San Diego, 9500 Gilman Drive, La Jolla, CA 92093, USA. E-mail: mihir@cs.ucsd.edu. URL: http://www-cse.ucsd.edu/users/mihir.

[2] Department of Computer Science & Engineering, Mail Code 0114, University of California at San Diego, 9500 Gilman Drive, La Jolla, CA 92093, USA. E-mail: markus@cs.ucsd.edu.

[3] CertCo, New York, NY, USA. E-mail: moti@certco.com.

Abstract. We fill a gap in the theory of zero-knowledge protocols by presenting NP-arguments that achieve negligible error probability and computational zero-knowledge in four rounds of interaction, assuming only the existence of a one-way function. This result is optimal in the sense that four rounds and a one-way function are each individually necessary to achieve a negligible error zero-knowledge argument for NP.

1 Introduction

In a zero-knowledge (ZK) protocol, a prover P wants to "convince" a verifier V that some claim is true, without "revealing" any extra information [GMR]. In the theory of ZK protocols, researchers have looked at the complexity assumptions based on which protocols can be constructed, and the resources necessary to do so. Here we fill a gap in this area. Let us begin by explaining the various dimensions of such protocols.

1.1 The big picture

The interaction between P and V takes place on some common input x, and P is trying to convince V that x belongs to some underlying language L. The length of x is denoted n and one measures complexity in terms of n. The verifier is always a (probabilistic) polynomial time algorithm. Typically (and here) L is in NP. The system has two dimensions: "conviction" and "zero-knowledge." Each can be formalized in one of two ways, a weak and a strong, depending on whether or not we restrict the adversary involved to polynomial time. To describe these dimensions, we use a terminology from [BCY] (which they credit to Chaum).

DEGREES OF CONVICTION. Conviction is about "soundness." If $x \notin L$ we ask that no matter how the prover behaves, it cannot convince V to accept, except with low probability (called the error probability, and denoted $\epsilon(\cdot)$). This has been formalized in two ways:

W. Fumy (Ed.): Advances in Cryptology - EUROCRYPT '97, LNCS 1233, pp. 280-305, 1997.
© Springer-Verlag Berlin Heidelberg 1997

- Statistical conviction: This is the notion of [GMR]. Even a computationally unrestricted prover should be unable to make the verifier accept $x \notin L$, except with probability $\epsilon(n)$. Protocols providing this strong degree of conviction are usually called "proofs."

- Computational conviction: This is the notion of [BrCr, BCC]. A prover restricted to (randomized) polynomial time should be unable to make the verifier accept $x \notin L$, except with probability $\epsilon(n)$.[4] (But a more powerful prover might succeed in making the verifier accept with high probability.) Although weaker, this kind of soundness is good enough for cryptographic protocols. The soundness will typically depend on the assumed intractability of some computational problem, like factoring or computing discrete logarithms. Protocols meeting this condition are usually called "arguments."

DEGREES OF ZERO-KNOWLEDGE. Roughly, the zero-knowledge condition of [GMR] asks that when $x \in L$, the transcript of an interaction between the prover and a verifier yield no information (other than the fact that $x \in L$) to an adversary who gets to examine the transcript. Again, this adversary may be weak or strong:

- Statistical ZK: Even a computationally unrestricted adversary will not get useful information out of a transcript, except with low (negligible) probability. Protocols meeting this are usually called SZK.

- Computational ZK: A (randomized) polynomial time adversary will not get useful information out of a transcript. (But a computationally unrestricted adversary might.) This will be the case when the transcript contains encryptions of sensitive data, which are useless to a polynomial time adversary, but can be opened by an unrestricted one. This type of ZK is usually called CZK and, although weaker, is good enough for cryptographic protocols.

We clarify that this discussion is very informal. The definitions talk of the indistinguishability of ensembles. (See Section 2.4.) We also don't make perfect ZK a special case, considering it included as a sub-case of statistical.

A NOTE ON COMPLETENESS. In addition, a basic completeness condition is always required. It asks that if $x \in L$ then there is a strategy via which the prover can make V accept. The definition of [BrCr, BCC] asks (as appropriate for a cryptographic protocol) that this be efficiently achievable: if P is given a witness for the membership of x in the NP language L then it can make V accept in polynomial time. The definition of [GMR] does not make such a requirement. However, all known proofs (statistically convincing) for NP languages do meet this efficient completeness requirement, so we won't discuss it further, assuming it always to be true.

A NOTE ON PROOFS OF KNOWLEDGE. One usually also wants that when $x \in L$, the ability of a prover to convince V to accept should be indicative of "knowledge" of a witness. Like soundness, in proofs it holds for arbitrary provers and in arguments for polynomial time ones. (The notion was suggested in [GMR],

---

[4] This description masks some subtleties. See Definition 2 and the following discussion.

and an appropriate formalization has emerged in [BeGo]. See Section 2.3 for more.) Again, we will not discuss it further here, concentrating just on the two dimensions mentioned above.

FOUR KINDS OF PROTOCOLS. Since the dimensions discussed above are orthogonal, we get four kinds of protocols:

- CZK arguments: *Computationally convincing, computational ZK*. The weakest kind, but still adequate for cryptographic protocols. For example the arguments for all of NP in [BrCr, BCC] when a standard bit commitment is used.

- CZK proofs: *Statistically convincing, computational ZK*. For example the proofs for all of NP in [GMW].

- SZK arguments: *Computationally convincing, statistical ZK*. For example the arguments for all of NP in [BrCr, BCC] when a discrete logarithm based bit commitment is used; also [NOVY].

- SZK proofs: *Statistically convincing, statistical ZK*. The strongest kind, but not possible for all of NP unless the polynomial time hierarchy collapses [Fo]. But there are examples for special languages: quadratic residuosity and its complement [GMR]; graph isomorphism and its complement [GMW]; constant round SZK proofs for quadratic residuosity and graph isomorphism [BMO1].

1.2 Complexity measures and optimality

Recall that the error-probability is the probability $\epsilon(\cdot)$ in the soundness condition, whether in a proof or an argument. Most atomic ZK protocols have constant error. But one really wants low error. A standard goal is to make the error negligible. (That is, a function vanishing faster than the reciprocal of any polynomial.) We will have the same goal.

COMPLEXITIES TO MINIMIZE. Theoretical research in ZK proofs has focused on achieving this low error while trying to minimize other complexity measures. Two main ones are:

- Rounds: The round complexity is the number of messages exchanged, or rounds of interaction in the protocol.[5]

- Assumptions: The complexity assumption underlying the protocol, it underlies either the computational ZK or the computational conviction (or both). For example it may be an algebraic assumption like the hardness of factoring or discrete log computation, or a general assumption like the existence of claw-free pairs, trapdoor permutations, one-way permutations, or one-way functions.

---

[5] There may be some danger of confusion in terminology. We call each sending of a message by a party a round. Some works like [FeSh] call this a move, and say a round is two consecutive moves. In their terminology, our four round protocols would be four move or two round protocols.

Rounds	Assumption	Reference	Type
poly(n)	One-way function	Combine [GMW, HILL, Na]	CZK proof
$\omega(\log n)$	Algebraic	[BrCr, BCC]	SZK argument
poly(n)	One-way permutation	[NOVY]	SZK argument
6	Claw-free pairs	[BCY]	SZK argument
6	Claw-free pairs	[GoKa]	CZK proof
5	One-way function	[FeSh]	CZK argument
4	Algebraic	[FeSh]	CZK argument
4	Trapdoor perm. + Algebraic	Combine [Bl, FLS, BeYu]	CZK argument
4	One-way function	This paper	CZK argument

Fig. 1. *Negligible error ZK protocols for NP.* We list round complexity, complexity assumption used, and type (CZK or SZK, proof or argument). Remember four rounds is optimal.

LOWER BOUNDS. We know that things can't go too low. Four rounds and a one-way function are each individually necessary to get low-error ZK:

- Four rounds needed: Goldreich and Krawczyk [GoKr] show that there do not exist *three round, negligible error* (whether proof or argument) ZK (whether computational or statistical) protocols for NP unless NP $\subseteq$ BPP. (There is a technical condition saying the ZK must be of a certain form called black-box. But all known ZK protocols are of this type. In this paper whenever we talk of ZK we always mean black box. See Definition 6.) Accordingly, four is the minimal number of rounds required to achieve ZK with low error. (The result also holds if the protocol is not sound but just a proof of knowledge, so that four rounds is also necessary for negligible knowledge error [IS1].)

- One-way function needed: ZK arguments can be used to implement many kinds of cryptographic schemes, whence by [ImLu] require a one-way function to implement. Even for the proof case with a computationally unbounded prover, it is known that for "hard" languages some kind of "one-way function" is necessary [OsWi]. Thus, a one-way function is a minimal assumption required to achieve ZK.

THE PROBLEM. There are many so-called "atomic" ZK protocols for NP that achieve constant error-probability in constant (three or four) rounds. Serial repetition lowers the error and preserves ZK [GoOr, ToWo], but at the cost of increasing the number of rounds to non-constant. So we would like to do parallel repetition. However, this is ruled out: first, we have the above mentioned results of [GoKr]; second, the latter also showed that in general parallel repetition does not preserve ZK. So one must build low error ZK protocols directly.

PREVIOUS WORK. A good deal of effort has gone into this, and a variety of ingenious constructions have been proposed. We summarize the known results in Figure 1. (One that may need elaboration is the protocol of [Bl, FLS, BeYu]. We discuss it briefly in Appendix A.)

Notice that prior to our work optimality had not been achieved in any protocol category. That is, neither for CZK arguments, SZK arguments or CZK proofs did we have four round, low error protocols based on any one-way function. In this paper we have filled the first of these gaps.

We also clarify that we are only tabulating ZK protocols for all of NP (ie. for NP-complete languages). There is also a lot of work on constant round ZK (especially statistical ZK) for special languages which we don't get into.

1.3 Our result

RESULT. We look at low error CZK arguments for all of NP. Figure 1 tells us that it is possible to do it in four rounds using an algebraic assumption (hardness of discrete log) [FeSh]; or in five rounds using a one-way function [FeSh]. This leaves a (small but noticeable) gap, which we fill: we provide an optimal protocol, that uses only four rounds and a one-way function.

Theorem 1. *Suppose there exists a one-way function. Then for any language in NP, there exists a protocol which has four rounds of interaction; is computationally convincing (ie. an argument) with negligible error probability; is computational zero-knowledge; and is a computational proof of knowledge (for the underlying NP-relation) with negligible knowledge-error.*

TECHNIQUES. Our protocol is for the NP-complete language SAT. Let φ be the input formula. We use the idea of Feige and Shamir [FeSh] of ORing to φ some formula Φ which represents some choices of the verifier, and then having the prover run a standard ZK proof on input $\Theta = \varphi \vee \Phi$. However, Feige and Shamir [FeSh] begin their protocol by having the verifier give a witness indistinguishable proof of knowledge of something underlying Φ. Instead, we work directly with the one-way function, having the verifier give a cut-and-choose type proof that Φ meets some conditions. This is interleaved with a standard ZK proof run on Θ. To implement the latter with a one-way function we use Naor's bit commitment scheme [Na] which can be based on a one-way function via [HILL].

The tricky part is getting the protocol to be ZK. When the protocol is finally designed, however, the ZK is not hard to see. It turns out the technically more challenging part is to prove computational soundness. We introduce what seems to be a new technique, proving the soundness by using proofs of knowledge, relying on the strong formulation of the latter given in [BeGo].

1.4 Open problems

We have filled the (small) existing gap between upper and lower bounds for CZK arguments. For other protocol categories, the existing gap is larger and still unfilled. For CZK proofs, it is not known whether constant error can be achieved with a one-way function (let alone with what value of the constant). For SZK arguments, it is not known whether it can be done at all (ie. in polynomially many rounds) with a one-way function.

2 Definitions

We provide definitions for zero-knowledge arguments and computational proofs of knowledge.

2.1 Preliminaries

NP-RELATIONS. Let $\rho(\cdot,\cdot)$ be a binary relation. We say that ρ is an NP-relation if it is polynomial time computable and, moreover, there exists a polynomial p such that $\rho(x,w) = 1$ implies $|w| \leq p(|x|)$. For any $x \in \{0,1\}^*$ we let $\rho(x) = \{ w \in \{0,1\}^* : \rho(x,w) = 1 \}$ denote the witness set of x. We let $L_\rho = \{ x \in \{0,1\}^* : \rho(x) \neq \emptyset \}$ denote the language defined by ρ. Note that a language L is in NP iff there exists an NP-relation ρ such that $L = L_\rho$. We say that ρ is NP-complete if L_ρ is NP-complete.

The example we will concentrate on is satisfiability. Let φ be a boolean formula (circuit) and T an assignment of 0/1 values to its variables. We let $Satisfy(\varphi, T) = 1$ if T satisfies φ (makes it true) and 0 otherwise. This is an NP-relation, and the corresponding language $L_{Satisfy}$ is of course just $SAT = \{ \varphi : \varphi$ is a satisfiable boolean formula $\}$.

NEGLIGIBILITY. Recall that a function $\delta \colon \mathsf{N} \to \mathsf{R}$ is *negligible* if for every polynomial $p(\cdot)$ there exists an integer n_p such that $\delta(n) \leq 1/p(n)$ for every $n \geq n_p$.

INTERACTIVE ALGORITHMS. Parties in our protocols (provers and verifiers) are modeled as interactive functions. An interactive function A takes input x (the common input), the conversation $M_1 \ldots M_i$ so far, and coins R to output $A(x, M_1 \ldots M_i, R)$, which is either the next message, or some indicator to stop, perhaps accepting or rejecting in the process. Probabilities pertaining to this function are over the choice of R. We let $A_x(\cdot,\cdot) = A(x,\cdot,\cdot)$ and $A_{x,R}(\cdot) = A(x,\cdot,R)$. Typically we will have fixed x and will be talking about A_x; sometimes we will also have fixed R and are talking about the deterministic function $A_{x,R}$. A may also take an auxiliary input w (when A is the prover, this is a witness $w \in \rho(x)$) and we write A^w for this algorithm. Thus we can have A_x^w or $A_{x,R}^w$.

The transcript of a conversation between a pair of interactive functions is the entire sequence of messages exchanged between them until one of them halts. We let $\mathsf{Acc}(A_x, B_x)$ denote the probability (over the coins of both parties) that B accepts when talking to A on common input x. We let $\mathsf{Acc}(A_x, B_x, M_1 \ldots M_i)$ denote the conditional probability that B accepts in talking to A on common input x when the conversation so far is $M_1 \ldots M_i$.

We refer to the sending of a message by one party as a round of interaction. So the number of rounds is the total number of messages sent.

2.2 Arguments, or computationally convincing proofs

The protocol must satisfy a standard completeness condition saying that a prover knowing a witness for $x \in L_\rho$ can convince the verifier to accept x. Soundness pertains to what happens when $x \notin L_\rho$. We want to say that it is unlikely that one can make the verifier accept, even if one is allowed to modify the strategy of the prover. The error-probability measures how unlikely. For the purpose of this paper we are interested in arguments of negligible error, but the definition that follows is for any error.

Definition 2. Let P, V be polynomial time interactive algorithms and let ρ be an NP-relation. We say that (P, V) is a computationally convincing proof (or argument) for ρ, with error-probability $\epsilon(\cdot)$, if the following two conditions are met:

(1) EFFICIENT COMPLETENESS: For every $x \in L_\rho$ and every witness $w \in \rho(x)$ it is the case that $\text{Acc}(P_x^w, V_x) = 1$.

(2) COMPUTATIONAL SOUNDNESS: For every polynomial time interactive algorithm $\widehat{P}$ there is a constant $N_{\widehat{P}}$ such $\text{Acc}(\widehat{P}_x, V_x) \le \epsilon(|x|)$ for all $x \notin L_\rho$ which have length at least $N_{\widehat{P}}$.

If ϵ is negligible then we say that the error-probability is negligible.

We highlight the case of negligible error: the system has negligible error as long as there is *some* negligible function $\epsilon(\cdot)$ such that the error is $\epsilon(\cdot)$.

Notice one difference with defining interactive proofs: we ask that the point at which the error goes down to $\epsilon(\cdot)$ depend on the prover $\widehat{P}$. This is necessary, as the discussion below explains.

ISSUES IN COMPUTATIONAL SOUNDNESS. In the interactive proof setting [GMR], the error-probability of a protocol (P, V) is $\epsilon(\cdot)$ if for any $x \notin L$ and any interactive algorithm $\widehat{P}$ playing the role of the prover, $\text{Acc}(\widehat{P}_x, V_x) \le \epsilon(|x|)$. The question of what is the error-probability of a computationally sound proof (argument) is more subtle. The first thought is that we say the same thing, except restricting our attention to polynomial time prover algorithms. Namely, the error-probability is $\epsilon(\cdot)$ if $\text{Acc}(\widehat{P}_x, V_x) \le \epsilon(|x|)$ for any polynomial time interactive algorithm $\widehat{P}$ and any $x \notin L$. But this is not right. Underlying the argument is some computationally hard problem like inverting a one-way function. The size of this problem is proportional to $|x|$. So for any *fixed* x there is *some* polynomial time prover who can convince the verifier with *high* probability, by solving the underlying computational problem. In other words, we cannot, for a fixed $x \notin L$, hope that the probability of convincing the verifier is at most $\epsilon(|x|)$ for *all* polynomial time provers. (Unless the argument is in fact a proof.) However, for any fixed polynomial time prover, as $|x|$ grows, the probability of convincing the verifier decreases, because the size of the underlying hard computational problem is increasing. In particular it is reasonable to ask that for each $\widehat{P}$ the error eventually goes below the desired error-probability $\epsilon(n)$, which is what we did above.

In particular, the probability of convincing the verifier to accept $x \notin L$ in a computationally convincing proof cannot be reasonably expected to be exponentially small. It is restricted by the probability of solving the underlying computational problem. Since the typical assumption is that the latter is negligible (not but less), the error of the argument too is negligible but not less. In particular, independent repetition will not lower the error to exponentially small.

Another way to resolve the issue is to have a security parameter k that is separate from the input x and measures the size of the underlying hard problem. For any fixed x, the error-probability still goes down as we increase k. This formulation is probably better for protocol design, but in the current theoretical setting, we stick, for simplicity, to just one input, and adopt the definition above.

2.3 Computational proofs of knowledge

We want to say that if an interactive algorithm can convince V to accept $x \in L$ then it must actually "know" a witness $w \in \rho(x)$. This notion of a "proof of knowledge" was suggested in [GMR]. It was formalized in [BeGo] both for the standard interactive proof setting and the argument, or computationally convincing setting. (They discuss the latter in [BeGo, Section 4.7].) We adopt their notion. It comes in two equivalent forms. We present both.

Recall an oracle algorithm E is an algorithm that can be equipped with an oracle. An invocation of the oracle counts as one step. We will talk of an "extractor" E which will be given an oracle for $\widehat{P}_x$, a prover algorithm on input x, and will then try to find a witness w to the membership of x in L_ρ. The first definition below is what [BeGo] refer to as the "alternative form of validity."

Definition 3. [BeGo] We say that verifier V defines a computational proof of knowledge for NP-relation ρ, with knowledge-error $\kappa(\cdot)$, if there is a an expected polynomial time oracle algorithm E (the extractor) such that for every polynomial time interactive algorithm $\widehat{P}$ there is a constant $N_{\widehat{P}}$ such that if $x \in L_\rho$ has length at least $N_{\widehat{P}}$ then

$$\Pr\left[E^{\widehat{P}_x}(x) \in \rho(x) \right] \geq \mathsf{Acc}(\widehat{P}_x, V_x) - \kappa(|x|) .$$

If $\kappa(\cdot)$ is negligible then we say the proof has negligible knowledge-error.

In other words, if E has oracle access to $\widehat{P}$ then it can output a witness for membership of x in L_ρ with a probability only slightly less than the probability that $\widehat{P}$ would convince V to accept x. Again, note negligible knowledge error means the above is true for *some* negligible function $\kappa(\cdot)$.

In the next formulation (the main one of [BeGo]) the extractor must find a witness with probability one. It is not limited to (expected) polynomial time, but must run in time inversely proportional to the excess of the accepting probability over the knowledge error.

Definition 4. [BeGo] We say that verifier V defines a computational proof of knowledge for NP-relation ρ, with knowledge-error $\kappa(\cdot)$, if there is a an oracle algorithm E (the extractor) and a constant c such that for every polynomial time interactive algorithm $\widehat{P}$ there is a constant $N_{\widehat{P}}$ such that if $x \in L_\rho$ has length at least $N_{\widehat{P}}$ and satisfies $\mathrm{Acc}(\widehat{P}_x, V_x) > \kappa(x)$, then $E^{\widehat{P}_x}(x) \in \rho(x)$, and moreover this computation halts in an expected number of steps bounded by

$$\frac{|x|^c}{\mathrm{Acc}(\widehat{P}_x, V_x) - \kappa(x)}.$$

If $\kappa(\cdot)$ is negligible then we say the proof has negligible knowledge-error.

See [BeGo] for the proof that these two notions are equivalent. Sometimes it is convenient to use one, sometimes the other.

2.4 Zero-knowledge

ENSEMBLES AND COMPUTATIONAL INDISTINGUISHABILITY. We recall these notions of [GoMi, GMR]. An *ensemble* indexed by $L \subseteq \{0,1\}^*$ is a collection $\{E(x)\}_{x \in L}$ of probability spaces (of finite support), one for each $x \in L$. Let $\mathcal{E}_1 = \{E_1(x)\}_{x \in L}$ and $\mathcal{E}_2 = \{E_2(x)\}_{x \in L}$ be ensembles over a common index set L. A *distinguisher* is a polynomial sized family of circuits $D = \{D_x\}_{x \in L}$, with one circuit for each $x \in L$. We say that $\mathcal{E}_1, \mathcal{E}_2$ are *(computationally) indistinguishable* if there is a negligible function $\delta(\cdot)$ such that for every distinguisher D there is a constant N_D such that if $x \in L$ has length at least N_D then

$$\left| \Pr\left[D_x(v) = 1 : v \xleftarrow{R} E_1(x) \right] - \Pr\left[D_x(v) = 1 : v \xleftarrow{R} E_2(x) \right] \right| \leq \delta(|x|).$$

ZERO-KNOWLEDGE. Let P, V be interactive algorithms. The definition of a zero-knowledge interactive proof [GMR] refers to a language L. It begins by defining a probability space, the view of a cheating verifier $\widehat{V}$ in talking to P on input $x \in L$. (And then says there is a simulator that on input x produces an "indistinguishable" view.) The basic idea is the same in the argument setting, but one must be careful about a couple of things. Recall prover P begins with a witness w to x. The view generated by P and V depends not just on P but on w. An elegant way to bring this into the picture is via the notion of a witness selector [BeYu].

Definition 5. [BeYu] A *witness selector* for an NP-relation ρ is a map $W: L_\rho \to \{0,1\}^*$ with the property that $W(x) \in \rho(x)$ for each $x \in L_\rho$.

That is, a witness selector is just a way of fixing an association of a particular witness to each input. When $\rho = Satisfy$ and $L_\rho = SAT$ this just means associating to any formula $x = \varphi \in SAT$ a particular satisfying assignment to it, out of all the possible satisfying assignments.

Now we can define the view. Let P, V be interactive algorithms, ρ an NP-relation, and W a witness selector for ρ. We let $\text{VIEW}(P, W, V, x)$ be the probability space whose points are of the form (R, τ), where R is a random tape for V_x and τ is a transcript of an interaction between $P_x^{W(x)}$ and $V_{x,R}$. The associated probability is that over the choice of R and the coins of $P_x^{W(x)}$. The collection $\{\text{VIEW}(P, W, \widehat{V}, x)\}_{x \in L_\rho}$ becomes an ensemble.

We define zero-knowledge in a strong "black-box" simulation form. The simulator S is an oracle algorithm given input x and oracle access to $\widehat{V}_{x,R}$ where R has been chosen at random. (The simulator does not have to pick R. It is done automatically and the simulator only sees the interface to the oracle $\widehat{V}_{x,R}$.) It will output a transcript τ of a conversation between P_x and $\widehat{V}_{x,R}$. We let $\overline{S}^{\widehat{V}_\bullet}(x)$ denote the probability space of pairs (R, τ) where R was chosen at random and $\tau \leftarrow S^{\widehat{V}_{\bullet,R}}(x)$.

Definition 6. We say that (P, V) is a (computational) zero-knowledge protocol for NP-relation ρ if there exists an expected polynomial time oracle algorithm S (the simulator) such that for every polynomial time interactive algorithm $\widehat{V}$ (the cheating verifier) and every witness selector W for ρ, the ensembles $\{\overline{S}^{\widehat{V}_\bullet}(x)\}_{x \in L_\rho}$ and $\{\text{VIEW}(P, W, \widehat{V}, x)\}_{x \in L_\rho}$ are computationally indistinguishable.

Note formally, zero-knowledge is no longer a property of the language L_ρ but of the relation ρ itself.

Under this definition of zero-knowledge, we know that any negligible error probability zero-knowledge argument for an NP-complete relation ρ must have at least four rounds, assuming NP is not in BPP [GoKr]. We want to meet this bound given only a one-way function.

REMARK. The above notion of black-box simulation zero-knowledge is stronger than those of [GoOr, GoKr, BMO2] in the following sense. In our notion, the simulator has no control over the coins R of $\widehat{V}_x$: they are automatically chosen (at random) and then fixed. The simulator does not even have direct access to them: it just gets an oracle for $\widehat{V}_{x,R}$. In the notions of [GoOr, GoKr], the simulator could choose these coins as it liked, even try running $\widehat{V}_x$ on many different random tapes. In the notion of [BMO2] it could not choose them, but did have direct access to them, and could try several random tapes. However, since our results are positive, making a more stringent definition only strengthens them. Also, all known zero-knowledge protocols do meet our definition.

For simplicity we do not talk of non-uniform verifiers, but of course the above definition could be extended to include them.

3 Building blocks for our protocol

Our protocol uses one-way functions, satisfiability, and a standard bit commitment based atomic ZK protocol for satisfiability.

3.1 One-way functions

Let $f: \{0,1\}^* \to \{0,1\}^*$ be some length-preserving function. An *inverter* for f is a family $I = \{I_n\}_{n \geq 1}$ where each I_n is a circuit, taking n bit inputs and yielding n bit outputs, and having size at most $p(n)$ for some polynomial $p(\cdot)$. We let

$$\mathsf{Inv}_f^I(n) = \Pr\left[f(x') = y : x \xleftarrow{R} \{0,1\}^n ; y \leftarrow f(x) ; x' \leftarrow I_n(y) \right]$$

denote the probability that I_n successfully inverts f at the point $y = f(x)$, taken over a random choice of $x \in \{0,1\}^n$.

Definition 7. Let $f: \{0,1\}^* \to \{0,1\}^*$ be a polynomial time computable, length-preserving function. We say f is one-way if there is a negligible function $\delta(\cdot)$ such that for every inverter I there is an integer N_I such that $\mathsf{Inv}_f^I(n) \leq \delta(n)$ for all $n \geq N_I$.

Hereafter we fix a one-way function f, and the notation f will always refer to this fixed function.

3.2 Formulas and satisfiability

We will present ZK arguments for the NP-complete language SAT. More precisely let *Satisfy* be the NP-relation defined by $Satisfy(\varphi, T) = 1$ if assignment T satisfies formula φ. The corresponding language $L_{Satisfy}$ is of course $SAT = \{ \varphi : \varphi \text{ is a satisfiable boolean formula } \}$. We will present ZK arguments for the NP-relation *Satisfy* meeting the definitions in Section 2. (In terms of those definitions, the NP-relation here is $\rho = Satisfy$, the common input is $x = \varphi$, a boolean formula, and the witness w is a satisfying assignment T to φ.)

We will be encoding statements about the one-way function f as formulas, and need some standard features of the Cook-Levin theorem. The NP-completeness of SAT as proved in this theorem implies the following. There is a polynomial time computable transformation $\mathrm{FORMULA}_f(\cdot)$ such that for any $y \in \{0,1\}^*$ it is the case that $\mathrm{FORMULA}_f(y)$ is a boolean formula which is satisfiable iff there exists an $x \in \{0,1\}^*$ such that $f(x) = y$. More important, there are polynomial time computable maps $t_{f,1}, t_{f,2}$ (called witness transformations) with the following properties. Given x, map $t_{f,1}$ outputs a satisfying assignment $T = t_{f,1}(x)$ to $\mathrm{FORMULA}_f(f(x))$. Conversely, given a satisfying assignment T to $\mathrm{FORMULA}_f(y)$, map $t_{f,2}$ outputs a point $x = t_{f,2}(T)$ such that $f(x) = y$. We will refer to both the transformation $\mathrm{FORMULA}_f$ and to the accompanying witness transformations in what follows. What is important to remember is that knowledge of a satisfying assignment T to $\mathrm{FORMULA}_f(y)$ is tantamount to knowledge of a pre-image x of y under f.

3.3 Naor's commitment scheme

We will use Naor's commitment scheme [Na] which can be based on any one-way function via [HILL]. Some special properties of the scheme are important for us.

It work like this. Suppose A has some data $d \in \{0,1\}^m$ that she wants to commit to B. First, B must send A a random string R, which we call the *commitment setup* string, and which has length polynomial in the security parameter n and the data length m. Then, A picks at random some string s to use as coins, and computes a function $\alpha = \text{COMMIT}_f(R, d, s)$. (This function depends on a pseudorandom bit generator [BlMi, Ya], constructed out of f via [HILL], but we don't need to know that.) This α is A's commitment to d and is sent to B. At a later stage, B can ask A to "open" the commitment, at which point A sends d and s, and B checks that $\alpha = \text{COMMIT}_f(R, d, s)$.

The protocol must have two properties. First is *privacy:* α gives B no information about d. Second is *soundness:* A can't create commitments which she can open in more than one way.

In Naor's scheme [Na], the privacy is true in a computational sense. That is, as long as B cannot invert the underlying one-way function f, it gets no partial information about d. Soundness however is true in a strong, unconditional sense, and since this is important for us, we need to discuss it further.

A *de-committal* of α is a pair (d, s) such that $\alpha = \text{COMMIT}_f(R, d, s)$. We say that A opens α as d if she provides a de-committal (d, s) of α. We say that a commitment setup string R is *bad* if there exists a pair $(d_1, s_1), (d_2, s_2)$ of de-committals of α such that $d_1 \neq d_2$. We say R is *good* if it is not bad. Naor's scheme has the property that a randomly chosen commitment setup string is bad with probability exponentially small in n [Na, Claim 3.1]. For our purposes we set the parameters of the scheme so that this probability is 2^{-2n}. (The length of R required to make this true depends not only on n but also on the data length m. In what follows, we assume R is of the right length to make this true with respect to whatever data length we have.) It follows that the probability that even one out of n random commitment setup strings $R_1, \ldots, R_n$ is bad is at most $n \cdot 2^{-2n} \leq 2^{-n}$. This will be used repeatedly in what follows.

3.4 The atomic protocol

We use as a primitive a atomic four round ZK argument achieving error $1/2$. We now specify the properties we want of it and the notation used to describe it. To avoid depending on the details of any specific protocol, it is described via generic components and steps.

THE PROTOCOL. In the literature there are several commitment-based three round ZK arguments with error $1/2$. For concreteness, take the one of Brassard, Crépeau and Chaum [BCC], or the one based on general commitment in [ImYu]. To set it up using one-way function based commitment, we first have the verifier send a commitment setup string, and then run a protocol such as the ones in [BCC, ImYu], so that we have four rounds.

To avoid depending on the details of any specific underlying protocol, we describe the protocol via generic components and steps. Let Θ denote the boolean formula which is the common input. The prover is assumed to have a satisfying assignment T for Θ. We now specify the instructions for the parties, with the nomenclature to be explained later:

(1) Verifier picks at random a commitment setup string R and sends it to the prover.

(2) Prover picks a random string ρ and computes an encapsulated circuit $C = \text{ENCCIRC}_f(\Theta, T, R, \rho)$. This is sent to the verifier.

(3) Verifier picks a random challenge bit c and sends it to the prover.

(4) Prover computes an answer $D = \text{ANSWER}_f(\Theta, T, R, \rho, c)$ and sends it to the verifier.

(5) Verifier checks that $\text{CHECK}_f(\Theta, R, C, c, D) = 1$. If this is true it accepts, else rejects.

Now let us explain the components. In the second step, the prover computes an object C we call an "encapsulated circuit." This step will involve a number of bit commitments which is proportional to the size of Θ, and they are performed, here, using the scheme of Section 3.3, which can be implemented given f. The commitment setup string used (for all the commitments) is R, and ρ represents some random choices that underlie the encapsulation. (Roughly, the prover will first create a randomized version of Θ that is annotated with the values given by the truth assignment T. This annotated circuit, call it d, would reveal T, but the prover does not send it directly. Instead, he commits to it, sending $\text{COMMIT}_f(R, d, s)$ where s is part of ρ. But the details, such as what is d, will not matter: later we will summarize all the properties we need.) As in a typical cut-and-choose protocol, the verifier then poses a random challenge question, which is the bit c, and prover must "open" the encapsulated circuit in one of two ways. This "answer" of the prover, denoted D, is computed as a function of the truth assignment, the challenge, and the random choices underlying the original encapsulation. It consists of de-committing certain parts of C. The answer being sent to the verifier, the latter checks that it is correct. The check is a function of the encapsulated circuit, the commitment setup string, the challenge, and the answer provided.

PROPERTIES. We assume certain properties of this protocol. The standard example protocols (eg. [BCC]) do have these properties.

We assume that if an encapsulated circuit C is successfully "opened" in both ways, ie. for both a 0-challenge and a 1-challenge, then one can obtain the truth assignment underlying Θ. This is true no matter how C was constructed, and is the technical fact underlying the protocol being a (computational) proof of knowledge with knowledge error $1/2$.

More precisely, there is a polynomial time algorithm EXTRACT_f such that the following is true. Suppose R is a good commitment setup string. Let C be some string sent by the prover in the first step. (It purports to be a correctly computed encapsulated circuit.) Let D_0, D_1 be strings such that $\text{CHECK}_f(\Theta, R, C, 0, D_0) =$

CHECK$_f(\Theta, R, C, 1, D_1) = 1$. Then EXTRACT$_f(\Theta, R, C, D_0, D_1) = T'$ is a truth assignment that satisfies Θ.

We stress that this requires the commitment setup string R to be good as defined in Section 3.3. We are using the fact that when this happens, it is impossible (not just computationally infeasible) for the commiter (here the prover) to open a commitment in two different ways.

We will need (to show our protocol is ZK) that one can compute ENCCIRC$_f(\Theta, T, R, \rho)$ for any T, not just a T that satisfies ρ. The underlying annotated circuit d will be non-sensical in this case, but the verifier will not know, because the annotated circuit is provided in committed form. (Of course, a prover providing such an encapsulated circuit will be hard put to answer the challenges, but that will not matter for us.)

Finally, of course, we also need that the protocol is ZK. (Actually, all we will use is that it is witness indistinguishable in the sense of [FeSh], something which follows from its being ZK.)

4   Protocol 4R-ZK and its properties

We now describe our protocol and its properties. We call the protocol 4R-ZK for "four round ZK."

4.1   Protocol description

We give instructions for the prover P and the verifier V to execute protocol 4R-ZK. The common input is a formula φ of size n, and the prover is assumed in possession of a satisfying assignment T to φ. Refer to Section 3 for the notation and components referred to below.

(1) The verifier's message $M_1 = M_{1,1}M_{1,2}$ consists of two parts computed as we now describe.

    (1.1) For $i = 1, \ldots, n$ and $j = 0, 1$ the verifier chooses $x_{i,j} \xleftarrow{R} \{0,1\}^n$ and sets $y_{i,j} = f(x_{i,j})$. These points are hereafter called the "Y-values." It lets $M_{1,1}$ consist of these $2n$ strings.

    (1.2) The verifier picks at random commitment setup strings $R_1, \ldots, R_n$. It is thereby initiating n parallel runs of the atomic protocol: R_i will play the role of the commitment setup string for the i-th run. (But the input formula Θ for these runs has however not yet been defined! That will appear later.) It sets $M_{1,2} = (R_1, \ldots, R_n)$.

The verifier **sends** $M_1 = M_{1,1}M_{1,2}$ to the prover. Now for $i = 1, \ldots, n$ and $j = 0, 1$ we let $\Phi_{i,j} = \text{FORMULA}_f(y_{i,j})$ as per Section 3.2. This is a formula both parties can now compute.

(2) The prover receives M_1. Its reply $M_2 = M_{2,1}M_{2,2}$ consists of two parts computed as we now describe.

    (2.1) The prover picks bits $b_1, \ldots, b_n \xleftarrow{R} \{0,1\}$ and sets $M_{2,1} = (b_1, \ldots, b_n)$. The bit b_i is viewed as selecting the Y-value y_{i,b_i}, and the verifier is being asked to reveal the pre-image of this value, which he will do in the next step.

(2.2) We now set $\Phi = \Phi_{1,1-b_1} \vee \ldots \vee \Phi_{n,1-b_n}$. (This is the OR of all formulas corresponding to Y-values which the prover has *not* asked be revealed. As long as f is one-way, the prover has very little chance of knowing a satisfying assignment to Φ.) We then set $\Theta = \Phi \vee \varphi$. Notice that T (the satisfying assignment to φ that the prover has) is also a satisfying assignment to Θ, so the prover has a satisfying assignment to Θ (even though he does not have one for Φ). Viewing $R_1, \ldots, R_n$ as commitment setup strings initiating n parallel runs of the atomic protocol on common input Θ, the prover will now perform the second step for each of these executions of the atomic protocol. Namely, for $i = 1, \ldots, n$ it picks at random a string ρ_i to be used as coins in the encapsulated circuit computation, and computes $C_i = \text{ENCCIRC}_f(\Theta, T, R_i, \rho_i)$ for $i = 1, \ldots, n$. He now sets $M_{2,2} = (C_1, \ldots, C_n)$.

The prover **sends** $M_2 = M_{2,1} M_{2,2}$ to the verifier.

(3) The verifier receives $M_2 = M_{2,1} M_{2,2}$. Its reply $M_3 = M_{3,1} M_{3,2}$ consists of two parts computed as we now describe:

(3.1) It sets $M_{3,1} = (x_{1,b_1}, \ldots, x_{n,b_n})$, meaning it returns the pre-images for the Y-values selected by the bits $b_1, \ldots, b_n$ that the prover sent in $M_{2,1} = (b_1, \ldots, b_n)$.

(3.2) Having $b_1, \ldots, b_n$, the verifier knows Φ and hence Θ, these formulas being as defined above. It now picks challenges $c_1, \ldots, c_n \stackrel{R}{\leftarrow} \{0, 1\}$, one for each run of the atomic protocol on input Θ, and sets $M_{3,2} = (c_1, \ldots, c_n)$.

The verifier **sends** $M_3 = M_{3,1} M_{3,2}$ to the prover.

(4) The prover receives $M_3 = M_{3,1} M_{3,2}$.

(4.1) Say $M_{3,1} = (x_1, \ldots, x_n)$. The prover checks that $f(x_i) = y_{i,b_i}$ for $i = 1, \ldots, n$, and if this check fails then it aborts the protocol. Else it goes on to the next step.

(4.2) Say $M_{3,2} = (c_1, \ldots, c_n)$. The prover computes the answers to these challenges. Namely for $i = 1, \ldots, n$ it sets $D_i = \text{ANSWER}_f(\Theta, T, R_i, \rho_i, c_i)$. (Recall ρ_i was the coins used to produce the encapsulated circuit C_i, so that here the prover is opening this encapsulated circuit according to challenge c_i.)

The prover **sends** $M_4 = (D_1, \ldots, D_n)$ to the verifier.

(5) The verifier receives M_4 and makes its final check. For $i = 1, \ldots, n$ it checks that $\text{CHECK}_f(\Theta, R_i, C_i, c_i, D_i) = 1$. (Recall the verifier received the encapsulated circuit C_i in $M_{3,2}$ and the opening D_i in M_4.) If this is true it accepts, else it rejects.

Notice that the protocol is indeed of four rounds. Next we address its properties.

4.2 Result

Our claims about the above protocol are summarized in the following theorem. Refer to Section 2 for definitions of the various notions.

Theorem 8. *Assume f is a one-way function. Then protocol 4R-ZK is:*

(1) *A computationally convincing proof (ie. an argument) with negligible error probability,*

(2) *A computational proof of knowledge with negligible knowledge error, and*

(3) *A (computational) zero-knowledge protocol,*

all for the NP-relation Satisfy corresponding to the NP-complete language SAT.

We will prove these items in turn. As one might imagine, the difficulty in the protocol design was making sure it was ZK. Having done the design to make this work out, however, it will be relatively easy to show. The other claims turn out to be more non-trivial. In particular the soundness is shown via a novel use of proofs of knowledge. We begin with a technical lemma that underlies the first two claims above.

4.3 The Θ-Extraction Lemma

The first two claims about the protocol are that it is computationally convincing and a computational proof of knowledge. The first says that if φ is unsatisfiable then a polynomial time prover has little chance of convincing the verifier to accept, and the second says that if φ is satisfiable then any prover convincing the verifier to accept actually "knows" a satisfying assignment to φ. Both these claims pertain to the input formula φ. Yet our main technical lemma is a claim not about φ but about the formula Θ constructed in the protocol. Remember this formula (a random variable depending on other choices in the protocol) is the one on which the atomic protocol is actually run. The crucial property of this formula is that (as long as the verifier is honest, namely is V) it is *always satisfiable*: whether or not φ is satisfiable, Θ is, because Φ is always satisfiable.

We claim that if a prover A convinces V to accept φ then we can extract a satisfying assignment for Θ, regardless of whether or not φ is satisfiable. Furthermore, this extraction can be done to meet the kinds of conditions asked in the definition of [BeGo]. This will help prove both the above mentioned claims, and, as motivation, it may help to say why. Roughly, an assignment to Φ corresponds to knowledge of inverses of f on random points. But remember $\Theta = \varphi \vee \Phi$. So if φ is unsatisfiable, then an assignment to Θ must be an assignment to Φ, and this will enable us to say in Lemma 10 that significant success in making the verifier accept when φ is unsatisfiable translates to inverting the one-way function f. On the other hand, if φ is satisfiable then an assignment to Θ will with high probability be one to φ since otherwise someone is inverting f. Now let us state and prove the lemma.

Lemma 9. *There is an expected polynomial time oracle algorithm E (the extractor) such that for any prover A and formula φ the following is true. Let R be a random tape for A_φ and $M_1 M_2 M_{3,1}$ a partial transcript of an interaction between $A_{\varphi,R}$ and V_φ. (The transcript includes the first two messages of the protocol and the first part of V's third message). Assume the commitment setup strings in M_1 are good. Let $n = |\varphi|$. Let $p = \text{Acc}(A_{\varphi,R}, V_\varphi, M_1 M_2 M_{3,1})$ be the probability that V accepts given the current partial transcript. Then on input $\varphi, M_1 M_2 M_{3,1}$ and with oracle access to $A_{\varphi,R}$, algorithm E outputs a satisfying assignment to the formula Θ defined by the above partial transcript as in the description of our protocol, and this with probability at least $p - 2^{-n}$.*

Proof. Let $\mathbf{R} = (R_1, \dots, R_n)$ be the sequence of commitment setup strings in M_1. We know that $M_2 = (\mathbf{b}, \mathbf{C})$ where $\mathbf{C} = (C_1, \dots, C_n)$ and C_i is (supposed to be) an encapsulated circuit as per an execution of the atomic protocol on input Θ. Say $\mathbf{c} = (c_1, \dots, c_n)$ is a challenge vector playing the role of message $M_{3,2}$ in the protocol, and $\mathbf{D} = (D_1, \dots, D_n) = M_4$ is some response. It is useful to let

$$\text{CHECK}_f^n(\Theta, \mathbf{R}, \mathbf{C}, \mathbf{c}, \mathbf{D}) = \bigwedge_{i=1}^n \text{CHECK}_f(\Theta, R_i, C_i, c_i, D_i)$$

be the final evaluation predicate of our verifier. We first describe a different oracle algorithm E_1. It takes the same inputs as E should. It always returns a satisfying assignment to Θ, and this within an expected number of steps bounded by $\text{poly}(n)/(p-2^{-n})$. (We can assume $p > 2^{-n}$ since otherwise there is nothing to show.) Algorithm E_1 will sample responses of $A_{\varphi,R}$ for different random challenge vectors $\mathbf{c}$, keeping other information fixed, until it finds a pair of challenge vectors that are accepted by V but are different in at least one component. Namely, repeat the following steps:

(1) Pick $\mathbf{c}_t = (c_{t,1}, \dots, c_{t,n}) \overset{R}{\leftarrow} \{0,1\}^n$ and let $M_{t,3}^* = M_{3,1} \cdot \mathbf{c}_t$
(2) Let $\mathbf{D}_t = (D_{t,1}, \dots, D_{t,n}) \leftarrow A_{\varphi,R}(M_1 M_2 M_{t,3}^*)$

until $\exists\, l, m \in [t]$ such that $\mathbf{c}_l \neq \mathbf{c}_m$ but

$$\text{CHECK}_f^n(\Theta, \mathbf{R}, \mathbf{C}, \mathbf{c}_l, \mathbf{D}_l) = \text{CHECK}_f^n(\Theta, \mathbf{R}, \mathbf{C}, \mathbf{c}_m, \mathbf{D}_m) = 1 \,.$$

Now let l, m satisfy the halting condition. Let $i \in [n]$ be such that $c_{l,i} \neq c_{m,i}$. By definition of CHECK_f^n it must be that $\text{CHECK}_f(\Theta, R_i, C_i, c_{l,i}, D_{l,i}) = \text{CHECK}_f(\Theta, R_i, C_i, c_{m,i}, D_{m,i}) = 1$, meaning encapsulated circuit C_i of the atomic protocol has been successfully opened both for a 0-challenge and 1-challenge. But then, we know from the properties of the atomic protocol described in Section 3.4, that we can compute a satisfying assignment for Θ via $\text{EXTRACT}_f(\Theta, R_i, C_i, D_{l,i}, D_{m,i})$. (We use here the assumption, made in the lemma statement, that the commitment setup strings in M_1 are good. See Sections 3.3 and 3.4.)

Now we need to analyze the running time of E_1. Say $\mathbf{c}$ is good if $\text{CHECK}_f^n(\Theta, \mathbf{R}, \mathbf{C}, \mathbf{c}, \mathbf{D}) = 1$ where $\mathbf{D} = A_{\varphi,R}(M_1 M_2 M_{3,1} \cdot \mathbf{c})$. The probability that a random $\mathbf{c}$ is good is p so one is found in expected $1/p$ tries. Another different one is then found in expected $1/(p-2^{-n})$ tries. So the pair is found within $2/(p-2^{-n})$ tries. Each try being $\text{poly}(n)$ time, we have the claimed time bound on the expected running time of E_1.

Finally, we need to specify the extractor E claimed in the lemma. We apply a trick used in [BeGo] to prove the equivalence of Definitions 3 and 4. On input $\varphi, M_1 M_2 M_{3,1}$ and with oracle access to $A_{\varphi,R}$, algorithm E produces $M_{3,2}$ as V would (this consists of just picking n random challenges), sets $M_3 = M_{3,1} M_{3,2}$, and runs $A_{\varphi,R}$ to get the response $M_4 = A_{\varphi,R}(M_1 M_2 M_3)$. If the resulting transcript is rejecting (as can be determined by running the verifier's check) then E just aborts. If not, it nonetheless aborts with probability exactly 2^{-n}. If neither of these aborts happens, it runs E_1. Since it runs E_1 with probability $p - 2^{-n}$, it finds the satisfying assignment with this probability, and moreover its expected running time is $\text{poly}(n) + (p - 2^{-n}) \cdot \text{poly}(n)/(p - 2^{-n})$ which is $\text{poly}(n)$. ∎

4.4 Protocol 4R-ZK is computationally convincing

We will justify the first claim of Theorem 8 by proving the following:

Lemma 10. *Assume f is a one-way function. Then protocol 4R-ZK is a computationally sound proof for the NP-relation Satisfy, achieving negligible error-probability.*

We first remark and explain that there is indeed something (non-trivial) to be proven here. Typically, error-reduction is done by (serial or parallel) repetition. Firstly, that's not what we are doing; there is some repetition in the protocol, but the protocol itself does not consist of independently repeating some atomic protocol. Moreover, even when the input φ is unsatisfiable, the atomic sub-protocols are actually being run on a *satisfiable* formula (namely Θ). So we are not counting on the soundness of the atomic protocol to prove the soundness of our protocol!

As mentioned earlier, our approach is to use proofs of knowledge, and in particular Lemma 9. Let us now provide the proof.

Proof of Lemma 10. It is easy to see that the specified polynomial time prover strategy P in 4R-ZK will meet the efficient completeness condition of Definition 2. The issue is to show that computational soundness is achieved, and with the claimed negligible error.

Let us assume protocol 4R-ZK does not have negligible error-probability. As per Definition 2 this means there is no negligible function ϵ such that 4R-ZK meets the computational soundness condition of Definition 2 with error set to ϵ. We will show this contradicts the assumption that f is one-way.

So we want to show that f is not one-way. As per Definition 7, this means we are given an arbitrary negligible function δ and must show that there is an inverter I and an infinite set K of integers such that $\text{Inv}_f^I(n) > \delta(n)$ for all $n \in K$. Let us set $\epsilon(n) = \delta(n) \cdot 64n$. This is still a negligible function. So by the above assumption, 4R-ZK does not achieve error-probability ϵ. Hence there exists a polynomial time prover $\widehat{P}$ and an infinite set F of unsatisfiable boolean formulae such that $\text{Acc}(\widehat{P}_\varphi, V_\varphi) \geq \epsilon(|\varphi|)$ for all $\varphi \in F$. Let K be the set of all integers n for which F contains a formula φ of length n. For each $n \in K$ we fix (arbitrarily) some formula $\varphi_n \in F$. Before describing the inverter I for f we need to isolate certain executions of the interaction between $\hat{P}_\varphi$ and V_φ, where $\varphi = \varphi_n$.

GOOD EXECUTIONS. Let $n \in K$ and let $\varphi = \varphi_n$. Let R be a random tape for $\hat{P}_\varphi$ and $M_1 M_2 M_{3,1}$ a partial transcript of an interaction between $P_{\varphi,R}$ and V_φ. (The transcript includes the first two messages of the protocol and the first part of V's third message.) We say that $R, M_1 M_2 M_{3,1}$ is good if the commitment setup string in M_1 is good (as defined in Section 3.3) and also $\text{Acc}(P_{\varphi,R}, V_\varphi, M_1 M_2 M_{3,1}) \geq \epsilon(n)/2$ (the probability here is only over the choice of the verifier's challenge vector c, since all other quantities are fixed). Since $\text{Acc}(\hat{P}_\varphi, V_\varphi) \geq \epsilon(n)$ it must be that the probability (over R and the coins of V leading to $M_1 M_2 M_{3,1}$) that $\text{Acc}(P_{\varphi,R}, V_\varphi, M_1 M_2 M_{3,1}) \geq \epsilon(n)/2$ is at least $1/2$. On the other hand the probability that the commitment setup string in M_1 is bad is 2^{-n} (cf. Section 3.3). So the probability that $R, M_1 M_2 M_{3,1}$ is good is at least, say, $\epsilon(n)/4$. (This is because we can assume wlog that $\delta(n) = \epsilon(n)/(64n)$ is, say, at least $2^{-n/2}$, whence $2^{-n} \leq \epsilon(n)/2$.) In the sequel we will focus on these good transcript prefixes.

STRUCTURE OF INVERTER. We now describe an inverter I for f. The inverter I is a polynomial sized collection of circuits $\{ I_n : n \geq 1 \}$ as described in Section 3.1. (Meaning there is a polynomial $p_2(\cdot)$ such that the size of I_n is a most $p_2(n)$ for all $n \geq 1$.) We will show that that for all $n \in K$ we have $\text{Inv}_f^I(n) > \delta(n) = \epsilon(n)/(64n)$. I_n has embedded into it the formula φ_n (which by assumption is unsatisfiable). The input to I_n is a n-bit string $y = f(x)$ where x was chosen at random from $\{0,1\}^n$. I_n wants to output a pre-image of y under f. We describe I_n as a randomized algorithm. (The coins can always be later eliminated by using the non-uniformity). Think if I_n as having oracle access to $\hat{P}_\varphi$ where $\varphi = \varphi_n$. (Meaning it will feed it messages and run it, sometimes "backing it up" and so forth. It implements this by running $\hat{P}$ as a subroutine with the common input fixed to φ. It is important here that $\hat{P}$ is polynomial time). It begins by picking a random string R for $\hat{P}_\varphi$ and initializing the latter with that.

FIRST MOVE. I_n will mimic the first move of V, with a slight twist. It picks $\alpha \xleftarrow{R} [n]$ and $\beta \xleftarrow{R} \{0,1\}$. Then for $i = 1, \ldots, n$ and $j = 0, 1$ it does the following: If $(i,j) = (\alpha,\beta)$ then set $y_{i,j} = y$, else pick $x_{i,j} \xleftarrow{R} \{0,1\}^n$ and set $y_{i,j} = f(x_{i,j})$. We let $\Phi_{i,j} = \text{FORMULA}_f(y_{i,j})$ be the boolean formula resulting from applying Cook's theorem to the "$f(\cdot) = \cdot$" relation on input $y_{i,j}$, as described in Section 3.2. Now I_n also picks random strings $R_1, \ldots, R_n$, of appropriate length, as setup strings for the bit commitment to be used in the atomic protocol. It lets M_1 consist of the strings $y_{i,j}$ for $i = 1, \ldots, n$ and $j = 0, 1$, together with $R_1, \ldots, R_n$. This, thought of as the first message of V to $\hat{P}_\varphi$, is then "sent" to $\hat{P}_\varphi$.

SECOND MOVE. I_n runs $\hat{P}_\varphi$ to get its response $M_2 = \hat{P}_\varphi(M_1 ; R)$ to the verifier message M_1. This response has the form $M_2 = M_{2,1} M_{2,2}$ where $M_{2,1} = (b_1, \ldots, b_n)$ and $M_{2,2} = (C_1, \ldots, C_n)$. Here C_i is (supposed to be) a committal for a run of the atomic protocol on input $\Theta = \varphi \vee \Phi$, where $\Phi = \Phi_{1,1-b_1} \vee \ldots \vee \Phi_{n,1-b_n}$.

OPENING. Recall that V_φ is supposed to return x_{i,b_i} to $\widehat{P}_\varphi$ for all $i = 1, \ldots, n$. I_n would like to do the same. But if $b_\alpha = \beta$ then this means it must return a pre-image of $y_{\alpha,\beta}$ under f, and it does not know such a pre-image. (Indeed, the goal of I_n is to find one). So in this case I_n aborts. But this can only happen with probability $1/2$ since β was a random bit. In case $b_\alpha \neq \beta$, our I_n sets $M_{3,1} = \mathbf{x} = (x_{1,b_1}, \ldots, x_{n,b_n})$. This is the first part of a verifier message M_3 to be sent to $\widehat{P}_\varphi$.

FINDING A WITNESS FOR Φ. Now comes the important step. I_n will run an "extractor" for the protocol which consists of n parallel runs of the atomic protocol on input Θ and find a satisfying assignment for Φ. Specifically, we apply Lemma 9. Let E be as in that lemma and let $p_1(\cdot)$ be the polynomial which is its expected running time. I_n runs E on input $\varphi, M_1 M_2 M_{3,1}$, giving it oracle access to $\widehat{P}_{\varphi,R}$. However, this execution is halted in $2p_1(n)$ steps. (Recall E has an expected polynomial running time, but I_n needs to halt within a fixed polynomial amount of time.) If E finds, within this time, a satisfying assignment T to $\Theta = \varphi \vee \Phi$, then I_n will be able to find what it wants, namely a point x satisfying $f(x) = y$. The crucial observation is that since φ is unsatisfiable, the assignment T *must satisfy* Φ. Hence it must satisfy $\Phi_{i,1-b_i}$ for some $i \in [n]$. Since α was chosen at random from $[n]$ it will be the case that $i = \alpha$ with probability at least $1/n$. We know $b_\alpha \neq \beta$ (since otherwise we aborted above) meaning $b_\alpha = 1 - \beta$. So we have an assignment to $\Phi_{\alpha,\beta}$. Now recall that $\Phi_{\alpha,\beta} = \text{FORMULA}_f(y)$. Applying the witness transformation $t_{f,2}$ discussed in Section 3.2, we can compute a string x such that $f(x) = y$. I_n does this and outputs x.

ANALYSIS. The running time of I_n is clearly poly(n). We must analyze its success probability. We assume $R, M_1 M_2 M_{3,1}$ is good in the sense defined above: we saw this happens with probability at least $1/4$. This means the commitment setup strings in M_1 are good and $p = \text{Acc}(P_{\varphi,R}, V_\varphi, M_1 M_2 M_{3,1}) \geq \epsilon(n)/2$. Now Lemma 9 says that E would find a satisfying assignment to Θ with probability at least $p - 2^{-n} \geq \epsilon(n)/2 - 2^{-n} > \epsilon(n)/4$. (Recall we assumed wlog that $\delta(n) = \epsilon(n)/(64n)$ is at least $2^{-n/2}$, whence the last inequality.) Since we halt E within twice its expected running time, Markov's inequality says we find the assignment with at least half the original probability. So I_n finds x with probability at least $\epsilon(n)/8$. Putting this together with the other probability losses, all together, I_n succeeds with probability at least $\epsilon(n)/(64n) = \delta(n)$, as desired. ∎

4.5 Protocol 4R-ZK is a computational proof of knowledge

The second claim of Theorem 8 is justified by the following lemma.

Lemma 11. *Assume f is a one-way function. Then protocol 4R-ZK is a computational proof of knowledge (with negligible knowledge error) for the NP-relation Satisfy.*

Before proving it let us discuss the issues. Given a satisfiable formula φ and oracle access to a polynomial time prover $\widehat{P}$, the goal is to extract a satisfying assignment to φ, with a success probability only marginally less than the probability that $\widehat{P}_\varphi$ convinces V_φ to accept. We can easily run the extractor of Lemma 9 to find a satisfying assignment T, but for Θ, not φ. But $\Theta = \varphi \vee \Phi$. Our worry is that T satisfies Φ, not φ. However, intuitively not, because a satisfying assignment to Φ corresponds to the ability to invert f, and thus should appear only with negligible probability. To capture this intuition we must show that were T to satisfy Φ too often then there would be a way to invert f. We can do this similarly to the proof of Lemma 10.

Proof of Lemma 11. We will exhibit an extractor E_1 such that the conditions of Definition 3 are met for some negligible function $\kappa(\cdot)$. (Recall Definition 3 and Definition 4 are equivalent.) E_1 has input satisfiable formula φ, and has oracle access to $\widehat{P}_{\varphi,R}$ where R is some (randomly chosen and then fixed) random tape for prover $\widehat{P}$. E_1 first picks a random a tape R' for V. It now plays the role of V, invoking $\widehat{P}$ for the role of the prover, and generates a partial transcript $M_1 M_2 M_{3,1}$ of the interaction between $\widehat{P}_\varphi$ and $V_{\varphi,R'}$. If the commitment setup strings in M_1 are not good then E_1 aborts. Else it runs the knowledge extractor E of Lemma 9 on input $\varphi, M_1 M_2 M_{3,1}$, giving it oracle access to $\widehat{P}_{\varphi,R}$. Whatever the latter outputs (hopefully an assignment T to Θ) is what E_1 outputs.

Since E runs in expected polynomial time, it is easy to see that E_1 does too. Similarly, given Lemma 9, it is easy to see that with probability at least $p - 2^{-n+1}$, algorithm E_1 outputs a satisfying assignment T to Θ (not φ!), where $p = \mathrm{Acc}(\widehat{P}_\varphi, V_\varphi)$. (We loose the additional 2^{-n} over the success probability of E because the commitment setup strings are bad with probability at most 2^{-n} (cf. Section 3.3) and E_1 aborts in this case.)

But our goal is to find a satisfying assignment to φ. Remember $\Theta = \Phi \vee \varphi$. Our worry is that T satisfies Φ rather than φ. Intuitively, however, not, because we know that the ability to find an assignment to Φ corresponds to the ability to invert f. Thus it might happen, but only negligibly often. We must now capture this.

We must show there exists a negligible function $\kappa(\cdot)$ such that T is a satisfying assignment to φ with probability $p - \kappa(n)$, for all φ of size at least $N_{\widehat{P}}$, where $N_{\widehat{P}}$ is an integer depending on $\widehat{P}$. Assume towards a contradiction that there is no such κ. So given any negligible function κ there is a polynomial time prover $\widehat{P}$ and an infinite set F of formulas such that when $\varphi \in F$, the assignment T output by E_1 satisfies Φ (rather than φ) with probability at least $(p - 2^{-n}) - (p - \kappa(n)) = \kappa(n) - 2^{-n}$. We must show that this implies f is not one-way.

We will not give the construction and proof for this last statement in full because the idea is essentially the same as in the proof of Lemma 10. We use the composite of $E_1^{\widehat{P}}$ as an algorithm to construct an inverter for f. Like in the proof of Lemma 10, we are given a value y and want to find a pre-image of y under f. We put y into the first message of the verifier in the same way as before. Eventually when $E_1^{\widehat{P}}$ gives us an assignment T to Φ, it has some probability of satisfying $\mathrm{FORMULA}_f(y)$ and then we get a pre-image of y under f, just as before. The details can be filled in by looking at the proof of Lemma 10. ∎

4.6 Protocol 4R-ZK is zero-knowledge

The third claim of Theorem 8 is justified by the following lemma.

Lemma 12. *Assume f is a one-way function. Then protocol 4R-ZK is a (computational) zero-knowledge protocol.*

Proof. We must specify a simulator $\widehat{S}$ for which Definition 6 is met. S has input φ and oracle access to $\widehat{V}_{\varphi,R}$ where $\widehat{V}$ is any (possibly cheating) polynomial time verifier algorithm and R is a randomly chosen random tape for $\widehat{V}_\varphi$. It must produce a transcript τ such that (R, τ) is distributed like random members of the view of the real interaction between P_φ and $\widehat{V}_\varphi$. Before describing the algorithm let us sketch the intuition.

S will be trying to produce the prover moves in a conversation with $\widehat{V}_{\varphi,R}$. Of course, not knowing a satisfying assignment for φ, it can't really play the prover. But recall the atomic protocol is run not on input φ but on input $\Theta = \varphi \vee \Phi$. The trick is that it suffices to know a satisfying assignment for Θ.

Indeed, suppose we know some satisfying assignment for Θ. This is not necessarily a satisfying assignment for φ. Still, we can "mimic the prover" by using this assignment in the atomic protocol. The verifier will never know it was not an assignment to φ, because the proof is ZK and hence witness indistinguishable [FeSh]: views of the verifier for different witnesses held by the prover are indistinguishable.

So if the simulator can find a satisfying assignment to Θ it can complete a simulation. How can it find one? It can force $\widehat{V}_{\varphi,R}$ to give it one! It will do this by forcing the verifier to reveal a pre-image $x_{i,1-b_i}$ of $y_{i,1-b_i}$ for some $i \in [n]$. This corresponds effectively to a satisfying assignment to $\Phi_{i,1-b_i}$ and hence to a satisfying assignment to Φ and hence to a satisfying assignment to Θ.

But how does it get $x_{i,1-b_i}$? What $\widehat{V}_{\varphi,R}$ reveals is x_{i,b_i}, exactly to prevent the prover from getting $x_{i,1-b_i}$, because if the prover had the latter, it could cheat. But the simulator has an advantage: it can backup the verifier and run it twice for different choices of $b_1, \ldots, b_n$. First it runs it in a normal way on some "dummy" challenges $b'_1, \ldots, b'_n$, gets back the corresponding pre-images, and then claims that the real challenges $b_1, \ldots, b_n$ were different, in particular have $b_\alpha = 1 - b'_\alpha$ for some $\alpha \in [n]$. For the new challenges, it has the pre-image.

Let us now specify all this in full. Here is the algorithm for S with input φ and oracle access to $\widehat{V}_{\varphi,R}$:

(1) S runs $\widehat{V}_{\varphi,R}$ to get the first message $M_1 = M_{1,1} M_{1,2}$. Here $M_{1,1}$ consists of strings $y_{i,j} \in \{0,1\}^n$ for $i = 1, \ldots, n$ and $j = 0, 1$, and $M_{1,2} = (R_1, \ldots, R_n)$ consists of n strings to play the role of commitment setup strings. We let $\Phi_{i,j} = \text{FORMULA}_f(y_{i,j})$ be the formula corresponding to $y_{i,j}$ via Cook's theorem, as explained in Section 3.2.

(2) S picks at random $b'_1, \ldots, b'_n \in \{0, 1\}$ and lets $\Phi' = \Phi_{1,1-b'_1} \vee \ldots \vee \Phi_{n,1-b'_n}$. It then lets $\Theta' = \varphi \vee \Phi'$ and picks at random an assignment T' to the variables of Θ'. (This assignment is extremely unlikely to satisfy Θ', but that does not matter!) For each $i = 1, \ldots, n$ it then picks at random some coins ρ_i' and computes an encapsulated circuit $C_i' = \text{ENCCIRC}_f(\Theta', T', R_i, \rho_i')$ for Θ'. We let $M_{2,1}' = (b_1', \ldots, b_n')$ and $M_{2,2}' = (C_1', \ldots, C_n')$. We view $M_2' = M_{2,1}' M_{2,2}'$ as the second protocol message (from the prover).

(3) S runs $\widehat{V}_{\varphi,R}(M_1 M_2')$ to get back its response $M_3' = M_{3,1}' M_{3,2}'$. Here $M_{3,1}'$ consists of values $x_{i,b_i'}$ for $i = 1, \ldots, n$ and $M_{3,2}'$ is a challenge vector. S checks that $f(x_{i,b_i'}) = y_{i,b_i'}$ for $i = 1, \ldots, n$. If this fails, it outputs the current partial conversation and halts. Else it continues.

(4) S now picks at random another sequence of bits $b_1, \ldots, b_n \in \{0, 1\}$. If $(b_1, \ldots, b_n) = (b_1', \ldots, b_n')$ then it aborts (but this happens only with probability 2^{-n}). Else it fixes an index $\alpha \in [n]$ such that $b_i \neq b_i'$. It lets $\Phi = \Phi_{1,1-b_1} \vee \ldots \vee \Phi_{n,1-b_n}$ and $\Theta = \varphi \vee \Phi$. Now, notice that $1 - b_\alpha = b_\alpha'$ and S knows $x_{\alpha,b_\alpha'}$, a pre-image of $y_{\alpha,b_\alpha'}$, from the previous step. Because of this, it can compute a satisfying assignment T to the formula $\Phi_{\alpha,b_\alpha'}$. (This is via the properties of Cook's reduction as explained in Section 3.2.) But then T also satisfies Φ and hence Θ, so S has in its possession a satisfying assignment to Θ. Now the idea is to act like the real prover on input this assignment. (Note this assignment does not satisfy φ, but the verifier will never be able to tell, because it does satisfy the formula Θ on which the atomic protocol is performed, and the bit commitments are secure.) So for each $i = 1, \ldots, n$ the simulator picks at random some coins ρ_i and computes an encapsulated circuit $C_i = \text{ENCCIRC}_f(\Theta, T, R_i, \rho_i)$ for Θ. We let $M_2 = M_{2,1} M_{2,2}$ where $M_{2,1} = (b_1, \ldots, b_n)$ and $M_{2,2} = (C_1, \ldots, C_n)$. We view M_2 as a second protocol message (from the prover).

(5) Backing up $\widehat{V}_{\varphi,R}$, the simulator S computes $\widehat{V}_{\varphi,R}(M_1 M_2)$ to get back its response $M_3 = M_{3,1} M_{3,2}$. Here $M_{3,1}$ consists of values x_{i,b_i} for $i = 1, \ldots, n$ and $M_{3,2}$ is a challenge vector $c_1, \ldots, c_n$. S checks that $f(x_{i,b_i}) = y_{i,b_i}$ for $i = 1, \ldots, n$. If this check fails S *cannot* abort or output this conversation. (One can check this would lead to an incorrect simulation.) Instead, it must return to Step 4 and try again, continuing this loop until the check does pass. (This is a standard procedure, used for example in [BMO1], and as there one can show that the expected number of tries in this process is at most 2.) So we go on assuming the check did pass.

(6) Having a satisfying assignment T to Θ, the simulator (now in guise of the prover) is able to answer the challenges $c_1, \ldots, c_n$ by opening the appropriate parts of the encapsulated circuits $C_1, \ldots, C_n$ just as the prover would. Namely S can compute $D_i = \text{ANSWER}_f(\Theta, T, R_i, \rho_i, c_i)$ for $i = 1, \ldots, n$ and let M_4 consist of $D_1, \ldots, D_n$.

(7) Finally, S can output $\tau = M_1 M_2 M_3 M_4$ as a transcript of the interaction between the prover and $\widehat{V}_{\varphi,R}$.

Fix some witness selector $W: SAT \to \{0, 1\}^*$ for the relation $Satisfy(\cdot, \cdot)$. That is, $W(\varphi)$ is a satisfying assignment to φ for every $\varphi \in SAT$. As per Definition 6 we want to show that the probability ensembles $\mathcal{E}_1 = \{\overline{S}^{\widehat{V}_\varphi}(\varphi)\}_{\varphi \in SAT}$ and $\mathcal{E}_2 = \{\text{VIEW}(P, W, \widehat{V}, \varphi)\}_{\varphi \in SAT}$ are computationally indistinguishable. (Refer to Section 2.4 for the definition of $\overline{S}$.) We will do this under the assumption that f is a one-way function. We will provide here only a brief outline of the intuition behind this proof.

The function f shows up in two places in the protocol. First, f is used in the construction of Y-values underlying the formula Φ. Second, f underlies the bit commitment scheme of the atomic protocol. The first use of f is not a concern for the zero-knowledge, in the sense that the protocol would be ZK (but not computationally convincing or a computational proof of knowledge!) even if the function used to produce the Y-values was not one-way. The ZK depends however on the security of the bit commitment scheme, and hence indirectly on the one-wayness of f.

The privacy (cf. Section 3.3) of the bit commitment scheme means that when S, in Step (2), forms an encapsulated circuit using a dummy truth assignment T', the verifier $\widehat{V}$ has no feasible way to detect it, and its behavior can change "only negligibly." Now, in Step (4) the simulator uses a satisfying assignment for Θ that is different from the one the prover would use. But since the atomic protocol is ZK it is also witness indistinguishable in the sense of [FeSh]. Furthermore, they show that witness indistinguishability is preserved under parallel repetition, so the protocol consisting of n parallel repetitions of the atomic protocol is also witness indistinguishable. So the transcripts produced for the two different witnesses in protocol 4R-ZK have (computationally) indistinguishable distributions.

The formal proof would be by contradiction. We assume the ensembles are not computationally indistinguishable. So for any negligible function $\delta(\cdot)$ there is a distinguisher $D = \{D_\varphi\}_{\varphi \in SAT}$ and an infinite set F of satisfiable boolean formulae such that

$$\left| \Pr\left[D_\varphi(v) = 1 \; : \; v \xleftarrow{R} \overline{S}^{\widehat{V}_\varphi}(\varphi) \right] - \Pr\left[D_\varphi(v) = 1 \; : \; v \xleftarrow{R} \text{VIEW}(P, W, \widehat{V}, \varphi) \right] \right|$$

is at least $\delta(|\varphi|)$ whenever $\varphi \in F$. Using D we would do one of the following. Either construct a polynomial sized circuit family that defeated the privacy of the bit commitment scheme, which would contradict the security of this scheme as proven in [Na, HILL]. Or, build a distinguisher that would contradict the witness indistinguishability of n parallel repetitions of the atomic protocol. We omit these proofs from this abstract. ∎

Acknowledgments

We thank the (anonymous) referees of Eurocrypt 97 for comments which improved the presentation of the paper.

Mihir Bellare is supported in part by NSF CAREER Award CCR-9624439 and a Packard Foundation Fellowship in Science and Engineering.

References

[BeGo]   M. BELLARE AND O. GOLDREICH. On Defining Proofs of Knowledge. *Advances in Cryptology – Crypto 92 Proceedings*, Lecture Notes in Computer Science Vol. 740, E. Brickell ed., Springer-Verlag, 1992.

[BMO1] M. BELLARE, S. MICALI AND R. OSTROVSKY. Perfect Zero-Knowledge in Constant Rounds. *Proceedings of the 22nd Annual Symposium on the Theory of Computing*, ACM, 1990.

[BMO2] M. BELLARE, S. MICALI AND R. OSTROVSKY. The true complexity of statistical zero-Knowledge. *Proceedings of the 22nd Annual Symposium on the Theory of Computing*, ACM, 1990.

[BeYu]   M. BELLARE AND M. YUNG. Certifying permutations: Non-interactive zero-knowledge based on any trapdoor permutation. *Journal of Cryptology*, Vol. 9, No. 1, pp. 149–166, Winter 1996.

[Bl]   M. BLUM. Coin Flipping over the Telephone. IEEE COMPCON 1982, pp. 133–137.

[BDMP] M. BLUM, A. DE SANTIS, S. MICALI, AND G. PERSIANO. Non-Interactive Zero-Knowledge Proof Systems. *SIAM Journal on Computing*, Vol. 20, No. 6, December 1991, pp. 1084–1118.

[BlMi]   M. BLUM AND S. MICALI. How to generate cryptographically strong sequences of pseudo-random bits. *SIAM Journal on Computing*, Vol. 13, No. 4, pp. 850–864, November 1984.

[BrCr]   G. BRASSARD AND C. CRÉPEAU. Non-transitive Transfer of Confidence: A perfect Zero-knowledge Interactive protocol for SAT and Beyond. *Proceedings of the 27th Symposium on Foundations of Computer Science*, IEEE, 1986.

[BCC]   G. BRASSARD, D. CHAUM AND C. CRÉPEAU. Minimum Disclosure Proofs of Knowledge. *J. Computer and System Sciences*, Vol. 37, 1988, pp. 156–189.

[BCY]   G. BRASSARD, C. CRÉPEAU AND M. YUNG. Constant round perfect zero knowledge computationally convincing protocols. *Theoretical Computer Science*, Vol. 84, No. 1, 1991.

[FFS]   U. FEIGE, A. FIAT, AND A. SHAMIR. Zero-Knowledge Proofs of Identity. *Journal of Cryptology*, Vol. 1, 1988, pp. 77–94.

[FLS]   U. FEIGE, D. LAPIDOT, AND A. SHAMIR. Multiple Non-Interactive Zero-Knowledge Proofs Based on a Single Random String. *Proceedings of the 31st Symposium on Foundations of Computer Science*, IEEE, 1990.

[FeSh]   U. FEIGE AND A. SHAMIR. Witness Indistinguishable and Witness Hiding Protocols. *Proceedings of the 22nd Annual Symposium on the Theory of Computing*, ACM, 1990.

[Fo]   L. FORTNOW. The Complexity of Perfect Zero-Knowledge. In *Advances in Computing Research*, Ed. S. Micali, Vol. 18, 1989.

[GoKa]   O. GOLDREICH AND A. KAHAN. How to Construct Constant-Round Zero-Knowledge Proof Systems for NP. *Journal of Cryptology*, Vol. 9, No. 3, 1996, pp. 167–190.

[GoKr]   O. GOLDREICH AND H. KRAWCZYK. On the Composition of Zero Knowledge Proof Systems. *SIAM J. on Computing*, Vol. 25, No. 1, pp. 169–192, 1996.

[GMW]   O. GOLDREICH, S. MICALI AND A. WIGDERSON. Proofs that yield nothing but their validity or all languages in NP have zero knowledge proof systems. *Journal of the Association for Computing Machinery*, Vol. 38, No. 1, July 1991.

[GoOr] O. GOLDREICH AND Y. OREN. Definitions and properties of zero-knowledge proof systems. *Journal of Cryptology*, Vol. 7, No. 1, 1994, pp. 1–32.

[GoMi] S. GOLDWASSER AND S. MICALI. Probabilistic Encryption. *J. Computer and System Sciences*, Vol. 28, 1984, pp. 270–299.

[GMR] S. GOLDWASSER, S. MICALI AND C. RACKOFF. The knowledge complexity of interactive proof systems. *SIAM J. on Computing*, Vol. 18, No. 1, pp. 186–208, February 1989.

[HILL] J. HÅSTAD, R. IMPAGLIAZZO, L. LEVIN AND M. LUBY. Construction of a pseudo-random generator from any one-way function. Manuscript. Earlier versions in STOC 89 and STOC 90.

[ImLu] R. IMPAGLIAZZO AND M. LUBY. One-way Functions are Essential for Complexity-Based Cryptography. *Proceedings of the 30th Symposium on Foundations of Computer Science*, IEEE, 1989.

[ImYu] R. IMPAGLIAZZO AND M. YUNG. Direct Minimum-Knowledge Computations. *Advances in Cryptology – Crypto 87 Proceedings*, Lecture Notes in Computer Science Vol. 293, C. Pomerance ed., Springer-Verlag, 1987.

[IS1] T. ITOH AND K. SAKURAI. On the complexity of constant round ZKIP of possession of knowledge. *IEICE Transactions on Fundamentals of Electronics, Communications and Computer Sciences*, Vol. E76-A, No. 1, January 1993.

[Na] M. NAOR. Bit Commitment using Pseudo-Randomness. *Advances in Cryptology – Crypto 89 Proceedings*, Lecture Notes in Computer Science Vol. 435, G. Brassard ed., Springer-Verlag, 1989.

[NOVY] M. NAOR, R. OSTROVSKY, R. VENKATASAN, M. YUNG. Perfect zero knowledge arguments for NP can be based on general complexity assumptions. *Advances in Cryptology – Crypto 92 Proceedings*, Lecture Notes in Computer Science Vol. 740, E. Brickell ed., Springer-Verlag, 1992.

[OsWi] R. OSTROVSKY AND A. WIGDERSON. One-way functions are essential for nontrivial zero-knowledge. *Proceedings of the Second Israel Symposium on Theory and Computing Systems*, IEEE, 1993.

[ToWo] M. TOMPA AND H. WOLL. Random Self-Reducibility and Zero-Knowledge Interactive-Proofs of Possession of Information. *Proceedings of the 28th Symposium on Foundations of Computer Science*, IEEE, 1987.

[Ya] A. C. YAO. Theory and Applications of Trapdoor functions. *Proceedings of the 23rd Symposium on Foundations of Computer Science*, IEEE, 1982.

A Constant round ZK via coin flipping plus NIZK

The protocol stated in Figure 1 as obtained by combining [Bl, FLS] is folklore. First use Blum's coin flipping in the well protocol [Bl] to get a common random string, then do a NIZK [BDMP] proof, which can be done with a trapdoor permutation [FLS, BeYu]. In somewhat more detail, the first move is the verifier committing. For a four round ZK protocol we need a "certified one-way permutation." (Based on algebraic assumption, e.g. Discrete Logarithm. An arbitrary trapdoor permutation won't suffice.) After this the prover sends bits in the clear, the verifier de-commits, and the XOR of the prover bits and the verifier's de-comitted bits is declared to be the common random string. The non-interactive ZK (NIZK) proof is run on the latter. The reason the full protocol is an argument, not a proof, is that the verifier's first round committals are done using a computational assumption.

Efficient Cryptographic Protocols
Based on Noisy Channels

Claude Crépeau*

Département d'Informatique et R.O.,
Université de Montréal,
C.P. 6128, succursale centre-ville,
Montréal (Québec), Canada H3C 3J7.
e-mail: crepeau@iro.umontreal.ca.

Abstract. The Wire-Tap Channel of Wyner [19] shows that a Binary Symmetric Channel may be used as a basis for exchanging a secret key, in a cryptographic scenario of two honest people facing an eavesdropper. Later Crépeau and Kilian [9] showed how a BSC may be used to implement Oblivious Transfer in a cryptographic scenario of two possibly dishonest people facing each other. Unfortunately this result is rather impractical as it requires $\Omega(n^{11})$ bits to be transmitted through the BSC to accomplish a single OT. The current paper provides efficient protocols to achieve the cryptographic primitives of Bit Commitment and Oblivious Transfer based on the existence of a Binary Symmetric Channel. Our protocols respectively require sending $O(n)$ and $O(n^3)$ bits through the BSC. These results are based on a technique known as Generalized Privacy Amplification [1] that allow two people to extract secret information from partially compromised data.

1 Introduction

The cryptographic power of a noisy channel has been demonstrated by Wyner [19] who showed that two honest parties, say A and B, can exchange a secret key on which an eavesdropper $\mathcal{E}$ may obtain only a small fraction of the information as long as A and B are connected by a Binary Symmetric Channel of better quality than a similar Channel connecting them to $\mathcal{E}$. More recently, a result of Bennett, Brassard, Crépeau and Maurer [1] provides a technique called Generalized Privacy Amplification to ensure that $\mathcal{E}$'s information is an arbitrary small fraction of a bit under the same conditions.

But cryptography is no longer interested solely in protecting communications. As a result of public-key cryptography, a large number of other cryptographic tasks have emerged. Examples of such tasks are Coin-flipping by telephone [3] and Mental Poker. These may involve two or more parties, some of which may be dishonest. The general concept of Distributed Function Evaluation was first introduced by Yao [20] and later extended to "Mental Games" by Goldreich, Micali and Wigderson [12].

---

* Supported in part by Québec's FCAR and Canada's NSERC.

Distributed Function Evaluation and Mental Games are multi-party algorithms which involve secret data that the parties want to keep from one another. In the model where we are ready to accept computational assumptions, such general tasks can be achieved from basic assumptions such as the existence of a One-Way Trapdoor Function [12].

The lesson derived in the computational model is that very simple protocols are sufficient to achieve the general ones. The two primitives known as Bit Commitment (defined in Section 3) and Oblivious Transfer (defined in Section 4) are elementary protocols that are sufficient in general to accomplish any Mental Games, even in a non-computational scenario [14, 8].

The current paper considers a scenario where only two people, $\mathcal{A}$ and $\mathcal{B}$, are involved and where we put no limitation on their computing power. If we made no further assumption, it would be impossible to accomplish Mental Games. Thus, the extra assumption we make is that $\mathcal{A}$ and $\mathcal{B}$ are connected by a Binary Symmetric Channel ($\mathbf{BS}_\epsilon$), that is a channel that will change the value of a bit b with probability ϵ as it travels from one party to the other.

A first protocol to accomplish Oblivious Transfer from a Noisy Channel was presented in [9]. Unfortunately, that protocol is quite complex and requires $\Omega(n^{11})$ bits sent through the BSC to perform a single Oblivious Transfer, where n is a security parameter that specifies the reliability of the protocol. As a consequence, any two-party computations may be performed from the assumption that there exists a reliable BSC. The current solution is by far more efficient than those suggested earlier. The current paper provides a protocol for Bit Commitment that uses $O(n)$ times the $\mathbf{BS}_\epsilon$ and a protocol for Oblivious Transfer that uses $O(n^3)$ times that primitive, where n is a security parameter that specifies the probabilities of failure of the protocols. These probabilities are all exponentially small in n.

2 General Tools

2.1 Error Channel

We consider a standard error model: the *binary symmetric channel*. In the binary symmetric channel $\mathcal{A}$ sends a bit to $\mathcal{B}$ that is flipped with probability ϵ

$$\mathbf{BS}_\epsilon(x) = \begin{cases} \bar{x} \text{ with prob. } \epsilon \\ x \text{ with prob. } 1 - \epsilon. \end{cases}$$

By extension, we also write $\mathbf{BS}_\epsilon(w)$ as a shorthand for $\mathbf{BS}_\epsilon(w_1)\mathbf{BS}_\epsilon(w_2)...\mathbf{BS}_\epsilon(w_n)$ when $w = w_1w_2...w_n$ is an n-bit word. Let $\mathbf{H}(\epsilon) = -\epsilon\lg\epsilon - (1 - \epsilon)\lg(1 - \epsilon)$ be the binary entropy function. We define the channel capacity of the $\mathbf{BS}_\epsilon$ to be $C_\epsilon = 1 - \mathbf{H}(\epsilon)$.

A nice property of the binary symmetric channel is that it is totally symmetrical between the participants: if $\mathcal{B}$ wants to send a bit x via $\mathbf{BS}_\epsilon(x)$ to $\mathcal{A}$ when it is only available from $\mathcal{A}$ to $\mathcal{B}$, they can do as follows:

Protocol 2.1 ($\overline{\mathbf{BS}}_\epsilon(x)$)

    1: $\mathcal{A}$ picks $r \in_R \{0,1\}$ and runs $\mathbf{BS}_\epsilon(r)$ with $\mathcal{B}$ who gets r',
    2: $\mathcal{B}$ announces $y \leftarrow x \oplus r'$ to $\mathcal{A}$,
    3: $\mathcal{A}$ returns $y \oplus r$.

In general, for the binary symmetric channel, any protocol may be inverted by permuting $\mathcal{A}$ and $\mathcal{B}$ and replacing $\mathbf{BS}_\epsilon$ by $\overline{\mathbf{BS}}_\epsilon$. Therefore the protocols of sections 3 and 4 may be achieved from a noisy channel running either way. This is not the case with all channels. The following is an example of the opposite type.

An alternative to the binary symmetric channel would have been to consider the *erasure channel* where bits are either received without errors, or completly lost with probalities $1 - \epsilon$ and ϵ. However this situation has been previously analyzed since the erasure channel is the same as Rabin's Oblivious Transfer [16]. Protocols for Bit Commitment and $\binom{2}{1}$–OT using Rabin's O.T. are available in [14] and [7].

2.2 Coding theory

An $[n, k, d]$ linear code $\mathcal{C}$ is a linear subspace of $\{0,1\}^n$ of dimension k (and cardinality 2^k) such that no two words c_1, c_2 from $\mathcal{C}$ are such that $d_H(c_1, c_2) < d$, except if $c_1 = c_2$, where $d_H(x, y)$ is the Hamming distance between x and y: the number of positions where they differ.

Such a code is defined as the linear combinations of the rows of a generating matrix G of dimension $k \times n$. Alternatively, $\mathcal{C}$ may be defined as the kernel of a parity check matrix H of dimension $n \times (n - k)$. Knowledge of G or H is computationally equivalent as it is easy to get one from the other.

For section 3 we need the well known fact [15, chap. 17, prob. (30)] that there exists a constant $\rho > 1$ such that a random binary matrix G of size $Rn \times n$ defines a binary linear code with minimal distance at least ϵn except with probability not greater than $\rho^{(R-C_\epsilon)n}$, for values of $R < C_\epsilon$.

For section 4 we need codes that are efficiently decodable with high correction rate and high dimension. For this purpose we use concatenated codes defined in [11] that are efficiently encoded and decoded. Asymptotically, very long $[n, Rn, d]$ concatenated codes may be constructed in such a way that for every $\epsilon > 0$ there exists a constant $\rho > 1$ such that the codes fail to correct ϵn errors except with probability not greater than $\rho^{(R-C_\epsilon)n}$, for values of $R < C_\epsilon$ (although the minimum distance d may be somewhat smaller than ϵn). Please consult [11] for more information on asymptotic performances of concatenated codes.

In some situations the information transmitted is not a codeword. In such a case, as long as the syndrome $syn(w) = H^\top w$ of a word w is known the decoding algorithm may be used to recover w from a noisy version of that word and the value of $syn(w)$. Please consult [15] for more information on coding theory.

2.3 Generalized Privacy Amplification

Let W be a random variable uniformly distributed over $\{0,1\}^n$ and let $\mathbf{BS}_\epsilon(W)$ be another random variable obtained from W through a binary symmetric channel of error rate ϵ, i.e.

$$\text{Prob}\left[\mathbf{BS}_\epsilon(W) = v | W = w\right] = (1 - \epsilon)^{n - d_H(w,v)} \epsilon^{d_H(w,v)}.$$

Let G be a random variable taking values $g : \{0,1\}^n \to \{0,1\}^r$ uniformly distributed from a *universal*$_2$ class of hash functions [6]. It is shown in [1] that

Theorem 1. *For any $\delta > 0$ and all sufficiently large n, for $s = n(\mathbf{H}(\epsilon) - \delta) - r$*

$$H(G(W)|\mathbf{BS}_\epsilon(W), G) \geq r - \frac{2^{-s}}{\ln 2}.$$

Moreover, according to [2, 5, 1] for the special case where we have a linear function $syn : \{0,1\}^n \to \{0,1\}^t$

Theorem 2. *For any $\sigma \in \{0,1\}^t$, $\delta > 0$ and all sufficiently large n, for $s = n(\mathbf{H}(\epsilon) - \delta) - r$*

$$H(G(W)|syn(W) = \sigma, \mathbf{BS}_\epsilon(W), G) \geq r - \frac{2^{t-s}}{\ln 2}.$$

Since $H(G(W)|syn(W) = \sigma, \mathbf{BS}_\epsilon(W), G) = r$ means that no information about $G(W)$ is given by $syn(W) = \sigma, \mathbf{BS}_\epsilon(W), G$, the above result is exponentially close to the best possible: the latter contains almost no information about $G(W)$.

3 Bit Commitment

Assume that a party, $\mathcal{A}$, has a bit b in mind, to which she would like to be committed toward another party, $\mathcal{B}$. That is, $\mathcal{A}$ wishes, through a procedure $\mathbf{BC}(b)$, to provide $\mathcal{B}$ with a piece of evidence w that she has a bit b in mind and that she cannot change it (*binding*). Meanwhile, $\mathcal{B}$ should not be able to tell from that evidence what b is (*concealing*). At a later time, $\mathcal{A}$ can reveal, through an unveiling procedure $\mathbf{UN}(b,p)$, the value of b and prove through p to $\mathcal{B}$ that the piece of evidence sent earlier (w) really corresponded to that bit.

Bit commitment schemes have several applications in the field of cryptographic protocols. In particular one can implement *zero-knowledge proofs* of a variety of statements using bit commitment schemes [13, 4]. The first implementations of bit commitment schemes were given in a computational complexity scenario [3]. Unfortunately, proofs of their (computational) security have always required an unproven assumption since otherwise they would imply very strong results such as $\mathcal{P} \neq \mathcal{NP}$.

This section is inspired by that work of [5] to achieve Bit Commitment in the model of Quantum Cryptography.

3.1 Bit Commitment from Binary Symmetric Channel

Intuition behind Protocols BC & UN After establishing a proper error-correcting code, $\mathcal{A}$ sends a codeword from that code to $\mathcal{B}$ through the BS_ϵ. The code is such that $\mathcal{B}$ should have many candidates for $\mathcal{A}$'s codeword after seeing it through the BS_ϵ. The secret bit of $\mathcal{A}$ is given by applying a random function from a *universal*$_2$ class to the codeword. To unveil her bit, $\mathcal{A}$ discloses her codeword. She should not be able to announce two codewords that $\mathcal{B}$ will find close enough to the word he received to believe her.

Formal Protocol Let ϵ be the error probability of the channel, and $\gamma < 1$ be a positive number. Let $\delta > 0$ be such that $\mathbf{H}(\epsilon) - \delta > \mathbf{H}(\gamma\epsilon)$ and such that $(\mathbf{H}(\epsilon) - \delta)n$ is an integer. The following protocols work *for any value of* ϵ such that $0 < \epsilon < 1/2$, in contrast to the protocols of Section 4.

Protocol 3.1 (BC(b))

1: $\mathcal{B}$ chooses and announces to $\mathcal{A}$ a binary linear $[n, k, d]$-code $\mathcal{C}$ with parameters $k = (1 - \mathbf{H}(\epsilon) + \delta)n$ and $d \geq \gamma\epsilon n$.

2: $\mathcal{A}$
- picks a random n-bit string m and announces it to $\mathcal{B}$,
- picks a random codeword $c \in \mathcal{C}$ such that $c \odot m = b$,
- $\overset{n}{\underset{i=1}{\mathbf{DO}}}$ runs $BS_\epsilon(c_i)$ with $\mathcal{B}$ who receives c'_i,
- returns c, b.

3: $\mathcal{B}$ sets $c' \leftarrow (c'_1 c'_2 \ldots c'_n)$ and returns $(\mathcal{C}, m, c')$.

$\mathcal{B}$ keeps c' secret forever, whereas $\mathcal{A}$ keeps b and c secret until (and if) unveiling takes place. If $\mathcal{A}$ subsequently decides to unveil her commitment, she initiates the next protocol with $\mathcal{B}$. There exists a positive number $\lambda < \gamma(1/2 - \epsilon)/2$ such that an honest $\mathcal{A}$ is likely to satisfy the following with overwhelming probability while a dishonest $\mathcal{A}$ is unable to open the commitment as both bits with overwhelming probability.

Protocol 3.2 (UN(c, b), ($\mathcal{C}, m, c'$))

1: **if** $(c \in \mathcal{C}) \wedge (b = c \odot m) \wedge (d_H(c, c') < \epsilon n + \lambda n)$
then $\mathcal{B}$ accepts **else** $\mathcal{B}$ rejects.

Details of the Protocol In the above Protocol **BC** we ask $\mathcal{B}$ to choose a code with specific parameters. The effect of these parameters on the security of the protocol explain why we require $\mathcal{B}$ to do this job and not $\mathcal{A}$: the bigger d is, the more unlikely it is for $\mathcal{A}$ to cheat and the bigger k is, the more unlikely it is for $\mathcal{B}$ to cheat. Coding theory give us limits on how big d and k can be at the same time. In order to have them as large as possible at the same time, the best construction known to this day is to pick the generating matrix of the code at random. Nevertheless, in this case the value of k is easy to figure out from the matrix (the rank of the matrix) while the exact value of d is more difficult to determine. All we know is that it is likely to be high.

As discussed in Sect. 2.2, a random binary matrix G of size $Rn \times n$ defines a binary linear code with minimal distance at least ϵn except with probability $\rho^{(R-C_\epsilon)n}$, thus $\mathcal{B}$ has an exponentially small probability of having d too small when he picks a $k \times n$ matrix at random. $\mathcal{A}$ can easily verify that the value of k is correct.

The random vector m is used to define a Privacy Amplification Function of $\{0,1\}^n$ to $\{0,1\}$.

3.2 Analysis of the Protocol

Concealing Let C and M be the random variables describing $\mathcal{B}$'s possibilities for c and m. Before c is sent through the BSC, C is uniformly distributed among all the possible codewords of $\mathcal{C}$ and M among all possible n–bit strings. Let $0 < \delta' < \delta$. We are in the scenario of Theorem 2 with $r = 1$, $t = (\mathbf{H}(\epsilon) - \delta)n$, and $s = (\mathbf{H}(\epsilon) - \delta')n - 1$. We therefore conclude that seeing a codeword c through a BSC and learning m is not enough to know much about $c \odot m$:

Theorem 3. *For any all sufficiently large* n

$$H(C \odot M | syn(C) = (0, 0, ..., 0), \mathbf{BS}_\epsilon(C), M) \geq 1 - \frac{2^{(\delta' - \delta)n + 1}}{\ln 2}.$$

Binding An honest $\mathcal{A}$ sends a random codeword c through the channel. Consider the random variable $d_H(c, \mathbf{BS}_\epsilon(c))$. It is clear that $E(d_H(c, \mathbf{BS}_\epsilon(c))) = \epsilon n$ and by Bernstein's law of large numbers [17, Chap. VII, Sect. 4, Theorem 2] Prob $[d_H(c, \mathbf{BS}_\epsilon(c)) > \epsilon n + \lambda n]$ is exponentially small in n for all λ sufficiently small, and all sufficiently large n. A dishonest $\mathcal{A}$ sends any word w through the channel and later would like to claim c_0 or c_1 to unveil as 0 or 1. One of these, say c_z, is such that $d_H(c_z, w) > \gamma \epsilon n / 2$. Consider the random variable $d_H(c_z, \mathbf{BS}_\epsilon(w))$. It is easy to calculate that $E(d_H(c_z, \mathbf{BS}_\epsilon(w))) \geq \epsilon n + \gamma(1/2 - \epsilon)n$ and by Bernstein's law of large numbers Prob $[d_H(c_z, \mathbf{BS}_\epsilon(w)) < \epsilon n + \gamma(1/2 - \epsilon)n - \lambda n]$ is exponentially small in n for all λ sufficiently small, and all sufficiently large n.

Thus any $\lambda < \gamma(1/2 - \epsilon)/2$ will satisfy our requirements that an honest $\mathcal{A}$ succeeds except with probability exponentially small in n, while a dishonest $\mathcal{A}$ succeeds to open both ways only with probability exponentially small in n.

4 Oblivious Transfer

One-out-of-two Oblivious Transfer, denoted $\binom{2}{1}$–OT, is a primitive that originates with [18] (under the label of "multiplexing"). According to this primitive, one party $\mathcal{A}$ owns two secret strings w_0 and w_1, and another party $\mathcal{B}$ wants to learn w_c for a secret bit c of his choice. $\mathcal{A}$ is willing to collaborate provided that $\mathcal{B}$ does not learn any information about $w_{\bar{c}}$, but $\mathcal{B}$ will only participate if $\mathcal{A}$ cannot obtain information about c.

Similarly, in an Oblivious Transfer [16], $\mathcal{A}$ sends a message to $\mathcal{B}$ that is received with probability ϵ (this fact is out of their control) while the message is otherwise lost. $\mathcal{A}$ does not find out what happened. $\mathcal{B}$ knows if he got the message or nothing. We note this protocol $\mathbf{OT}_\epsilon$. Independently from [18] but inspired by [16], $\binom{2}{1}$–OT was introduced subsequently in [10] with applications to contract signing protocols.

These two simple cryptographic tools have been extensively studied by several researchers because they turned out to be elementary blocks to build more elaborate cryptographic tasks known as "secure computations". This idea introduced by Yao [20] allows $\mathcal{A}$ and $\mathcal{B}$ to compute a two-argument function on data they would like to keep secret from one another. They find out the output of the function but not their respective inputs. It was shown in a *computational* model that One-out-of-two Oblivious Transfer suffices to perform general secure computations by Goldreich, Micali and Wigderson [12] and later in an abstract (not necessarily computational) model by Kilian [14]. Crépeau showed [7] that indeed Rabin's Oblivious Transfer can also do the job by describing a general technique to turn an Oblivious Transfer into a One-out-of-two Oblivious Transfer. The result of the current section is an extension of that technique.

4.1 Oblivious Transfer from Binary Symmetric Channel

Basic Idea For $\epsilon > 1/2$, simulate $\mathbf{OT}_\epsilon(b)$ with protocol $\widehat{\mathbf{OT}}_\epsilon(b)$ obtained by sending b twice through the BSC of error probability $\varphi = \frac{1-\sqrt{2\epsilon-1}}{2}$ and then reduce $\binom{2}{1}$–OT to $\widehat{\mathbf{OT}}_\epsilon(b)$ with a Protocol similar to that of [7].

Protocol 4.1 ($\widehat{\mathbf{OT}}_\epsilon(b)$)

   1: $\mathcal{A}$ runs $\mathbf{BS}_\varphi(b)\mathbf{BS}_\varphi(b)$ with $\mathcal{B}$ who receives $b_0 b_1$, for $\varphi = \frac{1-\sqrt{2\epsilon-1}}{2}$.

   2: if $b_0 = b_1$ then $\mathcal{B}$ returns b_0 else $\mathcal{B}$ returns ε.

The problems with this approach are that $\widehat{\mathbf{OT}}_\epsilon(b)$ makes errors and that $\mathcal{A}$ can send bad pairs $\bar{b}b$: if $\mathcal{A}$ is honest and sends bb through the binary symmetric channel then

$$\text{Prob}\left[\widehat{\mathbf{OT}}_\epsilon(b) = x\right] = \begin{cases} (1-\varphi)^2 & \text{if } x = b \\ \varphi^2 & \text{if } x = \bar{b} \\ 2\varphi(1-\varphi) & \text{if } x = \varepsilon \end{cases}$$

$\mathcal{B}$ receives a bit with probability $\epsilon = \varphi^2 + (1 - \varphi)^2$. If instead $\mathcal{A}$ is dishonest and sends $\bar{b}b$ or $b\bar{b}$ through the binary symmetric channel then the probability that $\mathcal{B}$ receives a bit is $1 - \epsilon = 2\varphi(1 - \varphi)$. If no extra checks are performed, $\mathcal{A}$ could send bad pairs and figure out in Protocol 4.2 which set is *good* and which set is *bad* by the fact that good pairs are more likely to have been received.

The errors are first solved (in Protocol 4.2) by the same trick as in [2] using codes to fix them, while the cheating by $\mathcal{A}$ is later taken care of (in Protocol 4.3) by running statistics on the frequency of bb pairs. Protocol 4.2 introduces another kind of cheating $\mathcal{A}$ could perform that is also solved in Protocol 4.3.

Intuition behind Protocol $\binom{2}{1}$-$\widehat{\mathrm{OT}}$ For this first protocol we assume $\mathcal{A}$ behaves honestly and will remove this assumption in the final protocol. The idea of the first protocol is that $\mathcal{A}$ sends $2n$ random bits $r_1, r_2, ..., r_{2n}$ to $\mathcal{B}$ using $\widehat{\mathrm{OT}}_\epsilon$. $\mathcal{B}$ should receive roughly $2\epsilon n$ of these and lose $2(1 - \epsilon)n$. $\mathcal{B}$ forms two sets I_0, I_1 of size n and thus defines two strings r'_{I_0}, r'_{I_1} of size n (r' restricted to I_0 and I_1). String r_{I_c} should be entirely known by $\mathcal{B}$, while string $r_{I_{\bar{c}}}$ should be partially unknown by $\mathcal{B}$. Nevertheless, because $\widehat{\mathrm{OT}}_\epsilon$ is imperfect, we expect an average of $\frac{\varphi^2}{\epsilon}n$ differences between r_{I_c} and r'_{I_c}.

A code is established between the parties to correct more than $\frac{\varphi^2}{\epsilon}n$ errors except with exponentially small probability in n.

The errors are corrected by having $\mathcal{A}$ send the syndrome of the two words $syn(r_{I_0}), syn(r_{I_1})$. Using r'_{I_c} and $syn(r_{I_c})$, $\mathcal{B}$ may recover r_{I_c} except with small probability of failure. Nevertheless, this correction information is not sufficient to find out both words $r_{I_c}, r_{I_{\bar{c}}}$ accurately, as long as the dimension of the code is somewhat greater than ϵn.

A privacy amplification function is finally used to extract one secret bit per string, so that one bit may be recovered by $\mathcal{B}$ but not both. This function is the scalar product by a random n-bit word m.

Incomplete Protocol Let γ be a number greater than 1.

Protocol 4.2 ($\binom{2}{1}$-$\widehat{\mathrm{OT}}(b_0, b_1)(c)$)

1: $\overset{2n}{\underset{i=1}{\mathrm{DO}}}$ $\mathcal{A}$ picks a random bit r_i and runs $\widehat{\mathrm{OT}}_\epsilon(r_i)$ with $\mathcal{B}$ who gets r'_i.

2: $\mathcal{B}$ picks and sends two random disjoint sets I_0, I_1 s.t. $|I_0| = |I_1| = n$, and ($\forall i \in I_c \, [r'_i \neq \epsilon]$).

3: $\mathcal{A}$ and $\mathcal{B}$ agree on a parity check matrix H of a concatenated code $\mathcal{C}$ with parameters $[n, k > (\epsilon + \delta)n, d]$ correcting $\gamma \frac{\varphi^2}{\epsilon}n$ errors.

4: $\mathcal{A}$
- computes and sends $s_0 \leftarrow syn(r_{I_0})$ and $s_1 \leftarrow syn(r_{I_1})$,
- picks and sends a random n-bit word m,
- computes and sends $\hat{b}_0 \leftarrow b_0 \oplus (m \odot r_{I_0})$ and $\hat{b}_1 \leftarrow b_1 \oplus (m \odot r_{I_1})$.

5: $\mathcal{B}$
- recovers r_{I_c} using r'_{I_c}, s_c and the decoding algorithm of $\mathcal{C}$,
- computes and returns $\hat{b}_c \oplus (m \odot r_{I_c})$.

Details and discussion of Protocol $\binom{2}{1}$–$\widehat{OT}$ The code used for this protocol requires the extra property that it must be efficiently decodable. This can be done by using concatenated codes. For $\varphi < 0.1982$ the conditions of Step **3** can be satisfied. Therefore, contrary to Protocol **BC**, this new protocol works only for reliable enough channels $\mathbf{BS}_\varphi$ (not for all φ).

$\mathcal{B}$ is unable to cheat this protocol because whatever way he splits the "good" bits $(r_i' \neq \varepsilon)$ between I_0, I_1, he will not be able to put more $(\epsilon + \delta/2)n$ good bits in at least one of I_0 or I_1. Since $k > (\epsilon + \delta)n$ then $syn(r_{I_0}), syn(r_{I_1})$ each contain $n - k$ bits of information, i.e. no more than $(1 - \epsilon - \delta/2)n$ bits. Thus, at least one of the two words r_{I_0}, r_{I_1} will be undetermined by at least $\delta n/2 = n - (1+\delta)n/2 - (1/2 - \delta)n$ bits. Using privacy amplification, this word will contain an exponentially small amount of information about its related bit. Therefore, $\mathcal{B}$ cannot learn both of $\mathcal{A}$'s bits.

Unfortunately, $\mathcal{A}$ can cheat this protocol in two different ways that allow her to figure out $\mathcal{B}$'s secret input c: at Step **2** $\mathcal{A}$ can send "bad" pairs $r_i \bar{r}_i$ or $\bar{r}_i r_i$ instead of $r_i r_i$ increasing the probability that it is lost $(r_i' = \varepsilon)$ by $\mathcal{B}$ and at Step **4** she can send a "bad" syndrome leading $\mathcal{B}$ to a decoding error. In the first cheat, "bad" pairs are more likely to end up in the "bad" set thus indicating to $\mathcal{A}$ which one is more likely to be the "good" and "bad" sets. In the second cheat, if $\mathcal{A}$ makes only one syndrome bad then $\mathcal{B}$ might have to abort depending on which bit he is trying to get. Protocol 4.3 solves these two problems.

Intuition behind Protocol $\binom{2}{1}$–OT The general idea of this new protocol is to repeat Protocol $\binom{2}{1}$–$\widehat{OT}$ several times for random $b_{\ell,0}, b_{\ell,1}$ and c_ℓ and combine these instances in such a way to prevent $\mathcal{A}$'s cheating as above.

More precisely, Protocol $\binom{2}{1}$–$\widehat{OT}$ is repeated n^2 times. We combine the n^2 instances of $\binom{2}{1}$–$\widehat{OT}$ in such a way that $\mathcal{A}$ must cheat in each instance if she wants to discover the value of c. Protocol $\widehat{OT}$ is used a total of $2n^3$ times. In order to obtain information $\mathcal{A}$ must send at least n^2 bad pairs in these protocols. This will make a statistical difference that will be detected with probability almost 1. If $\mathcal{A}$ uses less than n^2 bad pairs, she finds out nothing about c. Similarly, if $\mathcal{A}$ sends bad syndromes in protocol $\binom{2}{1}$–$\widehat{OT}$ with probability $1/2$ she will be detected by $\mathcal{B}$ because he reads according to a random choice. If she uses $O(n)$ such syndromes it is almost certain that $\mathcal{B}$ will detect her cheating.

Let n be an odd number. The instances are combined by requesting that
$$b_{\ell,0} \oplus b_{\ell,1} = b_0 \oplus b_1 \text{ for } 1 \leq \ell \leq n^2. \text{ Let } b_{0,0} = \bigoplus_{\ell=1}^{n^2} b_{\ell,0} \text{ and } b_{0,1} = \bigoplus_{\ell=1}^{n^2} b_{\ell,1}. \text{ These}$$
requirements cause that $\bigoplus_{\ell=1}^{n^2} b_{\ell,c_\ell} = b_{0,z}$ for $z = \bigoplus_{\ell=1}^{n^2} c_\ell$. Thus in order to find out which of $b_{0,0}$ or $b_{0,1}$ $\mathcal{B}$ is trying to get, $\mathcal{A}$ must find out all the c_ℓ.

Full Protocol Let γ be a number greater than 1 and n be an odd number. An extra index ℓ is added to each variable of the ℓth iteration of $\binom{2}{1}$–$\widehat{OT}$.

Protocol 4.3 ($\binom{2}{1}$–OT$(b_0, b_1)(c)$)

1: $\mathcal{A}$ picks n^2 random bits $b_{1,0}, b_{2,0}, ..., b_{n^2,0}$ and sets $b_{\ell,1} \leftarrow b_0 \oplus b_1 \oplus b_{\ell,0}$, for $1 \le \ell \le n^2$.

2: $\mathcal{B}$ picks n^2 random bits $c_1, c_2, ..., c_{n^2}$.

3: $\overset{n^2}{\underset{\ell=1}{\text{DO}}}$

   1. $\mathcal{A}$ runs $\binom{2}{1}$–$\widehat{\text{OT}}(b_{\ell,0}, b_{\ell,1})(c_\ell)$ with $\mathcal{B}$ who gets b'_ℓ,

   2. if $d_H(r_{\ell,I_{\ell,c_\ell}}, r'_{\ell,I_{\ell,c_\ell}}) > \gamma\frac{\varphi^2}{\epsilon}n$ then $\mathcal{B}$ aborts.

4: if $\left(\#\{\ell, i \mid r'_{\ell,i} \ne \varepsilon\} < 2\epsilon n^3 - \frac{(1-2\varphi)^2}{2}n^2 \right)$ then $\mathcal{B}$ aborts

else $\mathcal{B}$ computes and sends $c' \leftarrow c \oplus \left(\bigoplus_{\ell=1}^{n^2} c_\ell \right)$.

5: $\mathcal{A}$ computes and sends $\hat{b}_0 \leftarrow b_0 \oplus \left(\bigoplus_{\ell=1}^{n^2} b_{\ell,c'} \right)$ and $\hat{b}_1 \leftarrow b_1 \oplus \left(\bigoplus_{\ell=1}^{n^2} b_{\ell,\bar{c'}} \right)$

to $\mathcal{B}$.

6: $\mathcal{B}$ computes and returns $\hat{b}_c \oplus \left(\bigoplus_{\ell=1}^{n^2} b'_\ell \right)$.

Details of the Protocol The test of Step **3.2** is to decide if the syndrome sent by $\mathcal{A}$ was valid. The value $\gamma\frac{\varphi^2}{\epsilon}n$ is the scope of the decoding algorithm of the concatenated code. If the decoded word was further than this distance then clearly the syndrome was wrong. If the test of Step **4** is negative then $\mathcal{B}$ is almost certain that $\mathcal{A}$ has not cheated n^2 times over the $2n^3$ transmissions.

4.2 Analysis of the protocol

Let $z_{i,j} = \begin{cases} 0 \text{ if } r'_{i,j} = \varepsilon \\ 1 \text{ if } r'_{i,j} \ne \varepsilon \end{cases}$. When $\mathcal{A}$ sends valid pairs $r_{i,j}r_{i,j}$ in Protocol 4.3 clearly

we have $E\left(\sum_{i=1}^{n^2}\sum_{j=1}^{2n} z_{i,j} \right) = 2\epsilon n^3$. On the other hand, if $\mathcal{A}$ wants to take advantage of this kind of cheating, she must cheat in each of the n^2 iterations of the protocol (if not she will loose completely one of the c_ℓ and thus c). In that case

we get $E\left(\sum_{i=1}^{n^2}\sum_{j=1}^{2n} z_{i,j} \right) \le \epsilon(2n^3 - n^2) + (1-\epsilon)n^2 = 2\epsilon n^3 - (1-2\varphi)^2 n^2$.

Theorem 4. *There exists a constant $\rho < 1$ with the following properties: when $\mathcal{A}$ does not use "bad" pairs then*

$$\text{Prob}\left[\sum_{i=1}^{n^2}\sum_{j=1}^{2n} z_{i,j} < 2\epsilon n^3 - \frac{(1-2\varphi)^2}{2}n^2\right] < \rho^n$$

whereas, when she cheats n^2 times,

$$\text{Prob}\left[\sum_{i=1}^{n^2}\sum_{j=1}^{2n} z_{i,j} > 2\epsilon n^3 - \frac{(1-2\varphi)^2}{2}n^2\right] < \rho^n.$$

Proof (sketch). Follows from Bernstein's law of large numbers.

Thus, except with exponentially small probability, an honest $\mathcal{A}$ will pass the test of Step 4 while a dishonest $\mathcal{A}$ will fail that same test.

If $\mathcal{A}$ is honest, the probability that more than $\gamma \frac{\varphi^2}{\epsilon} n$ errors occur during transmission by accident is exponentially small. Thus an honest $\mathcal{A}$ who sends correct syndromes, is unlikely to fail the test of Step **3.2** while a dishonest $\mathcal{A}$ who deliberately sends a wrong syndrome will be detected with probability $1/2$, if $\mathcal{B}$ happens to use that syndrome at random.

Finally, for the same reasons discussed in Sect. 3.2, because of Privacy Amplification $\mathcal{B}$ cannot obtain information about both b_0 and b_1 through the instances of protocol $\binom{2}{1}$-$\widehat{\text{OT}}$.

5 Conclusion and Open Question

We have obtained two new protocols for the cryptographic primitives of Bit Commitment and One-out-of-Two Oblivious Transfer based on the existence of a BSC using Privacy Amplification. The protocol for BC requires $O(n)$ uses of the BSC, while the protocol for $\binom{2}{1}$-OT requires $O(n^3)$ uses of the BSC. If we combine these protocols with the protocol of Crépeau, van de Graaf and Tapp [8] for Private Multi-Party Computation to achieve any two-party function evaluation which requires $O(n^2)$ BCs and $O(n)$ $\binom{2}{1}$-OT per gate, we end up with a protocol requiring a total of $O(n^4)$ uses of the BSC per gate of the computation. Our main open question is to obtain $\binom{2}{1}$-OT with only $O(n^2)$ uses of the BSC and thus any two-party computation at a cost of $O(n^3)$ uses of the BSC per gate. Another open question is to find an equally efficient protocol for $\binom{2}{1}$-OT using a BS_ϵ for values of ϵ above 0.1982.

6 Acknowledgments

We thank Gilles Brassard, Jeroen van de Graaf, Joe Kilian, Ueli Maurer, Alain Tapp, and Louis Salvail for support, suggestions and comments on this work.

References

1. C.H. Bennett, G. Brassard, C. Crépeau, and U.M. Maurer. Generalized Privacy Amplification. *IEEE Transaction on Information Theory*, Volume 41, Number 6, November 1995, pp. 1915–1923.
2. C.H. Bennett, G. Brassard, C. Crépeau, and M.-H. Skubiszewska. Practical quantum oblivious transfer. In *Advances in Cryptology: Proceedings of Crypto '91*, Lecture Notes in Computer Science, Vol. 576, pages 351–366. Springer-Verlag, 1992.
3. M. Blum. Coin flipping by telephone. In *Proceedings of IEEE Spring Computer Conference*, pages 133–137. IEEE, 1982.
4. G. Brassard, D. Chaum, and C. Crépeau. Minimum disclosure proofs of knowledge. *Journal of Computer and System Sciences*, 37:156–189, 1988.
5. G. Brassard, C. Crépeau, R. Jozsa and D. Langlois, "A quantum bit commitment scheme provably unbreakable by both parties," *Proceedings of 34th IEEE Symposium on Foundations of Computer Science*, 1993, pp. 362–371.
6. J.L. Carter and M.N. Wegman, "Universal classes of hash functions", *Journal of Computer and System Sciences*, Vol. 18, 1979, pp. 143–154.
7. C. Crépeau. Equivalence between two flavours of oblivious transfers (abstract). In C. Pomerance, editor, *Advances in Cryptology: Proceedings of Crypto '87*, pages 350–354, Springer-Verlag, 1988.
8. C. Crépeau, J. van de Graaf and A. Tapp. Committed Oblivious Transfer and Private Multi-Party Computations. *Advances in Cryptology: Proceedings of Crypto '95*, August 1995, pp. 110–123.
9. C. Crépeau and J. Kilian. Achieving oblivious transfer using weakened security assumptions. In 29th *Symposium on Foundations of Computer Science*, pages 42–52. IEEE, 1988.
10. S. Even, O. Goldreich, and A. Lempel. A randomized protocol for signing contracts. In R.L. Rivest, A. Sherman, and D. Chaum, editors, *Proceedings CRYPTO 82*, pages 205–210, Plenum Press, New York, 1983.
11. Forney, G.D., *Concatenated Codes*, The M.I.T. Press, 1966.
12. O. Goldreich, S. Micali and A. Wigderson, *How to play any mental game, or: A completeness theorem for protocols with honest majority* In *Proc. 19th ACM Symposium on Theory of Computing*, pages 218–229, ACM, 1987.
13. O. Goldreich, S. Micali, and A. Wigderson. Proofs that yield nothing but their validity, or All languages in $\mathcal{NP}$ have zero-knowledge proof systems. *Journal of the ACM*, 38:691–729, 1991.
14. J. Kilian, *Founding cryptography on Oblivious transfer*, 20th ACM Symposium on Theory of Computation, 1988, pp. 20–31.
15. F.J. MacWilliams and N.J.A. Sloane. *The Theory of Error-Correcting Codes*. North-Holland, 1977.
16. M.O. Rabin, How to exchange secrets by oblivious transfer. Technical Memo TR-81, Aiken Computation Laboratory, Harvard University, 1981.
17. A. Rényi, *Probability Theory*, North Holland, 1970.
18. S. Wiesner. Conjugate coding. *SIGACT News*, 15(1):78–88, 1983. Manuscript written *circa* 1970, unpublished until it appeared in SIGACT News.
19. A.D. Wyner, "The wire-tap channel", *Bell System Technical Journal*, Vol. 54, no. 8, 1975, pp. 1355–1387.
20. YAO, A.C.-C., "Protocols for secure computations", In *Proceedings of the 23rd Annual IEEE Symposium on Foundations of Computer Science*, November 1982, pp. 160–164.

Rapid Demonstration of Linear Relations Connected by Boolean Operators

Stefan Brands

DigiCash, Kruislaan 419, NL-1098 VA Amsterdam, The Netherlands.
E-mail: brands@digicash.com

Abstract. Consider a polynomial-time prover holding a set of secrets. We describe how the prover can rapidly demonstrate any satisfiable boolean formula for which the atomic propositions are relations that are linear in the secrets, without revealing more information about the secrets than what is conveyed by the formula itself. Our protocols support many proof modes, and are as secure as the Discrete Logarithm assumption or the RSA/factoring assumption.

1 Introduction

Consider a polynomial-time prover that has committed to a vector of secrets and wants to demonstrate that the secrets satisfy some satisfiable formula from propositional logic, where the atomic propositions are relations that are linear in the secrets. An example formula is

$$((5x_1 - 3x_2 = 5) \text{ AND } (2x_2 + 3x_3 = 7)) \text{ OR } (\text{NOT}(x_1 + 4x_3 = 5)),$$

where $(x_1, \ldots, x_k)$ is the prover's vector of secrets. The prover does not want to reveal any more information about its secrets than what is conveyed by the formula itself. Can a truly practical protocol for this task be constructed?

In this paper we will show that truly practical protocols exist, assuming the intractability of the Discrete Logarithm problem or the RSA/factoring problem. Our protocols can be performed in all manner of proof modes, including four-move zero-knowledge proofs, three-move witness-hiding proofs, interactive or non-interactive signed proofs that are provably secure in the random oracle model, limited-show proofs, multi-prover proofs, and blinded and restrictively blinded signed proofs.

Our results are organized as follows. Section 2 discusses preliminary notions and reviews basic results. Related work is discussed in Section 3. In Section 4 we introduce our techniques for rapidly demonstrating linear relations connected by boolean operators. We conclude in Section 5.

2 Preliminaries

Throughout this paper, the polynomial-time prover and the (not necessarily polynomial-time) verifier are denoted by $\mathcal{P}$ and $\mathcal{V}$, respectively. The symbol "$\leftarrow$" is used to denote assignment, and $|\cdot|$ denotes binary length. The symbol "$\in_{\mathcal{R}}$" and the word "random" indicate an independent and uniformly random

W. Fumy (Ed.): Advances in Cryptology - EUROCRYPT '97, LNCS 1233, pp. 318-333, 1997.

choice, and we allow distributions that are computationally indistinguishable for polynomially bounded $\mathcal{V}$ and statistically indistinguishable for unbounded $\mathcal{V}$. Whenever we say that $\mathcal{P}$ is able to prove knowledge, we imply the existence of a knowledge extractor that outputs a witness when having oracle access to $\mathcal{P}$.

Our techniques can be based either on the Discrete Logarithm assumption or on the RSA/factoring assumption. We now discuss preliminary notions and basic cryptographic results for these two settings.

2.1 Discrete Logarithm Setting

Set-up. $\mathcal{P}$ and $\mathcal{V}$ initially agree on a cyclic group of order q, denoted by G_q, where q is an integer. Efficient algorithms must be available for recognizing, testing equivalence of, and multiplying numbers in G_q. Without loss of generality it is assumed that q uniquely identifies G_q. Additionally, $k \geq 1$ generators, $g_1, \ldots, g_k$, of G_q, are agreed on; we call $(g_1, \ldots, g_k)$ a generator-tuple. From now on, an integer in the Discrete Log setting is said to be "small" if it is polynomial in $|q|$, and "large" otherwise.

Using the terminology of [6, 7], a representation of a number $h \in G_q$ with respect to $(g_1, \ldots, g_k)$ is a vector of numbers, $(x_1, \ldots, x_k)$, such that

$$h = \prod_{i=1}^{k} g_i^{x_i},$$

where $x_1, \ldots, x_k$ are in $\mathbb{Z}_q$.

Intractability of Collision-Finding. For the security of $\mathcal{V}$ it is important that $\mathcal{P}$ cannot know more than one representation of the same number, since it serves as a commit on $\mathcal{P}$'s secrets. For the purpose of the following proposition, which has been proved by Chaum, van Heijst and Pfitzmann [14] for constant k, and by [6, page 16] more generally[1] for all small k, we assume that q is generated according to a probabilistic polynomial-time algorithm (the "DL-instance generator") that, on input a security parameter, outputs a triple (q, g, h), where g and h are generators of G_q.

Proposition 1. *Consider the case that q, as output by the DL-instance generator, is always a prime, and k is small. Assuming that the Discrete Logarithm problem is intractable over the DL-instance generator, there cannot exist a polynomial-time algorithm that, on input q (having a distribution that is indistinguishably close to that of q induced by the DL-instance generator) and a randomly chosen generator-tuple $(g_1, \ldots, g_k)$, outputs with non-negligible probability of success a number $h \in G_q$ and two different representations of h.*

---

[1] Bellare, Goldreich and Goldwasser [2] noted that the reduction can be modified to achieve a success probability for the Discrete-Logarithm finder that is within a constant factor of that of the collision-finding oracle, instead of being inversely proportionate to k. Specifically, their modification achieves a constant factor $1/2$, instead of $2/k$. Note, however, that the optimization mentioned in [6, page 17] already achieves this (the constant factor is $1/2 + 1/(2k)$).

In case the invulnerable DL-instance generator outputs composite q's, it may be easy to find collisions. In particular, as noted by Chaum et al. [13, page 13], if r is a small prime factor of q then one can easily find collisions, by raising the generators in the generator-tuple to the power q/r and computing their relative Discrete Logarithms in the subgroup of order r. As in Stinson [32, page 239], one can alleviate this situation if q/r is a large prime (or a composite that is infeasible to factor) by restricting the elements $x_1, \ldots, x_k$ of a representation to be in $\mathbb{Z}_{q/r}$. Alternatively, we can consider a DL-instance generator that outputs q's that have only large prime factors; finding collisions then requires one to break the Discrete Logarithm problem in G_q or to factor q. In addition, as in Brickell and McCurley [12], one can let the g_i's be generators of a non-trivial subgroup of G_q.

To guarantee $\mathcal{V}$'s security one should generate the elements of the set-up in accordance with the probability distributions of the appropriate DL-instance generator, depending on which form of q's one is interested in. The set-up should be generated by $\mathcal{V}$ itself, in a mutually random fashion between $\mathcal{V}$ and $\mathcal{P}$, by a party trusted by $\mathcal{V}$ or liable for security breaks by $\mathcal{P}$, or in any other manner that ensures that $\mathcal{P}$ cannot find collisions for the generated instance.

Proving Knowledge. Our results in Section 4 can be based on *any* proof of knowledge (see Bellare and Goldreich [1]) of a representation. For practical purposes we are interested in highly efficient protocols that offer a wide range of *proof modes*. The following generic protocol enables $\mathcal{P}$, for any m with $1 \leq m \leq k$, to demonstrate knowledge of a representation, $(x_1, \ldots, x_m)$ (its *secret key*), of a number $h \in G_q$ (its *public key*) with respect to a generator-tuple, $(g_1, \ldots, g_m)$. We assume for the moment that q is a prime.

Step 1. $\mathcal{P}$ generates at random m numbers $w_1, \ldots, w_m \in_\mathcal{R} \mathbb{Z}_q$, and sends $a \leftarrow \prod_{i=1}^m g_i^{w_i}$ to $\mathcal{V}$.

Step 2. $\mathcal{P}$ computes m responses, responsive to a challenge $c \in \mathbb{Z}_{2^t}$, according to $r_i \leftarrow cx_i + w_i \bmod q$, for $i = 1, \ldots, m$, and sends them to $\mathcal{V}$. The process of generating c and the size of t determine the proof mode of the protocol; in the appendix several proof modes of particular relevance are discussed.

Step 3. $\mathcal{V}$ accepts if and only if $h^{-c} \prod_{i=1}^m g_i^{r_i} = a$.

One can also consider the above protocol for q's that are not prime, and in particular for all the forms discussed in the preceding subsubsection. Of course, this has ramifications with respect to the proof modes and/or the intractability assumptions needed for security.

Rapid Computations. Rapid evaluation of $g_1^{x_1} \cdots g_m^{x_m}$ can be performed using simultaneous repeated squaring (see Knuth [23, exercise 27, page 465]). For efficiency one can, for all 2^m subsets of $\{g_1, \ldots, g_m\}$, precompute the product of the g_i's in the subset, and store the products in a table. With $1 < l \leq m$, the product $\prod_{i=1}^m g_i^{x_i}$ can be computed using $d = \lceil m/l \rceil$ precomputed tables, using simultaneous repeated squaring for each of the d sub-products and multiplying the sub-product results. Several variations and optimizations of this basic technique are known in the literature. For example, one can process $j > 1$ exponent bits at once; the size of the precomputed table then increases by a factor $2^{(j-1)m}$, while the workload decreases by approximately a factor j.

2.2 RSA Setting

Set-up. $\mathcal{P}$ and $\mathcal{V}$ initially agree on a group $\mathbb{Z}_n^*$, where $n = pq$ and p and q are distinct primes. They also agree on an integer, v. Additionally, $k \geq 1$ numbers, $g_1, \ldots, g_k$, all in $\mathbb{Z}_n^*$, are agreed on. From now on, an integer in the RSA setting is said to be "small" if it is polynomial in $|n|$, and "large" otherwise.

A *representation* of a number h in $\mathbb{Z}_n^*$, *with respect to* $(g_1, \ldots, g_k, v)$, is a vector of numbers, $(x_1, \ldots, x_k, x_{k+1})$, such that

$$h = (\prod_{i=1}^{k} g_i^{x_i}) x_{k+1}^v \bmod n,$$

where x_{k+1} is in $\mathbb{Z}_n^*$ and $x_1, \ldots, x_k$ are in $\mathbb{Z}_v$.

Intractability of Collision-Finding. For the security of $\mathcal{V}$ it is important that at least $\mathcal{P}$ cannot know more than one representation of the same number. For the purpose of the following two propositions, we assume that n is generated according to a probabilistic polynomial-time algorithm (the "RSA-instance generator") that, on input a security parameter and an integer v, outputs a pair $(n, y \in \mathbb{Z}_n^*)$. For any integer $v \geq 2$, we can consider the problem of extracting v-th roots modulo n; this is called the RSA problem for that particular v.

Proposition 2. *Suppose that v is a prime that is co-prime to $\varphi(n)$, and that k is small. Assuming that the RSA problem for v is intractable over the RSA-instance generator, there cannot exist a polynomial-time algorithm that, on input n (having a distribution that is indistinguishably close to that of n induced by the RSA-instance generator) and randomly chosen $(g_1, \ldots, g_k)$, outputs with non-negligible probability of success a number $h \in \mathbb{Z}_n^*$ and two different representations of h with respect to $(g_1, \ldots, g_k, v)$.*

Sketch of proof. To compute $y^{1/v} \bmod n$, on input $y \in \mathbb{Z}_n^*$, construct each g_i as $y^{r_i} s_i^v \bmod n$, for $r_i \in_{\mathcal{R}} \mathbb{Z}_v$ and $s_i \in_{\mathcal{R}} \mathbb{Z}_n^*$. If the oracle output is correct, a relation of the form $y^t = u^v \bmod n$ can be computed, for known $t \in \mathbb{Z}_v$ and $u \in \mathbb{Z}_n^*$, and from this $y^{1/v} \bmod n$ can be computed.

Note that if p and q are random primes of equal size, then a random element in $\mathbb{Z}_n^*$ has small order with negligible probability; see Håstad, Schrift and Shamir [21, Proposition 1].

Other choices of v are possible as well. For example, in case v is small and *not* co-prime to $\varphi(n)$, according to Ohta and Okamoto [25, Theorem 1] it is as hard to compute v-th roots as to factor the modulus. By restricting the g_i's and x_{k+1} in the definition of a representation to v-th residues, one can prove the difficulty of finding collisions for v's of this particular form in a likewise manner. The following result shows that this also holds for large v of a special form.

Proposition 3. *Consider the case in which $v = 2^l$, for any integer l, and the RSA-instance generator always outputs Blum integers (i.e., p and q are congruent to 3 mod 4). Furthermore, restrict the number x_{k+1} in a representation to be a quadratic residue. Assuming that the factoring problem is intractable over the*

RSA-instance generator, there cannot exist a polynomial-time algorithm that, on input n (having a distribution that is indistinguishably close to that of n induced by the RSA-instance generator) and randomly chosen quadratic residues $(g_1, \ldots, g_k)$, *outputs with non-negligible probability of success a number* $h \in \mathbb{Z}_n^*$ *and two different representations of h with respect to* $(g_1, \ldots, g_k, v)$.

Proving Knowledge. Our results in Section 4 can be based on any proof of knowledge of a representation. The following generic protocol is very efficient and offers a wide range of proof modes. For any m with $0 \le m \le k$, the protocol enables $\mathcal{P}$ to demonstrate knowledge of a representation, $(x_1, \ldots, x_m, x_{m+1})$, of a number $h \in \mathbb{Z}_n^*$ with respect to $(g_1, \ldots, g_m, v)$. For the moment, we assume that v is a prime that is co-prime to $\varphi(n)$.

Step 1. $\mathcal{P}$ generates at random m numbers $w_1, \ldots, w_m \in_\mathcal{R} \mathbb{Z}_v$, and a number $w_{m+1} \in_\mathcal{R} \mathbb{Z}_n^*$. $\mathcal{P}$ computes $a \leftarrow g_1^{w_1} \cdots g_m^{w_m} w_{m+1}^v \bmod n$, and sends a to $\mathcal{V}$.

Step 2. $\mathcal{P}$ computes $m+1$ responses, responsive to a challenge $c \in \mathbb{Z}_{2^t}$, according to $r_i \leftarrow cx_i + w_i \bmod v$, for $1 \le i \le m$, and

$$r_{m+1} \leftarrow (\prod_{i=1}^{m} g_i^{cx_i + w_i \text{ div } v}) \cdot x_{m+1}^c w_{m+1} \bmod n,$$

and sends them to $\mathcal{V}$.

Step 3. $\mathcal{V}$ accepts if and only if $(\prod_{i=1}^{m} g_i^{r_i}) \cdot r_{m+1}^v = h^c a \bmod n$.

All the proof modes for the proof of knowledge discussed in the Discrete Logarithm setting apply here as well, with the obvious modifications. If $m = 0$ we have the Guillou-Quisquater protocol [20] and if $m = 1$ the Okamoto protocol [26, page 39].

By making minor adjustments to the above protocol, we can use v's that are not prime and/or not co-prime to $\varphi(n)$. In all these cases, one has to restrict the set from which the g_i's and x_{m+1} and w_{m+1} are chosen, to avoid leakage of information about $\mathcal{P}$'s representation; similar adjustments as discussed in the preceding subsubsection can be made. Note, however, that if v is a large composite with a small prime factor, u, and it is feasible to randomly generate u-th residues without knowing a u-th root, then $\mathcal{P}$ can convince $\mathcal{V}$ in the three-move protocol with non-negligible success probability (specifically, $1/u$ if v/u has no small prime factors, and larger otherwise) without knowing a representation of h with respect to $(g_1, \ldots, g_k, v)$; for these v's, another protocol should be used.

3 Related Work

A constant-round zero-knowledge argument for our task can be constructed by properly reducing the boolean formula that is to be demonstrated to an instance of the NP-complete language Directed Hamiltonian Cycle, and applying the zero-knowledge argument of knowledge of Feige and Shamir [18]. However, techniques such as this are not practical, because they amount to encoding the statement into a boolean circuit and using commitments for each gate.

By restricting q in the Discrete Logarithm setting to be a prime, one can define a relation, $R = R_{q,(g_1,\ldots,g_k),(\alpha_1,\ldots,\alpha_k)}$, for any q, for any generator-tuple $(g_1,\ldots,g_k)$ and any vector of coefficients $(\alpha_1,\ldots,\alpha_k) \in (\mathbb{Z}_q)^k \setminus \{0\}^k$, as follows:

$$((h,b),(x_1,\ldots,x_k)) \in R \quad \Leftrightarrow \quad h = \prod_{i=1}^{k} g_i^{x_i} \text{ and } b = \sum_{i=1}^{k} \alpha_i x_i \bmod q$$

The corresponding language is easily seen to be random self-reducible. In the RSA setting, for v a prime that is co-prime to $\varphi(n)$, a random self-reducible language can be defined in a similar manner. Now, by applying the construction of Tompa and Woll [33] one gets a perfect zero-knowledge proof of knowledge for both languages, but these protocols use binary-valued challenges and require polynomially many rounds. For the special case $k = 3$ and $b = x_1 + x_2 + mx_3 \bmod q$, and b is an undeniable signature of $\mathcal{P}$ on a message m of the (unlimited powerful) $\mathcal{V}$, this construction has been used for signature confirmation by Chaum et al. [14]. We remark that it is straightforward to improve the protocols, by using a large challenge domain and prepending a move in which the verifier commits to its challenge (note that this improvement has been overlooked by Chaum et al. [14]), but the resulting protocols remain less efficient than ours. Moreover, our protocols facilitate many other proof modes.

De Santis, Di Crescenzo, Persiano and Yung [17] show how to prove any monotone formula over a random self-reducible language (Cramer, Damgård and Schoenmakers [15] independently discovered virtually the same technique). If a monotone formula has m logical connectives, then this technique requires the prover to perform m proofs of knowledge, one for each sub-formula. In contrast, our "AND" technique has the property that the communication complexity for both the prover and the verifier slightly *decreases* as the number of "AND" connectives increases, and the computational complexity is virtually unaffected. Moreover, the technique of De Santis et al. uses binary-valued challenges, and thus polynomially many repetitions are needed.

Furthermore, the technique of De Santis et al. and Cramer et al. applies only to monotone boolean formula, while we have a very efficient "NOT" technique. Chaum et al. [14] describe a perfect zero-knowledge protocol for the "NOT" of their special relation, $b = x_1 + x_2 + mx_3 \bmod q$, but this protocol inherently works for binary-valued challenges only. Moreover, in each iteration the signer must compute seven commitments, requiring many exponentiations in G_q, and it is unclear how to efficiently construct other proof modes.

Another important difference is in the scenario that is considered. Namely, De Santis et al. and Cramer et al. consider a situation in which there are many different public keys, and $\mathcal{P}$ demonstrates (in zero-knowledge) that it knows the secret keys corresponding to some of these. In contrast, we are concerned with the situation in which the prover knows a *single* public key, and demonstrates that its secret key satisfies a certain formula.

4 Demonstrating Boolean Formulae for Linear Relations

In this section we describe our proof techniques for "AND," "NOT" and "OR" connectives, respectively, and then show how to combine them in order to demonstrate arbitrary boolean formulae. Without loss of generality we base our discussions on the Discrete Logarithm setting, and for the RSA setting describe only the necessary adaptations. Note that if $k = 1$, then $\mathcal{V}$ can verify for any boolean formula directly whether the secret of $\mathcal{P}$ satisfies it, and so from now on we only consider the case $k \geq 2$.

4.1 Formulae with only "AND" connectives

We first consider the situation in which $\mathcal{P}$ has to demonstrate a satisfiable formula with zero or more "AND" connectives. At the outset, $\mathcal{P}$ has committed to a set of secrets, $(x_1, \ldots, x_k)$, by sending a number $h \in G_q$ to $\mathcal{V}$, where $(x_1, \ldots, x_k)$ is a representation of h with respect to $(g_1, \ldots, g_k)$. Without loss of generality, we assume that $\mathcal{P}$ has to demonstrate to $\mathcal{V}$ that this representation satisfies the following system of $l \geq 1$ independent linear relations:

$$
\begin{pmatrix}
\alpha_{11} & \cdots & \alpha_{1,k-l} & 1 & 0 & \cdots & 0 \\
\alpha_{21} & \cdots & \alpha_{2,k-l} & 0 & 1 & \cdots & 0 \\
\vdots & \vdots & \vdots & \vdots & \vdots & \ddots & \vdots \\
\alpha_{l1} & \cdots & \alpha_{l,k-l} & 0 & 0 & \cdots & 1
\end{pmatrix}
\begin{pmatrix}
x_{\pi(1)} \\
x_{\pi(2)} \\
\vdots \\
x_{\pi(k)}
\end{pmatrix}
=
\begin{pmatrix}
b_1 \\
b_2 \\
\vdots \\
b_l
\end{pmatrix}
\bmod q. \tag{1}
$$

The coefficients α_{ij}, for $1 \leq i \leq l$ and $1 \leq j \leq k - l$, are elements of $\mathbb{Z}_q$, and $\pi(\cdot)$ is a permutation of $\{1, \ldots, k\}$. The corresponding boolean formula is:

$$
(b_1 = \alpha_{11} x_{\pi(1)} + \cdots + \alpha_{1,k-l} x_{\pi(k-l)} + x_{\pi(k-l+1)} \bmod q) \text{ AND } \ldots
$$
$$
\ldots \text{ AND } (b_l = \alpha_{l1} x_{\pi(1)} + \cdots + \alpha_{l,k-l} x_{\pi(k-l)} + x_{\pi(k)} \bmod q). \tag{2}
$$

Note that the atomic proposition is the special case $l = 1$.

Our technique for demonstrating formula (2) is based on the following result.

Proposition 4. *$\mathcal{P}$ can demonstrate knowledge of a representation of*

$$
h \Big(\prod_{i=1}^{l} g_{\pi(k-l+i)}^{b_i} \Big)^{-1}
$$

with respect to

$$
\Big(g_{\pi(1)} \prod_{i=1}^{l} g_{\pi(k-l+i)}^{-\alpha_{i1}}, \quad \ldots, \quad g_{\pi(k-l)} \prod_{i=1}^{l} g_{\pi(k-l+i)}^{-\alpha_{i,k-l}} \Big)
$$

if and only if it knows a set of secrets that satisfies the formula (2).

The proof follows straightforwardly, by considering the relations that are satisfied by the output of the knowledge extractor. Note that the tuple in Proposition 4 is a generator-tuple with overwhelming probability in case q is a prime and the prover selects the matrix entries, α_{ij}, and can always be guaranteed to be so when the matrix entries are determined by $\mathcal{V}$.

We can efficiently implement the protocol by using the proof of knowledge for Discrete Logarithm representations described in Section 2. An important benefit of using this protocol is that one can expand the resulting expressions, so that $\mathcal{P}$ and $\mathcal{V}$ can use a single precomputed table for simultaneous repeated squaring, *independent* of the particular formula that is demonstrated. The resulting (generic) protocol steps are as follows:

Step 1. $\mathcal{P}$ generates at random $k - l$ numbers, $w_1, \ldots, w_{k-l} \in_{\mathcal{R}} \mathbb{Z}_q$, and computes

$$a \leftarrow \prod_{i=1}^{k-l} g_{\pi(i)}^{w_i} \prod_{i=1}^{l} g_{\pi(k-l+i)}^{-\sum_{j=1}^{k-l} \alpha_{ij} w_j}.$$

$\mathcal{P}$ then sends a to $\mathcal{V}$.

Step 2. $\mathcal{P}$ computes a set of responses, responsive to a challenge number c in $\mathbb{Z}_{2^t}$, as follows:

$$r_i \leftarrow c x_{\pi(i)} + w_i \bmod q, \quad \forall i \in \{1, \ldots, k-l\}.$$

$\mathcal{P}$ then sends $(r_1, \ldots, r_{k-l})$ to $\mathcal{V}$.

Step 3. $\mathcal{V}$ computes

$$r_{k-l+i} \leftarrow c b_i - \sum_{j=1}^{k-l} \alpha_{ij} r_j \bmod q, \quad \forall i \in \{1, \ldots, l\}.$$

and accepts if and only if

$$a = h^{-c} \prod_{i=1}^{k} g_{\pi(i)}^{r_i}.$$

The particular proof mode of the protocol is "inherited" from the mode in which the underlying proof of knowledge is performed, and a further discussion is therefore omitted here. Note, however, that special care must be taken for signed proofs: the transcript of a protocol execution is always convincing of the fact that $\mathcal{P}$ knows a set of secrets corresponding to h, but convinces of its conformity with the demonstrated formula only when a uniquely identifying description of the demonstrated formula is hashed along (or when the α_{ij}'s and the b_i's are all restricted to sets that are negligible in the range of the hash function).

To base the proof on the RSA/factoring problem, consider $\mathcal{P}$ having to prove the system of linear relations (1), but with "mod v" replacing "mod q." We assume that $\mathcal{P}$ has committed to a set of secrets, $(x_1, \ldots, x_k)$, by sending $h \leftarrow g_1^{x_1} \cdots g_k^{x_k} x_{k+1}^v \bmod n$ to $\mathcal{V}$, for some x_{k+1} in $\mathbb{Z}_n^*$.

Proposition 5. *For any integer $v \geq 2$, $\mathcal{P}$ can prove knowledge of a representation of*

$$h \left(\prod_{i=1}^{l} g_{\pi(k-l+i)}^{b_i} \right)^{-1} \bmod n$$

with respect to

$$\left(g_{\pi(1)} \prod_{i=1}^{l} g_{\pi(k-l+i)}^{-\alpha_{i1}} \bmod n, \; \ldots, \; g_{\pi(k-l)} \prod_{i=1}^{l} g_{\pi(k-l+i)}^{-\alpha_{i,k-l}} \bmod n, \; v \right).$$

if and only if it knows a set of secrets that satisfies the formula.

By using the efficient proof of knowledge for the RSA setting described in Section 2, again expanding the resulting expressions, a single precomputed table can be used for simultaneous repeated squaring; of course, one then also inherits the limitations in the range of choices for v.

4.2 Formulae with only "NOT" connectives

We next study the situation in which $\mathcal{P}$ has to demonstrate that a linear relation does *not* hold, without revealing more information than required. The situation at the outset is as in Subsection 4.1. This time, $\mathcal{P}$ has to demonstrate to $\mathcal{V}$ that its representation satisfies the formula

$$\text{NOT}\left(x_{\pi(1)} = \alpha_1 + \alpha_2 x_{\pi(2)} + \cdots + \alpha_k x_{\pi(k)} \bmod q\right). \tag{3}$$

The coefficients α_i, for $1 \leq i \leq k$, are elements of $\mathbb{Z}_q$. Clearly, the permutation $\pi(\cdot)$ can always be defined to interchange at most two elements and leave the rest unchanged.

Our technique for demonstrating formula (3) is based on the following result.

Proposition 6. *Let q be a prime. $\mathcal{P}$ can prove knowledge of a representation of $g_{\pi(1)}$ with respect to*

$$\left(g_{\pi(1)}^{\alpha_1} h^{-1}, g_{\pi(1)}^{\alpha_2} g_{\pi(2)}, \ \ldots, \ g_{\pi(1)}^{\alpha_k} g_{\pi(k)}\right)$$

if and only if it knows a set of secrets that satisfies the formula (3).

Sketch of proof. With $(y_1, \ldots, y_k)$ denoting the representation output by the knowledge extractor, if $y_1 = 0$ then a non-trivial representation of 1 has been found and hence the Discrete Logarithm problem is tractable, and if $y_1 \neq 0$ then the representation satisfies formula (3).

Proposition 7. *If the proof of knowledge performed by $\mathcal{P}$ in the preceding Proposition is witness indistinguishable, then it is impossible for $\mathcal{V}$ (even with unlimited computing power) to learn any information about the difference between $x_{\pi(1)}$ and $\alpha_1 + \alpha_2 x_{\pi(2)} + \cdots + \alpha_k x_{\pi(k)} \bmod q$.*

Sketch of proof. Denoting the representation known to $\mathcal{P}$ by $(z_1, \ldots, z_k)$ and the difference by ϵ, observe that $z_1 = 1/\epsilon \bmod q$, and so information about ϵ is leaked if and only if information about z_1 is leaked. Since $k \geq 2$, for each $z_1 \in \mathbb{Z}_q$ there is a representation containing that z_1; and because there are equally many (namely, q^{k-2}) such representations for each z_1 and the protocol is witness-indistinguishable, no information about ϵ leaks.

If q is not a prime, then the inverse of the difference number, ϵ, is not guaranteed to exist. If q is a composite that is hard to factor then zero-divisors cannot be found and so nothing is lost, and in other cases we can force the existence of an inverse by making additional assumptions about the coefficients in (3) and/or about the representation of $\mathcal{P}$.

Applying the efficient proof of knowledge for the Discrete Logarithm setting described in Section 2, the following practical protocol results:

Step 1. $\mathcal{P}$ generates at random k numbers, $w_1, \ldots, w_k \in_{\mathcal{R}} \mathbb{Z}_q$, and computes

$$a \leftarrow h^{-w_1} g_{\pi(1)}^{\sum_{i=1}^{k} \alpha_i w_i} \prod_{i=2}^{k} g_{\pi(i)}^{w_i}.$$

$\mathcal{P}$ then sends a to $\mathcal{V}$.

Step 2. Let ϵ denote $(\alpha_1 + \sum_{i=2}^{k} \alpha_i x_{\pi(i)}) - x_{\pi(1)}$ mod q, and let $\delta = \epsilon^{-1}$ mod q. $\mathcal{P}$ computes a set of responses, responsive to a challenge number c in $\mathbb{Z}_{2^t}$, as follows:

$$r_1 \leftarrow c\delta + w_1 \bmod q,$$
$$r_i \leftarrow cx_{\pi(i)}\delta + w_i \bmod q, \quad \forall i \in \{2, \ldots, k\}.$$

$\mathcal{P}$ then sends $(r_1, \ldots, r_k)$ to $\mathcal{V}$.

Step 3. $\mathcal{V}$ accepts if and only if

$$a = h^{-r_1} g_{\pi(1)}^{-c+\sum_{i=1}^{k} \alpha_i r_i} \prod_{i=2}^{k} g_{\pi(i)}^{r_i}.$$

As before, this protocol inherits the proof modes of the protocol described in Section 2. Note that signed proofs convince only of the demonstrated formula if the α_i's are hashed along or if they are restricted to be in small sets.

Our technique can also be based on the RSA/factoring problem. Consider $\mathcal{P}$ having to prove formula (3), with "mod v" replacing "mod q," and having committed to $(x_1, \ldots, x_k)$ using $h \leftarrow g_1^{x_1} \cdots g_k^{x_k} x_{k+1}^v \bmod n$, for x_{k+1} in $\mathbb{Z}_n^*$.

Proposition 8. *If v is a prime (or a composite that is hard to factor), then $\mathcal{P}$ can prove knowledge of a representation of $g_{\pi(1)}$ with respect to*

$$\left(g_{\pi(1)}^{\alpha_1} h^{-1} \bmod n, g_{\pi(1)}^{\alpha_2} g_{\pi(2)} \bmod n, \ldots, g_{\pi(1)}^{\alpha_k} g_{\pi(k)} \bmod n, v\right),$$

if and only if it knows a set of secrets that satisfies the formula.

Of course, if $v = 2$ then all boolean formula are monotone and one can do without this technique.

4.3 Formulae with only "OR" connectives

We now show how $\mathcal{P}$ can demonstrate that at least one of two linear relations holds, without revealing which one; this technique is an application of the "OR" technique of De Santis et al. and Cramer et al., although the scenario is different. The situation at the outset is again as in Subsection 4.1. This time $\mathcal{P}$ has to demonstrate to $\mathcal{V}$ that the representation known to it satisfies the formula

$$(x_{\pi(1)} = \alpha_1 + \sum_{i=2}^{k} \alpha_i x_{\pi(i)} \bmod q) \text{ OR } (x_{\rho(1)} = \beta_1 + \sum_{i=1}^{k} \beta_i x_{\rho(i)} \bmod q). \quad (4)$$

The coefficients α_i and β_i, for $1 \leq i \leq k$, are elements of $\mathbb{Z}_q$, and $\pi(\cdot)$ and $\rho(\cdot)$ are permutations of $\{1, \ldots, k\}$ that can always be defined to interchange at most two elements each.

If (and only if) the first linear relation holds, then $\mathcal{P}$ can compute, for any challenge c_1, responses $(r_2, \ldots, r_k)$ such that

$$a_1 = h^{-c_1} g_{\pi(1)}^{c_1 \alpha_1 + \sum_{i=2}^{k} \alpha_i r_i} g_{\pi(2)}^{r_2} \cdots g_{\pi(k)}^{r_k},$$

where

$$a_1 = g_{\pi(1)}^{\sum_{i=2}^{k} \alpha_i w_i} \prod_{i=2}^{k} g_{\pi(i)}^{w_i}$$

for random $w_2, \ldots, w_k$ in $\mathbb{Z}_q$. Likewise, if (and only if) the second linear relation holds, then $\mathcal{P}$ can compute, for any challenge c_2, responses $(s_2, \ldots, s_k)$ such that

$$a_2 = h^{-c_2} g_{\rho(1)}^{c_2 \beta_1 + \sum_{i=2}^{k} \beta_i s_i} g_{\rho(2)}^{s_2} \cdots g_{\rho(k)}^{s_k},$$

where

$$a_2 = g_{\rho(1)}^{\sum_{i=2}^{k} \beta_i v_i} g_{\rho(2)}^{v_2} \cdots g_{\rho(k)}^{v_k},$$

for random $v_2, \ldots, v_k$ in $\mathbb{Z}_q$. To demonstrate formula (4), we have $\mathcal{P}$ choose one of the two challenges, c_1 or c_2, at random by itself, so that it can anticipate that challenge by calculating a suitable a_i from the self-chosen challenge and a set of randomly self-chosen "responses." To ensure that $\mathcal{P}$ cannot choose the other challenge by itself as well, $\mathcal{P}$ must use challenges c_1 and c_2 such that, say, the bitwise exclusive-or of c_1 and c_2 is equal to the supplied challenge, c. (Of course, "simulation" is needed only for those sub-formulae that do not hold; if both sub-formulae would be true, $\mathcal{P}$ can do without a self-chosen challenge.)

This technique can straightforwardly be generalized to a formula with more than one "OR" connective, and as before an efficient implementation can be obtained by using the proof of knowledge of Section 2. A description based on the RSA/factoring problem is straightforward, and hence omitted.

4.4 Putting it all together

We now show how to combine the basic demonstration techniques, in order to demonstrate arbitrary satisfiable formulae from propositional logic, where the atomic propositions are linear relations over $\mathbb{Z}_q$. We hereto first show how to combine the techniques of Subsections 4.1 and 4.2 in order to demonstrate any satisfiable formula from propositional logic that has zero or more "AND" connectives and at most one "NOT" connective; these formulae play a central role in combining the basic techniques.

A consistent system consisting of linear relations and one linear inequality can be written as a system of linear relations by introducing a difference term, denoted by ϵ. By appropriate substitution, the system can then be represented by the matrix equation

$$\begin{pmatrix} \alpha_{11} & \cdots & \alpha_{1,k-l} & 1 & 0 & \cdots & 0 \\ \alpha_{21} & \cdots & \alpha_{2,k-l} & 0 & 1 & \cdots & 0 \\ \vdots & & \vdots & \vdots & \vdots & \ddots & \vdots \\ \alpha_{l1} & \cdots & \alpha_{l,k-l} & 0 & 0 & \cdots & 1 \end{pmatrix} \begin{pmatrix} x_{\pi(1)} \\ x_{\pi(2)} \\ \vdots \\ x_{\pi(k)} \end{pmatrix} = \begin{pmatrix} b_1 - f_1 \epsilon \\ b_2 - f_2 \epsilon \\ \vdots \\ b_l - f_l \epsilon \end{pmatrix} \bmod q, \qquad (5)$$

where $f_1, \ldots, f_l$ are numbers in $\mathbb{Z}_q$. (Clearly, one of the f_i's can always be 1.)

Our technique for demonstrating the boolean formula that corresponds to the system (5) is based on the following result.

Proposition 9. *$\mathcal{P}$ can prove knowledge of a representation of*

$$\prod_{i=1}^{l} g_{\pi(k-l+i)}^{f_i}$$

with respect to

$$\left(h^{-1} \prod_{i=1}^{l} g_{\pi(k-l+i)}^{b_i}, \quad g_{\pi(1)} \prod_{i=1}^{l} g_{\pi(k-l+i)}^{-\alpha_{i1}}, \quad \cdots, \quad g_{\pi(k-l)} \prod_{i=1}^{l} g_{\pi(k-l+i)}^{-\alpha_{i,k-l}} \right).$$

if and only if it knows a set of secrets that satisfies the system (5).

As in Proposition 7, ϵ is information-theoretically hidden if the proof is witness-indistinguishable, provided that $l < k$. If $k = l$, then $\mathcal{V}$ can check the validity of the system (5) directly from $\mathcal{P}$'s public key, without interacting with $\mathcal{P}$; computing ϵ then is as hard as breaking the Discrete Logarithm problem, and ϵ has at least $O(\log |q|)$ bits that are simultaneously hard-core.

We are now prepared to describe our general technique. Any boolean formula, F, can be expressed in the form

$$F = Q_1 \text{ AND } \cdots \text{ AND } Q_m, \tag{6}$$

where each sub-formula, Q_i, has the format R_{i1} OR $\cdots$ OR R_{i,m_i}, and each subsub-formulae, R_{ij}, is a formula from propositional logic that connects linear relations over $\mathbb{Z}_q$ by zero or more "AND" connectives, at most one "NOT" connective and no other logical connectives. We have just seen how to demonstrate R_{ij}, and by using the technique of Subsection 4.3 we can have $\mathcal{P}$ demonstrate a single sub-formula, Q_i. To prove the formula F, $\mathcal{P}$ needs to demonstrate the validity of all m sub-formulae, $Q_1, \ldots, Q_m$. Hereto the corresponding m proofs can all be performed in parallel, responsive to the same challenge.

An optimization is sometimes possible, depending on the complexity of F. Namely, a system of the form (5) can be interpreted as corresponding to an atomic proposition. To demonstrate knowledge for this atomic proposition, our techniques have $\mathcal{P}$ demonstrate knowledge of a secret key corresponding to a "distorted" public key, with respect to a "distorted" generator tuple. We can now apply the monotone formula technique of De Santis et al. and Cramer et al. to prove monotone boolean formulae over these atomic propositions. In particular, the restrictions according to which $\mathcal{P}$ generates its self-chosen challenges from the supplied challenge can be dictated in accordance with the secret-sharing construction of Benaloh and Leichter [4] for the access structure defined by the dual of the formula F (see Cramer et al. [15] for details). In other words, expressing F in a more compact form than (6) may lead to a more efficient protocol.

A further optimization is for $\mathcal{V}$ to batch-process verification relations that correspond to atomic formulae that are connected by "AND" operators; this can be done similarly to the technique of Naccache, M'Raïhi, Raphaeli and Vaudenay [24] for batch verification of DSA signatures.

A description of the above techniques based on the difficulty of factoring or computing RSA-roots poses no particular difficulties, and is hence omitted.

5 Conclusion

An interesting problem is to extend the set of atomic propositions beyond linear relations. True practicality requires constant-round proofs of knowledge for which the computation and communication complexity are linearly dependent on the number of secrets of $\mathcal{P}$ and the size of its public key, but independent of the parameters specifying the atomic proposition or anything else. The following approaches do not satisfy this criterion:

- The technique of Damgård [16] can be adapted in order to demonstrate atomic formulae of the form

$$x_1^{a_1} + \alpha_2 x_2^{a_2} + \cdots + \alpha_k x_k^{a_k} = \alpha_1 \bmod q,$$

  but this requires $\mathcal{P}$ to perform $O(\sum_{i=1}^{k} a_i)$ separate basic proofs of knowledge and proofs of equality of discrete logarithms;
- Brickell, Chaum, Dåmgard and Van de Graaf [11] showed how to prove that an exponent is in an interval, but their protocol inherently requires binary challenges (and thus polynomially many iterations), and moreover the proof must be performed for a substantially larger interval in order to avoid leakage of information; and
- The protocol of Pfitzmann [28] for demonstrating multiplications in zero-knowledge also inherently requires binary challenges.

Moreover, in all three cases the number of available proof modes is seriously limited. It is an open problem to construct truly practical protocols for atomic propositions of the above forms.

Our techniques have many practical applications. For example, they can be used to implement the confirmation and the disavowal protocols of Chaum et al. [14] more efficiently (the speed-up is polynomial). The main motivation, however, for devising the techniques in this paper has been to construct all manner of practical privacy-protecting credential mechanisms; this is the subject of a forthcoming paper.

References

1. M. Bellare and O. Goldreich. On defining proofs of knowledge. In E. F. Brickell, editor, *Advances in Cryptology–CRYPTO '92*, volume 740 of *Lecture Notes in Computer Science*, pages 390–420. Springer-Verlag, 1992.
2. M. Bellare, O. Goldreich, and S. Goldwasser. Incremental cryptography: The case of hashing and signing. In Y. G. Desmedt, editor, *Advances in Cryptology–CRYPTO '94*, volume 839 of *Lecture Notes in Computer Science*, pages 216–233. Springer-Verlag, 1994.
3. M. Bellare and P. Rogaway. Random oracles are practical: A paradigm for designing efficient protocols. In *First ACM Conference on Computer and Communications Security*, pages 62–73, Fairfax, 1993. ACM Press.
4. J. Benaloh and J. Leichter. Generalized secret sharing and monotone functions. In S. Goldwasser, editor, *Advances in Cryptology–CRYPTO '88*, volume 403 of *Lecture Notes in Computer Science*, pages 27–35. Springer-Verlag, 1988.

5. M. Blum, A. De Santis, S. Micali, and G. Persiano. Noninteractive zero-knowledge. *SIAM J. Computing*, 20(6):1084–1118, December 1991.
6. S. Brands. An efficient off-line electronic cash system based on the representation problem. Technical Report CS-R9323, Centrum voor Wiskunde en Informatica, April 1993.
7. S. Brands. Untraceable off-line cash in wallets with observers. In D. R. Stinson, editor, *Advances in Cryptology–CRYPTO '93*, volume 773 of *Lecture Notes in Computer Science*, pages 302–318. Springer-Verlag, 1994.
8. S. Brands. More on restrictive blind issuing of secret-key certificates in parallel mode. Technical Report CS-R9534, Centrum voor Wiskunde en Informatica, March 1995.
9. S. Brands. Restrictive blind issuing of secret-key certificates in parallel mode. Technical Report CS-R9523, Centrum voor Wiskunde en Informatica, March 1995.
10. S. Brands. Restrictive blinding of secret-key certificates. In L. C. Guillou and J.-J. Quisquater, editors, *Advances in Cryptology–EUROCRYPT '95*, volume 921 of *Lecture Notes in Computer Science*, pages 231–247. Springer-Verlag, 1995.
11. E. F. Brickell, D. Chaum, I. B. Damgård, and J. van de Graaf. Gradual and verifiable release of a secret. In C. Pomerance, editor, *Advances in Cryptology–CRYPTO '87*, volume 293 of *Lecture Notes in Computer Science*, pages 156–166. Springer-Verlag, 1988.
12. E. F. Brickell and K. S. McCurley. An interactive identification scheme based on discrete logarithms and factoring. *Journal of Cryptology*, 5(1):29–39, 1992.
13. D. Chaum, E. van Heijst, and B. Pfitzmann. Cryptographically strong undeniable signatures, unconditionally secure for the signer. Technical report, University of Karlsruhe, February 1991. Interner Bericht 1/91.
14. D. Chaum, E. van Heijst, and B. Pfitzmann. Cryptographically strong undeniable signatures, unconditionally secure for the signer. In J. Feigenbaum, editor, *Advances in Cryptology–CRYPTO '91*, volume 576 of *Lecture Notes in Computer Science*, pages 470–484. Springer-Verlag, 1992.
15. R. Cramer, I. Damgård, and B. Schoenmakers. Proofs of partial knowledge and simplified design of witness hiding protocols. In Y. G. Desmedt, editor, *Advances in Cryptology–CRYPTO '94*, Lecture Notes in Computer Science, pages 174–187. Springer-Verlag, 1994.
16. I. B. Damgård. Practical and provably secure release of a secret. In T. Helleseth, editor, *Advances in Cryptology–EUROCRYPT '93*, volume 765 of *Lecture Notes in Computer Science*, pages 200–217. Springer-Verlag, 1994.
17. A. De Santis, G. D. Crescenzo, G. Persiano, and M. Yung. On monotone formula closure of SZK. In *Proc. 35th IEEE Symp. on Foundations of Comp. Science*, pages 454–465, Santa Fe, 1994. IEEE Transactions on Information Theory.
18. U. Feige and A. Shamir. Zero-knowledge proofs of knowledge in two rounds. In G. Brassard, editor, *Advances in Cryptology–CRYPTO '89*, volume 435 of *Lecture Notes in Computer Science*, pages 526–544. Springer-Verlag, 1990.
19. A. Fiat and A. Shamir. How to prove yourself: Practical solutions to identification and signature problems. In A. Odlyzko, editor, *Advances in Cryptology–CRYPTO '86*, volume 263 of *Lecture Notes in Computer Science*, pages 186–194. Springer-Verlag, 1987.
20. L. C. Guillou and J.-J. Quisquater. A practical zero-knowledge protocol fitted to security microprocessors minimizing both transmission and memory. In C. Günther, editor, *Advances in Cryptology–EUROCRYPT '88*, Lecture Notes in Computer Science, pages 123–128. Springer-Verlag, 1988.
21. J. Håstad, A. Schrift, and A. Shamir. The discrete logarithm modulo a composite hides $o(n)$ bits. *JCSS*, 47(3):376–404, 1993.

22. M. Jakobsson, K. Sako, and R. Impagliazzo. Designated verifier proofs and their applications. In U. Maurer, editor, *Advances in Cryptology–EUROCRYPT '96*, volume 1070 of *Lecture Notes in Computer Science*, pages 143–154. Springer-Verlag, 1996.

23. D. E. Knuth. *Seminumerical Algorithms*, volume 2 of *The Art of Computer Programming*, pages 441–462. Addison-Wesley Publishing Company, 2 edition, 1981. ISBN 0-201-03822-6.

24. D. Naccache, D. M'Raïhi, S. Vaudenay, and D. Raphaeli. Can D.S.A. be improved? – complexity trade-offs with the digital signature standard. In A. D. Santis, editor, *Advances in Cryptology–EUROCRYPT '94*, volume 950 of *Lecture Notes in Computer Science*, pages 77–85. Springer-Verlag, 1995.

25. K. Ohta and T. Okamoto. A modification of the Fiat-Shamir scheme. In S. Goldwasser, editor, *Advances in Cryptology–CRYPTO '88*, volume 403 of *Lecture Notes in Computer Science*, pages 232–243. Springer-Verlag, 1988.

26. T. Okamoto. Provably secure and practical identification schemes and corresponding signature schemes. In E. F. Brickell, editor, *Advances in Cryptology–CRYPTO '92*, volume 740 of *Lecture Notes in Computer Science*, pages 31–53. Springer-Verlag, 1992.

27. T. Okamoto and K. Ohta. Divertible zero knowledge interactive proofs and communtative random self-reducibility. In J.-J. Quisquater and J. Vandewalle, editors, *Advances in Cryptology–EUROCRYPT '89*, volume 434 of *Lecture Notes in Computer Science*, pages 134–149. Springer-Verlag, 1989.

28. B. Pfitzmann. ZKP in $\mathbb{Z}_p$ or $\mathbb{Z}_{2^\sigma}$. Unpublished manuscript, April 1991.

29. D. Pointcheval and J. Stern. Provably secure blind signature schemes. In K. Kim and T. Matsumoto, editors, *Advances in Cryptology–ASIACRYPT '96*, 1163, pages 252–265. Springer-Verlag, 1996.

30. D. Pointcheval and J. Stern. Security proofs for signature schemes. In U. Maurer, editor, *Advances in Cryptology–EUROCRYPT '96*, volume 1070 of *Lecture Notes in Computer Science*, pages 387–398. Springer-Verlag, 1996.

31. C. P. Schnorr. Efficient signature generation by smart cards. *Journal of Cryptology*, 4:161–174, 1991.

32. D. R. Stinson. *Cryptography; theory and practice*. CRC Press, 1 edition, 1995. ISBN 0-8493-8521-0.

33. M. Tompa and H. Woll. Random self-reducibility and zero knowledge interactive proofs of possession of information. Technical Report RC 13207 (#59069), IBM, October 1987.

A Proof Modes

If c is chosen at random by $\mathcal{V}$, and 2^t is small, then the protocol must be repeated polynomially many times in order for $\mathcal{P}$'s proof to be convincing with overwhelming probability. Sequential repetitions result in a zero-knowledge proof, while parallel repetitions are not zero-knowledge unless preceded by an initial step in which $\mathcal{V}$ commits to its challenges (the commit must be unconditionally secure for $\mathcal{P}$ in case $\mathcal{V}$ is unbounded); alternatively, the challenges are determined in a mutually random fashion by $\mathcal{P}$ and $\mathcal{V}$.

If 2^t is large then $\mathcal{P}$ is convincing with overwhelming probability, without repetitions. The case $m = 1$ is the Schnorr proof of knowledge [31]; this is widely believed to be witness hiding, although no proof of this is known. In case $m \geq 2$ the protocol is non-trivially witness indistinguishable and provably witness hiding, and the case $m = 2$ is Okamoto's proof of knowledge [26, page 36]. The protocol can be made zero-knowledge in the manner described above.

The protocol can be performed as a signed proof, meaning that the transcript of the protocol execution is convincing evidence that $\mathcal{P}$ has performed a protocol execution. Following Fiat and Shamir [19], the challenge c is hereto computed as a one-way hash (implying that 2^t is large) of at least a. The hash-function must be such that it is infeasible to obtain more signed proofs than the number of protocol executions that $\mathcal{P}$ has engaged in ("unforgeability"). In addition, h and a message may be hashed along; in the latter case the signed proof serves as a digital signature of $\mathcal{P}$ on the message. The signed proof consist of $(r_1, \ldots, r_m)$, one of a and c, and any (other) information hashed in order to compute c; moreover, h must be included in case it is not associated with $\mathcal{P}$. The Schnorr signature scheme [31] (resp. the Okamoto signature scheme [26, page 46]) is the special case in which $\mathcal{P}$ determines c by itself, signed proofs serve as digital signatures, and $m = 1$ (resp. $m = 2$).

If we model the hash function as a random oracle (see Bellare and Rogaway [3]) and $\mathcal{P}$ computes c in Step 2 by itself, then the unforgeability of signed proofs is guaranteed for all $m \geq 1$, assuming the Discrete Logarithm assumption; see Pointcheval and Stern [30]. In particular, this holds also if $\mathcal{V}$ supplies a message, possibly adaptively chosen based on previous protocol executions, that is hashed along by $\mathcal{P}$.

In signed proof mode, it may be desirable to let $\mathcal{V}$ instead of $\mathcal{P}$ determine c, for example to enable $\mathcal{V}$ to obtain a blinded signed proof (it is straightforward to apply the blinding technique of Okamoto and Ohta [27]). In the random oracle model, the unforgeability of signed proofs for which c determines the challenge is guaranteed for all $m \geq 2$, assuming the Discrete Logarithm assumption and provided that $\mathcal{P}$ engages in no more than logarithmically many protocol executions; see Pointcheval and Stern [29].

Other proof modes are available as well. For example, one can perform the protocol as a non-interactive zero-knowledge proof (see Blum, De Santis, Micali and Persiano [5]), a limited-show proof, a designated verifier proof (see Jakobsson, Sako and Impagliazzo [22]), or a multi-prover proof. As an example of the latter proof mode, consider i parties that have each committed to their own secret, x_i, by publishing $h_i = g_1^{x_i} g_2^{y_i}$, for randomly chosen y_i; by taking h to be the product of appropriate powers of the h_i's, they can jointly demonstrate formulae pertaining to their secrets (without revealing them to any other party), by combining their responses in accordance with the formula that is demonstrated.

Finally, we note that the protocol can be modified in order to issue a signed proof that can be blinded only restrictively, by using the techniques of [10]. Hereto $\mathcal{P}$ and $\mathcal{V}$ perform the blinded signed proof with respect to a combination of $\mathcal{P}$'s public key and $\mathcal{V}$'s public key. In addition to the properties of unforgeability and independence of the signed proof and $\mathcal{P}$'s view, it can be proved under the Discrete Logarithm assumption that part of the representation of $\mathcal{V}$ remains invariant under $\mathcal{V}$'s blinding operations. In the random oracle model, this holds even if polynomially many verifiers, each with a different public key, conspire, provided that protocol executions are performed sequentially; for parallel executions, slight modifications are required, and the security can only be argued heuristically (see [9, 8]).

Oblivious Transfers and Privacy Amplification

Gilles Brassard[*] and Claude Crépeau[**]

Département IRO, Université de Montréal
C.P. 6128, succursale centre-ville
Montréal (Québec), Canada H3C 3J7
email: {brassard,crepeau}@iro.umontreal.ca

Abstract. Assume $\mathcal{A}$ owns two secret k–bit strings. She is willing to disclose one of them to $\mathcal{B}$, at his choosing, provided he does not learn anything about the other string. Conversely, $\mathcal{B}$ does not want $\mathcal{A}$ to learn which secret he chose to learn. A protocol for the above task is said to implement One-out-of-two String Oblivious Transfer, denoted $\binom{2}{1}$–OT^k. This primitive is particularly useful in a variety of cryptographic settings. An apparently simpler task corresponds to the case $k = 1$ of two one-bit secrets: this is known as One-out-of-two Bit Oblivious Transfer, denoted $\binom{2}{1}$–OT. We address the question of reducing $\binom{2}{1}$–OT^k to $\binom{2}{1}$–OT. This question is not new: it was introduced in 1986. However, most solutions until now have implicitly or explicitly depended on the notion of *self-intersecting codes*. It can be proved that this restriction makes it asymptotically impossible to implement $\binom{2}{1}$–OT^k with fewer than about $3.5277\,k$ instances of $\binom{2}{1}$–OT. The current paper introduces the idea of using *privacy amplification* as underlying technique to reduce $\binom{2}{1}$–OT^k to $\binom{2}{1}$–OT. This allows for more efficient solutions at the cost of an exponentially small probability of failure: it is sufficient to use slightly more than $2k$ instances of $\binom{2}{1}$–OT in order to implement $\binom{2}{1}$–OT^k. Moreover, we show that privacy amplification allows for the efficient implementation of $\binom{2}{1}$–OT^k from generalized versions of $\binom{2}{1}$–OT that would not have been suitable for the earlier techniques based on self-intersecting codes. An application of this more general reduction is given.

Key Words: Information-Theoretic Security, Reduction Between Protocols, Oblivious Transfer, Privacy Amplification.

[*] Supported in part by Canada's NSERC, The Canada Council and Québec's FCAR.
[**] Supported in part by Québec's FCAR and Canada's NSERC.

W. Fumy (Ed.): Advances in Cryptology - EUROCRYPT '97, LNCS 1233, pp. 334-347, 1997.

1 Introduction

One-out-of-two String Oblivious Transfer, denoted $\binom{2}{1}$-OT^k, is a primitive that originates with [Wie70] (under the name of "multiplexing"), a paper that marked the birth of quantum cryptography. According to this primitive, one party A owns two secret k–bit strings w_0 and w_1, and another party B wants to learn w_c for a secret bit c of his choice. A is willing to collaborate provided that B does not learn any information about $w_{\bar{c}}$, but B will only participate if A cannot obtain information about c. Independently from [Wie70] but inspired by [Rab81], a natural restriction of this primitive was introduced subsequently in [EGL83] with applications to contract signing protocols: One-out-of-two Bit Oblivious Transfer, denoted $\binom{2}{1}$-OT, concerns the case $k = 1$ in which w_0 and w_1 are single-bit secrets, generally called b_0 and b_1 in that case.

Techniques were introduced in [BCR86] and refined in [CS91b, BCS96] to reduce $\binom{2}{1}$-OT^k to $\binom{2}{1}$-OT: several two-party protocols were given to achieve One-out-of-two String Oblivious Transfer based on the assumption of the availability of a protocol for the simpler One-out-of-two Bit Oblivious Transfer. The fact that $\binom{2}{1}$-OT^k can be reduced to $\binom{2}{1}$-OT is not surprising because a number of authors [Kil88, Cré89, CGT95] have shown that $\binom{2}{1}$-OT is sufficient to implement *any* two-party computation. Our interest in direct reductions is their far greater efficiency. With the exception of [CS91a], all previous direct reductions that we are aware of [BCR86, CS91b, BCS96] are based on a notion called *zigzag functions*, whose construction is reduced to finding particular types of error-correcting codes called *self-intersecting codes*. In a nutshell, this approach consists in selecting once and for all a suitable function f from $\{0,1\}^n$ to $\{0,1\}^k$ for n as small as possible ($n > k$), so that if x_0 is a random preimage of w_0 and x_1 is a random preimage of w_1, and if B is given to choose via $\binom{2}{1}$-OT to see the ith bit of either x_0 or x_1, $1 \leq i \leq n$, then no information can be inferred on at least one of w_0 or w_1. This approach has led to various reductions with expansion factors β ranging from 4.8188 to 18, that is various polynomial-time constructible methods using $n = \beta k$ instances of $\binom{2}{1}$-OT to perform one $\binom{2}{1}$-OT^k on k–bit strings. Komlós proved that this approach cannot yield an expansion factor β that is asymptotically better than 3.5277 [CL85]. It was recently proven by Stinson that the same bound applies even to non-linear zigzags [Sti97].

The current paper exploits a new approach to this problem using *privacy amplification*, a notion first introduced in the context of key exchange protocols [BBR88]. The new approach allows for a solution requiring only slightly more than $2k$ instances of $\binom{2}{1}$-OT to perform one $\binom{2}{1}$-OT^k, and it can be extended to a whole range of generalizations of $\binom{2}{1}$-OT that could not be used with the reductions based on zigzag functions.

An application of the simplest of our generalizations is also considered: $\binom{2}{1}$-OT^k from A to B can be reduced to $\binom{2}{1}$-OT in the other direction (from B to A) by only doubling the cost of reducing to $\binom{2}{1}$-OT from A to B. This improves on an earlier result of [CS91a].

2 Privacy Amplification Method

Assume $\mathcal{A}$ knows a random n-bit string x about which $\mathcal{B}$ has partial information. *Privacy amplification* is a technique invented in [BBR88] and refined in [BBCM95] that allows $\mathcal{A}$ to shrink x to a shorter string w about which $\mathcal{B}$ has an arbitrarily small amount of information even if he knows the recipe used by $\mathcal{A}$ to transform x into w. Intuitively, this can be used to implement $\binom{2}{1}$-$\mathrm{OT}^k(w_0, w_1)(c)$ from $\binom{2}{1}$-OT because $\mathcal{A}$ can offer $\mathcal{B}$ to read one of two random strings x_0 or x_1 by a simple sequence of $\binom{2}{1}$-$\mathrm{OT}(x_0^i, x_1^i)(c_i)$. Subsequently, $\mathcal{A}$ tells $\mathcal{B}$ how to transform x_0 into w_0 and x_1 into w_1 by way of privacy amplification. An honest $\mathcal{B}$ who accessed all the bits of x_c can reconstruct w_c from this information. But a dishonest $\tilde{\mathcal{B}}$ who tried to access some of the bits of x_0 and some of the bits of x_1 will not have enough information on at least one of them to infer any information on the corresponding w_i or even joint information on both w_0 and w_1.

An important fact about the method based on zigzag functions considered in earlier papers is that there is no way for $\mathcal{B}$ to learn information about both w_0 and w_1 even though the zigzag function is known before he gets to choose which bits of x_0 and x_1 to obtain through the $\binom{2}{1}$-OT instances. In the new approach based on privacy amplification, $\mathcal{A}$ reveals the function to $\mathcal{B}$ *after* the necessary $\binom{2}{1}$-OT's have been performed. This allows for a protocol that is simpler, more general and more efficient, but at the cost of a vanishingly small probability of failure. A drawback of this approach is that a new function must be generated and transmitted at each run of the protocol.

The following table compares the efficiency of the earlier methods to that of privacy amplification. The column "expansion factor" gives a number β so that a $\binom{2}{1}$-OT^k can be achieved with βk instances of $\binom{2}{1}$-OT, s is a safety parameter, and ε is arbitrarily small in the limit of large k, Thus we see that the privacy amplification method is preferable provided a probability of failure can be tolerated.

Method	expansion factor	failure probability	construction time
Monte Carlo Zigzag[1]	$4.8188 + \varepsilon$	2^{-s}	$O(k^2)$
Las Vegas Zigzag[2]	$9.6377 + \varepsilon$	0	$O(k^2)$
Zigzag à la Justesen[3]	18	0	$O(k^4)$
Zigzag à la Goppa[4]	6.4103	0	$O(k^{32})$
Privacy Amplification	$2 + \varepsilon$	2^{-s}	$O(k^2)$

---

[1] Attributed to Cohen and Lempel in [BCS96].
[2] Attributed to Joe Kilian in [BCS96].
[3] From [BCS96].
[4] From [CZ94] based on a method of [CS91b].

3   The New Protocol

Let s be a security parameter chosen by $\mathcal{A}$ and $\mathcal{B}$ so that they agree to tolerate a probability 2^{-s} of failure. Let γ be a constant to be determined later, let $n = \gamma k + s$, and let $\mathcal{F}_2$ denote the field of integers modulo 2.

Privacy amplification is based on the general notion of universal classes of hash functions [CW79]. For sake of simplicity, we use a specific class of hash functions in our protocol to implement $\binom{2}{1}\text{-OT}^k$ from $\binom{2}{1}\text{-OT}$:

$$\{h \mid h(x) = Mx, \text{ for } M \text{ a } k \times n \text{ matrix over } \mathcal{F}_2\} \ .$$

Other, more efficient classes of hash functions can be used, but it is not known if the definition of universal classes is sufficient in general to make our protocol work.

Protocol 3.1 ($\binom{2}{1}\text{-OT}^k(w_0, w_1)(c)$)

1: $\mathcal{A}$ picks two random n-bit strings x_0 and x_1.

2: $\overset{n}{\underset{i=1}{\mathrm{DO}}}$ $\mathcal{A}$ transfers $t^i \leftarrow \binom{2}{1}\text{-OT}(x_0^i, x_1^i)(c)$ to $\mathcal{B}$.

3: $\mathcal{A}$ picks two random $k \times n$ matrices M_0 and M_1 over $\mathcal{F}_2$; she announces them to $\mathcal{B}$.

4: $\mathcal{A}$ sets $m_0 \leftarrow M_0 x_0$, $m_1 \leftarrow M_1 x_1$, $y_0 \leftarrow m_0 \oplus w_0$ and $y_1 \leftarrow m_1 \oplus w_1$; she announces y_0 and y_1 to $\mathcal{B}$.

5: $\mathcal{B}$ recovers w_c by computing $(M_c t) \oplus y_c$.

We postpone to Sect. 5 the proof that this protocol is private provided $\gamma \geq 2$ because we shall first generalize it to permit at no extra cost the use of another primitive called *XOR Oblivious Transfer*. (Informally, a protocol is *private* if $\mathcal{B}$ cannot learn information on both w_0 and w_1 except perhaps with negligible probability. In addition, $\mathcal{B}$ must not be able to obtain joint information on w_0 and w_1 except for what follows from his a priori knowledge and his learning one of the two strings. Conversely, $\mathcal{A}$ should learn nothing at all. See [BCS96] for a formal information-theoretic definition. We shall later relax the condition to allow $\mathcal{B}$ an exponentially small amount of unauthorized information.)

4   XOR Oblivious Transfer

A $\binom{2}{1}\text{-XOT}$ is an extension of $\binom{2}{1}\text{-OT}$ that enables a sender $\mathcal{A}$ to transfer to a receiver $\mathcal{B}$ either one bit among b_0 and b_1 or their exclusive-or, at $\mathcal{B}$'s choice. More formally, $\mathcal{A}$ inputs b_0 and b_1 into the protocol, $\mathcal{B}$ inputs $c \in \{0, 1, \oplus\}$, and $\mathcal{B}$ learns b_c while $\mathcal{A}$ learns nothing, where for convenience we use $b_\oplus$ to denote $b_0 \oplus b_1$. As usual, this is done in an all-or-nothing fashion: $\mathcal{B}$ cannot get more information about b_0 and b_1 than b_0, b_1 or $b_\oplus$, however malicious or computationally powerful he is. Note that in our application of $\binom{2}{1}\text{-XOT}$, which is to use it instead of

$\binom{2}{1}$–OT inside Protocol 3.1, an honest $\mathcal{B}$ would never requests $b_\oplus$. Therefore we can safely use any protocol in which it is merely *tolerated* that $\mathcal{B}$ might learn $b_\oplus$ in cheating attempts even though $\mathcal{A}$ is not required to provide it upon request.

The $\binom{2}{1}$–XOT comes naturally in a specific implementation of $\binom{2}{1}$–OT: in [BCR86a] a protocol for $\binom{2}{1}$–OT is given under the assumption that deciding quadratic residuosity modulo a composite number is hard. In that implementation, the possibility that $\tilde{\mathcal{B}}$ obtains $b_\oplus$ arises naturally and some effort is made to prevent it. The current paper shows that this effort was unnecessary if the final goal is to implement $\binom{2}{1}$–OT^k rather than simply $\binom{2}{1}$–OT.

5 Privacy

Consider a variation of Protocol 3.1 in which the transfers at step 2 are performed through $\binom{2}{1}$–XOT instead of $\binom{2}{1}$–OT. Even though this makes no difference if $\mathcal{B}$ follows the protocol honestly, it gives him additional opportunities for cheating if he so desires. Our goal is to show that whatever program $\tilde{\mathcal{B}}$ is ran by $\mathcal{B}$, he is not able to obtain information on both w_0 and w_1, except with a probability that is exponentially small in the security parameter s. Moreover, it is obvious from inspection of the protocol that a cheating $\mathcal{A}$ cannot obtain any information about $\mathcal{B}$'s secret parameter c. From now on, think of x_0 and x_1 as column-vectors of length n, and of $m_0 = M_0 x_0$ and $m_1 = M_1 x_1$ as column-vectors of length k, all over $\mathcal{F}_2$. First, we show that immediately after Step 3 of the protocol, whatever program $\tilde{\mathcal{B}}$ is ran by $\mathcal{B}$, he will have no information about one of m_0 or m_1 and no information allowing him to connect m_0 and m_1 (such as $m_0 \oplus m_1$ for instance), except with exponentially small probability. (Formally, there will be some bit $\tilde{c}$ such that the first three steps of the protocol would give no additional information to $\tilde{\mathcal{B}}$ about the pair (m_0, m_1) than if he were simply told the value of $m_{\tilde{c}}$; see [BCS96].) We conclude the result about w_0 and w_1 at the end of the protocol from the fact that m_0 and m_1 are used as one-time pads to transfer them.

Suppose $\tilde{\mathcal{B}}$ reads the bits $x^i_{c_i}$ with $c_i \in \{0, 1, \oplus\}$ at his choosing. Let g be a non-trivial linear function of m_0 and m_1. In other words, $g(m_0, m_1) = v_0 m_0 \oplus v_1 m_1$ for some line-vectors v_0 and v_1 of length k over $\mathcal{F}_2$ such that both v_0 and v_1 are non-zero [5].

Theorem 1. *Consider the knowledge that $\tilde{\mathcal{B}}$ has about m_0 and m_1 immediately after Step 3 of the protocol. Provided $\gamma \geq 2$,*

$$\text{Prob}\left(\exists \text{ non-trivial } g \text{ such that } \tilde{\mathcal{B}} \text{ knows } g(m_0, m_1)\right) < 2^{-s} .$$

---

[5] Note that by virtue of v_0 and m_0 being a line-vector and a column-vector, respectively, a "matrix" multiplication such as $v_0 m_0$ computes the scalar product; similarly, given that x_0 is also a column-vector, an expression such as $v_0 M_0 x_0$ makes sense: it is simply an element of $\mathcal{F}_2$. This notation is handy because $v_0 M_0 x_0$ can be thought of indifferently as either the scalar product of v_0 with $M_0 x_0$ or of $v_0 M_0$ with x_0.

Proof. We first describe the condition under which $\tilde{B}$ learns $g(m_0, m_1)$ at Step 3 of the protocol for some specific non-trivial linear function g. By definition

$$g(m_0, m_1) = v_0 m_0 \oplus v_1 m_1 = v_0 M_0 x_0 \oplus v_1 M_1 x_1 = z_0 x_0 \oplus z_1 x_1$$

where $z_0 = v_0 M_0$ and $z_1 = v_1 M_1$. Because x_0 and x_1 are random, $\tilde{B}$ cannot learn anything about $g(m_0, m_1)$ at Step 3 unless he is lucky enough that his choices c_i simultaneously follow

$$c_i = \begin{cases} 0 & \text{when } (z_0^i, z_1^i) = (1, 0) \\ 1 & \text{when } (z_0^i, z_1^i) = (0, 1) \\ \oplus & \text{when } (z_0^i, z_1^i) = (1, 1) \end{cases}$$

in all the instances of $\binom{2}{1}$–XOT such that z_0^i and z_1^i are not both 0. (The value of c_i is unimportant when $(z_0^i, z_1^i) = (0, 0)$ since neither x_0^i nor x_1^i is required in that case to compute $g(m_0, m_1)$.)

But remember that M_0 and M_1 are picked at random and neither v_0 nor v_1 is zero. Therefore $z_0 = v_0 M_0$ and $z_1 = v_1 M_1$ are random binary strings of length n chosen independently according to the uniform distribution. In particular, z_0 and z_1 are independent of $\tilde{B}$'s choices of c_i's. It follows that, for each i, the probability that either $(z_0^i, z_1^i) = (0, 0)$ or $\tilde{B}$ chose c_i appropriately according to the above case analysis is exactly $1/2$. Since $\tilde{B}$ must be lucky for each i, $1 \leq i \leq n$,

$$\text{Prob}\left(\tilde{B} \text{ learns } g(m_0, m_1)\right) = 2^{-n}$$

for each non-trivial linear function g, whatever choices $\tilde{B}$ makes for the c_i's. Finally, given that there are less than 2^{2k} such linear functions, we conclude that

$$\text{Prob}\left(\exists \text{ non-trivial } g \text{ such that } \tilde{B} \text{ learns } g(m_0, m_1)\right)$$
$$< 2^{2k-n} = 2^{(2-\gamma)k-s} \leq 2^{-s}$$

provided $\gamma \geq 2$. □

Theorem 2. *Protocol 3.1 is private even if the transfers at step 2 are performed through $\binom{2}{1}$-XOT instead of $\binom{2}{1}$-OT.*

Proof. We know from Theorem 1 that, except with probability at most 2^{-s}, B has not learned $g(m_0, m_1)$ by the end of Step 3 for any linear function g that involves both m_0 and m_1 in a non-trivial way.[6] It follows that there is a $d \in \{0, 1\}$ such that B learns no non-trivial linear function of m_d because if he could learn non-trivial linear functions $g_0(m_0)$ and $g_1(m_1)$, he would have learned $g_0(m_0) \oplus g_1(m_1)$, a non-trivial linear function of both m_0 and m_1. We can say something stronger: not only does B learn no non-trivial linear function of m_d, but he learns no information of any kind that involves m_d. This is true because

---

[6] Of course, it *is* possible for B to learn linear functions of m_0 or m_1 alone by setting all the $c_i = 0$ or $c_i = 1$ as in the honest protocol.

m_0 and m_1 are purely random and the only source of information that B has about them (up until Step 3) is given by linear functions of m_0 and m_1. Since m_d is used by A at Step 4 as one-time pad to transmit w_d to B, it follows that B learns no information of any kind that involves w_d. □

6 Application: Reversing Oblivious Transfer

Consider that A wants to send one of two words w_0 or w_1 to B when they only have an $\binom{2}{1}$–OT channel running from B to A. A very efficient protocol for sending one of two *bits* from A to B is given in [CS91a] provided A does not mind the possibility that B might learn the exclusive-or of her two bits: two instances of reversed $\binom{2}{1}$–OT are sufficient to implement $\binom{2}{1}$–XOT. No such efficient constructions are known that would implement $\binom{2}{1}$–OT from so few instances of reversed $\binom{2}{1}$–OT. In other words it is much easier to implement $\binom{2}{1}$–XOT than $\binom{2}{1}$–OT from A to B given an $\binom{2}{1}$–OT channel from B to A. This is fine because we just showed that $\binom{2}{1}$–XOT is just as good as $\binom{2}{1}$–OT for the purpose of implementing $\binom{2}{1}$–OT^k. Therefore, $\binom{2}{1}$–OT^k from A to B can be implemented from slightly more than $4k$ instances of $\binom{2}{1}$–OT from B to A. This is a three-fold improvement over [CS91a].

7 Generalized Oblivious Transfer

A $\binom{2}{1}$–GOT is a cryptographic protocol for two participants that enables a sender A to transfer a one-bit function evaluated on (b_0, b_1) to a receiver B who chooses secretly which one-bit function (f) he gets from her input bits. This is done in an all-or-nothing fashion: B cannot get more information about b_0 and b_1 than $f(b_0, b_1)$ for some f, however malicious or computationally powerful he is, and A finds out nothing about the choice f of B. As was the case with $\binom{2}{1}$–XOT in Sect. 4, one may think of a $\binom{2}{1}$–GOT protocol as merely tolerating the fact that a cheating B might learn $f(b_0, b_1)$ for some f rather than specifying that any such f can be learned at B's whim.

The following table enumerates all 14 possible non-constant functions from two bits to one. (We ignore the two constant function since they would yield no information if used.) The symbols used refer to the common boolean functions. Example: $\overline{\wedge}$ stands for $\overline{b_0 \wedge b_1}$. The notations 0 and 1 are used for the projection functions $b_0 0 b_1 = b_0$ and $b_0 1 b_1 = b_1$. We say that a function $f(b_0, b_1)$ is *biased* if the probability that $f(b_0, b_1) = 1$ is not $1/2$ when b_0 and b_1 are chosen randomly and independently according to the uniform distribution. The ordinary $\binom{2}{1}$–OT is a special case of $\binom{2}{1}$–GOT where B is limited to the functions 0 and 1.

b_0	b_1	$\vee$	$\Leftarrow$	1	$\Rightarrow$	0	$\oplus$	$\overline{\wedge}$	$\wedge$	$\overline{\oplus}$	0	$\rightarrow$	1	$\leftarrow$	$\vee$
0	0	1	0	1	0	1	0	1	0	1	0	1	0	1	0
1	0	0	1	1	0	0	1	1	0	0	1	1	0	0	1
0	0	0	0	0	1	1	1	1	0	0	0	0	1	1	1
1	0	0	0	0	0	0	0	0	1	1	1	1	1	1	1
biased		$\checkmark$	$\checkmark$		$\checkmark$		$\checkmark$	$\checkmark$		$\checkmark$		$\checkmark$		$\checkmark$	$\checkmark$	$\checkmark$

It has been shown in [BCR86] that $\binom{2}{1}$–GOT is a sufficient primitive to implement $\binom{2}{1}$–OT. The reduction they presented uses $\Theta(s)$ runs of $\binom{2}{1}$–GOT to achieve a single $\binom{2}{1}$–OT in such a way that the reduction may fail and give both bits to $\mathcal{B}$ with probability 2^{-s}. If this protocol is combined with a standard reduction of $\binom{2}{1}$–OT^k we obtain a global cost of $\Theta(ks)$ runs of $\binom{2}{1}$–GOT per $\binom{2}{1}$–OT^k. Contrary to reductions to $\binom{2}{1}$–OT, reductions to $\binom{2}{1}$–GOT *must* involve a failure probability since it is *always* possible to get all the information sent by $\mathcal{A}$ by selecting the appropriate biased function at each transfer by sheer luck. For example, if $\mathcal{B}$ requests $x_0^i \wedge x_1^i$ at step 2 of Protocol 3.1 for some i, and if he obtains the value 1, then he knows that both x_0^i and x_1^i are equal to 1. Using the new privacy amplification method we obtain a direct reduction of $\binom{2}{1}$–OT^k at a cost of only $\Theta(k + s)$ instances of $\binom{2}{1}$–GOT.

Consider a variation of Protocol 3.1 in which the transfers of step 2 are performed through $\binom{2}{1}$–GOT instead of $\binom{2}{1}$–OT. Our goal is to show that whatever program $\tilde{\mathcal{B}}$ is ran by $\mathcal{B}$, he is not able to obtain non-negligible information on both w_0 and w_1, except with a probability that is exponentially small in the security parameter s. Contrary to the analysis in Sect. 5, it will no longer suffice to take $n = \gamma k + s$ for some γ, but n will nevertheless remain in $\Theta(k + s)$—see the proof of Theorem 3 for details. First we show that immediately after Step 3 of the protocol, whatever program $\tilde{\mathcal{B}}$ is ran by $\mathcal{B}$, he will have negligible information about one of m_0 or m_1, and negligible information allowing him to connect m_0 and m_1. We conclude a similar result about w_0 and w_1 at the end of the protocol from the fact that m_0 and m_1 are used as one-time pads to transfer them.

Suppose $\tilde{\mathcal{B}}$ obtains bits $x_{c_i}^i$ with $c_i \in \{\nabla, \Leftarrow, \overline{1}, \Rightarrow, \overline{0}, \oplus, \overline{\wedge}, \wedge, \overline{\oplus}, 0, \rightarrow, 1, \leftarrow, \vee\}$ at his choosing. As before, let g be a non-trivial linear function of m_0 and m_1, that is $g(m_0, m_1) = v_0 m_0 \oplus v_1 m_1$ for some non-zero binary line-vectors v_0 and v_1 of length k. We say that $\mathcal{B}$ can α-*bias* a bit if he can guess it with probability better than $\frac{1}{2} + \alpha$ of being correct.

Theorem 3. *Consider the knowledge that $\tilde{\mathcal{B}}$ has about m_0 and m_1 immediately after Step 3 of the protocol.*

$$\text{Prob}\left(\exists \; \text{non-trivial } g \text{ such that } \tilde{\mathcal{B}} \text{ can } 2^{-k-1-s/2}\text{-bias } g(m_0, m_1)\right) < 2^{-s}$$

provided n is chosen appropriately in $\Theta(k + s)$.

Proof. Let γ and a be constants to be determined later and let $n = (a+1)(\gamma k + s)$. Let $Biased = \{i \,|\, c_i \in \{\triangledown, \Leftharpoondown, \Rightarrow, \overline{\wedge}, \wedge, \rightarrow, \leftarrow, \vee\}\}$, the set of positions where $\mathcal{B}$ uses a biased function. If $\#Biased < a(\gamma k + s)$ then Theorem 1 applies with $\gamma \geq 2$ and $n = \gamma k + s$. We thus get the desired result. Otherwise $\#Biased \geq a(\gamma k + s)$ is the more interesting case to consider. Consider the set of positions where $\tilde{\mathcal{B}}$ has used a biased function. As before, $\tilde{\mathcal{B}}$ would have learned $g(m_0, m_1)$ exactly if he had simultaneously obtained

$$x_0^i \text{ when } (z_0^i, z_1^i) = (1, 0)$$
$$x_1^i \text{ when } (z_0^i, z_1^i) = (0, 1)$$
$$x_\oplus^i \text{ when } (z_0^i, z_1^i) = (1, 1)$$

for all i for which z_0^i and z_1^i are not both 0.

Remember that M_0, M_1, x_0 and x_1 are picked at random. Thus z_0 and z_1 are random binary words of length n. Since $\mathcal{B}$ has used a biased function in position i, with probability $1/4$ he will have learned both x_0^i and x_1^i, and with probability $3/4$ he will be able to $1/6$-bias x_0^i, x_1^i and $x_\oplus^i$. (This is because each biased function has one output that uniquely defines a specific pair of inputs, while the other output leaves three pairs of inputs equally likely.) This means that in each such position i, $\tilde{\mathcal{B}}$ has obtained the bit he needs with probability $7/16$ and with probability $9/16$ he can only $1/6$-bias the bit he needs. Of the $a(\gamma k + s)$ such values of i, less than $a(\gamma k + s)/4$ of them will fall in the second case with probability at most $2e^{-25a(\gamma k+s)/1024} \approx 2^{-(\gamma k+s)}$ according to Bernstein's law of large numbers [Rén70, Chap. VII, Sect. 4, Theorem 2], for $a \approx 28$. When $7(\gamma k + s)$ of the bits involved in the calculation of $g(m_0, m_1)$ are $1/6$-biased, even if all the other bits are exactly known, $\mathcal{B}$ can only $(1/3)^{7(\gamma k+s)}/2$-bias the value of $g(m_0, m_1)$. (In general, δ-biasing each of $x_1, x_2, ..., x_l$ allows to $(2\delta)^l/2$-bias $x_1 \oplus x_2 \oplus ... \oplus x_l$ [Cré90].) It follows that for any set of choices $\{c_i\}$, and any $v_0, v_1 \neq 0^k$

$$\text{Prob} \left(\tilde{\mathcal{B}} \text{ can } 3^{-7(\gamma k+s)}/2\text{-bias } g(m_0, m_1) \right) < 2^{-(\gamma k+s)} .$$

Finally, given that there are less than 2^{2k} pairs v_0, v_1, taking $\gamma \geq 2$, and using the fact that $3^{-7(\gamma k+s)}/2 \leq 2^{-k-1-s/2}$, we conclude as desired that

$$\text{Prob} \left(\exists \text{ non-trivial } g \text{ such that } \tilde{\mathcal{B}} \text{ can } 2^{-k-1-s/2}\text{-bias } g(m_0, m_1) \right)$$
$$< 2^{2k} 2^{-(\gamma k+s)} \leq 2^{-s} .$$

$\square$

To conclude that, except with probability 2^{-s}, $\mathcal{B}$ has no more than 2^{-s} bit of information on at least one of m_0 or m_1 immediately after Step 3, and therefore no more than 2^{-s} bit of information on at least one of w_0 or w_1 at the end of the protocol (even if he is given the other string—see the Appendix for formal definitions), it suffices to apply the following theorem with $\varepsilon = 1/2^{k+1+s/2}$.

Theorem 4. *Let k be an integer and $\varepsilon \le 1/2^{k+1}$. Consider a k-bit string m so that B cannot ε-bias any non-trivial linear function of the bits of m. Then B's information on m in the sense of Shannon is less than $(2^{k+1}\varepsilon)^2$ bit.*

Proof sketch. Let X be the random variable over the binary strings of length k that corresponds to B's probability distribution on m. Consider the set G of all non-trivial linear functions on k-bit strings: there are exactly $2^k - 1$ such functions. For any $g \in G$, let p_g be the probability that $g(X) = 0$. We have $\frac{1}{2} - \varepsilon < p_g < \frac{1}{2} + \varepsilon$ for all $g \in G$ by assumption that B cannot ε-bias non-trivial linear functions of the bits of m.

It is easily shown that the probability that $X = x$ for any given string x is given by

$$\text{Prob}\,(X = x) \;=\; 2^{-k} + \frac{1}{2^k} \sum_{g \in G} s(g, x) \times (2p_g - 1)$$

for some function $s : G \times \{0,1\}^k \to \{-1,1\}$ whose detail does not concern us. It follows that $\text{Prob}\,(X = x)$ differs from 2^{-k} by less than the largest value of $2p_g - 1$ in absolute value, which is less than 2ε. The random variable X that would give the most information to B, yet respect the above constraint, would have half the strings with probability $2^{-k} - 2\varepsilon$ and the other half with probability $2^{-k} + 2\varepsilon$. Therefore,

$$\mathbf{H}(X) \le -2^{k-1}(2^{-k} - 2\varepsilon)\lg(2^{-k} - 2\varepsilon) - 2^{k-1}(2^{-k} + 2\varepsilon)\lg(2^{-k} + 2\varepsilon)$$

$$= \left(\frac{(2^{k+1}\varepsilon)^2}{1 \times 2} + \frac{(2^{k+1}\varepsilon)^4}{3 \times 4} + \frac{(2^{k+1}\varepsilon)^6}{5 \times 6} + \frac{(2^{k+1}\varepsilon)^8}{7 \times 8} + \cdots \right) / \ln 2$$

$$< (2^{k+1}\varepsilon)^2.$$

$\square$

8 Open Problems

The value of n used in our proof of Theorem 3 is in $\Theta(k + s)$ but we conjecture that it could be made significantly smaller in terms of the hidden constant, perhaps as small as $2k + s$.

As a further generalization, consider any $\alpha < 2$. An α-$\binom{2}{1}$–UOT is a cryptographic protocol for two participants that enables a sender A to transfer α bits of information, in the sense of Shannon, about two bits (b_0, b_1) to a receiver B who chooses secretly which information $\Omega_{(b_0,b_1)}$ he gets from her input bits. We require that $\Omega_{(x,y)}$ be a random variable such that $\mathbf{H}\left((B_0, B_1)|\Omega_{(B_0,B_1)}\right) \ge 2 - \alpha$ when B_0 and B_1 are uniformly distributed over $\{0,1\}$. This is done in an all-or-nothing fashion: B cannot get more information about b_0 and b_1 than a sample from $\Omega_{(b_0,b_1)}$ for some Ω, however malicious or computationally powerful he is, and that A finds out nothing about the choice Ω

of $\mathcal{B}$. To see that this is genuinely more general than $\binom{2}{1}$–GOT, consider the case in which $\mathcal{B}$ would request to see both bits through a binary symmetric channel with error rate 11%. Because $\mathbf{H}_2(11\%) \approx 0.5$, this would give $\mathcal{B}$ one bit of information about the two bits of $\mathcal{A}$. However, this scenario cannot be simulated with $\binom{2}{1}$–GOT.

Conjecture 5. *For all $\alpha < 2$ (or perhaps merely for all $\alpha \leq 1?$), Protocol 3.1 remains private even if occurrences of $\binom{2}{1}$–OT are replaced with α–$\binom{2}{1}$–UOT, provided $n \geq \beta_\alpha(k + s)$ for an appropriate constant β_α to be determined, where s is the safety parameter.*

Conjecture 6. *If conjecture 5 fails as stated, it works if Shannon entropy is replaced with Rényi entropy of order ρ in the definition of α–$\binom{2}{1}$–UOT for all $\rho > 1$ [Cac97] or perhaps merely for $\rho = 2$ [BBCM95].*

Acknowledgements

We thank Dominic Mayers and Louis Salvail for their help, comments, suggestions and support.

References

[BBCM95] C. H. Bennett, G. Brassard, C. Crépeau and U. M. Maurer, "Generalized privacy amplification", *IEEE Transaction on Information Theory*, Vol. 41, no. 6, November 1995, pp. 1915–1923.

[BBR88] C. H. Bennett, G. Brassard and J.-M. Robert, "Privacy amplification by public discussion", *SIAM Journal on Computing*, Vol. 17, no. 2, April 1988, pp. 210–229.

[BCR86] G. Brassard, C. Crépeau and J.-M. Robert, "Information theoretic reductions among disclosure problems", *Proceedings of 27th Annual IEEE Symposium on Foundations of Computer Science*, 1986, pp. 168–173.

[BCR86a] G. Brassard, C. Crépeau and J.-M. Robert, "All-or-nothing disclosure of secrets", *Advances in Cryptology: Proceedings of Crypto '86*, Springer-Verlag, 1987, pp. 234–238.

[BCS96] G. Brassard, C. Crépeau and M. Sántha, "Oblivious transfers and intersecting codes", *IEEE Transactions on Information Theory*, Vol. 42, no. 6, November 1996, pp. 1769–1780.

[Cac97] C. Cachin, "Smooth entropy and Rényi entropy", *Advances in Cryptology: Proceedings of Eurocrypt '97*, Springer-Verlag, 1997.

[CW79] J. L. Carter and M. N. Wegman, "New hash functions and their use in authentication and set equality", *Journal of Computer and System Sciences*, Vol. 22, 1981, pp. 265–279.

[CL85] G. D. Cohen and A. Lempel, "Linear intersecting codes", *Discrete Mathematics*, Vol. 56, 1985, pp. 35–43.

[CZ94]   G. D. Cohen and G. Zémor, "Intersecting codes and independent families", *IEEE Transactions on Information Theory*, Vol. 40, no. 6, November 1994, pp. 1872 – 1881.

[Cré89]   C. Crépeau, "Verifiable disclosure of secrets and application", *Advances in Cryptology: Proceedings of Eurocrypt '89*, Springer-Verlag, 1990, pp. 181 – 191.

[Cré90]   C. Crépeau, *Correct and Private Reductions Among Oblivious Transfers*, PhD thesis, Department of Electrical Engineering and Computer Science, Massachusetts Institute of Technology, 1990. Supervised by Silvio Micali.

[CGT95]   C. Crépeau, J. van de Graaf and A. Tapp, "Committed oblivious transfer and private multi-party computations", *Advances in Cryptology: Proceedings of Crypto '95*, Springer-Verlag, 1995, pp. 110 – 123.

[CS91a]   C. Crépeau and M. Sántha, "On the reversibility of oblivious transfer", *Advances in Cryptology: Proceedings of Eurocrypt '91*, Springer-Verlag, 1991, pp. 106 – 113.

[CS91b]   C. Crépeau and M. Sántha, "Efficient reductions among oblivious transfer protocols based on new self-intersecting codes", *Sequences II, Methods in Communications, Security and Computer Science*, Springer-Verlag, 1991, pp. 360 – 368.

[EGL83]   S. Even, O. Goldreich and A. Lempel, "A randomized protocol for signing contracts", *Proceedings of Crypto 82*, Plenum Press, New York, 1983, pp. 205 – 210.

[GMR89]   S. Goldwasser, S. Micali and C. Rackoff, "The knowledge complexity of interactive proof-systems", *SIAM Journal on Computing*, Vol. 18, 1989, pp. 186 – 208.

[Kil88]   J. Kilian, "Founding cryptography on oblivious transfer", *Proceedings of 20th Annual ACM Symposium on Theory of Computing*, 1988, pp. 20 – 31.

[Rab81]   M. O. Rabin, "How to exchange secrets by oblivious transfer", Technical Memo TR–81, Aiken Computation Laboratory, Harvard University, 1981.

[Rén70]   A. Rényi, *Probability Theory*, North Holland, 1970.

[Sti97]   D. R. Stinson, Private communication, 12 February 1997.

[Wie70]   S. Wiesner, "Conjugate coding", *Sigact News*, Vol. 15, no. 1, 1983, pp. 78 – 88. Original manuscript written circa 1970.

A Appendix: Information Theoretic Definition of Generalized Oblivious Transfer

A cryptographic protocol is a multi-party synchronous program that describes for each party the computations to be performed or the messages to be sent to some other party at each point in time. The protocol terminates when no party has any message to send or information to compute. The protocols we describe in this paper all take place between two parties $\mathcal{A}$ and $\mathcal{B}$. We denote by $\bar{A}$ and $\bar{B}$ the *honest* programs to be executed by $\mathcal{A}$ and $\mathcal{B}$: honest parties behave according to $\bar{A}$ and $\bar{B}$ and no other program. In the following definitions of *correctness* and *privacy* we also consider alternative *dishonest* programs $\tilde{A}$ and $\tilde{B}$ executed by $\mathcal{A}$ or $\mathcal{B}$ in a effort to obtain unauthorized information from one another. The definitions specify the result of honest parties interacting together through a specific protocol as well as the possible information leakage of an honest party facing a dishonest party. We are not concerned with the situation where both parties may be dishonest as they can do anything they like in that case; we are only concerned with protecting an honest party against a dishonest party. At the end of each execution of a protocol, each party will issue an "accept" or "reject" verdict regarding their satisfaction with the behaviour of the other party. Two honest parties should always issue "accept" verdicts at the end of their interactions. An honest party will issue a "reject" verdict at the end of a protocol if he received some message from the other party of improper format or some message not satisfying certain conditions specified by the protocol. We also implicitly assume certain time limits for each party to issue messages to each other: after a specified amount of time a party will give up interacting with the other party and issue a "reject" verdict.

As discussed in Sect. 7, a $\binom{2}{1}$–GOT is a cryptographic protocol for two participants that enables a sender $\mathcal{A}$ to transfer a one-bit function of two bits b_0 or b_1 to a receiver $\mathcal{B}$ who chooses secretly which function $f(b_0, b_1)$ he gets. This is done in an all-or-nothing fashion, which means that $\mathcal{B}$ cannot get partial information about b_0 and b_1 at the same time, however malicious or computationally powerful he is, and that $\mathcal{A}$ finds out nothing about the choice f of $\mathcal{B}$.

Formally speaking we describe a two-party protocol that satisfies the following constraints of *correctness* and *privacy*, similar to those introduced for $\binom{2}{1}$–OT in [BCS96].

Let $[P_0, P_1](a)(b)$ be the random variable (since P_0 and P_1 may be probabilistic programs) that describes the outputs obtained by $\mathcal{A}$ and $\mathcal{B}$ when they execute together the programs P_0 and P_1 on respective inputs a and b. Similarly, let $[P_0, P_1]^*(a)(b)$ be the random variable that describes the total information (including not only messages received and issued by the parties but also the result of any local random sampling they may have performed) acquired during the execution of protocol $[P_0, P_1]$ on inputs a and b. Let $[P_0, P_1]_P(a)(b)$ and $[P_0, P_1]_P^*(a)(b)$ be the marginal random variables obtained by restricting the above to only one party P. The latter is often called the *view* of P [GMR89].

In the following definition, the equality sign ($=$) means that the distributions on the l.h.s. and the r.h.s. are the same. When required, we shall use more flexible definitions that would allow an exponentially small probability of failure or amount of unauthorized information leakage. Details are left to the reader.

Definition 7 (Correctness). Protocol $[\bar{A}, \bar{B}]$ is *correct* for $\binom{?}{1}$–GOT if

- $\forall b_0, b_1 \in \{0, 1\}, f : \{0, 1\}^2 \to \{0, 1\}$

$$[\bar{A}, \bar{B}](b_0, b_1)(f) = (\epsilon, f(b_0, b_1)) \tag{1}$$

- for any program $\tilde{A}$ there exists a probabilistic program $\tilde{A}'$ s.t. $\forall b_0, b_1 \in \{0, 1\}, f : \{0, 1\}^2 \to \{0, 1\}$

$$\left[\tilde{A}, \bar{B}\right]_{\bar{B}} (b_0, b_1)(f) \mid \bar{B} \text{ accepts } = \left[\bar{A}, \bar{B}\right]_{\bar{B}} (\tilde{A}'(b_0, b_1))(f) \mid \bar{B} \text{ accepts .} \tag{2}$$

Intuitively, condition (1) means that if the protocol is executed as described, it will accomplish the task it was designed for: B receives bit $f(b_0, b_1)$ and A receives nothing. Condition (2) means that in situations in which B does not abort, A cannot induce a distribution on B's output using a dishonest $\tilde{A}$ that she could not induce simply by changing the input words and then being honest.

Let B_0, B_1 and F be the random variables taking values over $\{0, 1\}$ and $\{0, 1\}^2 \to \{0, 1\}$ that describe A's and B's inputs. We assume that both A and B are aware of the joint probability distribution of these random variables $P_{B_0, B_1, F}$. A sample b_0, b_1, f is generated from that distribution and b_0, b_1 is provided as A's secret input while f is provided as B's secret input.

Definition 8 (Privacy). Protocol $[\bar{A}, \bar{B}]$ is *private* for $\binom{?}{1}$–GOT if
$\forall B_0, B_1 \in \{0, 1\}, F : \{0, 1\}^2 \to \{0, 1\}$

- $\forall b_0, b_1 \in \{0, 1\}$ and for any program $\tilde{A}$

$$\mathbf{I}\left(F; \left[\tilde{A}, \bar{B}\right]_{\tilde{A}}^{*}(B_0, B_1)(F) \mid (B_0, B_1) = (b_0, b_1)\right) = 0 \tag{3}$$

- $\forall f : \{0, 1\}^2 \to \{0, 1\}$ and for any program $\tilde{B}$ there exists a random variable $\tilde{F} = \Omega(F) : \{0, 1\}^2 \to \{0, 1\}$ s.t.

$$\mathbf{I}\left((B_0, B_1); \left[\bar{A}, \tilde{B}\right]_{\tilde{B}}^{*}(B_0, B_1)(F) \mid F = f, \tilde{F}(B_0, B_1)\right) = 0. \tag{4}$$

The above two conditions are designed to guarantee that each party is limited to the information he or she should get according to the honest task definition. Condition (3) means that $\tilde{A}$ cannot acquire any information about F through the protocol. On the other hand, condition (4) means that $\tilde{B}$ may acquire only one bit of deterministic information about B_0, B_1 through the protocol. We do not require that $\tilde{B}$ be given $F(B_0, B_1)$ because there is no way to prevent him from obtaining any other $\tilde{F}(B_0, B_1)$ through otherwise honest use of the protocol.

SHA: A Design for Parallel Architectures?

Antoon Bosselaers, René Govaerts and Joos Vandewalle

Katholieke Universiteit Leuven, Dept. Electrical Engineering-ESAT
Kardinaal Mercierlaan 94, B-3001 Heverlee, Belgium

antoon.bosselaers@esat.kuleuven.ac.be

Abstract. To enhance system performance computer architectures tend to incorporate an increasing number of parallel execution units. This paper shows that the new generation of MD4-based customized hash functions (RIPEMD-128, RIPEMD-160, SHA-1) contains much more software parallelism than any of these computer architectures is currently able to provide. It is conjectured that the parallelism found in SHA-1 is a design principle. The critical path of SHA-1 is twice as short as that of its closest contender RIPEMD-160, but realizing it would require a 7-way multiple-issue architecture. It will also be shown that, due to the organization of RIPEMD-160 in two independent lines, it will probably be easier for future architectures to exploit its software parallelism.

Key words. Cryptographic hash functions, instruction-level parallelism, multiple-issue architectures, critical path analysis

1 Introduction

The current trend in computer designs is to incorporate more and more parallel execution units, with the aim of increasing system performance. However, available hardware parallelism only leads to increased software performance, if the executed code contains enough software parallelism to exploit the potential benefits of the multiple-issue architecture.

Cryptographic algorithms are often organized as an iteration of a common sequence of operations, called a round. Typical examples of this technique are iterated block ciphers and customized hash functions based on MD4. In many applications, encryption and/or hashing forms a computational bottleneck, and an increased performance of these basic cryptographic primitives is often directly reflected in an overall improvement of the system performance.

To increase the performance of round-organized cryptographic primitives it suffices to concentrate the optimization effort on the round function, knowing that each gain in the round function is reflected in the overall performance of the primitive multiplied by the number of rounds. Typical values for the number of rounds are between 8 and 32.

This paper confronts one class of cryptographic primitives, namely the customized hash functions based on MD4, with the most popular computer architectures in use today or in the near future. Although only the MD4-like hash functions are considered in the sequel, much of it also applies to other classes of

W. Fumy (Ed.): Advances in Cryptology - EUROCRYPT '97, LNCS 1233, pp. 348-362, 1997.

iterated cryptographic primitives. Our main aim is to investigate the amount of software parallelism in the different members of the MD4 hash family, and the extent to which nowadays RISC and CISC processors are able to exploit this parallelism. This approach differs of the one in [BGV96] in that we now take the hashing algorithms as a starting point, and investigate the amount of inherently available parallelism, while previously we took a particular superscalar processor as starting point, and investigated to which extent an implementation of the hashing algorithms could take advantage of that architecture.

The next section considers the basic requirements a processor has to meet to enable efficient implementations of MD4-like hash functions. Section 3 gives an overview of currently available processor architectures, and lists their, for our purposes, interesting characteristics. Section 4 introduces the notion of a critical path. The available amount of instruction-level parallelism in the MD4-like algorithms is determined in section 5, and confronted with the available hardware of section 3. Finally, section 6 formulates the conclusions.

2 Basic hardware requirements

The customized hash functions based on MD4 include MD4 [Riv92a], MD5 [Riv92b], SHA-1 [FIPS180-1], RIPEMD [RIPE95], RIPEMD-128 and RIPEMD-160 [DBP96]. It are all iterative hash functions using a compression function as their basic building block, the input to which consists of a 128 or 160-bit chaining variable and a 512-bit message block. The output is an update of the chaining variable. Internally, the compression function operates on 32-bit words. The conversion from external bit strings to internal word arrays uses a big-endian convention for SHA-1 and a little-endian convention for all the other hash functions. Depending on the algorithm the compression function consists of 3 to 5, possibly parallel, rounds, each made up of 20 (SHA-1) or 16 (all other) steps. Finally, a feedforward adds the initial value of the chaining variable to the updated value. Every round uses a particular non-linear function, and every step modifies one word of the chaining variable and possibly rotates another. Definitions of the round and step functions can be found in Tables 1 and 2, respectively.

Multiplexer	$(x \wedge y) \vee (\overline{x} \wedge z)$, $(x \wedge z) \vee (y \wedge \overline{z})$
Majority	$(x \wedge y) \vee (x \wedge z) \vee (y \wedge z)$
Xor	$x \oplus y \oplus z$
Or-Xor (OX)	$(x \vee \overline{z}) \oplus y$, $(x \vee \overline{y}) \oplus z$, $(y \vee \overline{z}) \oplus x$

Table 1. Definition of the Boolean round functions used in MD4-family algorithms.

This short overview allows us to conclude that an implementation of MD4-like hash functions will benefit from a processor that

Algorithm	Step function using Boolean function:	Mux	Maj	Xor	Or-Xor
MD4	$A := (A + f(B,C,D) + X_i + K)^{\lll s}$	1	2	3	
MD5	$A := B + (A + f(B,C,D) + X_i + K)^{\lll s}$	1,2		3	4
SHA-1	from step 17 onwards: $X_i := (X_i \oplus X_{i+2} \oplus X_{i+8} \oplus X_{i+13})^{\lll 1}$ $A := A + B^{\lll 5} + f(C,D,E) + X_i + K$ $C := C^{\lll 30}$	1	3	2,4	
RIPEMD	$A := (A + f(B,C,D) + X_i + K)^{\lll s}$	1	2	3	
RIPEMD-128	$A := (A + f(B,C,D) + X_i + K)^{\lll s}$	2L,3R 4L,1R		1L,4R	3L,2R
RIPEMD-160	$A := E + (A + f(B,C,D) + X_i + K)^{\lll s}$ $C := C^{\lll 10}$	2,4		1L,5R	5L,1R 3

Table 2. Definition of the step function used in MD4-family algorithms. Additions are modulo 2^{32}. Rotating x over s bits to the left is indicated as $x^{\lll s}$. A, B, C, D, E are the words of the chaining variable, K and s are constants, X_i is a message word or a combination thereof, and $f()$ is one of the functions defined in Table 1. The last 4 columns indicate in which rounds these functions are used, and, if different, whether in the left (L) or right (R) parallel line.

1. supports 32-bit operations.
2. can handle both little-endian and big-endian memory addressing.
3. has a rotate instruction, and, in addition to the standard logical instructions and, or, and xor, instructions like nand, nor, nxor, and-not, and or-not, where the latter two are defined as, respectively, the and and or of the first operand and the complement of the second. Remark that xor-not would be the same as nxor.
4. is able to keep all local variables in registers: 16 message words, 5 chaining words, and 2 auxiliary words. The RIPEMD-family, having two parallel lines, requires two copies of the last two items. So in total up to 30 registers are required.
5. supports parallel execution of arithmetic or logical (ALU) operations. This item will be further investigated in the next section.

3   Hardware parallelism

The basic implementation technique, applied by all nowadays processors, to improve CPU performance is pipelining. A pipeline is organized in a number of stages, each of which executes part of a CPU instruction. Multiple instructions can overlap in execution by letting each stage in the pipeline complete a part of a different instruction. Hence, this technique allows different parts of consecutive instructions to be executed in parallel. As a consequence, pipelining increases the CPU instruction throughput. The execution time of each instruction usually slightly increases due to pipeline control overhead, but this is more than compensated for by the increase in instruction throughput. The net effect is a substantial decrease in the number of clock cycles per instruction, ideally resulting in a speedup equaling the number of pipeline stages.

To enhance performance even further two approaches are available: increase the number of pipeline stages, or use a number of parallel pipelines. The former architecture is called superpipelined and emphasizes temporal parallelism, while the latter relies on spatial parallelism and comes in two flavors: superscalar or very long instruction word (VLIW). The aim of these techniques is to further increase the throughput. A superpipelined architecture achieves this by reducing the clock cycle time, while a superscalar/VLIW architecture tries to issue more than 1 instruction per clock cycle. However, there is a limit to what can be gained in terms of performance. This limit is determined by two factors: a software one and a hardware one. The software factor is the amount of parallelism in the instruction stream, i.e., the amount of data dependencies between the instructions. In the next section the available instruction-level parallelism in an instruction stream will be characterized by the its critical path. The hardware factor is the impact of the increase in the number of pipeline stages or pipelines on the clock cycle time.

In case of a superpipelined architecture limited parallelism in the instruction stream will eventually lead to so-called pipeline stalls due to data dependencies: the execution of an instruction has to be stalled until the data needed to complete it become available. But even in the absence of dependencies superpipelining will eventually run out of steam. The clock cycle time can never be lower than the overhead pipelining incurs on each stage: clock skew and pipeline register overhead [HePa96]. Therefore, increasing the number of pipeline stages beyond a critical point will result in performance degradation rather than performance gain.

Further increase in performance can then only be obtained by either going superscalar or using VLIWs.

- A superscalar processor has dynamic issue capability: a varying number of instructions is issued every clock cycle. The hardware dynamically decides which instructions are simultaneously issued and to which pipelines, based on issue criteria and possible data dependencies.
- A VLIW processor has fixed issue capability: every clock cycle a fixed number of instructions is issued, formatted as one large instruction (hence the name). The software (i.e., the compiler) is completely responsible for creating a package of instructions that can be simultaneously issued. No decisions about multiple issue are dynamically taken by the hardware.

An advantage of a VLIW over a superscalar is that the amount of required hardware can be reduced: choosing the instructions to be issued simultaneously is done at compile-time, and not at run-time. However, the superscalar has two major advantages: its code density is little affected by the available parallelism in the instruction stream, and it can be object-code compatible with a large family of non-parallel processors. The major challenge in the design of a superscalar processor will be to limit the impact on the clock cycle time of issuing and executing multiple instructions per cycle. This is illustrated by the fact that to date a factor of 1.5 to 2 in clock rate has consistently separated the highest clock rate processors and the most sophisticated multiple-issue processors [HePa96].

A final uniprocessor technique to exploit parallelism inherent in many algorithms is single-instruction, multiple-data (SIMD) processing, a term originally only used in the context of multiprocessor environments [Fly66]. A SIMD instruction performs the same operation in parallel on multiple data elements, packed into a single processor word. Tuned to accelerate multimedia and communications software, these instructions can be found in an increasing number of general-purpose processor architectures. Examples are Intel's MMX [PeWe96], UltraSPARC's VIS [TONH96], and PA-RISC 2.0 architecture's MAX [Lee96]. MMH [HaKr97] is an example of a cryptographic hash function taking advantage of this new technology. Remark that a combination of multiple-issue and SIMD techniques creates in effect a kind of multiple-issue, multiple-data (MIMD) parallelism, also called SIMD-MIMD parallelism [Lee95].

CPUs can be differentiated among based on the type of their internal storage: a stack, an accumulator, or a set of registers. Only the latter class of CPUs will be considered in the sequel, since virtually every processor designed after 1980 uses that architecture, called a (general-purpose) register architecture. A further division of this call can be made based on the way instructions can access memory and on the operands for a typical ALU instruction.

- In a register-memory architecture memory can be accessed as part of any instruction, while in a register-register architecture memory can only be accessed with load and store instructions, for which reason the latter is also called a load-store architecture.
- The maximum number of operands of an ALU instruction is either two or three. A three-operand instruction contains a destination and two source operands, while in a two-operand instruction one of the operands is both a source and a destination for the operation.
- The number of memory operands of an ALU instruction can vary from none to the maximum number of operands (2 or 3).

It turns out that two[1] combinations suffice to classify all the CPUs that will be considered:

class 1 - a tree-operand load-store architecture (no memory operands in ALU instruction): MIPS, Precision Architecture (PA-RISC), PowerPC, SPARC, Alpha.

class 2 - a two-operand register-memory architecture (at most one memory operand in ALU instruction): 80x86 (including Pentium and PentiumPro), 680x0.

Remark that the same division also distinguishes between RISC processors (class 1) and CISC processors (class 2).

---

[1] Three suffice to classify nearly all existing machines, see [HePa96, Section 2.2]

Table 3 summarizes the characteristics of these architectures with respect to the requirements formulated at the end of the previous section, including the available hardware parallelism for ALU instructions [Sta96, HePa96, Bha96]. The figures are for the most recent processors of each architecture. As far as RISC processors are concerned, these are all 64-bit, although compatibility with their 32-bit predecessors is retained. Since Alpha was designed as a 64-bit device, the support for 32-bit operations is limited. All RISC architectures include support for both little and big-endian addressing, but especially with PA-RISC and Alpha architectures an implementation is not required to implement both addressing modes. An Alpha implementation is not even required to support changing the convention during program execution, but only at boot time [Dig96]. The other RISCs can use either format, selectable in either software or hardware. Some architectures are more than 2-way superscalar, but none can issue more than 2 instructions in parallel of the ALU subset that interests us: add, logical operations, rotate/shift.

Architecture	MIPS IV	PA 2.0	PowerPC	SPARC V9	Alpha EV5	80x86	680x0
Word Size	64	64	64	64	64	32	32
Integer regs	31	31	32	31	31	7	8
Endianness	select.	select.	select.	select.	Little	Little	Big
AND	and	and, and-not	and,nand, and-not	and, and-not	and, and-not	and	and
OR	or,nor	or	or,nor, or-not	or, or-not	or, or-not	or	or
XOR	xor	xor	xor,nxor	xor,nxor	xor,nxor	xor	xor
ROT	No	Yes[a]	Yes	No	No	Yes	Yes
ALU pipe s	$1^b/2^c$	2	2	2	2	2	2
32-bit subset	Yes	Yes	Yes	Yes	No[d]	(Yes)	(Yes)
Processor	R4000, R10000	PA-8000	PowerPC 620	Ultra- SPARC	21164	Pentium PPro	68060

[a] The PA-RISC 2.0 instruction shrpw r1,r2,x,t shifts the concatenation of r1 and r2 to the right over x bits, and puts the result in t. By taking r1 = r2 = t it is in effect a rotate.

[b] The R4000 is superpipelined (but not superscalar) and its pipeline clock is twice the external clock frequency, so that 2 instructions can be issued per clock cycle.

[c] The R10000 is superscalar, but not superpipelined.

[d] The Alpha architecture has just 3 32-bit integer operations: add, subtract, multiply. In addition, it has a set of in-register manipulation instructions on 32-bit quantities, such as extract, insert, and mask.

Table 3. Overview of the latest designs of the most popular computer architectures. Only those characteristics are listed that are relevant when implementing MD4-like hash functions on these architectures.

From this table we can conclude that, with respect to the requirements of section 2, all listed RISC architectures fulfill requirement 4, while all of the first 3 requirements are only met by the PowerPC, and to a varying degree by the other RISCs. The most serious problem for a number of RISCs is certainly the absence of a rotate instruction, while CISCs are severely restricted by their small register set. In section 5 it is investigated whether such a two-way superscalar architecture suffices to exploit all the parallelism available in the MD4-like algorithms, using the analysis restricted to MD5 in [Tou95] as a starting point.

So far only superpipelined and/or superscalar processors have been considered. Nowadays most multiple-issue processors are superscalar, but VLIW is experiencing a comeback in popularity. An example of the latter is the recently introduced 32-bit VLIW processor TM-1 of Philips Trimedia [SRD96]. Up to 5 operations can be packed into a single VLIW-instruction and executed in a single clock cycle. Although intended for multimedia processing, its ability to execute 5 ALU operations in parallel creates new opportunities for fast implementations of existing and for the design of new cryptographic algorithms [Cla97].

4 Critical path length

To determine the amount of available instruction-level parallelism in the MD4-like hash functions, a critical path analysis is applied. To that end the algorithms are represented as a so-called activity-on-edge network, which is a directed graph with weighted edges.

Geometrically a graph G is defined as a set $V(G)$ of vertices v_i interconnected by a set $E(G)$ of edges e_i. In a directed graph or digraph an edge e_i is a directed pair $\langle v_i, v_j \rangle$ and represented by an arrow from the tail v_i to the head v_j. A directed path from v_p to v_q is a sequence of vertices $v_p, v_{i_1}, v_{i_2}, \ldots, v_{i_n}, v_q$ such that $\langle v_p, v_{i_1} \rangle, \langle v_{i_1}, v_{i_2} \rangle, \ldots, \langle v_{i_n}, v_q \rangle$ are edges in $E(G)$.

A network is a graph with weighted edges, i.e., to each edge e a weight $w(e)$ is assigned. In an activity-on-edge network (AOE-network) tasks to be performed are represented by directed edges. The vertices in the network represent events, signaling the completion of certain activities. Activities represented by edges leaving a vertex cannot be started until the event at that vertex has occurred. An event occurs only when all activities entering it have been completed. The weight $w(e)$ assigned to an edge e represents the time required to complete the activity associated with e.

The length of a path is then defined as $\sum_e w(e)$, where e runs over all edges on the path. It is the time it takes to complete the task represented by the path. Assuming the activities in an AOE network can be carried out in parallel, the minimum time to complete the overall task is the length of the longest path from the start vertex to the termination vertex. Such a path is called a critical path.

The evaluation of an arithmetic expression can be modeled as an AOE network. The start vertex corresponds to the availability of the input data, the activities represented by the edges correspond to the arithmetic operations constituting the expression, and the termination vertex corresponds to the result of the expression. The weight of an edge represents the time it takes to complete the corresponding arithmetic operation. Maximum performance in evaluating an arithmetic expression will therefore be obtained by making its critical path as short as possible, using, as much as possible, parallel execution of individual arithmetic operations. However, we must take into account that eventually the evaluation of the expression will take place on a multiple-issue architecture of the kind described in the previous section, i.e., all parallel execution units are pipelined, and all advance at the same rate. Unless out-of-order execution is supported, operations executed in parallel all deliver their result at the same moment, and therefore not faster than the time of the slowest operation. For this reason the critical path length will be expressed in terms of required pipeline stages, rather than in clock cycles. A measure similar to critical path length is depth, as used in the analysis of parallel algorithms [Ble96].

5 CPL analysis of the MD4-family

The critical path length (CPL) of the MD4-like compression functions is mainly determined by the CPL of the individual rounds: the CPL of the feedforward is at most 2. The CPL of each round is equal to the sum of the CPLs of each step, so that the CPL of the compression function is easily derived from the CPL of a step. Each step updates one of the chaining words, and this updated word is then input to the next step. It is this basic dependency between steps that will determine their CPL. An inspection of two consecutive steps of every MD4-family member (see Appendix A) learns us that, except for SHA-1, the chaining word updated in one step is input to the Boolean function of the next step. The chaining word updated in that step only becomes available after adding in the Boolean result, rotating the resulting sum, and, in case of MD5 and RIPEMD-160, adding in another chaining word. SHA-1, in contrast, inputs the updated chaining word to a simple rotate, and the next chaining word becomes available after only 1 more addition. These lower bounds on a step's CPL are summarized in Table 4.

Algorithm	Operations in CP	min. CPL
MD4, RIPEMD, RIPEMD-128	$f(), +, \lll$	3
MD5, RIPEMD-160	$f(), +, \lll, +$	4
SHA-1	$+, \lll$	2

Table 4. Lower bound on the CPL of a step for each of the MD4-family members, assuming that it takes a minimum of 1 stage to deliver the result of a Boolean function.

SHA-1 uses exactly the same kind and amount of operations as MD5 and RIPEMD-160 to update a chaining variable: 1 application of a Boolean function, 4 additions, and a rotate. However, the lower bound on a step's CPL is only half that of MD5 and RIPEMD-160. This is due to the fundamentally different way SHA-1's step function is organized compared to all the others:

1. The rotate is not applied to a sum of intermediate results, but to an individual chaining variable.
2. None of the arguments of the Boolean function are, except for a rotate, updated in the previous step, but in the step before that.

This in itself might be a coincidence, but it turns out that the lower bound is also the actual CPL of each SHA-1 step, while this is not the case for any of the other hash functions, as will be shown in the sequel. This seeming coincidence might well be a design principle.

For the other hash functions the Boolean function is part of the critical path. This results in an increase of the CPL if the result cannot be delivered within the 1 stage assumed for the lower bound. This is, e.g., the case for the multiplexer $(x \wedge y) \vee (\overline{x} \wedge z)$ used in all MD4-like hash functions. It would seem that from the moment x becomes available, and only using and, or, and xor, it takes three more stages to deliver the multiplexer result [Tou95]. However, using the mathematically equivalent expression $((y \oplus z) \wedge x) \oplus z$ [McC94, NMVR95], it only takes two more stages. Since this is still 1 more than the value assumed in the lower bound, this multiplexer lengthens the CPL of all steps using it by 1, except for SHA-1, where the Boolean function isn't necessarily part of the critical path. Remark that, as far as CPL is concerned, it doesn't always pay off to use the equivalent multiplexer expression. Consider the alternative multiplexer $(x \wedge z) \vee (y \wedge \overline{z})$ used in MD5, RIPEMD-128, and RIPEMD-160, and where the critical path runs through y. Without rewriting it only takes 2 stages to deliver the result from the point y becomes available, but using the equivalent expression $((x \oplus y) \wedge z) \oplus y$ the CPL increases to 3.

The results of this CPL analysis for the MD4-family of hash functions is given in Table 5. The analysis is done using both 3-operand and 2-operand instructions. With the exception of the first and third round steps of SHA-1, the shortest possible critical path is the same for both operand formats. However, for the same CPL a realization on a 2-operand architecture requires more parallel execution units than on a 3-operand one. This information can be derived from the last 4 columns, where for both formats the required number of parallel units and their efficiency is given. The efficiency is defined as

$$\frac{\text{number of instructions in a step}}{\text{CPL} \times \text{number of execution units}},$$

and is a measure of the average usage of the parallel execution units. The closer the value is to 1, the higher the degree of occupancy of the parallel units.

Table 5 also shows that if 3-operand instructions are used the shortest possible critical path of all SHA-1 steps is equal to the lower bound of Table 4: 2 stages. This is illustrated for the most involved case in Figure 1: the step function of the third round using the majority function. As a result the CPL of SHA-1's compression function is the shortest of all the MD4-like hash functions, as shown

Algorithm	Step function	CPL min.	CPL real	Regs state	Regs aux.	3-op pipe #	3-op pipe eff.	2-op pipe #	2-op pipe eff.
MD4	Mux	3	4	16+4	1	2	0.75	2	0.88
	Maj		4		2	2	1.00	3	0.83
	Xor		3		1	2	1.00	3	0.78
MD5	Mux1	4	5	16+4	1	2	0.80	2	0.90
	Mux2		5		2	2	0.90	3	0.73
	Xor		4		1	2	0.88	2	1.00
	Or-Xor		5		1	2	0.80	2	0.90
SHA-1	Mux[a]	2	2/3[b]	16+5	5	7	0.93	6	0.89
	Xor		2		4	6	1.00	7	1.00
	Maj		2/3[b]		5	7	1.00	6	1.00
	Xor		2		4	6	1.00	7	1.00
RIPEMD	Mux	3	4	16+8	2	4	0.81	4	0.94
	Maj		4		4	4	0.94	5	0.95
	Xor		3		2	4	1.00	5	0.93
RIPEMD-128	Xor/Mux2	3	3/4[c]	16+8	3	4	0.92	5	0.89
	Mux1/OX		4		2	4	0.88	4	1.00
RIPEMD-160	Xor/OX2	4	4	16+10	2	4	1.00	5	0.90
	Mux1/Mux2		5		3	4	0.95	5	0.88
	OX1/OX1		5		2	4	0.90	4	1.00

[a] The message expansion only starts at step 17. Therefore, the first 16 steps have only an efficiency of 0.64 and 0.61, respectively.

[b] 3-operand/2-operand figure

[c] Xor/Mux2 figure

Table 5. Results of the critical path analysis on the MD4-like steps. Listed are for each step the lower bound and the actual value of the CPL, the required number of state (message+chaining) and auxiliary registers, and the required number and efficiency of parallel ALU pipelines, both for 3-operand and 2-operand instruction formats. The figures for the last two rounds of RIPEMD-128 and RIPEMD-160 are not listed, since they are the same as those for the first two rounds.

Algorithm	CPL (stages)	#Regs	Pipes #	Pipes Eff.
MD4	176	22	2	0.91
MD5	304	22	2	0.84
SHA-1	160	26	7	0.85
RIPEMD	176	28	4	0.91
RIPEMD-128	240	27	4	0.90
RIPEMD-160	368	29	4	0.96

Table 6. The shortest possible CPLs of the MD4-like compression functions (without feedforward), and the required resources in terms of registers and parallel execution units. A 3-operand instruction format is assumed.

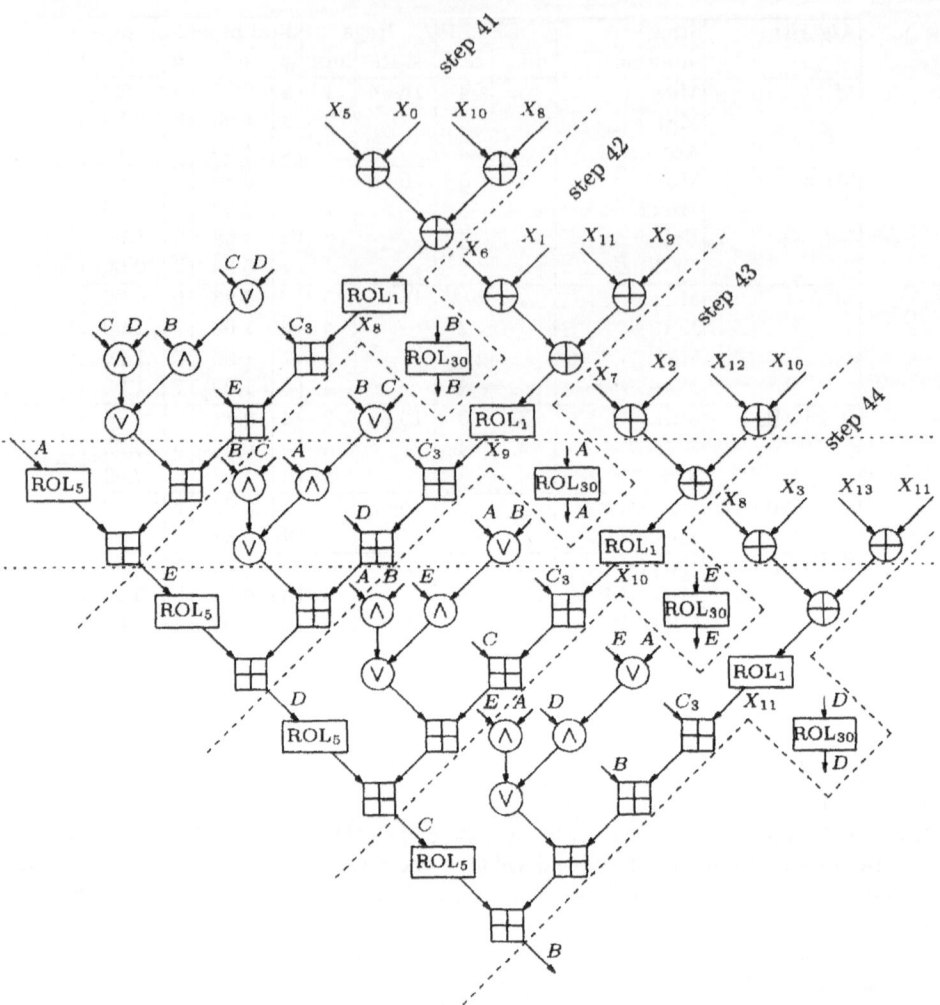

Fig. 1. The first 4 steps of SHA-1's round 3 on a 7-way multiple-issue architecture using a 3-operand instruction format. Instructions executed in parallel are drawn on the same horizontal level, while instructions belonging to the same step are shown between diagonal dotted lines. A CPL of 2 stages is realized by executing 7 instructions of up to 4 different steps in parallel, as shown between the 2 horizontal dotted lines.

in Table 6. To realize a CPL of 2 in round 1 and 3 of SHA-1, two parallel rotates of the same variable are required, see Figure 1. However, the rotate instruction is a unary operation, and hence its 2-operand format has equal source and destination, making a parallel execution on the same variable impossible. Comparing the requirements of Table 6 with the resources of Table 3 shows that current superscalar architectures are only able to exploit all the available instruction-level parallelism of MD4 and MD5, two algorithms that as collision-resistant hash functions can no longer be considered as secure [Dob96a, Dob96b, Rob96].

The natural question to ask is: how realistic are the prospects for a general-purpose processor issuing one day 7 ALU instructions in parallel? Issuing many instructions per clock is difficult due to an increasingly complex issuing logic having a negative impact on the clock cycle time. Therefore, a high issuing rate will only pay off if the parallel execution units are kept sufficiently busy, so that the increase in cycle time will be more than compensated for by an enhanced throughput. The CPL analysis of SHA-1 shows that some algorithms certainly contain enough instruction-level parallelism to sustain such an increased issuing rate, but it is doubtful whether this will be the case for an average instruction sequence.

The RIPEMD-family has, in contrast to SHA-1, two completely independent lines, leaving room for exploiting parallelism on a different level: the use of a multiprocessor system where the multiple-issue capability of each processor is limited, rather than a uniprocessor system with a single, very sophisticated processor capable of offering all the required parallelism on its own. In this respect [HePa96, Section 4.10] states that 'to date, computer architects do not know how to design processors that can effectively exploit instruction-level parallelism in a multiprocessor configuration.' The capability of placing two fully configured processors on a single die, which should be possible around the turn of the century, might result in a new type of architecture allowing processors to be more tightly coupled than before, and at the same time allowing them to achieve very high performance individually. Therefore, exploiting the instruction-level parallelism of the RIPEMD-family in the near future seems much more likely, since each of the independent lines only requires a two-way superscalar architecture, which is already a standard feature of most processors today.

Algorithms with more instruction-level parallelism than the hardware they are executed on can provide, will inevitably see their CPL increase. This is illustrated by means of the first step of MD4's round 2. Using a 3-operand instruction format two parallel units suffice two exploit all available instruction-level parallelism, as illustrated in the left diagram of Figure 2. Remark that the efficiency is 100%. Using a 2-operand instruction format will increase the number of instructions, as operations of the form $A \leftarrow B$ op C will require two instructions: $A \leftarrow B$ and $A \leftarrow A$ op C. Due to the already 100% efficiency of the 3-operand instruction stream, 3 parallel units are now required to realize the same CPL of 4. Therefore, an implementation using only 2 parallel units will inevitably have a longer critical path. This is illustrated in the right diagram of the same figure, showing an increase in CPL of 1 stage. The left diagram is expected to be found on e.g., a PowerPC 604 [SDC94] or a PA 7100LC [BKQW95], while the right diagram resembles the situation on a Pentium processor, except that a Pentium cannot execute a rotate over more than 1 bit in parallel with any other instruction, resulting in a further increase of the CPL.

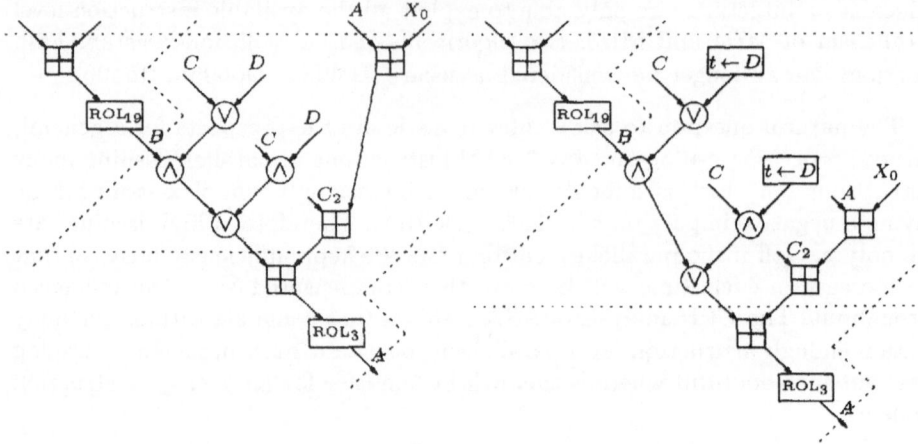

Fig. 2. The first step of MD4's round 2 implemented on a two-way superscalar architecture. Instructions executed in parallel are drawn on the same horizontal level, while instructions belonging to the same step are shown between diagonal dotted lines. The left diagram uses 3-operand instructions, and shows both instruction pipes already occupied for 100%. The use of 2-operand instructions increases the number of instructions by 2, either requiring an additional instruction pipe for the same CPL, or resulting in an increased CPL on the same architecture, as shown on the right.

6 Conclusion

The new generation of customized hash functions based on MD4 (RIPEMD-128, RIPEMD-160, SHA-1) contains more instruction-level parallelism than current general-purpose computer architectures are able to provide. The critical path of SHA-1 is shorter than any of the other MD4-like hash functions, but exploiting it would require a 7-way multiple-issue architecture. Exploiting the instruction-level parallelism of the RIPEMD-family in the near future seems more likely, due to their organization in two independent lines, each of which only requires a 2-way superscalar architecture. Opening up new perspectives is the recent introduction of a new 5-way VLIW processor, primarily intended for multimedia processing.

References

[BKQW95] M. Bass, P. Knebel, D.W. Quint, W.L. Walker, "The PA 7100LC microprocessor: a case study of IC design decisions in a competitive environment," *HP Journal*, Vol. 46, No. 2, April 1995, pp. 12–22.

[Bha96] D.P. Bhandarkar, *Alpha implementations and architecture*, Digital Press, Boston, MA, 1996.

[Ble96]   G.E. Blelloch, "Programming parallel algorithms," *Communications of the ACM*, Vol. 39, No. 3, 1996, pp. 85–97.

[BGV96]   A. Bosselaers, R. Govaerts, J. Vandewalle, "Fast hashing on the Pentium," *Advances in Cryptology, Proceedings Crypto'96, LNCS 1109*, N. Koblitz, Ed., Springer-Verlag, 1996, pp. 298–312.

[Cla97]   C. Clapp, "Optimizing a fast stream cipher for VLIW, SIMD, and superscalar processors," *Fast Software Encryption, LNCS*, E. Biham, Ed., Springer-Verlag, 1997, to appear.

[Dig96]   *Alpha architecture handbook, Version 3*, Digital Equipment Corp., Maynard, MA, 1996.

[Dob96a]   H. Dobbertin, "Cryptanalysis of MD4," *Fast Software Encryption, LNCS 1039*, D. Gollmann, Ed., Springer-Verlag, 1996, pp. 53–69.

.[Dob96b]   H. Dobbertin, "The status of MD5 after a recent attack," *CryptoBytes*, Vol. 2, No. 2, 1996, pp. 1–6.

[DBP96]   H. Dobbertin, A. Bosselaers, B. Preneel, "RIPEMD-160: A Strengthened Version of RIPEMD," *Fast Software Encryption, LNCS 1039*, D. Gollmann, Ed., Springer-Verlag, 1996, pp. 71–82. Final version available via ftp at `ftp.esat.kuleuven.ac.be/pub/COSIC/bosselae/ripemd/`.

[FIPS180-1] FIPS 180-1, "Secure hash standard," US Department of Commerce/NIST, Washington D.C., April 1995.

[Fly66]   M. Flynn, "Very high-speed computing systems," *Proceedings of the IEEE*, Vol. 54, No. 12, 1966, pp. 1901–1909.

[HaKr97]   S. Halevi and H. Krawczyk, "MMH: Software message authentication in the Gbit/second rates," *Fast Software Encryption, LNCS*, E. Biham, Ed., Springer-Verlag, 1997, to appear.

[HePa96]   J.L. Hennessy and D.A. Patterson, *Computer architecture: a quantitative approach, 2nd edition*, Morgan Kaufmann Publishers, San Francisco, 1996.

[Lee95]   R. Lee, "Accelerating multimedia with enhanced microprocessors," *IEEE Micro*, Vol. 15, No. 2, April 1995, pp. 22–32.

[Lee96]   R. Lee, "Subword parallelism with MAX-2," *IEEE Micro*, Vol. 16, No. 4, August 1996, pp. 51–59.

[NMVR95]   D. Naccache, D. M'Raïhi, S. Vaudenay, D. Raphaeli, "Can DSA be improved? Complexity trade-offs with the Digital Signature Standard," *Advances in Cryptology, Proceedings Eurocrypt'94, LNCS 950*, A. De Santis, Ed., Springer-Verlag, 1995, pp. 77–85.

[McC94]   K.S. McCurley, "A fast portable implementation of the secure hash algorithm, III," Technical Report SAND93-2591, Sandia National Laboratories, 1994.

[PeWe96]   A. Peleg and U. Weiser, "MMX technology extension to the Intel architecture," *IEEE Micro*, Vol. 16, No. 4, August 1996, pp. 42–50.

[RIPE95]   RIPE, *"Integrity Primitives for Secure Information Systems. Final Report of RACE Integrity Primitives Evaluation (RIPE-RACE 1040),"* LNCS 1007, A. Bosselaers and B. Preneel, Eds., Springer-Verlag, 1995.

[Riv92a]   R.L. Rivest, "The MD4 message-digest algorithm," Request for Comments (RFC) 1320, Internet Activities Board, Internet Privacy Task Force, April 1992.

[Riv92b]   R.L. Rivest, "The MD5 message-digest algorithm," Request for Comments (RFC) 1321, Internet Activities Board, Internet Privacy Task Force, April 1992.

[Rob96]   M. Robshaw, "On recent results for MD2, MD4 and MD5," Bulletin No. 4, RSA Laboratories, November 1996.

[SRD96]   G.A. Slavenburg, S. Rathnam, H. Dijkstra, "The Trimedia TM-1 PCI VLIW media processor," *Hot Chips VIII Conference*, Stanford University, Palo Alto, CA, 1996.

[SDC94]   S.P. Song, M. Denman, J. Chang, "The PowerPc 604 RISC microprocessor," *IEEE Micro*, Vol. 14, No. 5, October 1994, pp. 8–17.

[Sta96]   P.H. Stakem, *A practitioner's guide to RISC microprocessor architecture*, John Wiley & Sons, New York, 1996.

[Tou95]   J. Touch, "Performance analysis of MD5," *Proceedings of ACM SIG-COMM'95, Comp. Comm. Review*, Vol. 25, No. 4, 1995, pp. 77-86.

[TONH96]   M. Tremblay, J.M. O'Connor, V. Narayanan, L. He, "VIS speeds new media processing," *IEEE Micro*, Vol. 16, No. 4, August 1996, pp. 10–20.

A   Dependencies between consecutive steps

This appendix lists, for each member of the MD4-family, the first two steps of an arbitrary round.

- MD4

$$A := (A + f(B, C, D) + X_i + K)^{\lll s_1}$$
$$D := (D + f(A, B, C) + X_j + K)^{\lll s_2}$$

- MD5

$$A := B + (A + f(B, C, D) + X_i + K)^{\lll s_1}$$
$$D := A + (D + f(A, B, C) + X_j + K)^{\lll s_2}$$

- SHA-1

$$X_i := (X_i \oplus X_{i+2} \oplus X_{i+8} \oplus X_{i+13})^{\lll 1}$$
$$E := E + A^{\lll 5} + f(B, C, D) + X_i + K$$
$$B := B^{\lll 30}$$

$$X_{i+1} := (X_{i+1} \oplus X_{i+3} \oplus X_{i+9} \oplus X_{i+14})^{\lll 1}$$
$$D := D + E^{\lll 5} + f(A, B, C) + X_{i+1} + K$$
$$A := A^{\lll 30}$$

- RIPEMD

$$A := (A + f(B, C, D) + X_i + K)^{\lll s_1}$$
$$D := (D + f(A, B, C) + X_j + K)^{\lll s_2}$$

- RIPEMD-128

$$A := (A + f(B, C, D) + X_i + K)^{\lll s_1}$$
$$D := (D + f(A, B, C) + X_j + K)^{\lll s_2}$$

- RIPEMD-160

$$A := E + (A + f(B, C, D) + X_i + K)^{\lll s_1}$$
$$C := C^{\lll 10}$$
$$E := D + (E + f(A, B, C) + X_j + K)^{\lll s_2}$$
$$B := B^{\lll 10}$$

Fast Arithmetic Architectures for Public-Key Algorithms over Galois Fields $GF((2^n)^m)$

Christof Paar
(christof@ece.wpi.edu)

Pedro Soria-Rodriguez
(sorrodp@ece.wpi.edu)

ECE Department
Worcester Polytechnic Institute
Worcester, MA 01609, USA

Abstract. This contribution describes a new class of arithmetic architectures for Galois fields $GF(2^k)$. The main applications of the architecture are public-key systems which are based on the discrete logarithm problem for elliptic curves. The architectures use a representation of the field $GF(2^k)$ as $GF((2^n)^m)$, where $k = n \cdot m$. The approach explores bit parallel arithmetic in the subfield $GF(2^n)$, and serial processing for the extension field arithmetic. This mixed parallel-serial (hybrid) approach can lead to very fast implementations. The principle of these approach was initially suggested by Mastrovito. As the core module, a hybrid multiplier is introduced and several optimizations are discussed. We provide two different approaches to squaring which, in conjunction with the multiplier, yield fast exponentiation architectures.

The hybrid architectures are capable of exploring the time-space trade-off paradigm in a flexible manner. In particular, the number of clock cycles for one field multiplication, which is the atomic operation in most public-key schemes, can be reduced by a factor of n compared to all other known realizations. The acceleration is achieved at the cost of an increased computational complexity. We describe a proof-of-concept implementation of an ASIC for exponentiation in $GF((2^n)^m)$, m variable.

1 Introduction

Finite fields play an important role in public-key cryptography. Many public-key algorithms are either based on arithmetic in prime fields or on extension fields of $GF(2)$, denoted by $GF(2^k)$. Examples of schemes which can be based on Galois fields of characteristic two include the classical Diffie-Hellman key establishment protocol [1], the ElGamal encryption and digital signature scheme [2], and systems which use elliptic [3] and hyperelliptic curves [4]. Public-key algorithms which explore the assumed difficulty of the discrete logarithm (DL) in finite fields require extension degrees k of about 1000 bits in order to provide reasonable security [5, 6]. Schemes based on the DL problem over (non-supersingular) elliptic curves should have extension degrees of $k \geq 140$ [7]. These long word lengths required for public-key algorithms lead to relatively low performance which is widely recognized as a major shortcoming in practical applications. The provision of fast hardware architectures for arithmetic in Galois fields $GF(2^k)$ is thus of great interest.

W. Fumy (Ed.): Advances in Cryptology - EUROCRYPT '97, LNCS 1233, pp. 363-378, 1997.
© Springer-Verlag Berlin Heidelberg 1997

In the case of algorithms over $GF(2^k)$, addition can be realized with k bit-wise exclusive OR operations. Addition is thus a fast and relatively inexpensive operation. The other field operation, multiplication, on the other hand is very costly in terms of gate count and delay. Multipliers can be classified into bit parallel and bit serial architectures. The former ones compute a result in one clock cycle but have an area requirement of $\mathcal{O}(k^2)$. Bit serial multipliers compute a product in k clock cycles but have an area requirement of $\mathcal{O}(k)$. The two types of architectures are a typical example of the space-time trade-off paradigm. The main idea of this contribution is the introduction of a new class of Galois field arithmetic architectures which are faster than bit serial ones but with an area complexity which is considerably below the k^2 bound of bit parallel ones. It appears that avoiding the two extreme choices provided by bit parallel and bit serial architectures (very fast and large versus relatively slow and small) can lead to architectures with more optimized performance/cost characteristic for many applications. We will refer to the new arithmetic schemes as hybrid architectures. The name and the principle of the architecture was first introduced in [8, Chapter 6]. The reference, however, only describes a hybrid multiplier and does not address optimizations, hybrid squaring, exponentiation, and applications to cryptography as it is done here. The main application of the new architecture are systems based on elliptic curves. It is not recommended to use the architecture for public-key algorithms based on the DL in finite fields since it is based on subfields.

The outline of the remaining paper is as follows. Section 2 summarizes previous approaches of finite field architectures in general and of public-key architectures in particular. Section 3 introduces the general structure of the new multiplier architecture together with several optimizations. Section 4 describes two architecture options for hybrid squaring which enables the design of fast exponentiation units. Section 5 shows the design and results of a proof-of-concept ASIC implementation of an exponentiation unit. Section 6 concludes with a summary of results and a description of areas of application.

2 Previous Work

The use of composite fields $GF((2^n)^m)$ for public-key schemes, more specifically for elliptic curve systems, is described in [9, 10, 11]. All references deal with software implementations which explore table look-up for subfield arithmetic. Neither reference mentions the application to hardware architectures.

Computer architectures for finite field arithmetic have drawn considerable attention over the past decade. The majority of publications have concentrated so far on finite field architectures for relatively small fields, thus being mainly relevant for the implementation of channel codes. The focus in the research literature has been on architectures for the arithmetic operations multiplication [12, 13, 14], inversion [15, 16, 17], and exponentiation [18, 19, 20]. Multiplication in $GF(2^k)$ is usually considered the crucial operation which determines the speed or throughput of a cryptosystem. Finite field architectures can be classified into bit serial (one output bit per clock cycle) and bit parallel ones (all output bits are computed within one clock cycle.) All proposed schemes are based on either of these two types. Architectures which are of hybrid-type (partially serial, and partially parallel), as proposed here, have only be mentioned in the dissertation [8, Chapter 6].

Another classification of Galois field architectures is possible with respect to the basis representation of field elements. The most popular representations are standard (or polynomial or canonical), dual basis, and normal basis. Each basis representation has certain advantages; polynomial and dual basis representations are well suited for bit parallel multipliers, whereas normal basis representation allows for very efficient exponentiation. There have have been a few attempts to compare different types of arithmetic architectures for Galois fields. In [21, 22, 23] are multipliers in different basis representation compared. The focus is mainly on relatively small fields. Reference [18] compares normal and standard basis exponentiation architectures which are relevant for public-key algorithms.

There is a relatively small number of published work on Galois field architectures especially designed for cryptographic applications. Many of the bit serial architectures mentioned above, however, also extend to cryptographic applications. It should be noted that the $\mathcal{O}(k^2)$ complexity bound of parallel multiplier architectures would result in unrealistically large arithmetic units for most public-key algorithms. So far, polynomial basis and normal basis representation have been used for cryptographic applications.

There are two relevant reported implementations which gain their security from the discrete logarithm in finite fields. Reference [24] contains a detailed description of an implementation of an exponentiation unit in the field $GF(2^{593})$, using an optimal normal basis representation of field elements. Reference [25] deals with various aspects of bit serial architectures in Galois fields for cryptographic applications. An implementation of an exponentiation unit in $GF(2^{333})$ using polynomial basis representation is described. In addition, there is the early description of an implementation of a cryptosystem over $GF(2^{127})$ [26].

More recently, there have also been publications about successful implementations of elliptic curve systems in hardware. Reference [27] describes the realization of a non super-singular elliptic curve system over $GF(2^{155})$. Field elements are represented with respect to an optimal normal basis.

3   Hybrid Multipliers

3.1   General Architecture

This subsection describes the general structure of a hybrid multiplier architecture for Galois fields in standard basis. The critical operation in terms of system performance of almost all public-key algorithms is multiplication. Both exponentiation (in schemes based on the DL in finite fields) as well as inversion (in schemes based on the DL over elliptic curves) rely on finite field multiplication as elementary function. The new class of architecture for arithmetic in $GF(2^k)$ will be based on the following two principles:

1. Representation of the field $GF(2^k)$ as $GF((2^n)^m)$, where $nm = k$;
2. Application of bit parallel architectures to arithmetic in the subfield $GF(2^n)$ and of a bit serial structures to arithmetic in the extension field $GF((2^n)^m)$. The goal is to obtain an acceleration by reducing the number of clock cycles required for a field multiplication.

We consider arithmetic in an extension field of $GF(2^n)$. The extension degree is denoted by m, so that the field can be denoted by $GF((2^n)^m)$. This field is isomorphic to $GF(2^n)/(P(x))$, where $P(x)$ is an irreducible polynomial of degree m over $GF(2^n)$. In the following, a residue class will be identified with the polynomial of least degree in this class. For a standard basis multiplier we consider two field elements U, V:

$$U(x) = u_{m-1}x^{m-1} + \cdots + u_1 x + u_0,$$
$$V(x) = v_{m-1}x^{m-1} + \cdots + v_1 x + v_0,$$

where $u_i, v_i \in GF(2^n)$. Field multiplication with the two elements is performed by the operation $W(x) = U(x) \times V(x) \bmod P(x)$, with W being the product element, and $P(x) = x^m + \sum_{i=0}^{m-1} p_i x^i, p_i \in GF(2^n)$, is a monic irreducible polynomial. A possible hardware realization for this operation, polynomial multiplication modulo the field polynomial, is shown in Figure 1. At the kernel of the architecture is a linear feedback shift register (LFSR) of width n and length m. The registers of the LFSR hold the w_i coefficients. The coefficients p_i of the field polynomial are the feedback coefficients of the the LFSR.

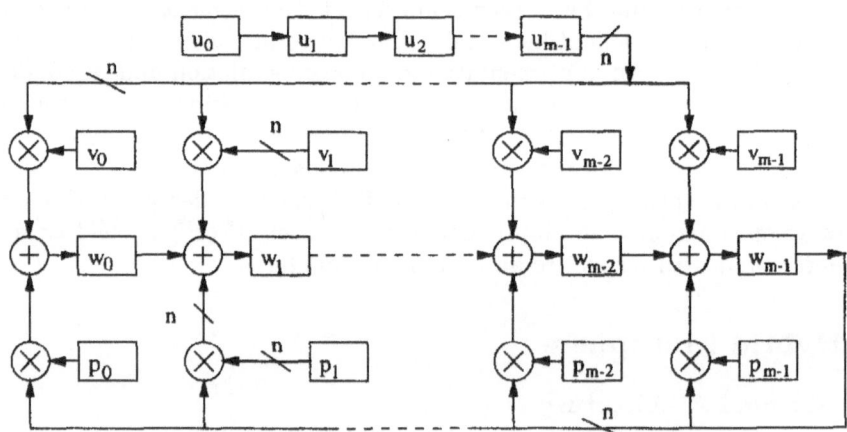

Fig. 1. General structure of a hybrid multiplier in $GF((2^n)^m)$

If $n = 1$, the structure degenerates into one of the classical bit-serial architectures for multiplication in the field $GF(2^m)$ (see, e.g., [28]). In this case all lines are one-bit connections. The U operand is fed into the architectures in a bit serial manner. The product coefficients w_i are available after m clock cycles, i.e., multiplication of m bit operands requires m clock cycles. All hardware implementations of public-key cryptosystems we are ware of are designed with $n = 1$, i.e., $m = k$. For the large m occurring in public-key algorithms, the resulting processing time can be considerable. The complexity of the classical architecture with $n = 1$ is given by:

$$\#AND = 2nm = 2k, \tag{1}$$

$$\#XOR = nm = k, \tag{2}$$

$$\#REG = 3k,$$

$$\#CLK = nm = k,$$

where we consider the number of $GF(2)$ multiplications (AND), additions (XOR), registers (in bits), an number of clock cycles for one multiplication, respectively.

However, if the field $GF(2^k)$ needed in a given cryptographic application allows a composite field extension $k = nm$, $n > 1$, application of the same principal structure leads to the new architecture. In that case, all connections are n bit wide buses and all arithmetic is performed in the subfield $GF(2^n)$. Assuming bit parallel architectures for the subfield multiplication and addition in the LFSR, the result is now computed in m clock cycles. We name this architecture a *hybrid* multiplier. The hybrid architecture reduces the number of clock cycles for one multiplication by a factor of $n = k/m$.

One attractive feature of the hybrid architecture is that it is still highly regular and modular which are very desirable features for VLSI realizations [29]. The multiplier can be built from m identical modules to which we will refer as "slices". Each slice consists of two subfield multipliers, one subfield adder, and three n-bit registers. The only global communication required is an n-bit feedback path which is common to all slices. The architecture allows also full flexibility with respect to the field polynomial $P(x)$. Any monic m degree polynomial over $GF(2^n)$ can be loaded into the architecture. The field polynomial can be changed during operation after each multiplication if desired. The complexity of the general hybrid architecture is given by:

$$\#AND = 2mn^2 = 2nk,$$

$$\#XOR = m(2(n^2 - 1) + n) = 2nk + k - \frac{2k}{n},$$

$$\#REG = 3k,$$

$$\#CLK = m,$$

where a bit parallel subfield multiplier with a complexity of n^2 AND gates and $n^2 - 1$ XOR gates [12] is assumed. It can be seen that the number of logic gates increases by roughly a factor of $2n$ compared to the traditional approach, whereas the number of registers is the same. The major advantage of the hybrid architecture is that the number clock cycles for one multiplication is reduced by a factor of n. The hybrid multiplier explores thus the time-space trade-off paradigm, where the degree of the trade-off (performance versus complexity) is determined by the field decomposition $n \cdot m$. The following section describes two optimizations of the general architecture which result in considerably reduced gate counts.

3.2 Optimizations

Binary Field Polynomials In many public-key algorithms, in particular for elliptic curves schemes, the extension degree m can be chosen such that $\gcd(n, m) = 1$. In this case a field polynomial $P(x)$ which is irreducible over $GF(2)$ is also irreducible over $GF(2^n)$ [30]. In particular, we can now chose a $P(x)$ with coefficients from $GF(2)$. Field polynomials with binary coefficients result in a hybrid multiplier with drastically improved complexity. A block diagram of the improved multiplier is shown in Figure 2.

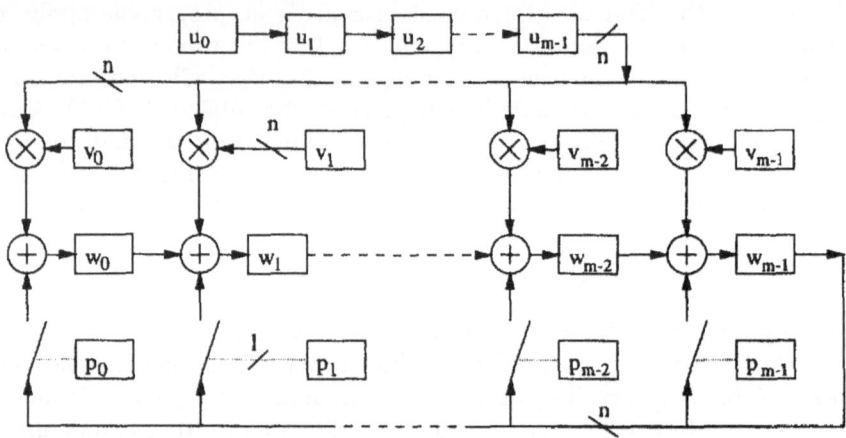

Fig. 2. Hybrid multiplier in $GF((2^n)^m)$ with binary field polynomial

In slice i, the signal from the feedback path is either passed through (coefficient $p_i = 1$) or not processed (coefficient $p_i = 0$). Hence, in each slice, the general multiplier with the polynomial coefficient p_i is now replaced by a binary n-bit switch. A switch can be realized efficiently in digital hardware. In a simple realization the switch can be built by n AND gates, but more efficient realizations, e.g., through transmission gates [29], are also possible. If we neglect the switch complexity relatively to the other components, an over-all complexity of

$$\#\text{AND} = nk, \tag{3}$$

$$\#\text{XOR} = k \left(n + 1 - \frac{1}{n} \right), \tag{4}$$

$$\#\text{REG} = 2k + \frac{k}{n},$$

$$\#\text{CLK} = m,$$

is achieved.

The architecture for binary field polynomials reduces the gate complexity roughly by half and the number of register bits by about one third, compared to the general hybrid multiplier from Section 3.1. It should be noted that the architecture still allows flexibility with respect to the field polynomial. Any irreducible polynomial with coefficients from $GF(2)$ of degree m can be loaded into the architecture and serve as the field polynomial. Also, the high degree of modularity and regularity is preserved with the optimization.

A further optimization is possible if the field polynomial is fixed. In this case switches are replaced by a connection or no connection. The optimum field polynomial for this option are trinomials

Low Complexity Subfield Multiplication The gate complexities in Equations (3) and (4) is now mainly determined by the bit parallel subfield multiplier. Applying a more efficient architecture to the subfield multiplication can thus be very beneficial to the over-all system complexity.

So far we assumed a subfield architecture with a complexity of n^2 AND gates and $n^2 - 1$ XOR gates. The vast majority of bit parallel architectures has at least these gate counts. However, the complexity can be further reduced by applying the bit parallel architecture described in [31, 32]. The multipliers are based on a representation of the subfield $GF(2^n)$ through another field decomposition $GF((2^o)^p)$, where $n = o \cdot p$. In particular, for values of $n = 6, \ldots, 14$, n even, a representation of $GF(2^n) \cong GF((2^{n/2})^2)$ will lead to highly efficient architectures. This range of values for n appears to be very attractive for practical applications such as elliptic curve cryptosystems. Elements A, B of the subfield are now represented as polynomials with a maximum degree of one over $GF(2^{n/2})$: $A(y) = a_1 y + a_0$ and $B(y) = b_1 y + b_0$, where $a_0, a_1, b_0, b_1 \in GF(2^{n/2})$. The complexity of one subfield multiplication can be reduced to $\#\text{AND} = 3/4 n^2$ and $\#\text{XOR} \approx 3/4 n^2 + 2n - 3$ [32].

It should be noted that for the specific value of $n = 8$, it has been shown [23] that a multiplier based on the decomposition $GF((2^4)^2)$ requires in fact a considerably smaller number of gate equivalences in an ASIC implementation than architectures based on $GF(2^8)$. At the same time, the former architecture was found to be faster than the latter ones which is an attractive feature since the subfield multiplier is in the critical path of the architecture.

The structure of a hybrid multiplier with a decomposed subfield $GF((2^{n/2})^2)$ is still given by Figure 2 but the complexity is reduced to:

$$\#\text{AND} = \frac{3}{4} nk,$$

$$\#\text{XOR} = k \left(\frac{3}{4} n + 3 - \frac{3}{n} \right)$$

$$\#\text{REG} = (2 + \frac{1}{n}) k.$$

$$\#\text{CLK} = m.$$

The introduction of the subfield decomposition has thus reduced the gate complexity by roughly 25%. If the complexity after the two optimization is compared with the one of the bit serial architecture given in (1) and (2), it can be seen that the time performance (clock cycles) improves by a factor of n, whereas the gate complexity increases only by a factor of $3/4n$.

3.3 Comparison to Normal Base Multipliers

Some implementations of public-key schemes over fields $GF(2^k)$ use a normal basis (NB) representation of field elements [24, 27]. The computational complexity for one multiplication depends heavily on the specific field polynomial [33]. A lower complexity bound, however, is given for irreducible polynomials which have a corresponding optimum normal basis [34]. Assuming an optimum normal basis for non-composite fields $GF(2^k)$, the complexity is given by #AND $= k$, #XOR $= 2k - 2$, and #CLK $= k$. The NB multiplier requires thus n times as many clock cycles as the hybrid multiplier but has a lower gate count. Other advantages of the hybrid architectures are that the field polynomial can be changed and that the field extension m can be alterable, as will be explained in Section 5.1. However, a major advantage of a NB is that squaring can be accomplished through a simple cyclic shift whereas squaring with the hybrid architectures is more costly as will be described in the following section.

4 Squaring and Exponentiation

Besides multiplication, the other arithmetic operation of central importance for the implementation of public-key algorithms is squaring. Systems based on the the DL problem for non-supersingular elliptic curves require two multiplications and one inversion, with respect to time-critical arithmetic, per group operation if non-projective coordinates are used. A popular method for inversion in hardware is based on Fermat's Little Theorem, according to which $A^{-1} = A^{2^k-2}$, $\forall A \in GF(2^k)$, $A \neq 0$. Although the extended Euclidean algorithm has a better theoretical performance, it requires more operands which in turn need more registers. It is thus less attractive for hardware implementations. The exponentiation in Fermat's Theorem can be realized using addition chains. The standard approach to exponentiation is the square-and-multiply algorithm or one of its derivatives (additions chains, sliding window, etc.) [35]. If the inputs to the algorithm are denoted by A and e and the output is the value A^e, each iteration stage of the algorithm performs one of the two operations:

1. Multiply result of previous iteration with A.
2. Square result of previous iteration.

The hybrid multiplier architecture from the previous section can be applied to the first operation. In this section, two architectures for squaring a result of a preceding multiplication (or squaring) will be developed which dovetail with the multiplier architecture.

4.1 Serial Squaring

The first architecture is based on the application of the general multiplier from Section 3 to squaring. We assume that the variable to be squared is contained in the registers $w_i, i = 0, 1, \ldots, m - 1$, of a hybrid multiplier as a result of a preceding multiplication or squaring. In order to square this variable we must assure that its coefficients are available as both inputs of the multiplier. This is achieved by the following two operations:

Preparation of operands: Before start of squaring, load values from register w_i into input register v_i for $i = 0, 1, \ldots, m - 1$. This can be performed simultaneously in all slices.

Squaring: Perform regular multiplication in m clock cycles. In clock cycle i, connect variable v_i as global input coefficient to all slices.

A corresponding hardware architecture is shown in Figure 3. It can be seen that the squaring functionality can be added to the hybrid multiplier with a modest amount of additional hardware. In every slice two switches must be added. Globally, a single control unit must be added to the system. The switches sp perform the initial parallel loading of the v_i registers. The switches ss allow for the generation of the global input coefficients during the multiplication cycles. The control logic assures that switch ss_{m-1-i} (and only switch ss_{m-1-i}) is closed during cycle i. The control logic can be realized as a counter with $\lceil \log_2 m \rceil$ bits and a $(\lceil \log_2 m \rceil)$-to-m decoder.

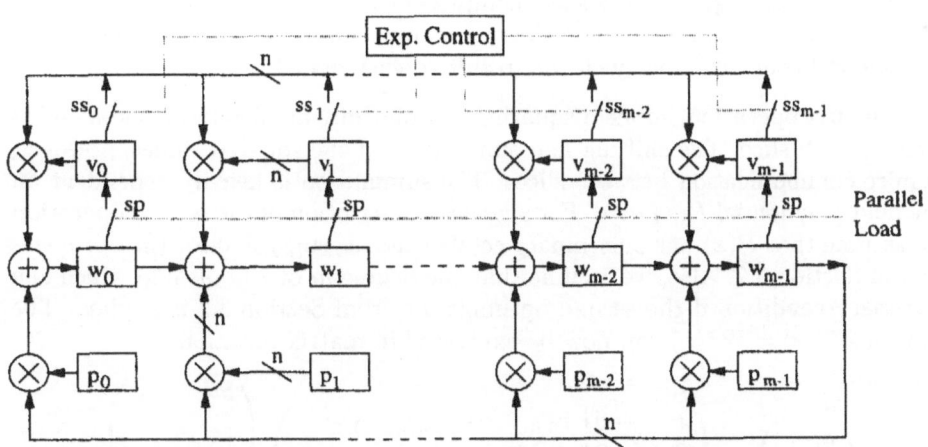

Fig. 3. Structure of a serial squarer for $GF((2^n)^m)$

The squaring functionality can be added to all three multiplier options discussed in Section 3. As stated earlier, switches can be realized very efficiently in ASIC implementations. The computational complexity of the expanded architecture is thus essentially the same as the hybrid multiplier complexity. It should be noted that a single squaring requires the same time as a general multiplication, namely m clock cycles. This is a major drawback compared to NB architectures which realize squaring in a single clock cycle by means of a cyclic shift. The following section introduces a much faster but more costly approach to squaring.

4.2 Parallel Squaring

Exponentiation with a k bit exponent requires $k - 1$ squarings and on average not more than $(k - 1)/2$ multiplications. Hence, a squaring architecture which requires fewer clock cycles than the one from the previous section can greatly improve the performance of a public-key system based on exponentiation. In the following we assume again that the input operand for the squaring is being held in the w registers of the hybrid multiplier. The architecture computes the result $T(x) = W^2(x)$ in one clock cycle, and puts the t_i coefficients in the w_i registers. For the development of a parallel squarer, i.e., squaring within one clock cycle, we note that [30]

$$T(x) = W^2(x) = \left(\sum_{i=0}^{m-1} w_i x^i \right)^2 = \sum_{i=0}^{m-1} w_i^2 x^{2i} = \sum_{i=0}^{m-1} t_i x^i \quad (\text{mod } P(x)).$$

A realization of this operation must provide the following two extensions to a hybrid multiplier:

Subfield squaring: In every slice, compute w_i^2.
Shift of coefficients and modulo reduction: Shift and summation of the squared coefficients w_i^2 yield the result coefficients t_i.

The first operation, subfield squaring, is uniformly applied to all slices and is local to each slice. The shifting and summation of the squared values, however, require communication between slices. The summation is heavily dependent on the field polynomial $P(x)$ used. For a general description of the second operation we assume that $P(x)$ has only binary coefficients, as suggested for the optimization of Section 3.2. Also, we assume that the degree m of $P(x)$ is odd which is a necessary condition if the second optimization from Section 3.2 is applied. The squaring $T(x) = W^2(x)$ can now be expressed in matrix notation as

$$\begin{pmatrix} t_0 \\ t_1 \\ \vdots \\ t_{m-1} \end{pmatrix} = \left(\begin{array}{cccc|ccc} 1 & 0 & \cdots & 0 & r_{0,0} & \cdots & r_{0,(m-3)/2} \\ 0 & 0 & \cdots & 0 & r_{1,0} & \cdots & r_{1,(m-3)/2} \\ 0 & 1 & \cdots & 0 & r_{2,0} & \cdots & r_{2,(m-3)/2} \\ \vdots & \vdots & & \vdots & \vdots & & \vdots \\ 0 & 0 & \cdots & 1 & r_{m-1,0} & \cdots & r_{m-1,(m-3)/2} \end{array} \right) \begin{pmatrix} w_0^2 \\ \vdots \\ w_{(m-1)/2}^2 \\ w_{(m+1)/2}^2 \\ \vdots \\ w_{m-1}^2 \end{pmatrix}, \quad (5)$$

where $r_{i,j} \in GF(2)$. This "reduction matrix" consists of two binary sub-matrices: A $m \times (m+1)/2$ matrix which describes the shift of the values $w_0^2, \ldots, w_{(m-1)/2}^2$, and a $m \times (m-1)/2$ matrix which describes the modulo reduction summation of the coefficients $w^2_{(m+1)/2}, \ldots, w^2_{m-1}$. The actual elements $r_{i,j}$ of the reduction matrix depend heavily on the specific field polynomial $P(x)$ used. In order to obtain low computational and connectivity complexities, it is desired to use an irreducible polynomial with low coefficient weight. In the following we will develop a complexity expression for field polynomials of the type $P(x) = x^m + x + 1$ which yield the lowest possible modulo reduction complexity.

If we assume a non-optimized binary field polynomial of degree n for the subfield, it can be shown that one subfield squaring requires on average $(n^2 + 2n-4)/4$ XOR gates. Using the trinomial $P(x) = x^m + x + 1$ results in a reduction matrix with $m-1$ "one" entries. The matrix-vector multiplication in (5) requires then exactly $(m-1)/2$ additions in $GF(2^n)$ or $n(m-1)/2$ XOR gates. Summing of the two complexity contributions, subfield squaring and modulo reduction, yields an over-all gate complexity for the parallel squarer of

$$\#\mathrm{XOR} = \frac{1}{4}\left(kn + 3k - 4\frac{k}{n} - 2n\right).$$

If this gate complexity is compared to the gate count of the hybrid multiplier given in (3) and (4) it can be seen that the parallel squarer has about $1/8$ of the hybrid multiplier gate count. Hence, adding a parallel squarer to a hybrid multiplier only modestly increase the computational complexity of the system if the field polynomial $P(x)$ is chosen with care. It should be stressed at this point that the architecture performs one squaring in $GF((2^n)^m)$ in one single clock cycle. The trade-off, however, is that the parallel squarer requires communication between slices in a relatively irregular manner, so that the connectivity complexity of the system would increase.

5   Proof-of-Concept Implementation

In order to gain further experience with the new class of finite field architectures, we performed a proof-of-concept hardware implementation. First we will show how a hybrid exponentiator for variable field extension m can be built.

5.1   Variable Field Order

In some cryptographic applications it is desirable to allow for an alterable order of the underlying finite field. If we impose the restriction that the subfield order 2^n is fixed, architectures with variable extension degree m can be designed from the hybrid multiplier and the serial squarer. We can essentially apply the architecture in Figure 1 for a design that allows for variable m.

With a modest amount of additional hardware, an exponentiation architecture with m slices can be programmed to use s slices, where $s \leq m$. In order to perform arithmetic in the field $GF((2^n)^s)$ with the m-slice architecture, the connection between slice $s-1$ and slice s is open, and the output of register w_{s-1} is redirected to the feedback data bus. This can be done with one switch st

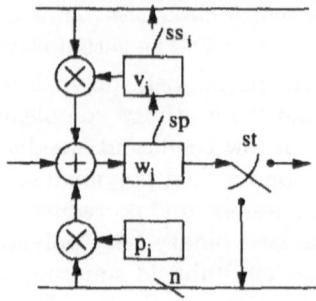

Fig. 4. General slice structure for variable extension degree m

(see Figure 4) which connects the output of slice $s-1$ (i in the figure) to either the next slice or to the feedback loop, but not both. Since the feedback happens now between slices $s-1$ and s, only s slices are used to perform a multiplication. Slices s and above are unused because there is no communication between them and the lower s slices.

As we mentioned, if only s slices are used, a connection has to be open and another connection made. This implies a need for digital switches between every pair of slices, and from the output of each slice to the feedback bus. Although switching can be done fast, this is not of major importance in this architecture because the actual switching from, say, s to t operative slices is done once to set up the desired configuration. However, during the actual computation of a product, these switches are located in the path of data flow, adding a delay to the propagation of the result from one slice to the next slice.

5.2 Prototype Implementation

We implemented the most general multiplier architecture described in Section 3.1 and the serial squarer from Section 4.1, with variable m and $n = 8$ [36]. One slice of our implementation has thus the structure shown in Figure 4. We applied a full-custom design approach using CMOS technology. The choice of this technology allowed us to area-optimize the implementation as opposed to, for instance, a VHDL-based realization. The fabrication facilities available for this research project enabled us to use $2\mu m$ technology. Obviously, this imposes a serious speed limitation on the prototype compared to current technologies. However, our goal was not to compete with commercial or semi-commercial implementations but rather to demonstrate the principle feasibility of the architectures and to obtain reliable area estimations. Similarly we did not implement an entire public-key system. Our main interest was to study the underlying arithmetic architectures.

For the two subfield multipliers in each design we used the architecture [12]. The implementation uses transmission gates (X-gates) to discern whether slice i is the final one in the chain or not. The transmission gates offer better switching characteristic than a pass transistor implementation [29], but still introduce a delay in passive mode. The coefficients of $V(x), W(x)$, and $P(x)$ are stored in 8-bit latches. This kind of memory requires less area than a SRAM implementation. While other types of memory elements can operate faster than a simple latch, they also consume more area.

The test ASICs that we implemented contains four slices, each of which requires 3040 transistors or 760 gate equivalences. Using the optimized architecture in Section 3.2, the gate count would roughly be reduced by a factor of 1/2 and only 400 gates would be needed per slice. This estimate indicates that hybrid architectures with a subfield of $n = 8$ are feasible even for relatively large field orders. An example of a field order which is well suited for an elliptic curve cryptosystem is $k = 152 = 8 \cdot 19$. For this field order we obtain an estimate of about 8000 gate equivalences for an exponentiation (or inversion) architecture. These would even allow for a realization with reprogrammable logic (FPGA, EPLD).

Our implementation allows a clock rate of 3.5 Mhz. A multiplication in $GF(2^k)$ with $k = 152 = 8 \cdot 19$ takes thus $5.4\mu\text{sec}$. Although we did not implement an elliptic curve system we can derive the following rough estimates: Using projective coordinates a point addition requires 7 multiplications or $70.2\mu\text{sec}$, and a point doubling 13 multiplications or $37.8\mu\text{sec}$. A point multiplication with a 152-bit integer would then take 5.3 msec on average, using the standard double-and-add algorithm. This estimate does not take any overhead into account, but also ignores possible improvement of the double-and-add algorithm (k-ary, sliding window). We would like to stress that the implementation is by no means speed optimized (but area optimized) and that we used relatively slow technology. We expect that the use of the parallel squaring architecture from Section 4.2, state-of-the-art technology ($0.8\mu\text{m}$ or smaller), and application of the faster subfield multiplier from Section 3.2 would lead to a very competitive performance of the exponentiation unit.

6 Conclusions and Applications

We developed new types of multipliers and squarers for Galois fields $GF(2^k)$. The multiplication and squaring architectures are designed such that exponentiation units can be built which are of central interest for public-key cryptosystems. The underlying idea is to represent the field $GF(2^k)$ by $GF((2^n)^m)$, $k = n \cdot m$, and to apply bit parallel architectures to arithmetic in the subfield and a serial approach to the extension field arithmetic. The main feature of the new hybrid architectures is that they have the potential of being considerably faster than previously reported public-key architectures for finite field arithmetic. Hybrid multiplication requires only m clock cycles as opposed to k in traditional fully bit serial approaches. The principal feasibility of the approach was demonstrated in an ASIC implementation for the field $GF((2^8)^m)$, m variable. It appears that hybrid architectures could result in improved performance for several important public-key schemes.

The most attractive public-key schemes for the new architecture are those based on elliptic curves. Most reported implementation of elliptic curve systems over Galois fields $GF(2^k)$ already use a composite field extension k (e.g., $k = 155$ [27], or $k = 176$ [10, 11]), although not all of them explore subfield arithmetic. This situation is ideally suited for hybrid architectures. Also, since values of $k = 140\ldots200$ provide high security against currently known attacks, hybrid architectures for elliptic curves can be realized with a moderate gate complexity.

Another type of cryptosystem which can be used in conjunction with our architecture are those based on hyperelliptic curves [4], practical aspects of which are described in [37, 38]. Secure one-way functions can be built with $k < 100$, where k can be composite. This range of field orders seems also very well suited for hybrid architectures.

Finally we would like to stress that the DL in finite fields appears to be insecure for composite Galois fields $GF((2^n)^m)$ [5]. Hence it is not recommended to apply the hybrid architecture to such schemes, which include, for instance, the classical Diffie-Hellman key exchange protocol.

References

1. W. Diffie and M. Hellman, "New directions in cryptography," *IEEE Transactions on Information Theory*, vol. IT-22, pp. 644–654, 1976.
2. T. ElGamal, "A public-key cryptosystem and a signature scheme based on discrete logarithms," *IEEE Transactions on Information Theory*, vol. IT-31, no. 4, pp. 469–472, 1985.
3. V. Miller, "Uses of elliptic curves in cryptography," in *Lecture Notes in Computer Science 218: Advances in Cryptology — CRYPTO '85*, pp. 417–426, Springer-Verlag, Berlin, 1986.
4. N. Koblitz, "Hyperelliptic cryptosystems," *Journal of Cryptology*, vol. 1, no. 3, pp. 129–150, 1989.
5. L. Adleman and J. DeMarrais, "A subexponential algorithm for discrete logarithms over all finite fields," in *Advances in Cryptography — CRYPTO '93*, pp. 147–158, Springer-Verlag, 1993.
6. D. Gordon and K. McCurley, "Massively parallel computation of discrete logarithms," in *Lecture Notes in Computer Science 453: Advances in Cryptology — CRYPTO '92* (E. Brickell, ed.), pp. 312 – 323, Springer-Verlag, Berlin, August 1993.
7. A. Menezes, *Elliptic Curve Public Key Cryptosystems*. Kluwer Academic Publishers, 1993.
8. E. Mastrovito, *VLSI Architectures for Computation in Galois Fields*. PhD thesis, Linköping University, Dept. Electr. Eng., Linköping, Sweden, 1991.
9. G. Harper, A. Menezes, and S. Vanstone, "Public-key cryptosystems with very small key lengths," in *Advances in Cryptology — EUROCRYPT '92*, pp. 163–173, May 1992.
10. E. D. Win, A. Bosselaers, S. Vandenberghe, P. D. Gersem, and J. Vandewalle, "A fast software implementation for arithmetic operations in $GF(2^n)$," in *Asiacrypt '96*, Springer Lecture Notes in Computer Science, 1996.
11. D. Beauregard, "Efficient algorithms for implementing elliptic curve public-key schemes," Master's thesis, ECE Dept., Worcester Polytechnic Institute, Worcester, Massachusetts, May 1996.

12. E. Mastrovito, "VLSI design for multiplication over finite fields $GF(2^m)$," in *Lecture Notes in Computer Science 357*, pp. 297–309, Springer-Verlag, Berlin, March 1989.

13. M. Hasan, M. Wang, and V. Bhargava, "Modular construction of low complexity parallel multipliers for a class of finite fields $GF(2^m)$," *IEEE Transactions on Computers*, vol. 41, pp. 962–971, August 1992.

14. S. Fenn, M. Benaissa, and D. Taylor, "$GF(2^m)$ multiplication and division over the dual base," *IEEE Transactions on Computers*, vol. 45, pp. 319–327, March 1996.

15. G. Feng, "A VLSI architecture for fast inversion in $GF(2^m)$," *IEEE Transactions on Computers*, vol. C-38, p. 1989, Oct 1989.

16. M. Morii and M. Kasahara, "Efficient construction of gate circuit for computing multiplicative inverses over $GF(2^m)$," *Transactions of the IEICE*, vol. E 72, pp. 37–42, January 1989.

17. S. Fenn, M. Benaissa, and D. Taylor, "Finite field inversion over the dual base," *IEEE Transactions on VLSI Systems*, vol. 4, pp. 134–136, March 1996.

18. W. Geiselmann and D. Gollmann, "VLSI design for exponentiation in $GF(2^n)$," in *Lecture Notes in Computer Science 453: Advances in Cryptology — AUSCRYPT '90* (J. Seberry and J. Pieprzyk, eds.), (Sydney, Australia), pp. 398–405, Springer-Verlag, Berlin, January 1990.

19. C. Wang and D. Pei, "A VLSI design for computing exponentiation in $GF(2^m)$ and its application to generate pseudorandom number sequences," *IEEE Transactions on Computers*, vol. C-39, pp. 258–262, February 1990.

20. M. Hasan and V. Bhargava, "Low complexity architecure for exponentiation in $GF(2^m)$," *Electronics Letters*, vol. 28, pp. 1984–86, October 1992.

21. I. Hsu, T. Truong, L. Deutsch, and I. Reed, "A comparison of VLSI architecture of finite field multipliers using dual-, normal-, or standard bases," *IEEE Transactions on Computers*, vol. 37, pp. 735–739, June 1988.

22. Y. Jeong and W. Burleson, "Choosing VLSI algorithms for finite field arithmetic," in *IEEE Symposium on Circuits and Systems, ISCAS 92*, 1992.

23. C. Paar and N. Lange, "A comparative VLSI synthesis of finite field multipliers," in *3rd International Symposium on Communication Theory and its Applications*, (Lake District, UK), July 10–14 1995.

24. G. Agnew, R. Mullin, I. Onyschuk, and S. Vanstone, "An implemenation for a fast public-key cryptosystem," *Journal of Cryptography*, vol. 3, pp. 63–79, 1991.

25. W. Gollmann, "Algorithmenentwurf in der Kryptographie." Habilitation, Fakultät für Informatik, Universität Karlsruhe, Germany, August 1990.

26. K. Yiu and K. . Peterson, "A single-chip VLSI implemenation of the discrete exponential public-key distribution system," *IBM Systems Journal*, vol. 15, no. 1, pp. 102–116, 1982.

27. G. Agnew, R. Mullin, and S. Vanstone, "An implementation of elliptic curve cryptosystems over $F_{2^{155}}$," *IEEE Journal on Selected areas in Communications*, vol. 11, pp. 804–813, June 1993.

28. S. Lin and D. Costello, *Error Control Coding: Fundamentals and Applications*. Englewood Cliffs, NJ: Prentice-Hall, 1983.

29. N. Weste and K. Eshraghian, *Principles of CMOS VLSI Design, A Systems Perspective*. Addison-Wesley Publishing Company, second ed., 1992.

30. R. Lidl and H. Niederreiter, *Finite Fields*, vol. 20 of *Encyclopedia of Mathematics and its Applications*. Reading, Massachusetts: Addison-Wesley, 1983.

31. V. Afanasyev, "On the complexity of finite field arithmetic," in *5th Joint Soviet-Swedish Intern. Workshop on Information Theory*, (Moscow, USSR), pp. 9–12, January 1991.

32. C. Paar, "A new architecture for a parallel finite field multiplier with low complexity based on composite fields," *IEEE Transactions on Computers*, vol. 45, pp. 856–861, July 1996.

33. W. Geiselmann, *Algebraische Algorithmenentwicklung am Beispiel der Arithmetik in Endlichen Körpern*. PhD thesis, Universität Karlsruhe, Fakultät für Informatik, Institut für Algorithmen und Kognitive Systeme, Karlsruhe, Germany, 1993.

34. R. Mullin, I. Onyszchuk, S. Vanstone, and R. Wilson, "Optimal normal bases in $GF(p^n)$," *Discrete Applied Mathematics, North Holland*, vol. 22, pp. 149–161, 1988/89.

35. D. Knuth, *The Art of Computer Programming. Volume 2: Seminumerical Algorithms*. Reading, Massachusetts: Addison-Wesley, 2nd ed., 1981.

36. M. Lehky, M. Nappi, and P. Soria-Rodriguez, "Coprocessor board for cryptographic applications." Major Qualifying Project (Senior Thesis), 1996. ECE Dept., Worcester Polytechnic Institute.

37. A.-M. Spallek, *Kurven vom Geschlecht 2 und ihre Anwendung in Public-Key-Kryptosystemen*. PhD thesis, Institute for Experimental Mathematics, University of Essen, Essen, Germany, July 1994.

38. S. Paulus, *Ein Algorithmus zur Berechnung der Klassengruppe quadratischer Ordnungen über Hauptidealringen*. PhD thesis, Institute for Experimental Mathematics, University of Essen, Essen, Germany, June 1996.

Finding Good Random Elliptic Curves for Cryptosystems Defined over F_{2^n}

Reynald Lercier

CELAR/CASSI, Route de Laillé, F-35170 Bruz, FRANCE
email: lercier@lix.polytechnique.fr

Abstract. One of the main difficulties for implementing cryptographic schemes based on elliptic curves defined over finite fields is the necessary computation of the cardinality of these curves. In the case of finite fields F_{2^n}, recent theoretical breakthroughs yield a significant speed up of the computations. Once described some of these ideas in the first part of this paper, we show that our current implementation runs from 2 up to 10 times faster than what was done previously. In the second part, we exhibit a slight change of Schoof's algorithm to choose curves with a number of points "nearly" prime and so construct cryptosystems based on random elliptic curves instead of specific curves as it used to be.

1 Introduction

It is well known that the discrete logarithm problem is hard on elliptic curves defined over finite fields F_q. This is due to the fact that the only known attacks (baby steps giant steps [Sha71], Pollard ρ [Pol78] and Pohlig-Hellman [PH78] methods) are still exponential in $\log q$. So, cryptosystems based on this problem can reach the same level of security as non elliptic versions with slightly higher computation rates and much smaller keys [SOOS95, HMV93].

The remaining difficulty to design elliptic cryptosystems is the computation of the cardinality of elliptic curves. Until recently, it was usually admitted that the cost needed to perform this task was too high for randomly chosen curves. To tackle this difficulty, one used to consider specific curves, for instance, supersingular curves [Mil87, Kob87, Kal86, BC89, MV90] or curves with complex multiplication [Mor91, Kob91, Miy91, Miy93, LZ94, CTT94]. Unfortunately, supersingular curves turned out to be disastrous and so, the use of specific curves seems to be quite compromised for cryptographical purposes [MOV93].

Thanks to recent theoretical as well as practical developments, the cost of computing the number of points on a randomly chosen curve is no longer prohibitive. For finite fields of characteristic two (specially attractive for industrial applications), the improvements of Schoof's algorithm due to Atkin, Elkies, Morain, Couveignes, Müller, Dewaghe, ... [Sch85, CM94, Mül95, Sch95, CDM96] were significantly speeded up by replacing the isogeny computation algorithm of Couveignes [Cou94] with a recent heuristic algorithm of the author [Ler96].

In this article, once briefly recalled some basic facts about elliptic curves in Section 2, we describe in Section 3 our current implementation of these ideas

W. Fumy (Ed.): Advances in Cryptology - EUROCRYPT '97, LNCS 1233, pp. 379-392, 1997.
© Springer-Verlag Berlin Heidelberg 1997

and we explain in Section 4 how we can take advantage of Schoof's algorithm for speeding up the search of an elliptic curve with a nearly prime number of points. Among others, it turns out that we are now able to compute the cardinality of any elliptic curve for sizes of finite fields recommended for cryptographical schemes in only a few seconds, that is to say a speed up factor from 2 up to 10 compared to our previous implementation [LM95a].

2   Elliptic Curves over $\mathbb{F}_{2^n}$

Following [Men93], we consider for our purposes elliptic curves over $\mathbb{F}_{2^n}$ defined by

$$E_a : y^2 + xy = x^3 + a, \ a \in \mathbb{F}_{2^n}^*. \tag{1}$$

Any non supersingular elliptic curve is isomorphic to a curve or the twist of a curve defined by this equation. Invariant J_a and discriminant Δ_a of E_a are equal to

$$J_a = 1/a \text{ and } \Delta_a = a.$$

Let us note that E_a can not be supersingular because, in $\mathbb{F}_{2^n}$, an elliptic curve is supersingular if and only if its invariant is equal to 0 (on the explicit determination of supersingular curves in finite fields of odd characteristic, see [Mor96]).
   The set of points of E_a over $\mathbb{F}_{2^n}$ is

$$E_a(\mathbb{F}_{2^n}) = \{O_{E_a}\} \cup \{(x, y) \in \mathbb{F}_{2^n}^2, y^2 + xy = x^3 + a\}.$$

This set is a finite group and the formulae of the abelian group law are:

- $\forall P = (x_P, y_P) \in E_a(\mathbb{F}_{2^n}), P + O_{E_a} = O_{E_a} + P = P, -P = (x_P, y_P + x_P);$
- if $P = (x_P, y_P), Q = (x_Q, y_Q), P \neq -Q$, then, if $P = Q$, let $\lambda = x_P + y_P/x_P$, otherwise let $\lambda = (y_Q + y_P)/(x_Q + x_P)$, and $R = P + Q = (x_{P+Q}, y_{P+Q})$ is obtained by

$$\begin{cases} x_{P+Q} = \lambda^2 + \lambda + x_P + x_Q, \\ y_{P+Q} = \lambda(x_P + x_{P+Q}) + x_{P+Q} + y_P. \end{cases}$$

Some endomorphisms will be of special interest in Section 3, namely $[m]_a$, multiplication by any integer m on E_a and ϕ_a, the Frobenius map. These endomorphisms are defined as follows.

$$[m]_a : \begin{array}{c} E_a(\overline{\mathbb{F}}_{2^n}) \longrightarrow E_a(\overline{\mathbb{F}}_{2^n}), \\ (x, y) \longmapsto m(x, y), \end{array} \text{ and } \phi_a : \begin{array}{c} E_a(\overline{\mathbb{F}}_{2^n}) \longrightarrow E_a(\overline{\mathbb{F}}_{2^n}), \\ (x, y) \longmapsto (x^{2^n}, y^{2^n}). \end{array}$$

In particular, multiplication by 2 is given by

$$[2]_a : \begin{array}{c} E_a(\overline{\mathbb{F}}_{2^n}) \longrightarrow E_a(\overline{\mathbb{F}}_{2^n}), \\ (x, y) \longmapsto \left(x^2 + \dfrac{a}{x^2}, \left(x + \dfrac{y}{x}\right)\left(x^2 + \dfrac{a}{x^2}\right) + \dfrac{a}{x^2}\right). \end{array} \tag{2}$$

Equation (2) shows that there exists a single point $P_a = (0, \sqrt{a})$ of order 2 on these curves and the formulae of the translation by P_a are

$$T_a : \begin{array}{l} E_a(\overline{\mathbb{F}}_{2^n}) \longrightarrow E_a(\overline{\mathbb{F}}_{2^n}), \\ P = (x, y) \longmapsto P + P_a = \left(\dfrac{\sqrt{a}}{x}, \sqrt{a} + \dfrac{\sqrt{a}}{x} + \dfrac{a}{x^2} + \sqrt{a}\dfrac{y}{x^2} \right). \end{array}$$

3 Counting Points on Elliptic Curves

The number of points of a *non supersingular* elliptic curve E_a defined over $\mathbb{F}_{2^n}$ satisfies Hasse's inequality [Sil86],

$$\#E_a(\mathbb{F}_{2^n}) = 2^n + 1 - t, \text{ with } |t| < 2\sqrt{2^n}. \tag{3}$$

Before 1985, the only known methods to compute this number consisted in testing all the possible integers t in Equation (3) with baby steps giant steps variants [Sha71]. The complexity of these algorithms is asymptotically $O(2^{n/4})$. With the work by Schoof [Sch85] and the numerous improvements that followed, it is now possible to compute this cardinality with a probabilistic complexity asymptotically equal to $O(n^6)$. We briefly describe this method in Section 3.1.

The heart of these algorithms is the computation of isogenies. In practice, the most efficient method to do that in $\mathbb{F}_{2^n}$ seems to be a heuristic algorithm due to the author [Ler96] and we overview it in Section 3.2. Thanks to this algorithm, first, we were able to speed up our previous implementation [LM95a] by a significant factor and secondly, compute the cardinality of an elliptic curve defined over $\mathbb{F}_{2^{1301}}$.

3.1 The Schoof-Elkies-Atkin Algorithm

The characteristic equation satisfied by the Frobenius map ϕ_a is

$$\phi_a^2 - [t]_a \circ \phi_a + [2^n]_a = 0, \tag{4}$$

where $2^n + 1 - t$ is the cardinality of $E_a(\mathbb{F}_{2^n})$. First, Schoof remarked, that once restricted to the ℓ^2 points of the kernel $E_a[\ell]$ of the multiplication $[\ell]_a$ (ℓ an odd prime), Equation (4) yields

$$\phi_{E_a[\ell]}^2 + [2^n \bmod \ell]_{E_a[\ell]} = [t \bmod \ell]_{E_a[\ell]} \circ \phi_{E_a[\ell]}. \tag{5}$$

Schoof's algorithm simply consists in computing left hand side of Equation (5) for a point P of $E_a[\ell]$ and then in computing $[k]_a \phi_a(P)$ for k in $0, \dots, \ell - 1$. When Equation (5) is satisfied for such an integer k, we have $t \bmod \ell = k$ and when $t \bmod \ell$ is known for enough primes ℓ, that is to say

$$\prod \ell \geq 4\sqrt{2^n},$$

we deduce t by using the Chinese Remainder Theorem.

The main drawback of this method is that we are virtually forced to work not only with one point P of $E_a[\ell]$, but with all the points of $E_a[\ell]$ because the x-coordinates of these points are basically defined in an extension of degree $(\ell^2 - 1)/2$ of $\mathbb{F}_{2^n}$.

Works by Atkin and Elkies improved largely this situation by noticing that, for half the primes ℓ (called Elkies primes), $E_a[\ell]$ contains at least one subgroup of ℓ points. Thus, x-coordinates of these points are defined in an extension of degree $(\ell - 1)/2$ of $\mathbb{F}_{2^n}$. Indeed, this subgroup is the kernel of an isogeny (morphism) I between the curve E_a and an isogenous curve E_b and, when such an isogeny exists, there exists another isogeny $\hat{I}$ from E_b to E_a (called the dual isogeny) such that $\hat{I} \circ I = [\ell]_a$. Therefore, $\mathrm{Ker} I \subset \mathrm{Ker}[\ell]_a$.

Elkies and Atkin gave a construction based on modular equations to obtain E_b for Elkies primes ℓ. This works in any finite field. Unfortunately, the nice analytical method that they proposed for computing explicitly the isogeny between E_a and E_b is only valid in finite fields of large characteristic [Sch95].

3.2 Isogenies between Elliptic Curves in $\mathbb{F}_{2^n}$.

Since the original method by Atkin and Elkies for computing isogenies between two elliptic curves E_a and E_b does not work in finite fields of small characteristic p [Sch95], only Schoof's algorithm was available during a while to count points [MVZ93]. Fortunately, the situation evolved quickly.

Known Algorithms. The first attempt to fill this gap is due to Couveignes [Cou94]. The computations take place in the formal group defined by E_a. The algorithm was successfully implemented by Morain and the author [LM95b] and we do not describe it here.

But the time needed to compute isogenies with this method turned out to be the major cost while counting points. We recently proposed another algorithm which performs much better in practice. It is specially designed for the characteristic two case and is only based on algebraic properties [Ler96].

Let us note that Couveignes proposed a third algorithm for finite fields of small characteristic p based on algebraic properties too. It consists in computing $E_a[p^k]$ and $E_b[p^k]$ and then, uses the fact that $I(E_a[p^k]) = E_b[p^k]$. But since the computations take place in extension of degree $p^i(p-1)/2 \simeq 2\ell$, it does not seem obvious to implement it efficiently in practice even if its asymptotical complexity is attractive [Cou96, Cou97].

Lercier's Approach. In finite fields of characteristic two, we exploited that there exists a unique point P_a of order 2 on E_a. Thus, an isogeny I must satisfy

$$I \circ T_a = T_b \circ I.$$

From this, we deduced the following characterization.

Theorem 1. *Let E_a and E_b be two elliptic curves defined over $\mathbb{F}_{2^n}$. Let ℓ be an odd integer, and $d = (\ell - 1)/2$. Let $\mathcal{I}$ be an isogeny of degree ℓ between E_a and E_b given by $(X, Y) \mapsto \left(\frac{G(X)}{Q^2(X)}, \frac{H(X)+YK(X)}{Q^3(X)} \right)$ where $Q(X), G(X), H(X), K(X)$ in $\mathbb{F}_{2^n}[X]$ with degrees at most d, ℓ, $3d$ and $2d$. Then $G(X) = XP^2(X)$ where $P(X)$ is a polynomial of degree d such that $\gcd(P(X), Q(X)) = 1$ and*

$$X^d Q(\sqrt{a}/X) = \frac{\sqrt[8]{a}}{\sqrt[8]{b}} \left(\sqrt[4]{a} \right)^d P(X),$$

or equivalently via $X \to \sqrt{a}/X$,

$$X^d P(\sqrt{a}/X) = \frac{\sqrt[8]{b}}{\sqrt[8]{a}} \left(\sqrt[4]{a} \right)^d Q(X). \tag{6}$$

In order to explicitly compute the isogeny I, it turns out that we have to find conditions satisfied by the polynomial $Q(X)$. This is achieved from the fact that $I \circ [2]_a = [2]_b \circ I$.

Corollary 2. *With the notations of theorem 1, polynomials $P(X)$ and $Q(X)$ must satisfy*

$$X^d \widehat{Q}(X + \sqrt{a}/X) = Q(X)P(X), \tag{7}$$

and

$$(X + \sqrt[4]{a}) X^d \widehat{P} \left(X + \sqrt{a}/X \right) = XP^2(X) + \sqrt[4]{b}Q^2(X), \tag{8}$$

where $\widehat{P}(X) = \sqrt{P(X^2)}$ and $\widehat{Q}(X) = \sqrt{Q(X^2)}$ (polynomials whose coefficients are square roots of coefficients of $P(X)$ and $Q(X)$).

Even if Equation (8) is a linear equation satisfied by $Q(X)$ over $\mathbb{F}_2$, asymptotic complexity to inverse this system is $O(\ell^3 n)$. This is too high in practice.

To decrease this complexity, we considered Equation (7) and replaced the resolution of this linear system over $\mathbb{F}_{2^n}$ by a quadratic system over $\mathbb{F}_2$. This yields an algorithm (we do not describe here) whose heurististic complexity is $O(\ell^3)$.

3.3 Results

We had an old implementation of the SEA (Schoof, Elkies, Atkin) algorithm including Couveignes's first algorithm to compute isogenies and using an "ad hoc" C arithmetic of $\mathbb{F}_{2^n}$ [LM95a]. We completely rewrote it with our approach and the formalism of ZEN library [CL96a, CL96b] which enables us to handle any finite field given recursively by a polynomial basis over a subfield (for instance, $\mathbb{F}_2$). Since we restrict ourselves to the case of the characteristic two in this article, we only give accurate timings for finite fields $\mathbb{F}_{2^n}$, even if this implementation allows us to compute the number of points of an elliptic curve defined over other finite fields [Ler97].

$\mathbb{F}_{2^{65}}$	min	max	avg	$\mathbb{F}_{2^{89}}$	min	max	avg	$\mathbb{F}_{2^{105}}$	min	max	avg
ℓ_{max}	31	31	31	ℓ_{max}	41	43	41	ℓ_{max}	47	47	47
#U	1	3	2	#U	1	5	2	#U	1	5	2
#L	8	10	9	#L	8	12	10	#L	10	14	12
#M	10^3	10^6	$3{\cdot}10^5$	#M	$3{\cdot}10^3$	$3{\cdot}10^8$	$3{\cdot}10^7$	#M	$6{\cdot}10^5$	$6{\cdot}10^9$	$5{\cdot}10^8$
X^{2^n}	2.8	2.9	2.9	X^{2^n}	7.3	9.5	7.5	X^{2^n}	15	16.5	15.7
$X^{2^n r}$	0.7	2.4	1.8	$X^{2^n r}$	3.5	6.8	5.1	$X^{2^n r}$	6.4	12.2	8.9
Schoof	0	0	0	Schoof	0	0	0	Schoof	0	0	0
g	0	0	0	g	0	0	0	g	0	0	0
k	0	0	0	k	0	0	0	k	0	0	0
M $-$ S	0.3	1.1	0.7	M $-$ S	0.4	5.8	2.2	M $-$ S	1.5	30.1	6.5
Total	4.2	6.1	5.4	Total	11.4	18.9	14.9	Total	24.9	53.3	31.1

Table 1. Statistics obtained with our first implementation for small finite fields $\mathbb{F}_{2^n}$.

All the timings (in seconds) are obtained on a DEC Alpha workstation 250 (266 MHz, 4th generation). First, we did the same benchmarks as in [LM95a]. That is to say, we measured the running times for 50 random curves $y^2 + xy = x^3 + a$ where $a \in \mathbb{F}_2[T]$ defined over $\mathbb{F}_{2^{65}} \simeq \mathbb{F}_2[T]/(T^{65} + T^4 + T^3 + T + 1)$, $\mathbb{F}_{2^{89}} \simeq \mathbb{F}_2[T]/(T^{89} + T^6 + T^5 + T^3 + 1)$ and $\mathbb{F}_{2^{105}} \simeq \mathbb{F}_2[T]/(T^{105} + T^4 + 1)$ with the so-called "dynamic strategy". Results are given in Table 2. For the sake of comparison, we also give statistics obtained with our previous implementation on this machine in Table 1.

$\mathbb{F}_{2^{65}}$	min	max	avg	$\mathbb{F}_{2^{89}}$	min	max	avg	$\mathbb{F}_{2^{105}}$	min	max	avg
ℓ_{max}	31	31	31	ℓ_{max}	41	41	41	ℓ_{max}	41	47	42
#U	0	5	1	#U	0	4	2	#U	1	6	3
#L	6	11	10	#L	9	13	11	#L	8	13	10
#M	10^3	$3{\cdot}10^6$	$2{\cdot}10^5$	#M	$6{\cdot}10^3$	$8{\cdot}10^7$	$6{\cdot}10^6$	#M	$5{\cdot}10^3$	$2{\cdot}10^8$	10^7
X^{2^n}	2.2	4.2	3.3	X^{2^n}	5.1	7.8	6.4	X^{2^n}	6.6	11.5	8.8
$X^{2^n r}$	0,3	0.9	0.6	$X^{2^n r}$	0.8	2.6	1.9	$X^{2^n r}$	1.0	3.8	2.6
Schoof	0	0	0	Schoof	0	0.4	0	Schoof	0	3.8	0.3
g	0	0	0	g	0	0.2	0	g	0	1.7	0.5
k	0	0	0	k	0.2	0.8	0.6	k	0.8	3.8	2.4
M $-$ S	0.6	1.5	1.0	M $-$ S	1.1	5.2	2.2	M $-$ S	1.1	9.8	2.9
Total	3.8	5.9	4.9	Total	9.2	14.6	11.2	Total	13.4	24.5	17.3

Table 2. Statistics for small finite fields $\mathbb{F}_{2^n}$.

We give: ℓ_{max}, the maximal prime used; the number of U (resp. L) primes; #M, the number of combinations; the cumulated time for X^{2^n}, $X^{2^n r}$, Schoof's algorithm; computing isogenies (g_ℓ) and $t \bmod \ell$ when ℓ is Elkies (k); the time for the match and sort program; the total time. For each category, minimal, maximal and average values are given.

Since for these "small" finite fields, the time needed to compute isogenies is negligible, we only gain a speed up factor from 1.1 up to 1.8 thanks in part to the arithmetic of ZEN which is faster than the arithmetic of our old implementation. We did the same experiments for three larger finite fields, $\mathbb{F}_{2^{155}} \simeq \mathbb{F}_2[T]/(T^{155} + T^7 + T^5 + T^4 + 1)$, $\mathbb{F}_{2^{196}} \simeq \mathbb{F}_2[T]/(T^{196} + T^3 + 1)$ and $\mathbb{F}_{2^{300}} \simeq \mathbb{F}_2[T]/(T^{300} + T^5 + 1)$ (note that our previous implementation is really too slow to provide similar statistics). Results are given in Table 3.

$\mathbb{F}_{2^{155}}$	min	max	avg	$\mathbb{F}_{2^{196}}$	min	max	avg	$\mathbb{F}_{2^{300}}$	min	max	avg
$\ell_{\max}$	59	71	60	$\ell_{\max}$	73	79	74	$\ell_{\max}$	97	157	113
$\#U$	4	11	7	$\#U$	7	13	10	$\#U$	11	19	16
$\#L$	7	15	10	$\#L$	8	15	11	$\#L$	9	20	14
$\#M$	$3 \cdot 10^4$	$7 \cdot 10^8$	$7 \cdot 10^7$	$\#M$	10^5	$7 \cdot 10^9$	$8 \cdot 10^8$	$\#M$	$5 \cdot 10^6$	$5 \cdot 10^{11}$	$5 \cdot 10^{10}$
X^{2^n}	30.4	56.1	40.6	X^{2^n}	113	475	147	X^{2^n}	744	1761	996
$X^{2^{nr}}$	4.4	13.9	7.8	$X^{2^{nr}}$	8.8	31.8	21.6	$X^{2^{nr}}$	46	387	119
Schoof	0	14.8	4.3	Schoof	0	55.5	17.9	Schoof	0	551	199
g	1.5	21.6	7.1	g	9.9	419	40.6	g	76	568	287
k	7.4	31.9	20.1	k	29.2	90.1	58.9	k	354	961	601
$M - S$	2.9	20.4	6.5	$M - S$	5	86.9	22.9	$M - S$	14	1510	230
Total	58.8	132	86.5	Total	212	1029	308	Total	1519	3686	2434

Table 3. Statistics for larger finite fields $\mathbb{F}_{2^n}$.

At this point, the advantage of our approach clearly appears. The time needed to compute isogenies is (completely) negligible while it used to be the main cost in [LM95a] and we gain a speed up factor from 4 up to 10 on the whole computation.

To compare Couveignes's and Lercier's approaches for two huge finite fields, we collected the same data in Table 4 for the curve

$$E_X : y^2 + xy = x^3 + T^{16} + T^{14} + T^{13} + T^9 + T^8 + T^7 + T^6 + T^5 + T^4 + T^3.$$

For the first finite field, $\mathbb{F}_{2^{1009}} \simeq \mathbb{F}_2[T]/(T^{1009} + T^{11} + T^4 + T^2 + 1)$, we first used Couveignes's and Lercier's algorithms (respectively noted JMC and RL). For the second field, $\mathbb{F}_{2^{1301}} \simeq \mathbb{F}_2[T]/(T^{1301} + T^{11} + T^{10} + T + 1)$, we could only use Lercier's (the current record, as of February 1997). The results are striking, the time needed to compute isogenies is completely negligible in the case of $\mathbb{F}_{2^{1301}}$ (3 days) while it was the main cost for $\mathbb{F}_{2^{1009}}$ (77 days).

To improve the SEA algorithm, future implementation should now optimize computations of $X^{2^n} \bmod \Phi$.

	X^{2^n}	$X^{2^{n''}}$	Schoof	g	k	M − S	Total
$\mathbb{F}_{2^{1009}}$ (JMC)	15d 3h	2d 21h	10d 14h	77d 21h	23d 3h	1h	121d 15h
$\mathbb{F}_{2^{1009}}$ (RL)	9d 16h	1d 9h	2h	1d 2h	7d 7h	2h	19d 11h
$\mathbb{F}_{2^{1301}}$ (RL)	51d 7h	8d 12h	2d 8h	3d 17h	36d 14h	2h	103d 5h

	ℓ_{max}	#U	#L	#M
$\mathbb{F}_{2^{1009}}$ (JMC)	577	57	46	$4 \cdot 10^9$
$\mathbb{F}_{2^{1009}}$ (RL)	547	48	47	$2 \cdot 10^{10}$
$\mathbb{F}_{2^{1301}}$ (RL)	673	88	50	$9 \cdot 10^{10}$

Table 4. Timings for huge finite fields (days/hours).

4 Finding Random Elliptic Curves with Nearly Prime Cardinality Efficiently

Since the best known attacks against the discrete logarithm problem on elliptic curves are

1. the Weil pairing reduction for supersingular curves,
2. the baby steps giant steps, Pollard-ρ and Pohlig Helman algorithms for other curves,

"good curves" for cryptographical purposes only have to be defined in a not too small finite field and to be of "nearly prime" cardinality (to avoid point 2.) different from 2^n, $2^n + 1 \pm \sqrt{2^n}$, $2^n + 1 \pm \sqrt{2^{n+1}}$ and $2^n + 1 \pm 2\sqrt{2^n}$ (to avoid point 1.) if defined over $\mathbb{F}_{2^n}$.

In Section 4.1, we describe an early abort strategy suggested by Morain that takes advantage of the SEA algorithm to quickly throw away most of the curves which do not meet this condition. For convenience, we explain it only in the case of elliptic curves E_a defined over $\mathbb{F}_{2^n}$. But this strategy obviously works in any finite field. Then we give timing and examples of "good curves" provided by this strategy.

4.1 Early Abort Strategy

The Algorithm. An elliptic curve E_a given by Equation (1) is non supersingular and has a point $Q_a = (\sqrt[3]{a}, \sqrt{a})$ of order 4. Thus, the previous condition can be reformulated as follows : "A good curve E_a is a curve defined over $\mathbb{F}_{2^n}$ with $n \geq 60$ whose cardinality is 4 times a prime".

To find such "good curves", we proceed as follows:

1. Choose an element $a \in \mathbb{F}_{2^n}^*$ at random.
2. As explained in Section 3, compute $t \bmod \ell$ with the SEA algorithm checking during the computation that, for each Elkies prime $\ell \neq 2$,

$$2^n + 1 - t \bmod \ell \neq 0.$$

Otherwise, this means that the number of points of the curve is divisible by ℓ. In this case, go to step 1.

3. Check that the cardinality of the curve is 4 times a prime, otherwise go to step 1.

First of all, let us note that when $2^n + 1 - t \bmod \ell = 0$ for a prime ℓ, this means there is a point of order ℓ in E_a. Therefore, there exists an isogeny of degree ℓ defined from E_a, and ℓ is necessarily an Elkies prime.

Let us observe too that it is better to test the primality of the cardinality at step 3., first, by a pseudo primality test, and then by an exact primality prover (for instance ECPP [Mor90]). But for practical reasons, we used MAPLE system [CGGW85].

In practice, this algorithm works well because most of the time a curve does not have a prime cardinality, we will see in Section 4.1 that this cardinality is divided by a small integer. Since we choose primes ℓ as small as possible in the SEA algorithm, we detect such a curve quickly.

Analysis. A theorem by Howe [How93], which extends works by Lenstra [Len87] (see also [Kob88]), gives the asymptotic behavior of the probability that a random elliptic curve over a finite field $\mathbb{F}_q$ has ℓ^k ($k \in \mathbb{N}^*$) dividing the number M of its points when $q \to \infty$.

Theorem 3. *There is a constant $C \leq 1/12 + 5\sqrt{2}/6 \simeq 1.262$ such that the following statement is true. Given a prime power q, let r be the multiplicative arithmetic function such that for all primes ℓ and positive integers k*

$$
r_q(\ell^k) = \begin{cases} \dfrac{1}{\ell^{k-1}(\ell - 1)} & \text{if } q \neq 1 \bmod \ell^\mu, \\[2mm] \dfrac{\ell^{\nu+1} + \ell^\nu - 1}{\ell^{\nu+\mu-1}(\ell^2 - 1)} & \text{if } q = 1 \bmod \ell^\mu, \end{cases}
$$

where $\mu = \lceil k/2 \rceil$ and $\nu = \lfloor k/2 \rfloor$. Then for all positive integers N, the probability $\pi_{q,N}$ that a random elliptic curve over $\mathbb{F}_q$ has N dividing the number of its $\mathbb{F}_q$-defined points satisfies

$$
|\pi_{q,N} - r_q(N)| \leq \frac{CN\chi(N)2^{\sigma(N)}}{\sqrt{q}},
$$

where $\chi(N) = \prod_{\lambda | N}(\lambda + 1)/(\lambda - 1)$ and $\sigma(N)$ denotes the number of prime divisors of N.

Let $g_q(\ell)$ be the probability that the smallest prime factor of M is ℓ. This probability is equal to

$$
g_q(\ell) = r_q(\ell) \prod_{\text{primes } \lambda < \ell} (1 - r_q(\lambda)).
$$

In our particular case, we test random curves E_a defined over $\mathbb{F}_{2^n}$ with cardinalities always divisible by 4, so, we make the strong assumption that Howe's theorem applies, except for $\ell = 2$, and the probabilities $r_{2^n}(\ell^k)$ become

$$\rho_n(\ell^k) = \begin{cases} 1 & \text{for } \ell = 2 \text{ and } k = 1, \\ \frac{1}{2^{k-2}} & \text{for } \ell = 2 \text{ and } k > 1, \\ r_{2^n}(\ell^k) & \text{for } \ell > 2. \end{cases}$$

Consequently, the probability $\gamma_n(\ell)$ we detect at step 2. of the algorithm that an odd prime ℓ divides the cardinality of E_a is equal to

$$\gamma_n(\ell) = \rho_n(\ell)(1 - \rho_n(2^3)) \prod_{\text{odd primes } \lambda < \ell} (1 - \rho_n(\lambda)).$$

This quantity can be easily computed for n fixed but for any n, one can only state that

$$\rho_n(2^3) = \frac{1}{2} \text{ and } \frac{\ell}{\ell^2 - 1} \leq \rho_n(\ell) \leq \frac{1}{\ell - 1},$$

and therefore, $3/16 \leq \gamma_n(3) \leq 1/4$, $5/96 \leq \gamma_n(5) \leq 5/64$, $7/256 \leq \gamma_n(7) \leq 95/2304 \ldots$

4.2 Results

The implementation described in Section 3.3 allows to compute a lot of such "good curves" defined over $\mathbb{F}_{2^{65}}$, $\mathbb{F}_{2^{89}}$, $\mathbb{F}_{2^{105}}$, $\mathbb{F}_{2^{155}}$ and $\mathbb{F}_{2^{196}}$ in a reasonable amount of time. Accurate statistics are given in Table 5.

In this table, it turns out that the theoretical estimations of Section 4.1 are in practice satisfied most of the time, except maybe for the number of cardinalities divisible by 5 in $\mathbb{F}_{2^{196}}$ (150 instead of $1000 \cdot 25/394 \simeq 65$). In any case, the probability that an elliptic curve has its number of points divisible by a small prime ℓ is quite high and thus we need to compute the cardinality of a curve completely in only a few case. Some of these "good curves" are given in Table 6 with the notation $a_0 + a_1 2 + \cdots + a_{n-1} 2^{n-1} = a_0 + a_1 T + \cdots + a_{n-1} T^{n-1}$.

5 Conclusion

Thanks to the contribution of many people in this field of research, computing the number of points of an elliptic curve defined over $\mathbb{F}_{2^n}$ can be performed quickly in practice. From this, we derived an efficient way for finding elliptic curves with nearly prime cardinality. Even if it is harder to obtain such curves when n increases (only 2 among 1000 for $n = 196$), we think this method is of special interest for cryptographic purposes.

Performances we obtained for $\mathbb{F}_{2^n}$ are now similar to the performances we already had for the case $\mathbb{F}_p$ with p, a large prime, and this, even when the size of the finite field increases. The only problem which remains in practice is the case p odd and small. But, as what was foreseen at the end of [LM95a] for $p = 2$, we hope that the situation might evolve very soon for these fields too.

	$\mathbb{F}_{2^{65}}$	$\mathbb{F}_{2^{89}}$	$\mathbb{F}_{2^{105}}$	$\mathbb{F}_{2^{155}}$	$\mathbb{F}_{2^{196}}$
# curves tested	1000	1000	1000	1000	1000
$1000\gamma_n(8)$	500	500	500	500	500
# cardinalities divisible by 8	491	507	500	509	490
$1000\gamma_n(3)$	250	250	250	250	187.5
# cardinalities divisible by 3	255	253	256	236	177
$1000\gamma_n(5)$	62.5	62.5	62.5	62.5	65.1
# cardinalities divisible by 5	63	73	74	68	150
$1000\gamma_n(7)$	31.2	31.2	27.4	31.2	41.2
# cardinalities divisible by 7	28	28	59	25	34
# cardinalities divisible by $\ell \geq 11$ and detected at step 2. of the algorithm	29	52	43	61	57
# cardinalities divisible by $\ell \geq 11$ and detected at step 3. of the algorithm	116	77	62	96	90
Number of "good curves"	18	10	6	5	2
Total time needed (s)	1277	1733	2231	14112	30254

Table 5. Statistics of the "early abort strategy".

	a	Cardinality
$\mathbb{F}_{2^{65}}$	2108463510029530717	$2^2 \cdot 9223372038308612213$
$\mathbb{F}_{2^{65}}$	15004298573160993787	$2^2 \cdot 9223372035176356667$
$\mathbb{F}_{2^{89}}$	362244896591784868971148794	$2^2 \cdot 154742504910673945144969913$
$\mathbb{F}_{2^{89}}$	578529593362960704429241468	$2^2 \cdot 154742504910669983358163303$
$\mathbb{F}_{2^{105}}$	654393540540047802571729043 2415	$2^2 \cdot 101412048018258375221758412\backslash$ 06867
$\mathbb{F}_{2^{105}}$	229598971637660130735605103979\ 54	$2^2 \cdot 101412048018258342875266703\backslash$ 03267
$\mathbb{F}_{2^{155}}$	838795043588789173323661086541\ 2790131341725747	$2^2 \cdot 114179815416476790484662819\backslash$ 27805319915233345669
$\mathbb{F}_{2^{155}}$	110027220687791685841747180597\ 77371906785324958	$2^2 \cdot 114179815416476790484662992\backslash$ 30130487707830550127
$\mathbb{F}_{2^{196}}$	250334701759594235393108283794\ 6490796156769623968851 1281965	$2^2 \cdot 251084069415467230553431576\backslash$ 92759220570140916154347737377983
$\mathbb{F}_{2^{196}}$	404284818812143036331788043458\ 3715482432038248020058 8296980	$2^2 \cdot 251084069415467230553431576\backslash$ 92813473113492187155697729606263

Table 6. Curves with a nearly prime cardinality.

Acknowledgments. I would like to thank François Morain for fruitful discussions. I also thank the referees for careful comments and for suggesting a title corresponding closer to the content of this article.

References

[BC89]   A. Bender and G. Castagnoli. On the implementation of elliptic curve cryptosystems. In G. Brassard, editor, *Advances in Cryptology*, volume 435 of *Lecture Notes in Comput. Sci.*, pages 186–192. Springer-Verlag, 1989. Proc. Crypto '89, Santa Barbara, August 20–24.

[CDM96]   J.-M. Couveignes, L. Dewaghe, and F. Morain. Isogeny cycles and the Schoof-Elkies-Atkin algorithm. Research Report LIX/RR/96/03, LIX, April 1996.

[CGGW85]   B. W. Char, K. O. Geddes, G. H. Gonnet, and S. M. Watt. *MAPLE Reference Manual, Fourth Edition*. Symbolic Computation Group, Department of Computer Science, University of Waterloo, 1985.

[CL96a]   F. Chabaud and R. Lercier. A new toolbox for finite extensions of finite fields. Rapport technique, Laboratoire d'Informatique de l'École polytechnique (LIX), 1996. In preparation.

[CL96b]   F. Chabaud and R. Lercier. *ZEN, User Manual.* Laboratoire d'Informatique de l'École polytechnique (LIX), 1996. Available at http://lix.polytechnique.fr/~zen/.

[CM94]   J.-M. Couveignes and F. Morain. Schoof's algorithm and isogeny cycles. In L. Adleman and M.-D. Huang, editors, *ANTS-I*, volume 877 of *Lecture Notes in Comput. Sci.*, pages 43–58. Springer-Verlag, 1994. 1st Algorithmic Number Theory Symposium - Cornell University, May 6-9, 1994.

[Cou94]   J.-M. Couveignes. *Quelques calculs en théorie des nombres*. Thèse, Université de Bordeaux I, July 1994.

[Cou96]   J.-M. Couveignes. Computing l-isogenies with the p-torsion. In H. Cohen, editor, *ANTS-II*, volume 1122 of *Lecture Notes in Comput. Sci.*, pages 59–65. Springer-Verlag, 1996.

[Cou97]   J. M. Couveignes. Isomorphisms between towers of artin-schreier exetensions over a finite fields. Draft, 1997.

[CTT94]   J. Chao, K. Tanada, and S. Tsujii. Design of elliptic curves with controllable lower boundary of extension degree for reduction attacks. In Y. Desmedt, editor, *Advances in Cryptology – CRYPTO '94*, volume 839 of *Lecture Notes in Comput. Sci.*, pages 50–55. Springer-Verlag, 1994. Proc. 14th Annual International Cryptology Conference, Santa Barbara, Ca, USA, August 21–25.

[HMV93]   G. Harper, A. Menezes, and S. Vanstone. Public-key cryptosystems with very small key length. In R. A. Rueppel, editor, *Advances in Cryptoloy – EUROCRYPT '92*, volume 658 of *Lecture Notes in Comput. Sci.*, pages 163–173. Springer-Verlag, 1993. Workshop on the Theory and Application of Cryptographic Techniques, Balatonfüred, Hungary, May 24–28, 1992, Proceedings.

[How93]   E. W. Howe. On the group orders of elliptic curves over finite fields. *Compositio Mathematica*, 85:229–247, 1993.

[Kal86]   B. S. Kaliski, Jr. A pseudo-random bit generator based on elliptic logarithms. In *Proc. Crypto 86*, volume 263 of *Lecture Notes in Comput. Sci.*, 1986. Proceedings Crypto '86, Santa Barbara (USA), August 11-15, 1986.

[Kob87]   N. Koblitz. Elliptic curve cryptosystems. *Math. Comp.*, 48(177):203-209, January 1987.

[Kob88]   N. Koblitz. Primality of the number of points on an elliptic curve over a finite field. *Pacific Journal of Mathematics*, 131(1):157-165, 1988.

[Kob91]   N. Koblitz. Elliptic curve implementation of zero-knowledge blobs. *Journal of Cryptology*, 4(3):207-213, 1991.

[Len87]   H. W. Lenstra, Jr. Factoring integers with elliptic curves. *Annals of Math.*, 126:649-673, 1987.

[Ler96]   R. Lercier. Computing isogenies in $GF(2^n)$. In H. Cohen, editor, *ANTS-II*, volume 1122 of *Lecture Notes in Comput. Sci.*, pages 197-212. Springer-Verlag, 1996.

[Ler97]   R. Lercier. *Courbes elliptiques et cryptographie*. Thèse, École polytechnique, 1997. Draft.

[LM95a]   R. Lercier and F. Morain. Counting the number of points on elliptic curves over finite fields: strategies and performances. In L. C. Guillou and J.-J. Quisquater, editors, *Advances in Cryptology – EUROCRYPT '95*, number 921 in Lecture Notes in Comput. Sci., pages 79-94, 1995. International Conference on the Theory and Application of Cryptographic Techniques, Saint-Malo, France, May 1995, Proceedings.

[LM95b]   R. Lercier and F. Morain. Counting the number of points on elliptic curves over F_{p^n} using Couveignes's algorithm. Rapport de Recherche LIX/RR/95/09, Laboratoire d'Informatique de l'École polytechnique (LIX), 1995. Available at http://lix.polytechnique.fr/~morain/Articles.

[LZ94]   G.-J. Lay and H. G. Zimmer. Constructing elliptic curves with given group order over large finite fields. In L. Adleman and M.-D. Huang, editors, *ANTS-I*, volume 877 of *Lecture Notes in Comput. Sci.*, pages 250-263. Springer-Verlag, 1994. 1st Algorithmic Number Theory Symposium - Cornell University, May 6-9, 1994.

[Men93]   A. J. Menezes. *Elliptic curve public key cryptosystems*. Kluwer Academic Publishers, 1993.

[Mil87]   V. Miller. Use of elliptic curves in cryptography. In A. M. Odlyzko, editor, *Advances in Cryptology*, volume 263 of *Lecture Notes in Comput. Sci.*, pages 417-426. Springer-Verlag, 1987. Proceedings Crypto '86, Santa Barbara (USA), August 11-15, 1986.

[Miy91]   A. Miyaji. On ordinary elliptic curve cryptosystems. In *Advances in Cryptology – ASIACRYPT '91*, volume 739 of *Lecture Notes in Comput. Sci.*, pages 50-55. Springer-Verlag, 1991.

[Miy93]   A. Miyaji. Elliptic curves over F_p suitable for cryptosystems. In J. Seberry and Y. Zheng, editors, *Advances in cryptology - AUSCRYPT '92*, volume 718 of *Lecture Notes in Comput. Sci.*, pages 479-491. Springer-Verlag, 1993. Workshop on the theory and application of cryptographic techniques, Gold Coast, Queensland, Australia, December 13-16, 1992.

[Mor90]   F. Morain. *Courbes elliptiques et tests de primalité*. PhD thesis, Université Claude Bernard-Lyon I, September 1990.

[Mor91]   F. Morain. Building cyclic elliptic curves modulo large primes. In D. Davies, editor, *Advances in Cryptology – EUROCRYPT '91*, volume 547 of

Lecture Notes in Comput. Sci., pages 328–336. Springer–Verlag, 1991. Proceedings of the Workshop on the Theory and Application of Cryptographic Techniques, Brighton, United Kingdom, April 8–11, 1991.

[Mor96] F. Morain. Classes d'isomorphismes des courbes elliptiques supersingulières en caractéristique ≥ 3. To appear in Utilitas Mathematica. Available at http://lix.polytechnique.fr/~morain/, March 1996.

[MOV93] A. Menezes, T. Okamoto, and S. A. Vanstone. Reducing elliptic curves logarithms to logarithms in a finite field. *IEEETIT*, 39(5):1639–1646, 1993.

[Mül95] V. Müller. *Ein Algorithmus zur Bestimmung der Punktanzahl elliptischer Kurven über endlichen Körpern der Charakteristik größer drei.* PhD thesis, Technischen Fakultät der Universität des Saarlandes, 1995.

[MV90] A. Menezes and S. A. Vanstone. The implementation of elliptic curve cryptosystems. In J. Seberry and J. Pieprzyk, editors, *Advances in Cryptology*, number 453 in Lecture Notes in Comput. Sci., pages 2–13. Springer–Verlag, 1990. Proceedings Auscrypt '90, Sysdney (Australia), January 1990.

[MVZ93] A. J. Menezes, S. A. Vanstone, and R. J. Zuccherato. Counting points on elliptic curves over F_{2^m}. *Math. Comp.*, 60(201):407–420, January 1993.

[PH78] S. Pohlig and M. Hellman. An improved algorithm for computing logarithms over gf(p) and its cryptographic significance. *IEEE Transactions on Information Theory*, 24:106–110, 1978.

[Pol78] J. M. Pollard. Monte Carlo methods for index computation (mod p). *Math. Comp.*, 32(143):918–924, July 1978.

[Sch85] R. Schoof. Elliptic curves over finite fields and the computation of square roots mod p. *Math. Comp.*, 44:483–494, 1985.

[Sch95] R. Schoof. Counting points on elliptic curves over finite fields. *J. Théor. Nombres Bordeaux*, 7:219–254, 1995. Available at http://www.emath.fr/Maths/Jtnb/jtnb1995-1.html.

[Sha71] D. Shanks. Class number, a theory of factorization, and genera. In *Proc. Symp. Pure Math. vol. 20*, pages 415–440. AMS, 1971.

[Sil86] J. H. Silverman. *The arithmetic of elliptic curves*, volume 106 of *Graduate Texts in Mathematics*. Springer, 1986.

[SOOS95] R. Schroeppel, H. Orman, S. O'Malley, and O. Spatscheck. Fast key exchange with elliptic curve systems. In Don Coppersmith, editor, *Advances in Cryptology – CRYPTO '95*, volume 963 of *Lecture Notes in Comput. Sci.*, pages 44–56. Springer-Verlag, 1995.

Incremental Cryptography and Memory Checkers

Marc Fischlin

Fachbereich Mathematik/Informatik
Johann Wolfgang Goethe-Universität Frankfurt am Main
PSF 111932
60054 Frankfurt/Main, Germany

e-mail: marc @ informatik.uni-frankfurt.de
URL: http://www.uni-frankfurt.de/~roessner/group/marc/marc.html

Abstract. We introduce the relationship between incremental cryptography and memory checkers. We present an incremental message authentication scheme based on the XOR MACs which supports insertion, deletion and other single block operations. Our scheme takes only a constant number of pseudorandom function evaluations for each update step and produces smaller authentication codes than the tree scheme presented in [BGG95]. Furthermore, it is secure against message substitution attacks, where the adversary is allowed to tamper messages before update steps, making it applicable to virus protection. From this scheme we derive memory checkers for data structures based on lists. Conversely, we use a lower bound for memory checkers to show that so-called message substitution detecting schemes produce signatures or authentication codes with size proportional to the message length.

1 Introduction

The notion of incremental cryptography has been introduced by Bellare, Goldreich and Goldwasser in [BGG94] and refined by the same authors in [BGG95]. Suppose that we are given a block-by-block message M and its cryptographic form μ, i.e. encryption, signature or authentication code. Let M' be a message that is obtained by applying a text modification from a set $\mathcal{M}$ of modifications to M. With an incremental scheme supporting the text modifications $\mathcal{M}$ a cryptographic form μ' for M' can be produced much faster from μ and M than it would take to compute it from scratch.

Our results. We present the incremental authentication scheme IncXMACC that supports single block insertion and deletion, and therefore other operations like replacement. To update an authentication code for inserting or deleting a single block at a given position, this scheme performs only a constant number of pseudorandom function evaluations. Additionally, insertion can be done without accessing the message and deletion merely needs the corresponding block.

W. Fumy (Ed.): Advances in Cryptology - EUROCRYPT '97, LNCS 1233, pp. 393-408, 1997.

Security against Message Substitution Attacks. Our scheme remains secure if an adversary is allowed to alter messages before applying the update algorithm — while the shorter authentication code must be kept on some secure medium. Security against these message substitution attacks implies application to virus protection. To protect a large file stored on some insecure medium against unauthorized alternation, authenticate this file and store the shorter authentication code in some incorruptible memory. Whenever an authorized user modifies the file, we can update the authentication code very fast using the incremental algorithm. Conversely, it is very unlikely that an attacker, e.g. a virus, will be able to produce a forgery even if he tampers the documents before update steps. In this sense, message substitution attacks lie between (total) substitution attacks, where both the message and signature can be tampered before update steps, and basic attacks, where the adversary isn't allowed to alter messages or signatures before updating.

Related Work. In [BGG94] a hash-and-sign scheme based on an incremental hash function was presented. The signature consists of the hash value h and a signature for h produced by an arbitrary non-incremental signature scheme. To update a signature, increment the hash value and sign this new hash value. Unfortunately, this scheme only supports single block replacement and it is provably not secure against message substitution attacks.

In [BGG95] the same authors present the tree scheme supporting single block operations like insertion and deletion (and the more powerful modifications cut and paste to devide a text into two documents resp. to append a document to another). The tree scheme takes $\Omega(\log n)$ verification and authentication steps for the abovementioned operations, where n is the number of blocks of the document. For the cut modification, the tree scheme is much faster than IncXMACC, while our scheme supports the insert, delete and paste modifications applying a pseudorandom function only a constant number of times. Moreover, our scheme produces considerably smaller authentication codes than the tree scheme, though the authentication code must be kept on a secure medium. In contrast to that, signatures and authentication codes produced by the tree scheme can be stored in the insecure memory. A randomized version of the tree scheme is given in [M97]. This scheme hides the fact whether the incremental or non-incremental algorithm has been used to produce a signature.

Our scheme IncXMACC refines the incremental authentication scheme presented in [BGG95], which is also based on the XOR MACs. This scheme has several disadvantages in comparison to our scheme: It doubles the key size by using two pseudorandom functions and it requires many random bits. For an update step the incremental algorithm reads more than the corresponding block and security has only been proven for basic attacks.

Memory Checkers. Using IncXMACC, we present a method to obtain memory checkers for lists and similar data structures. The memory checker model has been introduced by Blum et al. in [BEG+94] (a prelimary version appeared in [BEG+91]). Informally, a memory checker for a data structure $\mathcal{D}$ verifies that for a given sequence of operations, an implementation of $\mathcal{D}$ works correctly for this sequence. If not, the checker outputs some error message. There are two sources of errors: The program implementing the data structure can be buggy or the memory where the elements are stored can be tampered by an adversary, e.g. a virus. Intuitively, incremental schemes that are secure against message substitution attacks seem to provide a suitable method to design such checkers. To do so, keep a signature for the current memory content and update the signature accordingly for an operation for $\mathcal{D}$. Nevertheless, in some settings the checker should be able to update the signature given only the old signature and the element resp. block that for example shall be deleted or inserted, without accessing other parts of the memory content. IncXMACC has this property.

Making the connection between memory checkers and incremental schemes we transfer a lower bound for checkers to incremental schemes. Informally, an incremental scheme is message substitution detecting, if it detects when relevant parts of message have been altered before calling the update algorithm. We give a sufficient condition under which an incremental message substitution detecting scheme that is secure against basic attacks, is also secure against message substitution attacks. The lower bound states that the length of a signature produced by a substitution detecting scheme must be very large, roughly proportional to the size of the message.

For a discussion about the differences between the memory checker setting and the program checking model (which has been introduced by Blum and Kannan in [BK89]) resp. the software protection model of Goldreich and Ostrovsky [GO96] we refer the reader to [BEG+94].

Exact Security. We follow the paradigm presenting our results in terms of *exact security* [BKR94,BGR95]. Informally, the notion of exact security can be described as follows. Assume that we have an adversary for IncXMACC with running time[1] t that makes at most q signature queries for messages of length at most L and achieves success probability ϵ. Then we derive (in a constructive way) a distinguisher D for the underlying function family F with parameters t', q', ϵ', such that D can distinguish F and the family of all functions with running time t', making at most q' oracle queries and achieving advantage at least ϵ'. Here, t', q', ϵ' are determined by t, q, L, ϵ.

2 Incremental Cryptography

We briefly review the definitions of incremental cryptography. This part is mainly based on [BGG95]. See this work for further discussion. In section 2.2 we introduce the notion of message substitution attacks.

---

[1] To be precise, t describes the running time *and* the size of the adversary's algorithm. For simplicity, we will only deal with the issue of running time in this paper.

2.1 Incremental Schemes

Let $S = (\mathsf{Gen}, \mathsf{Sig}, \mathsf{Vf})$ be an ordinary (i.e. non-incremental) signature or message authentication scheme which allows to sign block messages. That is, on input a security parameter s and a block size b in unary, the Gen algorithm outputs in probabilistic polynomial time a pair of keys (e, d). For simplicity we assume that s and b are recoverable from e or d and that $b = \mathrm{poly}\,(s)$. On input the key d and an admissible message $M \in \Sigma^*$, where $\Sigma = \{0, 1\}^b$, the signer Sig outputs a signature or message authentication code (MAC) μ in probabilistic polynomial time in s (and b). The polynomial time verifier Vf outputs a bit a where $a = 1$ stands for "accept" and $a = 0$ for "reject". A scheme is called *complete*, if $\mathsf{Vf}(e, M, \mathsf{Sig}(d, M)) = 1$ for all keys produced with positive probability by Gen and all admissible messages M. We say that a signature μ for M is *valid*, if $\mathsf{Vf}(e, M, \mu) = 1$. Else it is called *invalid*.

To every document we associate a name $\alpha \in \{0, 1\}^*$ and a counter cnt_α. For the rest of this paper, we assume that the counter value is bounded above by 2^b and that the document name has length at most b, so that both values can be treated as message blocks, and that all messages $M \in \Sigma^i$ with $1 \leq i \leq \mathrm{poly}\,(s)$ are admissible. Let $\pi(M_1, \dots, M_m, y) \in \Sigma^*$ denote the message that is obtained by applying text modification π to messages $M_1, \dots, M_m$ with argument vector y. For example, $\pi(M, i, M_*) = \mathsf{replace}(M, i, M_*)$ for $y = (i, M_*)$ is the message where the ith block in M is replaced by $M_* \in \Sigma$. We only present the definition for incremental signature schemes. The definition for message authentication schemes is similar.

Definition 1. Let $S = (\mathsf{Gen}, \mathsf{Sig}, \mathsf{Vf})$ be a signature scheme and $\mathcal{M}$ a set of text modifications. An $\mathcal{M}$-incremental scheme is an interactive machine such that:

- The machine is initialized with a pair (e, d) of keys produced by Gen on input $(1^s, 1^b)$.
- For a create command with arguments $\alpha \in \{0, 1\}^*$ and $D \in \Sigma^*$, the machine initializes a counter cnt_α with 1 and produces a signature $\mathsf{Sig}(d, D)$. (The Sig algorithm might take as additional input the name α and cnt_α.) The machine stores the document D, the counter cnt_α and the signature with reference to name α. If a document for this name already exists, it is replaced by D and cnt_α is incremented instead of initialized before calling Sig.
- On an edit command for the text modification $\pi \in \mathcal{M}$ with argument vector y and document names $\alpha_1, \dots, \alpha_m$ and β, the machine works as follows:
  - The machine increments the counter of document β.
  - It updates the signature of the document for β.
  - It replaces the document specified by β by applying modification π with argument vector y to the documents defined by the values α_i.

The update step is done by applying the incremental algorithm IncSig to the documents D_{α_i} and signatures μ_{α_i} specified by the values α_i, the modification π with argument vector y and the key d.[2] The algorithm might take

---

[2] To be more precise, IncSig is passed a description of π, where we assume that $|\mathcal{M}|$ is constant.

as additional input all the counter values and document names, including β and cnt_β.

The incremental scheme is called complete, if S is complete and for all pairs (e, d) of keys which are produced by Gen with positive probability and all valid signatures μ_{α_i} for D_{α_i}, the output of IncSig satisfies

$$\mathsf{Vf}\left(e, D_\beta, \mathsf{IncSig}(d, D_{\alpha_1}, \dots, D_{\alpha_m}, \mu_{\alpha_1}, \dots, \mu_{\alpha_m}, \pi, y)\right) = 1,$$

where the verifier Vf might take β and cnt_β as additional input.

For simplicity, we also write $S = (\mathsf{Gen}, \mathsf{Sig}, \mathsf{IncSig}, \mathsf{Vf})$ for the incremental scheme. $S(b, s)$ denotes the incremental scheme with fixed parameters b and s.

2.2 Security

In this section we review the notion of security for incremental signature and authentication schemes. Basically, an adversary performs an adaptive chosen message attack [GMR88]. So far, all values are stored securely by the interactive machine. As done in [BGG95], we augment our model by an alter command that takes as arguments a document name α, a document $D \in \Sigma^*$ and a signature μ. For an alter command the interactive machine replaces the document with name α by D and the signature by μ regardless of the current values. The counter value cnt_α remains unchanged.

The alter command models the following settings: Suppose that the documents and signatures are kept on an insecure medium like a remote host. Then an adversary, e.g. a virus, might change the document before issuing an edit command. If the adversary doesn't use alter commands during his attack, we call it a *basic attack*. If he tampers only documents but no signatures, we call this a *message substitution attack*. This corresponds to the case when the possibly short signature is kept on a secure medium. If the adversary changes documents and signatures, it is called a *(total) substitution attack*.

In substitution attacks, we must associate the signature or authentication code to some document. [BGG95] therefore introduce *virtual documents*. To every document D we define the virtual document $\mathrm{virt}(D)$ as follows: If the document D was issued by a create command, let $\mathrm{virt}(D) = D$. If the document was obtained by an edit command applying π with argument vector y to documents $D'_1, \dots, D'_m$, let $\mathrm{virt}(D)$ be the document that is obtained by applying π with y to $\mathrm{virt}(D'_1)\dots, \mathrm{virt}(D'_m)$. If the document D was obtained by an alter command replacing document D', let $\mathrm{virt}(D) = \mathrm{virt}(D')$. An adversary is *successful*, if he produces a signature or authentication code for a document which hasn't appeared as a virtual document before. We define security in terms of exact security:

Definition 2. Let $S(b, s)$ be an incremental signature or message authentication scheme with block size b and security parameter s. A $(t, q_s, q_v, q_i, L_s, L_v, L_i, \epsilon)$-adversary E makes at most t steps (in a standard RAM model [AHU74]), queries

Sig, IncSig, Vf at most q_s, q_i, q_v times, each query with messages of no more than L_s, L_i, L_v blocks, and is successful with probability at least ϵ. $S(b, s)$ is said to be $(t, q_s, q_v, q_i, L_s, L_v, L_i, \epsilon)$-secure against basic/message substitution/total substitution attacks, iff there is no $(t, q_s, q_v, q_i, L_s, L_v, L_i, \epsilon)$-adversary performing the corresponding attack.

For the rest of this paper, we write $(t, \mathbf{q}, \mathbf{L}, \epsilon)$ for $\mathbf{q} = (q_s, q_i, q_v)$ and $\mathbf{L} = (L_s, L_i, L_v)$. In some settings, parameters may be irrelevant, for example q_v and L_v in signature schemes. It this case, it is understood that $\mathbf{q}$ and $\mathbf{L}$ abbreviate (q_s, q_i) and (L_s, L_i).

3 Incremental Message Authentication: IncXMACC

3.1 Notations and Definitions

For two strings $x, y \in \{0, 1\}^*$, let $x \cdot y$ be the concatenation of x and y. For $x, y \in \{0, 1\}^n$, $x \oplus y$ denotes the bitwise exclusive-or of x, y. For a number $i \in \{0, \dots, 2^m - 1\}$, let $\langle i \rangle_m$ denote the m-bit binary representation of i.

Let $\mathrm{Map}(X, Y)$ denote the set of all functions with domain X and range Y. A *function family* $F \subseteq \mathrm{Map}(X, Y)$ is a set of functions, where we associate a key a to each function $f \in F$. Let F_a be the function specified by key a. To draw a function $f \in F$ at random means to choose at random with equal probability a key a from the set of all keys of functions in F and to set $f := F_a$. For a function f from the family $\mathrm{Map}(X, Y)$ the associated key is the sequence of all $|X|$ function values in some fixed order.

Let $F, G \subseteq \mathrm{Map}(X, Y)$ be two function families and D be a probabilistic algorithm. Define the *advantage* of D distinguishing between F and G as

$$\mathrm{Adv}_D(F, G) = \mathrm{Prob}_{f \in F}\left[D^f = 1\right] - \mathrm{Prob}_{g \in G}\left[D^g = 1\right],$$

where the probabilities are taken over the random choice of $f \in F$ resp. $g \in G$ and the coin tosses of D. We say that D is a (t, q, ϵ)-*distinguisher* if it makes at most t steps (in a standard RAM model), makes at most q oracle queries and achieves $\mathrm{Adv}_D(F, \mathrm{Map}(X, Y)) \geq \epsilon$. We say that the family F is (t, q, ϵ)-*secure* if there exists no (t, q, ϵ)-distinguisher.

3.2 XOR Schemes

Bellare, Guérin and Rogaway [BGR95] introduced the XOR MAC schemes, a general framework for designing message authentication schemes. Let F be a function family with domain $\{0, 1\}^l$ and range $\{0, 1\}^L$ and let F_a be a function in F according to key a. Given a message $M = M[1] \cdots M[n]$ and some state information, e.g. a counter, an algorithm $\mathcal{R}$ outputs probabilistically some seed r. On input r and M, a deterministic algorithm $\mathcal{E}$ produces a set $Z \subseteq \{0, 1\}^l$. Both algorithms must not dependend on the key a. The message authentication code for M is (r, z), where $z = \bigoplus_{x \in Z} F_a(x)$. The verifier knowing the key a

works as follows: On input a MAC (r', z') and a message M', it runs $\mathcal{E}$ with input r' and M' to obtain a set $Z' \subseteq \{0,1\}^l$ and accepts iff $\bigoplus_{x \in Z'} F_a(x) = z'$.

Security of such schemes can be reduced to the algebraic problem that an associated matrix has full rank. For a set $Z \subseteq \{0,1\}^l$ let the characteristic 2^l-bit vector be the vector where the xth entry is 1 iff $x \in Z$. Assume that the underlying function family is $\mathrm{Map}(\{0,1\}^l, \{0,1\}^L)$. Then the probability that the verifier accepts one of the q_v queries for a new message is bounded above by $\delta := q_v \cdot 2^{-L} + \max_{M,r} \{\mathsf{NFRank}_{q_s}(M, r)\}$ with

$$\mathsf{NFRank}_{q_s}(M, r) := \mathrm{Prob}\left[\mathsf{Matrix}_{q_s}(M, r) \text{ hasn't full rank} \mid M \notin \{\mathsf{M}_1, \dots, \mathsf{M}_{q_s}\}\right]$$

Here, $\mathsf{Matrix}_{q_s}(M, r)$ describes the random matrix over GF[2], consisting of the $q_s + 1$ characteristic vectors, where the first q_s vectors for the signing queries are defined by $\mathcal{E}$'s output for the random messages M_i and seeds R_i, and the row vector $q_s + 1$ is specified by $\mathcal{E}$'s output for the possible forgery M and seed r in the first verify query. Note that these two values determine the MAC, since $\mathcal{E}$ is deterministic. Given an adversary A for such an XOR scheme based on a function family F such that A is successful with probability ϵ', one can derive a distinguisher for F with comparable running time and advantage $\epsilon \geq \epsilon' - \delta$. See for example [BGR95] or the proof of Theorem 4.

3.3 The Scheme IncXMACC

The scheme $\mathsf{IncXMACC}_{F,b}$ is based on a function family $F \subseteq \mathrm{Map}(\{0,1\}^l, \{0,1\}^L)$ and has block size $b \leq l^*$ where $l^* = \frac{1}{2}l - 1$. For notational convenience we assume that l is even. It supports the operations $\mathsf{insert}(M, i, M_*)$ and $\mathsf{delete}(M, j)$ for inserting block M_* at position i resp. deleting the jth block in message $M = M[1] \cdots M[n]$, where $1 \leq i \leq n+1$ and $1 \leq j \leq n$. Therefore, the scheme supports other operations like $\mathsf{replace}(M, i, M_*)$, $\mathsf{swap}(M, i, j)$ or $\mathsf{move}(M, i, j)$ to replace block i by M_*, to swap block i and j or to move block i to position j, respectively. We sometimes abbreviate $\mathsf{delete}(M, j)$ by $\mathsf{delete}(j)$ if the corresponding message M is clear from the context. Similar for the other operations.

We will first discuss the single document setting and then show how to proceed in the multi document case. In the single document model, the scheme holds two counters dcnt and bcnt, a document counter resp. a block counter, both initialized with 0. For technical reasons, only messages with more than two blocks are allowed. In the multi document setting, only message with more than four blocks are admissible. In both cases, the counter values are bounded above by 2^{l^*}. The underlying idea is that we link every message block to a unique block counter value and incorperate the order of the message blocks by chaining the counter values.

We define the algorithms Sig and IncSig. Assume that the user or adversary issues a create command for the document $M[1] \cdots M[n] \in \Sigma^n$. Then Sig increments dcnt by one and produces the MAC $(\mathrm{dcnt}, \mathrm{bcnt} +1, \dots, \mathrm{bcnt} +n, z)$, where $z = \bigoplus_{x \in Z} F_a(x)$ with

$$Z = \{0 \cdot \langle \text{dcnt} \rangle_{l-1}\} \cup \{10 \cdot \langle M[i] \rangle_{l^*} \cdot \langle \text{bcnt} + i \rangle_{l^*} \mid i = 1, \ldots, n\}$$
$$\cup \{11 \cdot \langle \text{bcnt} + i \rangle_{l^*} \cdot \langle \text{bcnt} + i + 1 \rangle_{l^*} \mid i = 1, \ldots, n-1\}$$

Finally, Sig increments bcnt by n. On an insert(i, M_*) command for the current document $M = M[1] \cdots M[n]$ and MAC $\mu = (d, c_1, \ldots, c_n, z)$ for M, the system works as follows: IncSig increments the counters dcnt and bcnt and outputs a new MAC $(\text{dcnt}, c_1, \ldots, c_{i-1}, \text{bcnt}, c_i, \ldots, c_n, z')$ for the document $M[1] \ldots M[i-1] M_* M[i] \cdots M[n]$, where

$$z' = z \oplus F_a(0 \cdot \langle d \rangle_{l-1}) \oplus F_a(0 \cdot \langle \text{dcnt} \rangle_{l-1})$$
$$\oplus F_a(10 \cdot \langle M_* \rangle_{l^*} \cdot \langle \text{bcnt} \rangle_{l^*}) \oplus F_a(11 \cdot \langle c_{i-1} \rangle_{l^*} \cdot \langle c_i \rangle_{l^*})$$
$$\oplus F_a(11 \cdot \langle c_{i-1} \rangle_{l^*} \cdot \langle \text{bcnt} \rangle_{l^*}) \oplus F_a(11 \cdot \langle \text{bcnt} \rangle_{l^*} \cdot \langle c_i \rangle_{l^*})$$

That is, the old document counter value of the document is replaced by the new one and the new block M_* is linked to its block counter value bcnt. Moreover, bcnt is put in the chain between c_{i-1} and c_i breaking up the link between c_{i-1} and c_i. For $i = 1$ (resp. $i = n + 1$) drop the fourth and fifth (resp. fourth and last) function value.

A delete(i) command for $1 \leq i \leq n$ is processed similarly. Having incremented dcnt, the new MAC for the document $M[1] \cdots M[i-1] M[i+1] \cdots M[n]$ is given by $(\text{dcnt}, c_1, \ldots, c_{i-1}, c_{i+1}, \ldots, c_n, z')$ where

$$z' = z \oplus F_a(0 \cdot \langle d \rangle_{l-1}) \oplus F_a(0 \cdot \langle \text{dcnt} \rangle_{l-1})$$
$$\oplus F_a(10 \cdot \langle M[i] \rangle_{l^*} \cdot \langle c_i \rangle_{l^*}) \oplus F_a(11 \cdot \langle c_{i-1} \rangle_{l^*} \cdot \langle c_i \rangle_{l^*})$$
$$\oplus F_a(11 \cdot \langle c_i \rangle_{l^*} \cdot \langle c_{i+1} \rangle_{l^*}) \oplus F_a(11 \cdot \langle c_{i-1} \rangle_{l^*} \cdot \langle c_{i+1} \rangle_{l^*})$$

In this case, the system doesn't increment bcnt. For $i = 1$ or $i = n$ adapt the last lines as above.

Finally, we define the verify procedure Vf. Given $M = M[1] \ldots M[n]$ and a MAC $(d', c'_1, \ldots, c'_{n'}, z')$, check that $n' = n$ and that all r'_j values are different and reject if one of these properties doesn't hold. Otherwise compute $z = \bigoplus_{x \in Z} F_a(x)$ with

$$Z = \{0 \cdot \langle d' \rangle_{l-1}\} \cup \{10 \cdot \langle M[i] \rangle_{l^*} \cdot \langle c'_i \rangle_{l^*} \mid i = 1, \ldots, n\}$$
$$\cup \{11 \cdot \langle c'_i \rangle_{l^*} \cdot \langle c'_{i+1} \rangle_{l^*} \mid i = 1, \ldots, n-1\}$$

Reject if $z \neq z'$, otherwise accept.

Security is proven as in [BGR95]. We first deal with the case $F = R = \text{Map}(\{0,1\}^l, \{0,1\}^L)$ and show an upper bound for the success probability. Due to space restriction we skip the rather technical proof. It will be given in the final version.

Theorem 3. *Let $R = \text{Map}(\{0,1\}^l, \{0,1\}^L)$ and $2b + 2 \leq l$. Let E be a computationally unbounded adversary attacking the incremental scheme $\text{IncXMACC}_{R,b}$ in a message substitution attack making at most q_v verify queries. The probability that E is successful is bounded above by $\delta_I := q_v \cdot 2^{-L}$.*

Obviously, this bound is tight. From this Theorem we derive:

Theorem 4. *Let $F \subseteq \mathrm{Map}(\{0,1\}^l, \{0,1\}^L)$ be a function family with $2b+2 \leq l$. If F is (t', q', ϵ')-secure then $\mathsf{IncXMACC}_{F,b}$ is $(t, q_s, q_v, q_e, L_s, L_v, \epsilon)$-secure, where*

$$t' = t + c(q_s + q_v + q_e)(L + l + b), \quad q' = 2q_v L_v + 2q_s L_s + 6q_e, \quad \epsilon' = \epsilon - q_v \cdot 2^{-L}$$

for a small constant $c \in \mathbb{N}$ depending only on the computational model.

Proof. (Sketch) Let E be an adversary for $\mathsf{IncXMACC}$ with the specified parameters and success probability at least ϵ. From E we construct a distinguisher D for F. D is given oracle access to a randomly chosen function g in F resp. R. D simulates E and $\mathsf{IncXMACC}$'s program by replacing each function evaluation F_a with the oracle values for g and outputs 1 iff E is successful. By Theorem 3, for $g \in R$ the adversary E is successful with probability at most $q_v \cdot 2^{-L}$. Therefore,

$$\mathrm{Prob}_{g \in F}[D^g = 1] - \mathrm{Prob}_{g \in R}[D^g = 1]$$
$$= \mathrm{Prob}_{g \in F}[E \text{ is successful}] - \mathrm{Prob}_{g \in R}[E \text{ is successful}] \geq \epsilon - q_v \cdot 2^{-L}.$$

Hence, D is a (t', q', ϵ')-distinguisher for F. $\qquad\square$

We compare $\mathsf{IncXMACC}$ and the tree scheme presented in [BGG95]. Our scheme is only secure when the MAC is kept on a secure medium, while the tree scheme is secure against total substitution attacks. The tree scheme can be applied with any secure signature or authentication scheme, but deleting or inserting a block takes $\Omega(\log n)$ evaluations of the ordinary signature scheme, where n is the number of message blocks of the document. Additionally the tree structure must be maintained. Nevertheless, the tree scheme supports the more powerful modifications paste and cut. The advantage of our scheme is that it takes only a constant number of function evaluations for insert and delete (below we'll show that this holds also for the paste modification), that it merely accesses the corresponding message block in update steps, and that the size of the MAC is considerably smaller. Namely, let s be the output length of the pseudorandom function used by $\mathsf{IncXMACC}$ and the output length of the ordinary authentication scheme used in the tree scheme. Moreover, assume that both schemes have block size b. If the block counter is bounded above by s^c, then $\mathsf{IncXMACC}$ produces MACs for messages of n blocks with bit size at most $s + c(n+1)\log s = O(s + n \log s)$, while MACs produced by the tree scheme have size at least $(\frac{3}{2}s + 1)n = \Omega(ns)$.

The scheme $\mathsf{IncXMACC}$ is provably not secure against (nonadaptive) total substitution attacks. The adversary queries Sig for the document ABCD, where A,B,C,D are different blocks in $\{0,1\}^b$. He alters the document to AABC and changes the MAC $(d, c_1, c_2, c_3, c_4, z)$ to $(d, c_1, c_1, c_2, c_3, z)$. Then he asks IncSig to delete the third symbol. Replacing this MAC $(d+1, c_1, c_1, c_3, z')$ by $(d+1, c_1, c_3, c_4, z')$, he obtains a valid MAC for the document ACD, which hasn't appeared as a virtual document.

We now adress the multi document setting. For every document we associate a name $\alpha \in \{0,1\}^b$. Additionaly, we keep a block counter bcnt_α and a document counter dcnt_α for each document. Signing a document is similar to IncXMACC but we use the value $00 \cdot \langle \text{dcnt}_\alpha \rangle_{l^*} \cdot \langle \alpha \rangle_{l^*}$ instead of $0 \cdot \langle \text{dcnt} \rangle_{l-1}$ for the source and $00 \cdot \langle \text{dcnt}_\beta + 1 \rangle_{l^*} \cdot \langle \beta \rangle_{l^*}$ instead of $0 \cdot \langle \text{dcnt} + 1 \rangle_{l-1}$ for the destination. Security follows as in Theorem 3 and Theorem 4.

Theorem 5. *Let $F \subseteq \text{Map}(\{0,1\}^l, \{0,1\}^L)$ be a function family with $2b + 2 \leq l$. If F is (t', q', ϵ')-secure then $\text{IncXMACC}_{F,b}$ is $(t, q_s, q_v, q_e, L_s, L_v, \epsilon)$-secure in the multi document setting with at most I documents, where*

$$t' = t + cI(q_s + q_v + q_e)(L + l + b), \quad q' = 2q_v L_v + 2q_s L_s + 6q_e, \quad \epsilon' = \epsilon - q_v \cdot 2^{-L}$$

for a small constant $c \in \mathbb{N}$.

In the multi document setting, we can allow a paste modification if we use *one* block counter for *all* documents. The paste command for documents M, M' with names α_1, α_2 and MACs $(d, c_1, \ldots, c_n, z)$, $(d', c_1', \ldots, c_{n'}', z')$ produces the MAC $(\text{dcnt}_\beta + 1, c_1, \ldots, c_n, c_1', \ldots, c_{n'}', \hat{z})$ with

$$\hat{z} = z \oplus z' \oplus F_a(00 \cdot \langle \text{dcnt}_\beta + 1 \rangle_{l^*} \cdot \langle \beta \rangle_{l^*}) \oplus F_a(00 \cdot \langle d \rangle_{l^*} \cdot \langle \alpha_1 \rangle_{l^*})$$
$$\oplus F_a(00 \cdot \langle d' \rangle_{l^*} \cdot \langle \alpha_2 \rangle_{l^*}) \oplus F_a(11 \cdot \langle c_n \rangle_{l^*} \cdot \langle c_1' \rangle_{l^*})$$

for the document $M \cdot M'$ with name β.

4 Memory Checkers

4.1 Definition

Let $\mathcal{D}$ be a data structure with a set of operations that define the behaviour of $\mathcal{D}$ on an initial configuration. Consider for example the data structure stack. The sequence push(a), push(b), pop, push(b), pop for an empty stack produces the output $-, -, b, -, b$, where $-$ stands for "no output".

We assume that all arguments for the operations are specified by a parameter n. To emphasize this dependence we write $\mathcal{D}_n$. We want to design a program C that checks whether an implementation D_n of $\mathcal{D}_n$ works correctly for a sequence of operations for this data structure. We call these operations *user* or *input operations*. C filters the interaction between the user and the data structure resp. memory, so that the user can interact with the data structure only via the checker. After having read the next user operation, the program C shall return the output of that operation to the user or BUGGY if an error occurs, e.g. D_n returns a different value than the expected one. Obviously, the worst case occurs if the user and the memory is totally under control of one adversary. Additionally, the adversary works adaptively, i.e. his next action depends on all previous steps.

To allow multiple instances, we extend every operation by an argument taking values between 0 and $I - 1$ in binary, where I stands for the maximal number of instances available. Let $\mathcal{D}_n^I$ be the augmented version of $\mathcal{D}_n$. The checker can use further instances to save additional information like time stamps to the insecure medium.

An execution is divided into rounds. Each round starts with the checker reading the next user operation. Then it performs some local computation and may interact arbitrarily with the data structure. After having finished this computation, the checker shall return the correct answer for the user operation to the user (or "−" if the operation doesn't produce an output) before reading the next operation. The checker shall output BUGGY if the data structure returns a faulty value at some point in the execution. On the other hand, it shall never output BUGGY if no error occurs. Before starting the first round, the checker might perform a preprocessing, and additionally, after having read the last user operation, it might do some "postprocessing" (and perhaps output BUGGY then).

We use the RAM model to define our checker. The space complexity is measured logarithmically, while time complexity can either be uniform or logarithmic. In this work, time will be meassured uniformly. We assume that the adversary's model of computation is a RAM, too, and that both RAM share a sufficient large number of registers to exchange information, while every other memory of each machine is private. See [GMR89,GO96] for a more formal treatment of interactive machines.

Definition 6. A $(t_{pre}, t_{post}, t_{op}, s, q, J)$-memory checker for a data structure $\mathcal{D}_n^I$ is a probabilistic RAM C such that for every execution with at most q user operations, C takes only t_{pre} preprocessing steps, at most t_{post} postprocessing steps and only t_{op} steps to process each user operation. Additionally, C's private memory is bounded above by s bits and the checker uses at most J instances of $\mathcal{D}_n$. A $(t_{pre}, t_{post}, t_{op}, s, q, J)$-memory checker for $\mathcal{D}_n^I$ is called (t, δ, ϵ)-secure if the following holds for every adversary A running in time t:

- Completeness: If the output of D_n^J is correct for all operations issued by C, then the probability that C returns BUGGY or that not all answers of C for the user operations are correct is at most δ, where the probability is taken over the coin tosses of C and A.
- Soundness: If the output of D_n^J is false for some operation, then C should output BUGGY with probability at least $1 - \epsilon$.

In most settings we are interested in checkers for which $\delta = 0$ holds. These checkers are called *complete*. Definition 6 doesn't rule out the trivial solution, that C simply keeps all values in his private memory. This would rather prevent errors and guarantee correct outputs than check the data structure. We are interested in checkers using only a few bits private memory and causing a small overhead.[3] So this trivial solution gives us an upper bound and a starting point to build more efficient solutions. A checker is called an *on-line checker* iff it

---

[3] Note that we don't charge the checker's running time e.g. for inserting or deleting an element using **insert** and **delete** commands passed to the implementation (except for the time to write the operation and to read the answer).

outputs BUGGY in that round in which an error occurs. Otherwise it is called an *off-line checker*. A checker is called *noninvasive* if at the end of each round, the insecure memory contains only values specified by the input operations when the checker reads the next operation. Otherwise it is called *invasive*. In particular, our checker based on lncXMACC is off-line and noninvasive with the additional property that the checker passes only user operations to the implementation.

4.2 Designing Checkers via Incremental Schemes

In this section we show how we can derive a memory checker from lncXMACC. We prove that we can check any data structure based on the structure List_n, where List_n represents a list with elements from $\{0,1\}^n$. The initial configuration is empty. List_n supports four operations: $\text{insert}(i,v)$ to insert element $v \in \{0,1\}^n$ at position i, $\text{delete}(i)$ to remove the element at position i and return this value to the user, $\text{replace}(i,v)$ to replace the ith value by v and return this element, and $\text{read}(i)$ to return the ith element to the user.

We can design checkers for other data structures based on List_n like stacks and queues. If the checker maintains a counter for the number m of elements currently in the list, the stack resp. queue commands pop, $\text{push}(v)$, dequeue and $\text{enqueue}(v)$ are equivalent to $\text{delete}(m)$, $\text{insert}(m+1,v)$, $\text{delete}(1)$ and $\text{insert}(m+1,v)$. If the data structure can be implemented with lists, we can combine the checker's program and the list implementation of the data structure to obtain a method to securely store the data of this structure on an insecure medium. The following notion of a sound scheme will help us to prove stronger security:

Definition 7. Let $S(b,s) = (\text{Gen}, \text{Sig}, \text{IncSig}, \text{Vf})$ be an $\mathcal{M}$-incremental authentication or signature scheme. $S(b,s)$ is called sound iff for all keys produced with positive probability by Gen the following holds: Let M be a message that is obtained by applying a text modification $\pi \in \mathcal{M}$ with argument y to documents $M_1, \ldots, M_m$ and let $\mu_1, \ldots, \mu_m$ and $\mu = \text{IncSig}(M_1, \ldots, M_m, \mu_1, \ldots, \mu_m, \pi, y)$ the corresponding (valid or invalid) signatures. If $\text{Vf}(M, \mu) = 1$, then $\text{Vf}(M_i, \mu_i) = 1$ holds for all $i = 1, \ldots, m$.

Informally, a sound scheme is a scheme such that applying IncSig with an invalid signature μ_i for some M_i doesn't yield a valid signature for M. Note that the soundness property doesn't guarantee security. It only states that one cannot produce a valid signature form invalid signatures *directly*. It may yet be possible to deduce a valid signature from an invalid one.

Lemma 8. *The* $\{\text{delete}, \text{insert}\}$*-incremental scheme* $\text{IncXMACC}_{F,b}$ *is sound.*

The proof is omitted. One can easily verify that the tree scheme is sound, too.

Theorem 9. *Let F be a function family with input length l, output length L and key length κ. Assume that $\text{IncXMACC}_{F,b}$ is $(t, \mathbf{q}, \mathbf{L}, \epsilon)$-secure against message substitution attacks for block size $b = n$. Then there exists a non-invasive*

$(t_{\mathrm{pre}}, t_{\mathrm{post}}, t_{\mathrm{op}}, s, q, I)$-off-line checker for List_n^I, which is $(t', 0, \epsilon)$-secure where

$$t_{\mathrm{pre}} = \mathrm{Time}(\mathsf{FGen}), \quad t_{\mathrm{post}} = c_1 q \cdot \mathrm{Time}(F), \quad t_{\mathrm{op}} = c_1 \cdot (\mathrm{Time}(F) + \log q),$$
$$s = c_2 \cdot (n + l + q \log q + IL + \mathrm{Space}(F)) + \kappa,$$
$$t' = t - c_3(q t_{\mathrm{op}} + t_{\mathrm{pre}} + t_{\mathrm{post}}), \quad q_i = q, \quad I = \min\{q_s, q_v\}.$$

for small constants $c_1, c_2, c_3 \in \mathbb{N}$. Here, $\mathrm{Time}(F)$ resp. $\mathrm{Space}(F)$ denotes the time resp. space to evaluate a function from F and $\mathrm{Time}(\mathsf{FGen})$ denotes the time to draw a key for a function in F.

A sketch of the proof is given in Appendix A. It is easy to see that we can derive an on-line checker for List_n from the tree scheme. Storing the signature in the checker's private memory is too expensive. Hence, we need additional instances to store the nodes of the signature tree on the insecure memory. In this case, security is provided by the fact that the tree scheme is secure against total substitution attacks. However, this checker is invasive and we cannot for example efficiently apply this construction to stacks, because in this case we cannot access all parts of the signature fast.

4.3 A Lower Bound for Substitution Detecting Schemes

First, we define a *normal form* for adversaries performing attacks on the message substitution detection property. Let $\mathcal{S}(b, s) = (\mathsf{Gen}, \mathsf{Sig}, \mathsf{IncSig}, \mathsf{Vf})$ an $\mathcal{M}$-incremental (signature or authentication) scheme. We assume that IncSig outputs the invalid signature $\perp$ if, for some reason, it refuses to produce a valid one. An attack on the detection property is a message substitution attack, such that each IncSig query $(\alpha_1, \ldots, \alpha_m, \beta, \pi, y)$ has the following form:

1. The adversary may replace any message M_{α_i} with $M_{\alpha_i}^*$ by alter commands. Let $M_{\alpha_i}^*$, $i = 1, \ldots, m$, be this sequence of messages (where we allow $M_{\alpha_i}^* = M_{\alpha_i}$). Additionally, the adversary stores the current content M_β.
2. The adversary queries IncSig for $(\alpha_1, \ldots, \alpha_m, \beta, \pi, y)$.
3. The adversary replaces all messages with name α_i by M_{α_i} again. If IncSig has returned $\perp$, the adversary replaces the document with name β by the former value.

Furthermore, the adversary doesn't use additional alter commands. It is easy to see that every adversary can be assumed w.l.o.g. to be in normal form. Therefore, we can associate each alter command uniquely to an IncSig query. If IncSig doesn't return $\perp$ in step 2, the adversary may either replace M_β again or not.

For notational convenience, let $M[i] = \star$ for the message $M[1] \cdots M[n]$ and $i > n$, where $\star$ denotes a special symbol $\star \notin \Sigma$. In particular, we have $M[i] \neq M'[i]$ for messages $M[1] \cdots M[n]$ and $M'[1] \cdots M'[n']$ with $n < i \leq n'$.

Definition 10. A (normal form) adversary for the detection property is successful, if IncSig returns in step 2 a signature different from $\perp$ for a query $(\alpha_1, \ldots, \alpha_m, \beta, \pi, y)$, such that for the blocks $M_{\alpha_{i_h}}^*[j_h]$, $h = 1, \ldots, k$, that IncSig has read to produce this signature, we have $M_{\alpha_{i_h}}^*[j_h] \neq M_{\alpha_{i_h}}[j_h]$ for some h.

Note that Definition 10 doesn't rule out the trivial solution that IncSig always outputs ⊥ resp. that IncSig never reads a block.

Definition 11. Let $S(b, s) = ($Gen, Sig, IncSig, Vf$)$ be an $\mathcal{M}$-incremental scheme. A $(t, \mathbf{q}, \mathbf{L}, \delta)$-adversary for the detection property is specified by the parameters in definition 2, where δ is the success probability. $S(b, s)$ is called $(t, \mathbf{q}, \mathbf{L}, \delta)$-detecting, if there exists no $(t, \mathbf{q}, \mathbf{L}, \delta)$-adversary for the detection property.

Thus, message substitution detecting schemes can be viewed as on-line checkers. To prove that a detecting scheme which is secure against basic attacks, is also secure against message substitution attacks, we need the following definition:

Definition 12. The $\mathcal{M}$-incremental scheme $S(b, s) = ($Gen, Sig, IncSig, Vf$)$ is a scheme with p-predictable IncSig-access, iff one can for all (with positive probability generated) keys, all messages M_{α_i} with $M_{\alpha_i} = M_i[1] \cdots M_i[n_i]$ and signatures μ_{α_i}, $i = 1, \ldots, m$, predict the message blocks, which IncSig accesses to update the signature in response to $(\alpha_1, \ldots, \alpha_m, \beta, \pi, y)$ in time $p(\max\{n_i\})$ (in the corresponding computational model) from μ_{α_i}, $i = 1, \ldots, m$, and π, y.

For simplicity, we have assumed that IncSig's access is predictable from μ_{α_i}, π, y in time $p(\max\{n_i\})$. Extensions to other parameters are straightforward. Clearly, the tree scheme is a detecting scheme with predictable IncSig-access.

Proposition 13. *Let* $S(b, s) = ($Gen, Sig, IncSig, Vf$)$ *be a* $(t, \mathbf{q}, \mathbf{L}, \delta)$-*detecting* $\mathcal{M}$-*incremental scheme with p-predictable* IncSig-*access, which is* $(t, \mathbf{q}, \mathbf{L}, \epsilon)$-*secure against basic attacks. Then* $S(b, s)$ *is* $(t', \mathbf{q}, \mathbf{L}, \epsilon')$-*secure against message substitution attacks, where* $t' = t - q_i p(L_i)$ *and* $\epsilon' = \epsilon + \delta$.

Proof. (Sketch) Let E be a normal form adversary with parameters $t, \mathbf{q}, \mathbf{L}$, which is successful with probability at least ϵ in a message substitution attack. From E we construct via black-box-simulation an adversary A performing a basic attack.

A simulates each query E to Sig and Vf by its oracle access to $S(b, s)$. If E issues an IncSig query without having used an associated alter command in step 1 of the normal form specification, then A passes this query to IncSig and returns the signature to E. Assume, that E tampers messages M_{α_i} to $M_{\alpha_i}^*$ before. Then A computes in time $p(L_i)$ from μ_{α_i}, $i = 1, \ldots, m$, and π, y the message blocks $M_{\alpha_{i_h}}^*[j_h]$, $h = 1, \ldots, k$, which IncSig would read. If $M_{\alpha_{i_h}}^*[j_h] \neq M_{\alpha_{i_h}}[j_h]$ for some h, A returns ⊥ to E without quering IncSig. Else A passes the query to IncSig without tampering the messages and returns the signature to E. In this case, the signature does not depend on other (altered or unaltered) blocks and the answer is correct.

As alter commands don't change virtual documents, every virtual document appearing in A's attack appears in E's attack as well. Let Detect be the event, that E isn't successful in an attack for the detection property. Furthermore, let Succ_A resp. Succ_E be the events that A resp. E performs a successful attack on the signature scheme. We have

$$\epsilon' \leq \text{Prob}\left[\text{Succ}_E\right] \leq \text{Prob}\left[\text{Succ}_E \mid \text{Detect}\right] + \text{Prob}\left[\neg\, \text{Detect}\right] \leq \text{Prob}\left[\text{Succ}_A\right] + \delta.$$

Hence, A is successful with probability at least ϵ. $\qquad\square$

We show that we cannot design detecting schemes producing small signatures:

Proposition 14. *Let $S(b,s)$ be a complete $(t, \mathbf{q}, \mathbf{L}, \delta)$-detecting scheme for $t = cbn$, $q_s = 1$, $q_i = n$, $L_s = L_i = n$, which supports the replace modification such that IncSig always accesses the ith block for valid replace(M_α, i, M_*) commands. Then for $\Delta := 1 - \delta > \frac{1}{2}$ the bit length of a signature for a message $M = M[1] \cdots M[n]$ must be at least*

$$(1 - \beta)\frac{n}{t_{max}} + \log_2 \gamma,$$

where $\beta = 1 - 2(\alpha - \frac{1}{2})^2 \log_2 e < 1$, $\gamma = \frac{\Delta - \alpha}{1 - \alpha} < 1$ for $\frac{1}{2} < \alpha < \Delta$. Here, t_{max} is the maximal number of blocks IncSig reads for an update step.

The proof is a variation of the proof given in [BEG+94] for on-line checkers and is omitted. If Δ and α are close to 1, we have $1 - \beta \approx \frac{1}{3}$ and $\gamma \approx 1$, i.e. a signature must have at least $\frac{n}{3t_{max}}$ bits.

Acknowledgements

We thank Roger Fischlin for pointing out the topic of memory checkers and C.P. Schnorr and the anonymous referees for their comments. We also thank Mihir Bellare and Daniele Micciancio for discussions about their works.

References

[AHU74] A.AHO, J.HOPCROFT, J.ULLMAN: The Design and Analysis of Computer Algorithms, *Addison Wesley*, 1974.

[BGG94] M.BELLARE, O.GOLDREICH, S.GOLDWASSER: Incremental Cryptography: The Case of Hashing and Signing, *Crypto '94, Lecture Notes in Computer Science, Vol. 839, Springer-Verlag*, pp. 216-233, 1994.

[BGG95] M.BELLARE, O.GOLDREICH, S.GOLDWASSER: Incremental Cryptography and Application to Virus Protection, *Proceedings of the 27th Annual ACM Symposium on the Theory of Computing*, pp. 45-56, 1995.

[BGR95] M.BELLARE, R.GUÉRIN, P.ROGAWAY: XOR MACs: New Methods for Message Authentication Using Finite Pseudorandom Functions, *Crypto '95, Lecture Notes in Computer Science, Vol. 963, Springer-Verlag*, pp. 15-29, extended version available at http://www.cs.ucdavis.edu/~rogaway/, 1995.

[BKR94] M.BELLARE, J.KILLIAN, P.ROGAWAY: On the Security of Cipher Block Chaining, *Crypto '94, Lecture Notes in Computer Science, Vol. 839*, pp. 341-358, 1994.

[BEG+91] M.BLUM, W.EVANS, P.GEMMELL, S.KANNAN, M.NAOR: Checking the Correctness of Memories, *Proceedings of the 32nd IEEE Symposium on Foundations of Computer Science*, pp. 90-99, 1991.

[BEG+94] M.BLUM, W.EVANS, P.GEMMELL, S.KANNAN, M.NAOR: Checking the Correctness of Memories, *Algorithmica, Volume 12*, pp. 225-244, 1994.

[BK89] M.BLUM, S.KANNAN: Designing Programs that Check Their Work, *Proceedings of the 21st Annual ACM Symposium on the Theory of Computing*, pp. 86-97, 1989.

[GGM86] O.GOLDREICH, S.GOLDWASSER, S.MICALI: How to Construct Random Fun-
tions, *Journal of ACM, Vol. 33(4), pp.* 792–807, 1986.

[GMR89] S.GOLDWASSER, S.MICALI, C.RACKOFF: The Knowledge Complexity of Interactive Proof Systems, *SIAM Journal on Computation, Vol. 18, pp.* 186–
208, 1989.

[GMR88] S.GOLDWASSER, S.MICALI, R.L.RIVEST: A Digital Signature Scheme Secure Against Adaptive Chosen Message Attacks, *SIAM Journal on Computation, Vol. 17(2), pp.* 281–308, 1988.

[GO96] O.GOLDREICH, R.OSTROVSKY: Software Protection and Simulation on
Oblivious RAM, *Journal of ACM, Vol. 43(3), pp.* 431–473, 1996.

[M97] D.MICCIANCIO: Oblivious Data Structures: Application to Cryptography,
*(to appear at) Proceedings of the 29th Annual Symposium on the Theory of
Computing,* 1997.

A   Sketch of Proof of Theorem 9

Clearly, the checker runs the incremental scheme IncXMACC to check the correctness. For every instance we'll have a signature for the content. Updating
this signature when inserting, deleting, replacing or reading an element will be
done with the insert, delete commands for the incremental scheme. To prevent
repetition attacks, we prepend every "message" with a time stamp which the
checker stores in its local memory, not in the insecure memory. This time stamp
is updated before processing insert, delete commands.

If no more operations are left, the checker empties the memory in a postprocessing phase: For each initialized instance it deletes the values in the instance
using delete commands and checks that the obtained signatures are accepted by
Vf. If some signature is not accepted, it outputs BUGGY, otherwise C accepts.

If all operations work correctly, the checker never outputs BUGGY since
IncXMACC is complete. Assume that there is a sequence of operations such
that the checker is fooled. We design a adversary E for IncXMACC. E works as
follows: Let A be the adversary for the checker. Then E first runs the whole
execution simulating C and A by black-box-simulation using the oracle access
for the incremental scheme. Moreover, E maintains the correct memory contents
and stores all signatures.

Since E has simulated the whole execution first, he knows the last user operation for which a wrong value has been returned. E builds a message M that
consists of the time stamp, the correct memory content (at this point) and replaces the corresponding block with the wrong value. E outputs this message M
and the signature μ for this message as a forgery. As the scheme is sound and the
checker doesn't output BUGGY, i.e. the signature for the final value has been
accepted, this signature μ is valid for M. Virtual documents are only changed by
insert and delete commands, therefore all virtual documents are defined by the
correct memory content and the counter values. Since there is some error in M,
and the time stamps make every virtual document unique, M hasn't appeared
as a virtual document during the execution. Hence, E is successful if A is.

Almost k-wise Independent Sample Spaces and Their Cryptologic Applications

Kaoru Kurosawa[1], Thomas Johansson[2], Douglas Stinson[3]

[1] Dept. of Computer Science
Graduate School of Information Science and Engineering
Tokyo Institute of Technology
2–12–1 O-okayama, Meguro-ku, Tokyo 152, Japan
kurosawa@ss.titech.ac.jp

[2] Dept. of Information Technology, Lund University,
PO Box 118, S-22100 Lund, Sweden
thomas@it.lth.se

[3] Dept. of Computer Science and Engineering
University of Nebraska
Lincoln NE 68588, USA
stinson@bibd.unl.edu

Abstract. An almost k-wise independent sample space is a small subset of m bit sequences in which any k bits are "almost independent". We show that this idea has close relationships with useful cryptologic notions such as multiple authentication codes (multiple A-codes), almost strongly universal hash families and almost k-resilient functions.

We use almost k-wise independent sample spaces to construct new efficient multiple A-codes such that the number of key bits grows linearly as a function of k (here k is the number of messages to be authenticated with a single key). This improves on the construction of Atici and Stinson [2], in which the number of key bits is $\Omega(k^2)$.

We also introduce the concept of ϵ-almost k-resilient functions and give a construction that has parameters superior to k-resilient functions.

Finally, new bounds (necessary conditions) are derived for almost k-wise independent sample spaces, multiple A-codes and balanced ϵ-almost k-resilient functions.

1 Introduction

An *almost k-wise independent sample space* is a probability space on m-bit sequences such that any k bits are almost independent. A ϵ-*biased sample space* is a space in which any (boolean) linear combination of the m bits has the value 1 with probability close to $1/2$. These notions were introduced by Naor and Naor [17] and further studied in [1] due to their applications to algorithms and complexity theory. However, there are also cryptographic applications: Krawczyk applied ϵ-biased sample spaces to the construction of authentication codes [13].

In this paper, we investigate several new relationships between almost k-wise independent sample spaces and useful cryptologic notions such as multiple

W. Fumy (Ed.): Advances in Cryptology - EUROCRYPT '97, LNCS 1233, pp. 409-421, 1997.
© Springer-Verlag Berlin Heidelberg 1997

authentication codes (multiple A-codes) [2] and k-resilient functions [10, 3, 11, 24, 4].

In a multiple A-code, $k \geq 2$ messages are authenticated with the same key. (In "usual" A-codes, just one message is authenticated with a given key.) Recently, Atici and Stinson [2] defined some new classes of almost strongly universal hash families which allowed the construction of multiple A-codes. Here, we prove that almost k-wise independent sample spaces are equivalent to multiple A-codes. This allows us to obtain a more efficient construction of multiple A-codes from the almost k-wise independent sample spaces of [1].

Next, we present a lower bound on the size of the keyspace in a multiple A-code. Numerical examples show that the multiple A-codes we construct are quite close to this bound. Further, from the above equivalence, a lower bound on the size of almost k-wise independent sample spaces is obtained for free. (While a lower bound on the size of ϵ-biased sample spaces was given in [1], no lower bound was known for the size of almost k-wise independent sample spaces.)

Finally, we generalize the idea of resilient functions. A function $\phi : \{0,1\}^m \rightarrow \{0,1\}^l$ is called k-resilient if every possible output l-tuple is equally likely to occur when the values of k arbitrary inputs are fixed by an opponent and the remaining $m - k$ input bits are chosen at random. This is a useful tool for achieving key renewal: an m-bit secret key $(x_1, \cdots, x_m)$ can be renewed to a new l-bit secret key $\phi(x_1, \cdots, x_m)$ about which an opponent has no information if the opponent knows at most k bits of $(x_1, \cdots, x_m)$.

We show that k can be made larger if the definition of resilient function is slightly relaxed. Thus, we define an ϵ-almost k-resilient function as a function ϕ such that every possible output l-tuple is almost equally likely to occur when the values of k arbitrary inputs are fixed by an opponent. (The statistical difference between the output distribution of a k-resilient function and an ϵ-almost k-resilient function is ϵ.) We prove that a large set of almost k-wise independent sample spaces is equivalent to a balanced ϵ-almost k-resilient function, generalizing a result of [24]. From this equivalence, we are able to obtain both efficient constructions and bounds for balanced ϵ-almost k-resilient functions.

2   Almost k-wise independent sample spaces

Let $S_m \subseteq \{0,1\}^m$, and let $X = x_1 \cdots x_m$ be chosen uniformly from S_m.

Definition 1. [1] We say that S_m is an (ϵ, k)-*independent sample space* if for any k positions $i_1 < i_2 < \cdots < i_k$ and any k-bit string α, we have

$$| \Pr[x_{i_1} x_{i_2} \cdots x_{i_k} = \alpha] - 2^{-k}| \leq \epsilon. \tag{1}$$

If $\epsilon = 0$, then S_m is equivalent to an *orthogonal array* $OA_\lambda(k, m, 2)$, where $\lambda = |S_m|/2^k$.

The following efficient construction for (ϵ, k)-independent sample spaces is proved in [1].

Proposition 2. *There exists an (ϵ, k)-independent sample space S_m such that*

$$\log_2 |S_m| = 2(\log_2 \log_2 m - \log_2 \epsilon + \log_2 k - 1).$$

In this section, we prove that almost k-wise independent sample spaces are equivalent to multiple authentication codes (more precisely, almost strongly universal-k hash families, as defined in [2]). This allows us to obtain more efficient multiple A-codes than were previously known.

2.1 Multiple A-codes and ASU-k hash families

We briefly review basic concepts of (multiple) authentication codes. In the usual Simmons model of authentication codes (A-codes) [21, 22], there are three participants, a *transmitter*, a *receiver* and an *opponent*. In an A-code *without secrecy*, the transmitter sends a *message* (s, a) to the receiver, where s is a *source state* (plaintext) and a is an *authenticator*. The authenticator is computed as $a = e(s)$, where e is a secret *key* shared between the transmitter and the receiver. The key e is chosen according to a specified probability distribution.

In a *multiple A-code*, we suppose that an opponent observes $i \geq 2$ messages which are sent using the same key. Then the opponent places a new bogus message (s', a') into the channel, where s' is distinct from the i source states already sent. This attack is called a *spoofing attack of order i*. P_{d_i} denotes the success probability of a spoofing attack of order i, see [15].

Almost strongly universal hash families are a very useful way of constructing practical A-codes. This idea was introduced by Wegman and Carter [26], and further developed and refined in papers such as [23, 5, 13, 12]. Atici and Stinson [2] generalized the definitions so that they could be applied to multiple A-codes. We review these definitions now.

Definition 3. *An $(N; m, n)$ hash family is a set F of N functions such that $f : A \to B$ for each $f \in F$, where $|A| = m, |B| = n$ and $m > n$.*

Definition 4. *An $(N; m, n)$ hash family F of functions from A to B is ϵ almost strongly universal-k (or ϵ-ASU $(N; m, n, k)$) provided that, for all distinct elements $x_1, x_2, \cdots, x_k \in A$, and for all (not necessary distinct) $y_1, y_2, \cdots, y_k \in B$, we have*

$$|\{f \in F : f(x_i) = y_i, 1 \leq i \leq k\}| \leq \epsilon \times |\{f \in F : f(x_i) = y_i, 1 \leq i \leq k - 1\}|.$$

The following result gives the connection between ϵ-ASU $(N; m, n, k)$ hash families and multiple A-codes.

Proposition 5. *[2] There exists an A-code without secrecy for m source states, having n authenticators and N equiprobable authentication rules and such that $P_{d_{k-1}} \leq \epsilon$, if and only if there exists an ϵ-ASU $(N; m, n, k)$ hash family F.*

2.2 Equivalence of hash families and sample spaces

We can can rephrase Definition 1 in terms of hash families, and generalize it to the non-binary case, as follows.

Definition 6. An $(N; m, n)$ hash family F of functions from A to B is (ϵ, k)-independent if for all distinct elements $x_1, x_2, \cdots, x_k \in A$, and for all (not necessary distinct) $y_1, y_2, \cdots, y_k \in B$, we have

$$| \Pr(f(x_i) = y_i, 1 \leq i \leq k) - n^{-k}| \leq \epsilon, \tag{2}$$

where $f \in F$ is chosen uniformly at random.

The following results are straightforward.

Proposition 7. An (ϵ, k)-independent sample space S_m is equivalent to an (ϵ, k)-independent $(|S_m|; m, 2)$ hash family.

Proposition 8. If there exists an (ϵ, k)-independent sample space S_m, then there exists an $(\epsilon, k/t)$-independent $(|S_m|; m/t, 2^t)$ hash family.

Now we show the equivalence of (ϵ, k)-independent sample spaces and almost strongly universal-k hash families.

Theorem 9. If F is an (ϵ, k)-independent $(N; m, n)$ hash family, then F is a δ-ASU $(N; m, n, k)$ hash family, where

$$\delta = \frac{(n^{-k} + \epsilon)}{n(n^{-k} - \epsilon)}.$$

Proof. Suppose that Eq. (2) holds. Then for any $y_1, \cdots, y_k \in B$, we have

$$\Pr[f(x_i) = y_i, 1 \leq i \leq k] \geq n^{-k} - \epsilon,$$

$$\sum_{y_k \in B} \Pr[f(x_i) = y_i, 1 \leq i \leq k] \geq \sum_{y_k \in B} (n^{-k} - \epsilon), \quad \text{and}$$

$$\Pr[f(x_i) = y_i, 1 \leq i \leq k - 1] \geq n(n^{-k} - \epsilon).$$

From the above inequality and Eq. (2), we have

$$\frac{\Pr[f(x_i) = y_i, 1 \leq i \leq k]}{\Pr[f(x_i) = y_i, 1 \leq i \leq k - 1]} \leq \frac{n^{-k} + \epsilon}{n(n^{-k} - \epsilon)}.$$

Let $\delta \triangleq (n^{-k} + \epsilon)/(n(n^{-k} - \epsilon))$. Then

$$|\{f \in F : f(x_i) = y_i, 1 \leq i \leq k\}| \leq \delta \times |\{f \in F : f(x_i) = y_i, 1 \leq i \leq k - 1\}|.$$

Hence, F is a δ-ASU $(N; m, n, k)$ hash family. $\qquad \square$

Definition 10. An $(N; m, n)$ hash family F of functions from A to B is *strongly* (ϵ, k)-*independent* if for any t such that $1 \leq t \leq k$ and for all distinct elements $x_1, x_2, \cdots, x_t \in A$, and for all (not necessary distinct) $y_1, y_2, \cdots, y_t \in B$, we have

$$|\Pr(f(x_i) = y_i, 1 \leq i \leq t) - n^{-t}| \leq \epsilon \tag{3}$$

where $f \in F$ is chosen uniformly at random.

Theorem 11. *If an* $(N; m, n)$ *hash family* F *is strongly* (ϵ, k)-*independent, then* F *is a* δ-*ASU* $(N; m, n, k)$ *hash family, where* $\delta = (n^{-k} + \epsilon)/(n^{-(k-1)} - \epsilon)$.

Proof. The proof is similar to the proof of Theorem 9. □

Lemma 12. *[2] Suppose that a hash family* F *of functions from* A *to* B *is* ϵ-*ASU* $(N; m, n, k)$. *Then for for all* $1 \leq j \leq k$, *for all distinct elements* $x_1, x_2, \cdots, x_j \in A$, *and for all (not necessary distinct)* $y_1, y_2, \cdots, y_j \in B$, *we have*

$$|\{f \in F : f(x_i) = y_i, 1 \leq i \leq j\}| \leq \epsilon^j \times N \tag{4}$$

Lemma 13. *[2] If a hash family* F *is* ϵ-*ASU* $(N; m, n, k)$, *then* $\epsilon \geq 1/n$.

Theorem 14. *If a hash family* F *is* ϵ-*ASU* $(N; m, n, k)$, *then* F *is* (δ, k)-*independent, where* $\delta = (n^k - 1)(\epsilon^k - n^{-k})$.

Proof. From Lemma 12, we have

$$\Pr[f(x_i) = y_i, 1 \leq i \leq k] \leq \epsilon^k \quad \text{and} \tag{5}$$
$$\Pr[f(x_i) = y_i, 1 \leq i \leq k] - n^{-k} \leq \epsilon^k - n^{-k}. \tag{6}$$

On the other hand, from eq.(5), we have

$$\sum_{(\hat{y}_1, \cdots, \hat{y}_k) \neq (y_1, \cdots, y_k)} \Pr[f(x_i) = \hat{y}_i, 1 \leq i \leq k] \leq (n^k - 1)\epsilon^k.$$

Therefore, we have

$$\Pr[f(x_i) = y_i, 1 \leq i \leq k] = 1 - \sum_{(\hat{y}_1, \cdots, \hat{y}_k) \neq (y_1, \cdots, y_k)} \Pr[f(x_i) = \hat{y}_i, 1 \leq i \leq k]$$
$$\geq 1 - (n^k - 1)\epsilon^k.$$

Hence,

$$\Pr[f(x_i) = \hat{y}_i, 1 \leq i \leq k] - n^{-k} \geq 1 - (n^k - 1)\epsilon^k - n^{-k}$$
$$= 1 - \epsilon^k n^k + \epsilon^k - n^{-k}$$
$$= -(n^k - 1)(\epsilon^k - n^{-k}).$$

From Lemma 13, we see that $\epsilon^k - n^{-k} \geq 0$. Hence,

$$-(n^k - 1)(\epsilon^k - n^{-k}) \leq \Pr[f(x_i) = \hat{y}_i, 1 \leq i \leq k] - n^{-k} \leq \epsilon^k - n^{-k}$$

Then the family is (δ, k)-independent, where

$$\delta = \max\{|\epsilon^k - n^{-k}|, |-(n^k - 1)(\epsilon^k - n^{-k})|\} = (n^k - 1)(\epsilon^k - n^{-k})$$

□

2.3 New multiple A-codes

By combining Propositions 2 and 8 with Theorem 9 or Theorem 11, we can obtain new multiple A-codes (ASU-k hash families) from an (ϵ, k)-independent sample space. Since the (ϵ, k)-independent sample spaces from [1] mentioned in Proposition 2 can be shown to be strong, we will apply Theorem 11.

Theorem 15. *There exists a δ-ASU $(N; m, n, k)$ hash family where*

$$\log_2 N = 2(\log_2 \log_2(m \log_2 n) + k \log_2 n - \log_2(n\delta - 1) + \log_2(k \log_2 n) - 1). \quad (7)$$

Proof. Define $l = k \log_2 n$, $u = m \log_2 n$, and

$$\epsilon = \frac{n^{-k}(\delta n - 1)}{\delta + 1} \approx n^{-k}(\delta n - 1).$$

Apply Proposition 2 and 8, constructing a strongly (ϵ, k)-independent (N, m, n) hash family, where $\log_2 N = 2(\log_2 \log_2 u - \log_2 \epsilon + \log_2 l - 1)$. Now apply Theorem 11, to obtain a δ-ASU $(N; m, n, k)$ hash family. We compute $\log_2 N$ as

$$\log_2 N = 2(\log_2 \log_2(m \log_2 n) - \log_2(n^{-k}(\delta n - 1)) + \log_2(k \log_2 n) - 1)$$
$$= 2(\log_2 \log_2(m \log_2 n) + k \log_2 n - \log_2(\delta n - 1) + \log_2(k \log_2 n) - 1).$$

$\square$

3   A lower bound

In this section, we present a lower bound on the size of ASU-k hash families and almost k-wise independent sample spaces.

Theorem 16. *If there exists an ϵ-ASU$(N; m, n, k)$ hash family such that*

$$\epsilon^k \leq 1/n, \quad (8)$$

then

$$N \geq \frac{1}{\epsilon^k} \left(\frac{\log\left(\frac{mn}{k-1}\right)}{\log\left(\frac{1-\epsilon^k}{\frac{1}{n}-\epsilon^k}\right)} - 1 \right).$$

Proof. Suppose F is an ϵ-ASU$(N; m, n, k)$ hash family from A to B, where $|A| = m$, $|B| = n$ and $k \geq 2$. Construct an $N \times mn$ binary matrix $G = (g_{ij})$, with rows indexed by the functions in F and columns indexed by $A \times B$, defined by the rule

$$g_{f,(x,y)} = \begin{cases} 1 \text{ if } f(x) = y \\ 0 \text{ if } f(x) \neq y. \end{cases}$$

Interpret the columns of G as incidence vectors of the N-set F. We obtain a set-system $(F, \mathcal{C} = \{C_{x,y} : x \in A, y \in B\})$, where

$$C_{x,y} = \{f \in F : f(x) = y\}$$

for all $x \in A$, $y \in B$. Let

$$t \triangleq \lfloor \epsilon^k N \rfloor + 1. \tag{9}$$

This set-system satisfies the following properties: (A) $|F| = N$, (B) $|C| = mn$, (C) $\sum_{C \in \mathcal{C}} |C| = Nm$, (D) there does not exist a subset of t points that occurs as a subset of k different blocks (see Lemma 12).

Property (D) says that $(F, \mathcal{C})$ is a *t-packing of index* $\lambda = k - 1$ (i.e., no t-subset of points occurs in more than λ blocks). Hence we obtain the following:

$$\lambda \binom{N}{t} \geq \sum_{C \in \mathcal{C}} \binom{|C|}{t}. \tag{10}$$

Property (C) implies that the average block size is $Nm/mn = N/n$. Define a real-valued function $f(x)$ as

$$f(x) = \begin{cases} 0 & \text{if } x < t \\ x(x-1)\dots(x-t+1) & \text{otherwise.} \end{cases}$$

Since $f(x)$ is convex, we have

$$\frac{\lambda}{mn}\binom{N}{t} \geq \frac{1}{mn}\sum_{C \in \mathcal{C}}\binom{|C|}{t} \geq \frac{f(N/n)}{t!} \tag{11}$$

from Jensen's inequality. We observe that $N/n \geq t - 1$ follows from Eq. (8) and Eq. (9). Then, we obtain

$$(k-1)\frac{N(N-1)\cdots(N-t+1)}{\frac{N}{n}\left(\frac{N}{n}-1\right)\cdots\left(\frac{N}{n}-t+1\right)} \geq mn, \tag{12}$$

and hence

$$(k-1)\left(\frac{N-t+1}{\frac{N}{n}-t+1}\right)^t \geq mn. \tag{13}$$

From Eq. (9), we have $t \leq \epsilon^k N + 1$. Then Eq. (13) can be simplified as follows.

$$(k-1)\left(\frac{1-\epsilon^k}{\frac{1}{n}-\epsilon^k}\right)^t \geq mn, \quad \text{and hence}$$

$$(\epsilon^k N + 1)\log\left(\frac{1-\epsilon^k}{\frac{1}{n}-\epsilon^k}\right) \geq \log\left(\frac{mn}{k-1}\right),$$

from which our bound is obtained. □

Corollary 17. *Suppose S_m is an (ϵ, k)-independent sample space. Denote $\delta = (2^{-k} + \epsilon)/(2(2^{-k} - \epsilon))$. If $\delta^k \leq 1/2$, then*

$$|S_m| \geq \frac{1}{\delta^k}\left(\frac{\log\left(\frac{2m}{k-1}\right)}{\log\left(\frac{1-\delta^k}{\frac{1}{2}-\delta^k}\right)} - 1\right).$$

Proof. This follows from Theorem 9. □

3.1 Some numerical examples of multiple A-codes

We give some numerical examples to compare the multiple A-codes constructed by Atici and Stinson in [2], our new multiple A-codes obtained from Theorem 15, and the lower bound of Theorem 16. Suppose we want an authentication code for $m = 2^{2^{128}}$ source states with deception probability $\delta = 2^{-40}$. We tabulate the number of key bits (i.e., $\log_2 N$) for $k = 3, 4, 10$. Note that we take $n = 2/\delta = 2^{41}$ in Theorem 15 and Theorem 16 (whereas in [2], $n > 2/\delta$).

k	[2]	Theorem 15	Lower bound
3	657	518	243
4	1043	602	283
10	5376	1096	523

A counter-based multiple authentication scheme would (of course) require less key bits than the proposed construction. For example, tabulated values from [2] show that the construction from [5] would for the parameters above and $k = 4$ require 447 key bits. Hence, the $602 - 447 = 155$ additional key bits we use can be thought of as the price payed for having a stateless multiple authentication scheme. An interesting property that can be verified through Theorem 15 is the following. When $k \to \infty$, the number of key bits required per message approaches $\log_2 n$, which is the same as for the counter-based multiple authentication scheme.

4 Almost resilient functions

In what follows, let $m \geq l \geq 1$ be integers and let $\phi : \{0,1\}^m \to \{0,1\}^l$.

Definition 18. ϕ is called an (m, l, k)-resilient function if

$$\Pr[\phi(x_1, \ldots, x_m) = (y_1, \ldots, y_l) \mid x_{i_1} x_{i_2} \cdots x_{i_k} = \alpha] = 2^{-l}$$

for any k positions $i_1 < \cdots < i_k$, for any k-bit string α and for any $(y_1, \cdots, y_l) \in \{0,1\}^l$, where the values x_j ($j \notin \{i_1, \ldots, i_k\}$) are chosen independently at random.

Resilient functions have been studied in several papers, e.g., [10, 3, 11, 24, 4]. We now introduce a generalization, which we call ϵ-almost resilient functions, in which the the output distribution may deviate from the uniform distribution by a small amount ϵ.

Definition 19. We say that ϕ is an ϵ-almost (m, l, k)-resilient function if

$$|\Pr[\phi(x_1, \ldots, x_m) = (y_1, \ldots, y_l) \mid x_{i_1} x_{i_2} \cdots x_{i_k} = \alpha] - 2^{-l}| \leq \epsilon$$

for any k positions $i_1 < \cdots < i_k$, for any k-bit string α and for any $(y_1, \cdots, y_l) \in \{0,1\}^l$, where the values x_j ($j \notin \{i_1, \ldots, i_k\}$) are chosen independently at random.

4.1 Relation with (ϵ, k)-independent sample space

It is well-known that a resilient function is equivalent to a large set of orthogonal arrays [24]. Here we prove a similar result for almost resilient functions that involves k-wise independent sample spaces.

Definition 20. A *large set of* (ϵ, k, m, t)-*independent sample spaces*, denoted $LS(\epsilon, k, m, t)$, is a set of 2^{m-t} (ϵ, k, m, t)-independent sample spaces, each of size 2^t, such that their union contains all 2^m binary vectors of length m.

Theorem 21. *If there exists an* $LS(\epsilon, k, m, t)$, *then there exists a* δ-*almost* $(m, m - t, k)$-*resilient function, where* $\delta = \epsilon / 2^{m-t-k}$.

Proof. There are 2^{m-t} (ϵ, k)-independent sample spaces in the set. Name the (ϵ, k)-independent sample spaces C_γ, $\gamma \in \{0, 1\}^{m-t}$. Then define a function $\phi : \{0, 1\}^m \to \{0, 1\}^{m-t}$ by the rule

$$\phi(x_1, \ldots, x_m) = \gamma \text{ if and only if } (x_1, \ldots, x_m) \in C_\gamma.$$

For any k positions $i_1 < \cdots < i_k$, any k-bit string α and any $\gamma \in \{0, 1\}^{m-t}$, let

$$L \overset{\triangle}{=} |\{(x_1, \ldots, x_m) : x_{i_1} \cdots x_{i_k} = \alpha, (x_1, \ldots, x_m) \in C_\gamma\}|.$$

Then

$$\Pr[\phi(x_1, \ldots, x_m) = \gamma \mid x_{i_1} x_{i_2} \cdots x_{i_k} = \alpha] = \frac{L}{2^{m-k}}. \tag{14}$$

From Definition 1, we have

$$2^{-k} - \epsilon \leq \frac{L}{2^t} \leq 2^{-k} + \epsilon. \tag{15}$$

Hence, from (14) and (15), we obtain

$$|\Pr[\phi(x_1, \ldots, x_m) = \gamma \mid x_{i_1} x_{i_2} \cdots x_{i_k} = \alpha] - 2^{-(m-t)}| \leq \frac{\epsilon}{2^{m-t-k}}.$$

$\square$

Definition 22. The function $\phi : \{0, 1\}^m \to \{0, 1\}^l$ is called *balanced* if we have

$$\Pr[\phi(x_1, \ldots, x_m) = (y_1, \ldots, y_l)] = 2^{-l}$$

for all $(y_1, \cdots, y_l) \in \{0, 1\}^l$.

For balanced functions, we can prove the converse of Theorem 21.

Theorem 23. *If there exists a balanced* ϵ-*almost* (m, l, k)-*resilient function,* ϕ, *then there exists an* $LS(\delta, k, m, m - l)$, *where* $\delta = \epsilon / 2^{k-l}$.

Proof. For $\gamma \in \{0,1\}^l$, let

$$C_\gamma \triangleq \{(x_1, \ldots, x_m) : \phi(x_1, \ldots, x_m) = \gamma\}.$$

Since ϕ is balanced, $|C_\gamma| = 2^{m-l}$. If each C_γ is an (ϵ, k)-independent sample space, then we automatically get a large set. For any k positions $i_1 < \cdots < i_k$, for any k-bit string α for and any $\gamma \in \{0,1\}^l$, let

$$L \triangleq |\{(x_1, \ldots, x_m) : x_{i_1} \cdots x_{i_k} = \alpha, (x_1, \ldots, x_m) \in C_\gamma\}|.$$

Then, within the sample space C_γ, we have

$$\Pr[x_{i_1} x_{i_2} \cdots x_{i_k} = \alpha] = \frac{L}{|C_\gamma|} = \frac{L}{2^{m-l}}. \tag{16}$$

From Definition 19, we get

$$2^{-l} - \epsilon \le \frac{L}{2^{m-k}} \le 2^{-l} + \epsilon. \tag{17}$$

Hence, from (16) and (17), we obtain

$$|\Pr(x_{i_1} x_{i_2} \cdots x_{i_k} = \alpha) - 2^{-k}| \le \frac{\epsilon}{2^{k-l}}.$$

$\square$

4.2 Constructions of ϵ-almost resilient functions

Definition 24. An (ϵ, k)-independent sample space S_m is *t-systematic* if $|S_m| = 2^t$, and there exist t positions $i_1 < \cdots < i_t$ such that each t-bit string occurs in these positions for exactly one m-tuple in S_m.

A t-systematic (ϵ, k)-independent sample space can be transformed into an $LS(\epsilon, k, m, t)$ by using the same technique as [25, Theorem 3]. We have the following result.

Theorem 25. *If there exists a t-systematic (ϵ, k)-independent sample space S_m, then there exists a balanced δ-almost $(m, m - t, k)$-resilient function, where $\delta = \epsilon/2^{m-t-k}$.*

Due to space limitations, we will present only a very brief summary of our construction for t-systematic (ϵ, k)-independent sample spaces. Our approach is similar to [12] (see also [18]), and depends on the Weil-Carlitz-Uchiyama bound. In what follows, let Tr denote the *trace* function from $GF(2^t)$ to $GF(2)$.

Proposition 26 Weil-Carlitz-Uchiyama bound. *[9] Let $f(x) = \sum_{i=1}^{D} f_i x^i \in GF(2^t)[x]$ be a polynomial that is not expressible in the form $f(x) = g(x)^2 - g(x) + \theta$ for any polynomial $g(x) \in GF(2^t)[x]$ and for any $\theta \in F_{2^t}$. Then*

$$\left| \sum_{\alpha \in GF(2^t)} (-1)^{Tr(f(\alpha))} \right| \le (D-1)\sqrt{2^t}.$$

Definition 27. A polynomial $h(x) \in GF(2^t)[x]$ is a $(2^t, D)$-*polynomial* if h has degree at most D and $a_i = 0$ for all even i, where $h = \sum_{i=0}^{D} a_i x^i$. Define $H(2^t, D, k)$ to be a set of $(2^t, D)$-polynomials such that any k polynomials in the set are independent over $GF(2)$.

For $h_{i_1}, h_{i_2}, \ldots, h_{i_k} \in H(2^t, D, k)$ and for any k elements $\alpha_1, \cdots, \alpha_k \in GF(2)$, define

$$N_{\alpha_1, \ldots, \alpha_k}(h_{i_1}, \ldots, h_{i_k}) \triangleq |\{x \in GF(2^t) : Tr(h_{i_1}(x)) = \alpha_1, \cdots, Tr(h_{i_k}(x)) = \alpha_k\}|.$$

Lemma 28. *[12]* $|N_{\alpha_1, \ldots, \alpha_k}(h_{i_1}, \ldots, h_{i_k}) - 2^{t-k}| \leq (D-1)\sqrt{2^t}$.

Proof. The proof is an application of Proposition 26. The case $k = 2$ can be found in [12] and the general case is proved similarly. □

Theorem 29. *Suppose that β is a primitive element of $GF(2^t)$, and $H(2^t, D, k)$ is chosen such that $\{x, \beta x, \beta^2 x, \ldots, \beta^{t-1} x\} \subseteq H(2^t, D, k)$. There exists a t-systematic (ϵ, k)-independent sample space S_m where $m = |H(2^t, D, k)|$ and $\epsilon = (D-1)/\sqrt{2^t}$.*

Proof. Let $H(2^t, D, k) = \{h_1, \cdots, h_m\}$. Construct a sample space S_m as follows: A binary string $X_\gamma = x_1 x_2 \cdots x_m \in S_m$ is specified by any $\gamma \in GF(2^t)$, where the ith bit of X_γ is $x_i = Tr(h_i(\gamma))$. The proof that S_m is (ϵ, k)-independent follows from Lemma 28. Further, S_m can be shown to be systematic using the fact that $\{x, \beta x, \beta^2 x, \ldots, \beta^{t-1} x\} \subseteq H(2^t, D, k)$ (the proof will be given in the final paper). □

4.3 An Application

In our approach, using Theorem 29, we need to construct a set of polynomials $H(2^t, D, k)$ such that any k of them are linearly independent over $GF(2)$. For this we can use linear error-correcting codes (see [14]). For a fixed (odd) degree D, we can express each polynomial as a linear combination of polynomials in the set

$$\{x, \beta x, \ldots, \beta^{t-1} x, x^3, \beta x^3, \ldots, \beta^{t-1} x^3, \ldots, x^D, \beta x^D, \ldots, \beta^{t-1} x^D\}.$$

Indexing the polynomials in $H(2^t, D, k)$ as $h_1, h_2, \ldots, h_m$ we obtain a binary $tD' \times m$ matrix, where $D' = (D+1)/2$, which is a parity check matrix of an $[m, l, d]$ error correcting code in which $m - l = tD'$ and $d = k + 1$. Conversely, given such a code, we obtain a t-systematic sample space, and hence a balanced ϵ-almost $(m, m - t, k)$-resilient function, as follows.

Theorem 30. *Suppose $D = 2D' - 1$ and there is a $[m, m - tD', k+1]$ code. Then there exists a balanced ϵ-almost $(m, m - t, k)$-resilient function such that*

$$\epsilon = \frac{(D-1)\sqrt{2^t}}{2^{m-k}}.$$

A suitable value of ϵ would be 2^{-m+t-1}. We obtain the following corollary of Theorem 30 by taking $D = 3$ and $k = (t/2) - 2$.

Corollary 31. *Suppose there is an $[m, m - 4k - 8, k + 1]$ code. Then there exists a balanced $2^{-m+2k+3}$-almost $(m, m - 2k - 4, k)$-resilient function.*

As a typical example, suppose we take $m = 160$ and $k = 18$. A $[160, 80, 23]$ code is known to exist see ([6]), so we obtain a balanced 2^{-121}-almost $(160, 120, 18)$-resilient function.

Let's compare the above result to the best-known $(160, 120, k)$-resilient function. The most important construction method for resilient functions [3, 10] uses linear error-correcting codes, as follows: Let G be a generator matrix for an $[m, l, d]$ linear code. Define a function $f : (GF(2))^m \mapsto (GF(2))^l$ by the rule $f(x) = xG^T$. Then f is an $(m, l, d - 1)$ linear resilient function. The maximum d for which a $[160, 120, d]$ code is known to exist is $d = 12$ (see [6]). Hence, the maximum k for which we can construct a $(160, 120, k)$-resilient function is $k = 11$.

5   Comments

The techniques of this paper can also be used to construct "almost" versions of other cryptographic tools. These include *correlation-immune functions* (see, for example, [19, 8, 7]) and *locally random pseudo-random number generators* (see [20, 16, 18]). Details will be given in the full version of the paper.

References

1. N. Alon, O. Goldreich, J. Hastad, and R. Peralta. Simple constructions of almost k-wise independent random variables. *Random Structures and Algorithms* **3** (1992), 289–304.
2. M. Atici and D. R. Stinson. Universal hashing and multiple authentication. *Lecture Notes in Computer Science* **1109** (1996), 16–30 (CRYPTO '96).
3. C. H. Bennett, G. Brassard, and J.-M. Robert. Privacy amplification by public discussion. *SIAM Journal on Computing* **17** (1988), 210–229.
4. J. Bierbrauer, K. Gopalakrishnan and D. R. Stinson. Bounds for resilient functions and orthogonal arrays. *Lecture Notes in Computer Science* **839** (1994), 247–257 (CRYPTO '94).
5. J. Bierbrauer, T. Johansson, G. Kabatianskii and B. Smeets. On families of hash functions via geometric codes and concatenation. *Lecture Notes in Computer Science* **773** (1994), 331–342 (CRYPTO '93).
6. A. E. Brouwer. Bounds on the minimum distance of binary linear codes. http://www.win.tue.nl/win/math/dw/voorlincod.html
7. P. Camion and A. Canteaut. Generalization of Siegenthaler inequality and Schnorr-Vaudenay multipermutations. *Lecture Notes in Computer Science* **1109** (1996), 372–386 (CRYPTO '96).
8. P. Camion, C. Carlet, P. Charpin and N. Sendrier. On correlation-immune functions. *Lecture Notes in Computer Science* **576** (1992), 86–100 (CRYPTO '91).

9. L. Carlitz and S. Uchiyama. Bounds on exponential sums. *Duke Math. Journal*, (1957), 37–41.

10. B. Chor, O. Goldreich, J. Hastad, J. Friedman, S Rudich and R. Smolensky. The bit extraction problem or *t*-resilient functions. *26th IEEE symposium on Foundations of Computer Science*, pages 396–407, 1985.

11. J. Friedman. On the bit extraction problem. *33rd IEEE symposium on Foundations of Computer Science*, pages 314–319, 1992.

12. T. Helleseth and T. Johansson. Universal hash functions from exponential sums over finite fields and Galois rings. *Lecture Notes in Computer Science* **1109** (1996), 31–44 (CRYPTO '96).

13. H. Krawczyk. New hash functions for message authentication. *Lecture Notes in Computer Science* **921** (1995), 301–310 (EUROCRYPT '95).

14. F. J. MacWilliams and N. J. A. Sloane. *The Theory of Error-Correcting Codes*. North-Holland, 1977.

15. J. L. Massey. Cryptography – A selective survey. *Digital Communications*, North-Holland (1986), 3–21.

16. U. M. Maurer and J. L. Massey. Perfect local randomness in pseudo-random sequences. *Lecture Notes in Computer Science* **435** (1990), 100–112 (CRYPTO '89).

17. J. Naor and M. Naor. Small bias probability spaces: efficient constructions and applications. *SIAM Journal on Computing* **22** (1993), 838–856.

18. H. Niederreiter and C. P. Schnorr. Local randomness in polynomial random number and random function generators. *SIAM Journal on Computing* **22** (1993), 684–694.

19. T. Siegenthaler. Correlation-immunity of nonlinear combining functions for cryptographic applications. *IEEE Trans. Inform. Theory* **30** (1984), 776–780.

20. C. P. Schnorr. On the construction of random number generators and random function generators. *Lecture Notes in Computer Science* **330** (1988), 225–232 (EUROCRYPT '88).

21. G.J. Simmons. A game theory model of digital message authentication. *Congressus Numeratium* **34** (1982), 413–424.

22. G.J. Simmons. Authentication theory/coding theory, *Lecture Notes in Computer Science*. **196** (1985), 411–431 (CRYPTO '84).

23. D. R. Stinson. Universal hashing and authentication codes. *Lecture Notes in Computer Science* **576** (1992), 74–85 (CRYPTO '91).

24. D. R. Stinson. Resilient functions and large set of orthogonal arrays. *Congressus Numerantium* **92** (1993), 105–110.

25. D .R. Stinson and J. L. Massey. An infinite class of counterexamples to a conjecture concerning nonlinear resilient functions. *Journal of Cryptology* **8** (1995), 167–173.

26. M. N. Wegman and J. L. Carter. New hash functions and their use in authentication and set equality. *Journal of Computer and System Sciences* **22** (1981), 265–279.

More Correlation-Immune and Resilient Functions over Galois Fields and Galois Rings

Claude Carlet

GREYC, Université de Caen
and
INRIA Projet Codes
Domaine de Voluceau, BP 105
78153 Le Chesnay Cedex
FRANCE
email: Claude.Carlet@inria.fr

Abstract. We show that the usual constructions of bent functions, when they are suitably modified, allow constructions of correlation-immune and resilient functions over Galois fields and, in some cases, over Galois rings.

1 Introduction

The functions used in a conventional cipher must provide both diffusion, for merging several inputs, and confusion, for hiding any structure (cf. [19]). These notions are respectively formalized through the properties of correlation-immunity [2, 3, 4, 5, 20, 22] and nonlinearity [15, 16].

Correlation-immune functions play an important role in several aspects of cryptography such as, for instance, the design of running-key generators in stream ciphers which resist the correlation attack [20] or the design of hash functions (cf. [21]). The most general definition (cf. [3]) defines them over finite alphabets (the original definition was given in [20] for binary functions): let $\mathcal{A}$ be a finite alphabet; a function f from $\mathcal{A}^n$ to $\mathcal{A}^m$ is t-th order correlation-immune if the probability distribution of the output vector $f(X_1, \ldots, X_n)$, where $X_1, \ldots, X_n$ are random input variables assuming values from $\mathcal{A}$ with independent equiprobable distributions, is unaltered when at most t of the variables $X_1, \ldots, X_n$ are fixed (i.e. replaced by constants).
In [22], Xiao Guo-Zhen and J. L. Massey give a convenient characterization of binary correlation-immune functions by means of characters. It is generalized in [3] by Camion and Canteaut to finite abelian groups. Recall that the group of characters on a finite abelian group G is isomorphic with G itself. For $x, u \in G$, we denote by $\langle x, u \rangle$ the image of x under the character associated to u via such an isomorphism. We have:

$$\sum_{x \in G} \langle x, u \rangle \neq 0 \Leftrightarrow u = 0. \tag{1}$$

W. Fumy (Ed.): Advances in Cryptology - EUROCRYPT '97, LNCS 1233, pp. 422-433, 1997.

Such an isomorphism being chosen, the characters on the group G^n $(n > 0)$ are:

$$\langle x, u \rangle_n = \prod_{i=1}^{n} \langle x_i, u_i \rangle, \; x = (x_1, \ldots, x_n), \; u = (u_1, \ldots, u_n).$$

A function f from G^n to G^m is t-th order correlation-immune if:

$$\forall v \in G^m, \forall u \in G^n, 1 \leq w_H(u) \leq t, \sum_{x \in G^n} \langle x, u \rangle_n \langle f(x), v \rangle_m = 0 \qquad (2)$$

where $w_H(u)$ denotes the Hamming weight of u.
According to property (1), the equality in (2) is satisfied for every $u \neq 0$ if $v = 0$.
Thus, v may be assumed to be nonzero in (2).
f is t-resilient if it is t-th order correlation-immune and balanced. It is a simple matter to show that, thanks to the characterization above, this is equivalent to:

$$\forall v \in G^m, v \neq 0, \forall u \in G^n, w_H(u) \leq t, \sum_{x \in G^n} \langle x, u \rangle_n \langle f(x), v \rangle_m = 0. \qquad (3)$$

In [4] is given a bound on the degree relative to each variable of the algebraic normal form of a t-th order correlation-immune (resp. t-resilient) function over a finite field: in each monomial, at most $n - t$ (resp. $n - t - 1$, provided $q^m \neq 2$ or $t \neq n - m$) of the variables have (maximum) degree $q - 1$.
This bound, that generalizes Siegenthaler inequality [20], shows that the functions over finite fields are better suited than binary ones to achieve high linear complexity, given the order of their correlation-immunity.

The bent functions [5, 6, 7, 9, 11, 13, 15, 17] are those Boolean functions whose nonlinearity is maximum. The notion has been first defined for Boolean functions over $GF(2)^n$ (cf. [17], recall that n must then be even) and later generalized to functions over residue class rings (cf. [13]): let q and n be any positive integers; we denote by $\mathbf{Z}_q$ the ring $\mathbf{Z}/q\mathbf{Z}$. A function f from $\mathbf{Z}_q^n$ to $\mathbf{Z}_q$ is called bent if, for any vector s, the character sum:

$$\sum_{x \in \mathbf{Z}_q^n} w_q^{f(x) - x \cdot s}$$

has magnitude $q^{\frac{n}{2}}$, where $w_q = e^{2i\pi/q}$. The function f is called regular-bent if there exists a function $\tilde{f}$ such that, for any s:

$$\sum_{x \in \mathbf{Z}_q^n} w_q^{f(x) - x \cdot s} = q^{\frac{n}{2}} w_q^{\tilde{f}(s)}.$$

There exists also a generalization of the notion to functions over finite fields (cf. [1]), that is not equivalent for prime fields. These definitions can be extended to definitions of (regular-) bent functions over a Galois ring $GR(p^k, m)$ (whose definition is recalled in subsection 2.1): the character sums to be considered in this wider framework are:

$$\sum_{x \in GR(p^k,m)^n} w_{p^k}^{Tr(f(x)-x \cdot s)}$$

where Tr is the trace function from $GR(p^k,m)$ to $\mathbf{Z}_{p^k}$.

These notions of correlation-immune and bent functions are very similar. The purpose of this paper is to show that various constructions of bent functions, when they are suitably modified, lead to constructions of correlation-immune functions. Some of these constructions will be primary, in the sense that they lead to new classes of correlation-immune functions without using known ones. Others, on the contrary, will be secondary constructions.

2 Primary constructions

2.1 A Maiorana-McFarland-like class

Maiorana-McFarland class (cf. [11]) is the set of all the (bent) Boolean functions on $GF(2)^n = \{(x,y), x,y \in GF(2)^{\frac{n}{2}}\}$ (n even) of the form : $f(x,y) = x \cdot \pi(y) + g(y)$ where π is any permutation on $GF(2)^{\frac{n}{2}}$ and g is any Boolean function on $GF(2)^{\frac{n}{2}}$.

In [5] is derived a construction of binary resilient functions:
let t and $n = r + s$ be any positive integers ($r > t > 0$, $s > 0$), g any boolean function on $GF(2)^s$ and ϕ a mapping from $GF(2)^s$ to $GF(2)^r$ such that every element in $\phi(GF(2)^s)$ has Hamming weight greater than t, then the function:

$$f(x,y) = x \cdot \phi(y) + g(y), \ x \in GF(2)^r, \ y \in GF(2)^s$$

is t-resilient.

We generalize this construction to any Galois ring in theorem 1. Before we state this theorem, we recall what are the definition and major properties of Galois rings.

For any prime p and any positive integers k and m, the Galois ring $GR(p^k,m)$ is the Galois extension of degree m of the ring $\mathbf{Z}_{p^k}$. When $m = 1$, $GR(p^k,m)$ is equal to $\mathbf{Z}_{p^k}$ and when $k = 1$, it is equal to the Galois field $GF(p^m)$. We refer to [14] for a general presentation of this notion and to [12] for the special case $p = k = 2$.

Galois rings share with Galois fields almost all their properties.

• Their elements can be described in two different forms by means of a primitive element ξ of order p^m:

- the "multiplicative" form (this term comes from field theory):

$$x = \sum_{i=1}^{k} p^{i-1} u_i, \ u_i \in \{0,1,\xi,\ldots,\xi^{p^m-2}\},$$

- the "additive" form:

$$x = \sum_{r=0}^{m-1} a_r \, \xi^r, \; a_r \in \mathbf{Z}_{p^k}.$$

- They admit a Frobenius automorphism:

$$\varphi : \sum_{i=1}^{k} p^{i-1} u_i \to \sum_{i=1}^{k} p^{i-1} u_i{}^p$$

and a trace map from $GR(p^k, m)$ to $\mathbf{Z}_{p^k}$:

$$Tr : x \to x + \varphi(x) + \ldots + \varphi^{m-1}(x),$$

where φ^{m-1} is $m-1$ times the composition of φ by itself.

The difference between Galois fields and general Galois rings is obviously that every nonzero element of $GR(p^k, m)$ is not necessarily a unit; the units of $GR(p^k, m)$ are the elements:

$$\sum_{i=1}^{k} p^{i-1} u_i, \; u_1 \in \{1, \xi, \ldots, \xi^{p^m-2}\}, \; u_2, \ldots, u_k \in \{0, 1, \xi, \ldots, \xi^{p^m-2}\}.$$

Their number is $p^{(k-1)m} \cdot (p^m - 1) = |GR(p^k, m)| \cdot \left(\frac{p^m-1}{p^m}\right)$.

We denote again by $x \cdot y$ the expression:

$$\sum_{j=1}^{n} x_j \, y_j, \; x = (x_1, \ldots, x_n) \in GR(p^k, m)^n, \; y = (y_1, \ldots, y_n) \in GR(p^k, m)^n.$$

The characters on $GR(p^k, m)^n$ are the functions: $x \to \langle x, y \rangle_n = w_{p^k}{}^{Tr(x \cdot y)}$, where $w_{p^k} = e^{2\pi i/p^k}$.

The construction given in [5] could be extended to general finite rings. In the case of Galois rings, it is easy to state:

Theorem 1. *Let G be any Galois ring, t and $n = r + s$ any positive integers ($r > t > 0$, $s > 0$), g any function from G^s to G and ϕ a mapping from G^s to G^r such that any element in $\phi(G^s)$ has more than t coordinates that are units, then the function:*

$$f(x, y) = x \cdot \phi(y) + g(y), \; x \in G^r, \; y \in G^s$$

is a t-resilient function on G^n.

Proof:

For any nonzero element v of G and any element (u, u') of G^n ($u \in G^r$, $u' \in G^s$), we have:

$$\sum_{x \in G^r, y \in G^s} \langle x, u \rangle_r \langle y, u' \rangle_s \langle f(x, y), v \rangle =$$

$$\sum_{x \in G^r, y \in G^s} w_{p^k}{}^{Tr\left(v[x \cdot \phi(y) + g(y)] + x \cdot u + y \cdot u'\right)} =$$

$$\sum_{y \in G^s} \left(\left(\sum_{x \in G^r} w_{p^k}{}^{Tr(x \cdot [v\phi(y) + u])} \right) w_{p^k}{}^{Tr(vg(y) + y \cdot u')} \right).$$

The sum:

$$\sum_{x \in G^r} w_{p^k}{}^{Tr(x \cdot [v\, \phi(y) + u])}$$

is equal to 0, unless $v\phi(y) + u = 0$, according to property (1). Therefore:

$$\sum_{x \in G^r, y \in G^s} \langle x, u \rangle_r \langle y, u' \rangle_s \langle f(x, y), v \rangle =$$

$$|G|^r \sum_{y \in G^s \,|\, v\phi(y) + u = 0} w_{p^k}{}^{Tr(vg(y) + y \cdot u')}.$$

If we assume that (u, u') has Hamming weight at most t, then u, whose Hamming weight is *a fortiori* at most t, cannot be equal to $-v\,\phi(y)$: according to the hypothesis on ϕ, $v\,\phi(y)$ has more than t nonzero coordinates. Thus, the sum $\sum_{x \in G^r, y \in G^s} \langle x, u \rangle_r \langle y, u' \rangle_s \langle f(x, y), v \rangle$ is equal to zero. f is t-resilient. $\square$

Example: if G is a Galois field and $\phi(y) = (\phi_1(y), \ldots, \phi_r(y))$ is such that:
. the sets $E_i = \{y \in G^s \,|\, \phi_i(y) = 0\}$, $i = 1, \ldots, r$ are disjoint each others;
. a monomial in the algebraic normal form of one of the functions ϕ_i has maximum degree $q - 1$ relative to each variable;
then $f(x_1{}^{q-2}, \ldots, x_r{}^{q-2}, y_1, \ldots, y_s)$ is $(r - 2)$-resilient (according to theorem 1 and to [3], prop. 9) and almost reaches the bound on the degrees recalled in the introduction.

2.2 A Partial-Spreads-like class

In [11] is also introduced the class of bent functions called $\mathcal{PS}_{ap}$ (a subclass of Partial-Spreads class), whose elements are defined the following way:
$GF(2)^{\frac{n}{2}}$ is identified to the Galois field $GF(2^{\frac{n}{2}})$; $\mathcal{PS}_{ap}$ is the set of all the functions of the form $f(x, y) = g(x\, y^{2^{\frac{n}{2}} - 2})$ (i.e. $g(\frac{x}{y})$ with $\frac{x}{y} = 0$ if $x = 0$ or $y = 0$) where g is a balanced Boolean function on $GF(2)^{\frac{n}{2}}$. We have then $\tilde{f}(x, y) = g(\frac{y}{x})$.
The idea of this construction may be used to obtain a construction of correlation-immune functions. We give this construction in its most general form (involving a Galois field $GF(q)$ where q is any prime power).
In the next theorem, we identify a power $\mathcal{F}^m$ of a Galois field $\mathcal{F} = GF(q)$ to the Galois field $GF(q^m)$. Such an identification is done the following way: we choose a basis $(\alpha_1, \ldots, \alpha_m)$ of the $\mathcal{F}$-vector space $GF(q^m)$ and we identify $x = (x_1, \ldots, x_m) \in \mathcal{F}^m$ to $\sum_{i=1}^m x_i \alpha_i \in GF(q^m)$. We know that a dot product on $\mathcal{F}^m$ is, via this identification $Tr_m(x\, y)$, where Tr_m is the trace map from

$GF(q^m)$ to $GF(q)$. But the notion of correlation-immune function depends on the choice of the dot product on $\mathcal{F}^m$. So, we assume that the basis $(\alpha_1, \ldots, \alpha_m)$ is self-dual (it is always possible to find such a basis when q is even or m is odd), so that:

$$Tr_m(x\,y) = \sum_{i=1}^{m} x_i\,y_i = x \cdot y.$$

Notice that if we do not have a self-dual basis, we still have, for any basis, $x \cdot y = Tr_m(a\,x\,y)$, $a \in GF(q^m)$.

We will use a well-known fact about linear mappings: let ϕ be a linear mapping from $GF(q^n)$ to $GF(q^m)$, there exists a linear mapping ϕ^* (called adjoint of ϕ) from $GF(q^m)$ to $GF(q^n)$ such that, for every $x \in GF(q^m)$ and every $y \in GF(q^n)$:

$$Tr_m(x\,\phi(y)) = Tr_n(\phi^*(x)\,y).$$

We state theorem 2 in the case we have self-dual basis in $GF(q^m)$ and $GF(q^n)$. It can be easily generalized to any case.

Theorem 2. *Let $\mathcal{F} = GF(q)$ ($q = p^s$) be a finite field and tr the trace function from $\mathcal{F}$ to its prime field $GF(p)$. Let n and m be two positive integers (n, m odd if q is odd), g a function from $GF(q^m)$ to $\mathcal{F}$, ϕ a linear mapping from $GF(q^n)$ to $GF(q^m)$ and a an element of $GF(q^m)$ such that $a + \phi(y) \neq 0$, $\forall y \in GF(q^n)$. Let f be the function from $\mathcal{F}^m \times \mathcal{F}^n$ to $\mathcal{F}$ defined by:*

$$f(x,y) = g\left(\frac{x}{a + \phi(y)}\right) + Tr_n(b\,y),$$

where $b \in GF(q^n)$ and where x, y are viewed as elements of $GF(q^m)$, $GF(q^n)$ respectively.

Assume that, for every z in $GF(q^m)$ and every $v \neq 0$ in $\mathcal{F}$, $\phi^*(z) + v\,b$ has weight greater than t, then f is t-resilient.

Proof:
We have, for any (u, u') in $\mathcal{F}^m \times \mathcal{F}^n$ and any nonzero v in $\mathcal{F}$:

$$\sum_{x \in \mathcal{F}^m,\, y \in \mathcal{F}^n} \langle u, x \rangle_m \langle u', y \rangle_n \langle v, f(x,y) \rangle = \sum_{x \in \mathcal{F}^m,\, y \in \mathcal{F}^n} w_p{}^{tr[u \cdot x + u' \cdot y + v\,f(x,y)]} =$$

$$\sum_{x \in GF(q^m),\, y \in GF(q^n)} w_p{}^{tr\left[Tr_m(ux) + Tr_n(u'y) + v\,g\left(\frac{x}{a+\phi(y)}\right) + v\,Tr_n(b\,y)\right]}.$$

Since, for every y, $a + \phi(y) \neq 0$, the element $z = \frac{x}{a + \phi(y)}$ ranges over the whole field $GF(q^m)$ when x does. We deduce:

$$\sum_{x \in \mathcal{F}^m,\, y \in \mathcal{F}^n} \langle u, x \rangle_m \langle u', y \rangle_n \langle v, f(x,y) \rangle_r =$$

$$\sum_{z \in GF(q^m),\, y \in GF(q^n)} w_p{}^{tr[Tr_m(u(az + z\phi(y))) + Tr_n(u'y) + v\,g(z)) + v\,Tr_n(b\,y)]} =$$

$$\sum_{z\in GF(q^m),\, y\in GF(q^n)} w_p{}^{tr[Tr_m(uaz)+Tr_n(y[\phi^*(uz))+u'+v\,b])+v\,g(z)]} =$$

$$\sum_{z\in GF(q^m)} w_p{}^{tr[Tr_m(uaz)+v\,g(z)]}\left(\sum_{y\in GF(q^n)} w_p{}^{tr[Tr_n(y[\phi^*(uz))+u'+v\,b])]}\right) =$$

$$q^n \sum_{z\in GF(q^m)\,|\,\phi^*(uz))+u'+v\,b=0} w_p{}^{tr[Tr_m(uaz)+v\,g(z)]},$$

according to property (1).
If $w_H(u,u') \le t$, then according to the hypothesis on ϕ^*, the set

$$\{z \in GF(q^m)\,|\,\phi^*(uz)) + u' + v\,b = 0\}$$

is empty, and this sum is equal to 0. Thus, f is t-resilient. □

Example: Let E be an $\mathcal{F}$-subspace of $\mathcal{F}^n$ of maximum weight $n - t - 1$ and ψ a linear mapping from $\mathcal{F}^m$ to E. Let b be a word of weight n in $\mathcal{F}^n$. Then the condition of theorem 2 is satisfied by $\phi = \psi^*$, provided that a does not belong to the image of ψ^* (which is always possible if $n < m$).

3   Secondary constructions

3.1   Modifying a correlation-immune function on a subgroup

Dillon proves in [11] that if a binary function f is bent on $GF(2)^n$ (n even) and if E is a $\frac{n}{2}$-dimensional flat on which f is constant, then, denoting by δ_E the indicator of E, the function $f + \delta_E$ is bent too.
We shall prove a similar result on correlation-immune functions.

Theorem 3. *Let G be any finite abelian group, t, m and n any positive integers and f a t-th order correlation-immune function from G^n to G^m.*
Assume there exists a subgroup E of G^n, whose minimum nonzero weight is greater than t and such that the restriction of f to the orthogonal of E (i.e. the subgroup of G^n: $E^\perp = \{u \in G^n\,|\,\forall x \in E, \langle u,x\rangle_n = 1\}$) is constant. Then f remains t-th order correlation-immune if we change its constant value on $E^\perp$ into any other one.

Proof:
Let a be the constant value of f on $E^\perp$ and b any element of G^m. Set $f'(x) = f(x)$ if $x \notin E^\perp$, $f'(x) = b$ if $x \in E^\perp$.
For any nonzero element v of G^m and any element u of G^n, we have:

$$\sum_{x\in G^n} \langle x,u\rangle_n\langle f'(x),v\rangle_m =$$

$$\sum_{x\in G^n} \langle x,u\rangle_n\langle f(x),v\rangle_m + \sum_{x\in E^\perp} \langle x,u\rangle_n\langle b,v\rangle_m - \sum_{x\in E^\perp} \langle x,u\rangle_n\langle a,v\rangle_m.$$

If u is nonzero and if its weight is at most equal to t, then:

$$\sum_{x \in G^n} \langle x, u \rangle_n \langle f'(x), v \rangle_m = \left(\sum_{x \in E^\perp} \langle x, u \rangle_n \right) (\langle b, v \rangle_m - \langle a, v \rangle_m).$$

The sum: $\displaystyle\sum_{x \in E^\perp} \langle x, u \rangle_n$ is equal to 0, since u does not belong to E. $\qquad\square$

3.2 Adapting a secondary construction known for bent functions

It is known, cf. [11, 17], that if g, h, k and $g + h + k$ are bent on $GF(2)^m$ (m even), then the function defined on any element (x_1, x_2, x) of $GF(2)^{m+2}$ by:

$$f(x_1, x_2, x) =$$

$$g(x)h(x) + g(x)k(x) + h(x)k(x) + [g(x) + h(x)]x_1 + [g(x) + k(x)]x_2 + x_1 x_2$$

is bent.

Theorem 4. *Let g, h and k be three functions from $GF(2)^m$ to $GF(2)$. If g is t-resilient, h and k are $(t-1)$-resilient and $g + h + k$ is $(t-2)$-resilient, then the function on $GF(2)^{m+2}$:*

$$f(x_1, x_2, x) =$$

$$g(x)h(x) + g(x)k(x) + h(x)k(x) + [g(x) + h(x)]x_1 + [g(x) + k(x)]x_2 + x_1 x_2$$

is t-resilient (the converse is true).

Proof:
We have:

$$\sum_{x_1, x_2 \in GF(2), x \in GF(2)^m} (-1)^{f(x_1, x_2, x) + a_1 x_1 + a_2 x_2 + a \cdot x} = \sum_{x_1, x_2 \in GF(2), x \in GF(2)^m}$$

$$(-1)^{g(x) + [x_1 + g(x) + k(x) + a_2][x_2 + g(x) + h(x) + a_1] + a_1[g(x) + k(x)] + a_2[g(x) + h(x)] + a_1 a_2 + a \cdot x}.$$

Changing x_1 into $x_1 + g(x) + k(x) + a_2$ and x_2 into $x_2 + g(x) + h(x) + a_1$, we obtain:

$$\sum_{x_1, x_2 \in GF(2), x \in GF(2)^m} (-1)^{g(x) + x_1 x_2 + a_1[g(x) + k(x)] + a_2[g(x) + h(x)] + a \cdot x + a_1 a_2}$$

that is equal to:

$$2 \sum_{x \in GF(2)^m} (-1)^{g(x) + a_1[g(x) + k(x)] + a_2[g(x) + h(x)] + a \cdot x + a_1 a_2}.$$

Assume that the word (a_1, a_2, a) has Hamming weight at most t. Then if $a_1 = a_2 = 0$, we obtain:

$$2 \sum_{x \in GF(2)^m} (-1)^{g(x) + a \cdot x},$$

that is equal to zero, according to the hypothesis and since a has Hamming weight at most t. If $a_1 = 0$ and $a_2 = 1$ (resp. $a_1 = 1$ and $a_2 = 0$), we obtain:

$$2 \sum_{x \in GF(2)^m} (-1)^{h(x)+a \cdot x} \quad (\text{resp. } 2 \sum_{x \in GF(2)^m} (-1)^{k(x)+a \cdot x}), \text{ that is also equal to zero,}$$

since a has Hamming weight at most $t - 1$. If $a_1 = a_2 = 1$, we obtain:

$$-2 \sum_{x \in GF(2)^m} (-1)^{g(x)+h(x)+k(x)+a \cdot x},$$

that is equal to zero too, since a has Hamming weight at most $t - 2$.
The converse is similar. ∎

Example: This result may be applied to functions g, h and k chosen in Maiorana-McFarland-like class (over $GF(2)$): $g(x, y) = x \cdot \phi(y) + g_1(y)$, $h(x, y) = x \cdot \phi'(y) + h_1(y)$, $k(x, y) = x \cdot \phi''(y) + k_1(y)$, where any element of $\phi(G^s)$ (resp. $\phi'(G^s)$, $\phi''(G^s)$, $(\phi + \phi' + \phi'')(G^s)$)) has more than t (resp. $t - 1$, $t - 1$, $t - 2$) nonzero coordinates.

Remark: It is possible to extend this result to general finite fields, but the hypothesis becomes hard to satisfy.

3.3 Constructing correlation-immune functions from bent functions

The construction of bent functions that is recalled in the previous subsection is generalized in [8]:
Let m and r be two positive even integers. Let f be a Boolean function on $GF(2)^{m+r}$ such that, for any element x' of $GF(2)^r$, the function on $GF(2)^m$:

$$f_{x'} : x \to f(x, x')$$

is bent. Then f is bent if and only if for any element u of $GF(2)^m$, the function

$$\varphi_u : x' \to \widetilde{f_{x'}}(u)$$

is bent on $GF(2)^r$ ($\widetilde{f_{x'}}$ always exists: every bent function on $GF(2)$ in even dimension is regular-bent). This result generalizes to functions f over $\mathbf{Z}_q^{m+r}$ (as stated in [8]) such that for every x', the function $f_{x'}$ is regular-bent.
It leads us to a construction of resilient functions from regular-bent functions:

Theorem 5. *Let r be a positive integer, m a positive even integer and p a prime. Let f be a function from $(GF(p))^{m+r}$ to $GF(p)$ such that, for any element x' of $(GF(p))^r$, the function on $(GF(p))^m$:*

$$f_{x'} : x \to f(x, x')$$

is regular-bent.
If, for every element u of $(GF(p))^m$ of Hamming weight at most t, the function

$$\varphi_u : x' \to \widetilde{f_{x'}}(u)$$

is $(t - w_H(u))$-resilient, then f is t-resilient (the converse is true).

Proof:
For every nonzero v in $GF(p)$, and every (u, u') in $GF(p)^{m+r}$, we have:

$$\sum_{(x,x')\in GF(p)^{m+r}} \langle u, x \rangle_m \langle u', x' \rangle_r \langle v, f(x, x') \rangle =$$

$$\sum_{(x,x')\in GF(p)^{m+r}} w_p{}^{vf_{x'}(x)+u\cdot x+u'\cdot x'}. \tag{4}$$

$f_{x'}$ being regular-bent, we have:

$$\sum_{x\in GF(p)^m} w_p{}^{f_{x'}(x)+u\cdot x} = p^{\frac{m}{2}} w_p{}^{\widetilde{f_{x'}}(-u)}, \forall u \in GF(p)^m. \tag{5}$$

Let us first prove that, for every nonzero v in $GF(p)$:

$$\sum_{x\in GF(p)^m} w_p{}^{vf_{x'}(x)+u\cdot x} = p^{\frac{m}{2}} w_p{}^{\widetilde{f_{x'}}(-\frac{u}{v})}, \forall u \in GF(p)^m :$$

let C_p be the cyclotomic field generated by w_p over the rationnals, i.e.

$$C_p = \mathbf{Q}(w_p);$$

we know (cf. [18], see also [13]) that its Galois group is the abelian group each element σ of which raises w_p to the v-th power, $v \in \{1, \dots, p-1\}$ (every element of $\mathbf{Q}$ being invariant under σ). Say $\sigma = \sigma_v$.
From equality (5) and since $p^{\frac{m}{2}} \in \mathbf{Q}$, we deduce:

$$\sigma_v \left(\sum_{x\in GF(p)^m} w_p{}^{f_{x'}(x)+u\cdot x} \right) = p^{\frac{m}{2}} \sigma_v \left(w_p{}^{\widetilde{f_{x'}}(-u)} \right),$$

thus:

$$\sum_{x\in GF(p)^m} w_p{}^{v(f_{x'}(x)+u\cdot x)} = p^{\frac{m}{2}} w_p{}^{v\widetilde{f_{x'}}(-u)}$$

and therefore:

$$\sum_{x\in GF(p)^m} w_p{}^{vf_{x'}(x)+u\cdot x} = p^{\frac{m}{2}} w_p{}^{v\widetilde{f_{x'}}(-\frac{u}{v})}. \tag{6}$$

From equalities (4) and (6), we deduce:

$$\sum_{(x,x')\in GF(p)^{m+r}} \langle u, x \rangle_m \langle u', x' \rangle_r \langle v, f(x, x') \rangle =$$

$$p^{\frac{m}{2}} \sum_{x'\in GF(p)^r} w_p{}^{v\widetilde{f_{x'}}(-\frac{u}{v})+u'\cdot x'} =$$

$$p^{\frac{m}{2}} \sum_{x'\in GF(p)^r} w_p{}^{v\varphi_{-\frac{u}{v}}(x')+u'\cdot x'}.$$

This completes the proof, since $w_H\left(-\frac{u}{v}\right) = w_H(u)$ and since $w_H(u,u') \le t$ implies $w_H(u) \le t$ and $w_H(u') \le t - w_H(u)$. The converse is similar. $\qquad\square$

Example: taking $f_{x'}$ in Partial Spreads class and φ_u in Partial Spreads-like class, we obtain that the function $f(x,y,x',y') = k\left(\frac{x}{y}, \frac{x'}{a+\phi(y')}\right) + Tr_n(by')$, where for every x', the function $x \to k(x,x')$ is balanced, for every y', $a + \phi(y') \ne 0$, and for every z and every $v \ne 0$, $\phi^*(z) + vb$ has weight greater than t, is t-resilient.

Remark: Theorem 5 could be generalized to functions $f(x,x')$ over a more general Galois field $GF(q)$ such that, for every $x' \in GF(q)^r$ and every nonzero $v \in GF(q)$:

. the function $f_{x',v} : x \to vf(x,x')$ is regular-bent,

. $\widetilde{f_{x',v}} = v\,\widetilde{f_{x',1}}$.

References

1. A.S. Ambrosimov. Properties of bent functions of q-valued logic over finite fields. *Discrete Math. Appl.* vol 4, N° 4, pages 341-350 (1994)
2. J. Bierbrauer, K. Gopalakrishnan and D.R. Stinson. Bounds for resilient functions and orthogonal arrays. *Advances in Cryptology, CRYPTO'94, Lecture Notes in Computer Sciences, Springer Verlag* n° 839, pages 247-256 (1994)
3. P. Camion and A. Canteaut. Construction of t-resilient functions over a finite alphabet, *Advances in Cryptology, EUROCRYPT'96, Lecture Notes in Computer Sciences, Springer Verlag* n° 1070, pages 283-293 (1996)
4. P. Camion and A. Canteaut. Generalization of Siegenthaler inequality and Schnorr-Vaudenay multipermutations. In N. Koblitz, editor, *Advances in Cryptology - CRYPTO'96*, number 1109 in Lecture Notes in Computer Science, pages 372–386 Springer-Verlag, 1996.
5. P. Camion, C. Carlet, P. Charpin and N. Sendrier. On correlation-immune functions. *Advances in Cryptology, CRYPTO'91, Lecture Notes in Computer Sciences, Springer Verlag* n° 576, pages 86-100 (1992)
6. C. Carlet. Two new classes of bent functions. *EUROCRYPT' 93, Advances in Cryptology, Lecture Notes in Computer Science* 765, pages 77-101 (1994)
7. C. Carlet, Generalized Partial Spreads, *IEEE Transactions on Information Theory* vol 41 pages 1482-1487 (1995)
8. C. Carlet. A construction of bent functions. *Finite Fields and Applications, London Mathematical Society, Lecture Series* 233, Cambridge University Press, pages 47-58 (1996)
9. C. Carlet and P. Guillot. A characterization of binary bent functions. *Journal of Combinatorial Theory, Series A*, Vol. 76, No. 2 pages 328-335 (1996)
10. C. Carlet. Hyperbent functions. *PRAGOCRYPT'97, Czech Technical University Publishing House*, pages 145-155 (1996).
11. J. F. Dillon. Elementary Hadamard Difference sets. Ph. D. Thesis, Univ. of Maryland (1974).
12. A. R. Hammons Jr., P. V. Kumar, A. R. Calderbank, N. J. A. Sloane and P. Solé. The Z_4-linearity of Kerdock, Preparata, Goethals and related codes. *IEEE Transactions on Information Theory*, vol 40, pages 301-320, (1994)

13. P. V. Kumar, R.A. Scholtz and L.R. Welch. Generalized bent functions and their properties. *Journal of Combinatorial Theory*, Series A 40, pages 90-107 (1985)
14. B.R. MacDonald. Finite rings with identity. *Marcel Dekker*, NY, 1974
15. W. Meier and O. Staffelbach. Nonlinearity Criteria for Cryptographic Functions. *Advances in Cryptology, EUROCRYPT' 89, Lecture Notes in Computer Science* 434, pages 549-562, Springer Verlag (1990)
16. K. Nyberg. Perfect non-linear S-boxes. *Advances in Cryptology, EUROCRYPT' 91*, Lecture Notes in Computer Science 547, pages 378-386, Springer Verlag (1992)
17. O. S. Rothaus. On bent functions. *J. Comb. Theory*, 20A, pages 300- 305(1976)
18. P. Samuel. Algebraic Theory of Numbers. Boston, Houghton Mifflin, 1970
19. C. E. Shannon. Communication theory of secrecy systems. in *Bell system technical journal*, vol. 28, pages 656-715 (1949)
20. T. Siegenthaler. Correlation-Immunity of Nonlinear Combining Functions for Cryptographic Applications. *IEEE Trans. on Inf. Theory*, vol IT-30, n° 5, pages 776-780 (1984)
21. C.P. Schnorr and S. Vaudenay. Black box cryptanalysis of hash networks based on multipermutations. *Advances in Cryptology, EUROCRYPT' 94*, Lecture Notes in Computer Science 950, pages 47-57, Springer Verlag (1995)
22. Xiao Guo-Zhen and J. L. Massey. A Spectral Characterization of Correlation-Immune Combining Functions. *IEEE Trans. Inf. Theory*, Vol IT 34, n° 3, pages 569-571 (1988).

Design of SAC/PC(*l*) of Order *k* Boolean Functions and Three Other Cryptographic Criteria

Kaoru Kurosawa[1] and Takashi Satoh [*][2]

[1] Dept. of Computer Science,
Graduate School of Information Science and Engineering,
Tokyo Institute of Technology
[2] Dept. of Physical Electronics, Faculty of Engineering, Tokyo Institute of Technology

2–12–1 O-okayama, Meguro-ku, Tokyo 152, Japan

`kurosawa@ss.titech.ac.jp, tsato@ss.titech.ac.jp`

Abstract. A Boolean function f satisfies PC(l) of order k if $f(x) \oplus f(x \oplus \alpha)$ is balanced for any α such that $1 \leq W(\alpha) \leq l$ even if any k input bits are kept constant, where $W(\alpha)$ denotes the Hamming weight of α. This paper shows the first design method of such functions which provides $\deg(f) \geq 3$. More than that, we show how to design "balanced" such functions. High nonlinearity and large degree are also obtained. Further, we present balanced SAC(k) functions which achieve the maximum degree. Finally, we extend our technique to vector output Boolean functions.

1 Introduction

The security of block ciphers is often studied by viewing their S-boxes (or F functions) as a set of Boolean functions. SAC [15] and PC(l) [11] are important cryptographic criteria of such Boolean functions. Let $W(\alpha)$ denote the Hamming weight of $\alpha \in \{0,1\}^n$. For a Boolean function $f(x) = f(x_1, \ldots, x_n)$, define

$$\frac{Df}{D\alpha} \triangleq f(x) \oplus f(x \oplus \alpha) \ .$$

$f(x)$ is said to satisfy

- SAC if $Df/D\alpha$ is balanced for any α such that $W(\alpha) = 1$.
- SAC(k) if any function obtained from f by keeping any k input bits constant satisfies SAC.
- PC(l) if $Df/D\alpha$ is balanced for any α such that $1 \leq W(\alpha) \leq l$.
- PC(l) of order k if any function obtained from f by keeping any k input bits constant satisfies PC(l).

[*] This author was supported by the Telecommunications Advancement Foundation, Japan.

Well known bent functions satisfy both SAC and PC(l) for all $l \leq n$, but not necessarily SAC(k) nor PC(l) of order k for $k \geq 1$.

On the other hand, balancedness, algebraic degree and nonlinearity are another important cryptographic criteria.

- Let deg(f) denote the degree of the highest degree term in the algebraic normal form of f. Then deg(f) must be large. Actually, Jacobsen and Knudsen showed an attack against block ciphers with small deg(f) recently [2].
- The nonlinearity of a Boolean function f, denoted by $N(f)$, is defined as the minimum distance of f from the set of affine functions.

$$N(f) \stackrel{\triangle}{=} \min_{a_0,\ldots,a_n} \left| \{x \mid f(x) \neq a_0 \oplus a_1 x_1 \oplus \cdots \oplus a_n x_n\} \right| .$$

$N(f)$ must be large to avoid the linear attack [7].
- Preneel et al. showed a balanced SAC($n-2$) function for n =odd [11]. Lloyd [5] showed a condition such that SAC($n-3$) functions are balanced. Balanced SAC functions with high nonlinearity were constructed by [14]. Recently, other balanced SAC functions were given by [16].

However,

(1) No general methods are known which design Boolean functions satisfying PC(l) of order k except deg(f) = 2. (For deg(f) = 2, see [11, 12].)
(2) Balanced SAC(k) functions are not known for $1 \leq k \leq n - 4$.
(3) Balanced functions satisfying PC(l) of order k are not known for any $l \geq 2$ and any k.

This paper shows a design method of PC(l) of order k functions. The proposed method is the first design method which provides deg(f) ≥ 3. We construct f as

$$f(x_1,\ldots,x_s,y_1,\ldots,y_t) \stackrel{\triangle}{=} [x_1,\ldots,x_s]Q[y_1,\ldots,y_t]^T \oplus g(x_1,\ldots,x_s) , \qquad (1)$$

where Q is an $s \times t$ binary matrix and $g(x_1,\ldots,x_s)$ is any function. Then f satisfies PC(l) of order k if Q satisfies the following conditions.

- $W(Q\gamma_1) \geq k+1$ for any $t \times 1$ vector γ_1 such that $1 \leq W(\gamma_1) \leq l$.
- $W(\gamma_2 Q) \geq k+1$ for any $1 \times s$ vector γ_2 such that $1 \leq W(\gamma_2) \leq l$.

Such a matrix Q is obtained by the product of two generator matrices of error correcting codes. Further, it is shown that balanced f can be obtained by choosing g appropriately in (1). We can also obtain large degree and high nonlinearity such that

- deg(f) = $s/2$ and $N(f) \geq 2^{t+s-1} - 2^{t+s/2-1}$ for s =even.
- deg(f) = $(s-1)/2$ and $N(f) \geq 2^{t+s-1} - 2^{t+(s-1)/2}$ for s =odd.

The above $N(f)$ is almost the maximum if t is small. (The $\deg(f)$ and $N(f)$ for SAC(k) are obtained by substituting $t = k+1$ and $s = n-k-1$.)

Next, SAC(k) functions with the maximum $\deg(f)$ are obtained for $k \le n/2 - 1$. This shows that an upper bound on $\deg(f)$ of SAC(k) functions given by Preneel et al. [11] is tight. Further, balanced SAC(k) functions with the same maximum degree are presented for $n - k - 1 = $ odd. This means that the bound of [11] is tight even for balanced SAC(k) functions if $k \le n/2 - 1$ and $n - k - 1 = $ odd. It will be a further work to find a tight upper bound on $\deg(f)$ of balanced SAC(k) functions for $n - k - 1 = $ even.

Finally, we extend our technique to vector output Boolean functions. Vector output PC(2) of order $2^{r-1} - 1$ functions and vector output SAC(k) functions are obtained which also possess high nonlinearity and large degree.

2 Preliminaries

$f(x_1, \ldots, x_n)$ denotes a mapping from $\{0,1\}^n$ to $\{0,1\}$. For a binary string α, $W(\alpha)$ denotes the Hamming weight of α. We use square brackets to denote vectors like $[a_1, \ldots, a_n]$ and round brackets to denote functions like $f(x_1, \ldots, x_n)$.

2.1 Balance and Algebraic Degree

We say that $f(x)$ is balanced if

$$\left|\{x \mid f(x) = 0\}\right| = \left|\{x \mid f(x) = 1\}\right| = 2^{n-1} \ ,$$

where $x = [x_1, \ldots, x_n]$.

Definition 1. We call $f(x) = c \oplus a_1 x_1 \oplus \cdots \oplus a_n x_n$ an affine function.

Proposition 2. *A non-constant affine function is balanced.*

Proposition 3. *[14] $f(x_1, \ldots, x_s) \oplus g(y_1, \ldots, y_t)$ is balanced if f is balanced or g is balanced.*

The following form is called the *algebraic normal form* of f.

$$f(x_1, \ldots, x_n) = a_0 \oplus \bigoplus_{i=1}^{n} a_i x_i \oplus \bigoplus_{1 \le i < j \le n} a_{ij} x_i x_j \oplus \cdots \oplus a_{12\ldots n} x_1 x_2 \ldots x_n \ .$$

$\deg(f)$ denotes the degree of the highest degree term in the algebraic normal form of f.

2.2 Bent Function and Nonlinearity

Bent functions are defined as follows.

Definition 4. [13] $f(x_1, \ldots, x_n)$ is a bent function if

$$\left| \sum_x (-1)^{f(x)} (-1)^{\omega_1 x_1 + \cdots + \omega_n x_n} \right| = 2^{n/2} \tag{2}$$

for any $[\omega_1, \ldots, \omega_n] \in \{0, 1\}^n$.

Define a distance between two Boolean functions $f(x)$ and $g(x)$ as

$$d(f, g) \triangleq \left| \{x \mid f(x) \neq g(x)\} \right| .$$

Definition 5. [10] The nonlinearity of a Boolean function f, denoted by $N(f)$, is defined as

$$N(f) \triangleq \min_{a_0, \ldots, a_n} d(f(x), a_0 \oplus a_1 x_1 \oplus \cdots \oplus a_n x_n) .$$

$N(f)$ is the distance of f from the set of affine functions and it should be large to avoid the linear attack. It is known that each bent function has the maximum $N(f)$.

Proposition 6. [8, 13] $N(f) \leq 2^{n-1} - 2^{n/2-1}$.

Proposition 7. [8, 13] The equality of Proposition 6 is satisfied if and only if f is a bent function.

2.3 SAC and SAC(k)

f satisfies SAC if complementing any single input bit changes the output bit with probability a half.

Definition 8. [1, 15]

(1) $f(x_1, \ldots, x_n)$ satisfies SAC (the strict avalanche criterion) if $f(x) \oplus f(x \oplus \alpha)$ is balanced for any $\alpha \in \{0, 1\}^n$ such that $W(\alpha) = 1$.
(2) $f(x)$ satisfies SAC(k) if any function obtained from $f(x)$ by keeping any k input bits constant satisfies SAC. We say that f is an SAC(k) function if $f(x)$ satisfies SAC(k).

Proposition 9. [1] There exist no SAC(n − 1) functions.

Proposition 10. [11]

(1) If $f(x_1, \ldots, x_n)$ satisfies SAC(n − 2), then $\deg(f)=2$.
(2) If $f(x_1, \ldots, x_n)$ satisfies SAC(k) for $0 \leq k \leq n − 3$, then

$$\deg(f) \leq n - k - 1 . \tag{3}$$

Preneel et al. showed a design method of SAC(k) functions for $\deg(f) = 2$.

Proposition 11. [11] Suppose that $\deg(f) = 2$ and $n > 2$. Then, f satisfies SAC(k) if and only if every variable x_i occurs in at least $k + 1$ second order terms of the algebraic normal form, where $0 \leq k \leq n - 2$.

2.4 PC(l) and PC(l) of Order k

f satisfies PC(l) if complementing any l or less input bits changes the output bit with probability a half.

Definition 12. [11]

(1) $f(x_1, \ldots, x_n)$ satisfies PC(l) if $f(x) \oplus f(x \oplus \alpha)$ is balanced for any $\alpha \in \{0,1\}^n$ such that $1 \leq W(\alpha) \leq l$.
(2) $f(x)$ satisfies PC(l) of order k if any function obtained from $f(x)$ by keeping any k input bits constant satisfies PC(l). We say that f is a PC(l) of order k function if $f(x)$ satisfies PC(l) of order k.

It is well known that f satisfies PC(n) if and only if f is a bent function [11]. Bent functions, however, do not necessarily satisfy PC(l) of order k.

PC(n) functions, therefore bent functions, exist only for $n =$even from (2). Preneel et al. [12] showed the following functions which have $\deg(f) = 2$.

Proposition 13. *There exists a PC($n-1$) of order 1 function for $n =$odd.*

Proposition 14. *[11] Let*

$$s_n(x_1, \ldots, x_n) \triangleq \bigoplus_{1 \leq i < j \leq n} x_i x_j \ .$$

Then s_n satisfies PC(l) of order k if $l + k \leq n - 1$ or if $l + k = n$ and l is even. Further,

(1) s_n is the only function which satisfies PC(1) of order $n - 2$ (or SAC($n-2$)).
(2) s_n is the only function which satisfies PC(2) of order $n - 2$.
(3) s_n is balanced if $n =$odd.

Proposition 15.

(1) There exists a balanced SAC($n - 2$) function if $n =$odd.
(2) There exist no balanced SAC($n - 2$) functions if $n =$even

Proof.

(1) From (1) and (3) of Proposition 14.
(2) From line 4 of p.171 of [11] and (1) of Proposition 14, a SAC($n - 2$) function is a bent function if $n =$even. Further, bent functions cannot be balanced [13].

□

3 How to Design PC(l) of Order k Functions

This section shows the first design method of PC(l) of order k functions which provides $\deg(f) \geq 3$. (For $\deg(f) = 2$, see Sect. 2.4.) The proposed method is also a design method of SAC(k) functions since SAC(k) is equivalent to PC(1) of order k.

3.1 Basic Theorem

Theorem 16. *For positive integers l and k, suppose that there exists an $s \times t$ binary matrix Q such as follows.*

(1) $s \geq \max\{l, k+1\}$ and $t \geq \max\{l, k+1\}$.
(2) $W(Q\gamma_1) \geq k+1$ for any $t \times 1$ vector γ_1 such that $1 \leq W(\gamma_1) \leq l$.
(3) $W(\gamma_2 Q) \geq k+1$ for any $1 \times s$ vector γ_2 such that $1 \leq W(\gamma_2) \leq l$.

Now define

$$f(x_1,\ldots,x_s,y_1,\ldots,y_t) \triangleq [x_1,\ldots,x_s]Q[y_1,\ldots,y_t]^T \oplus g(x_1,\ldots,x_s) , \qquad (4)$$

where $g(x_1,\ldots,x_s)$ is any function and $n = s + t$. Then f satisfies PC(l) of order k.

Proof. Keep any k input bits constant. Without loss of generality, we can assume that

$$x_1 = b_1,\ldots,x_u = b_u, \quad y_1 = c_1,\ldots,y_v = c_v,$$

where $u + v = k$, $u < s$ and $v < t$. Substitute these bits into f and let

$$\hat{f}(x_{u+1},\ldots,x_s,y_{v+1},\ldots,y_t) \triangleq f(b_1,\ldots,b_u,x_{u+1},\ldots,x_s,c_1,\ldots,c_v,y_{v+1},\ldots,y_t) .$$

We have to prove that $\hat{f}(x) \oplus \hat{f}(x \oplus \alpha)$ is balanced for any α such that $1 \leq W(\alpha) \leq l$. For simplicity, we show a proof for $l = 2$. The proof for $l \geq 3$ is similar.

For $W(\alpha) = 2$, define

$$\frac{D\hat{f}}{Dx_{u+i}x_{u+j}} \triangleq \hat{f}(x_{u+1},\ldots,x_s,y_{v+1},\ldots,y_t) \oplus \hat{f}(\ldots,x_{u+i} \oplus 1,\ldots,x_{u+j} \oplus 1,\ldots)$$

$$\frac{D\hat{f}}{Dy_{v+i}y_{v+j}} \triangleq \hat{f}(x_{u+1},\ldots,x_s,y_{v+1},\ldots,y_t) \oplus \hat{f}(\ldots,y_{v+i} \oplus 1,\ldots,y_{v+j} \oplus 1,\ldots)$$

$$\frac{D\hat{f}}{Dx_{u+i}y_{v+j}} \triangleq \hat{f}(x_{u+1},\ldots,x_s,y_{v+1},\ldots,y_t) \oplus \hat{f}(\ldots,x_{u+i} \oplus 1,\ldots,y_{v+j} \oplus 1,\ldots) .$$

Let q_i be the i-th column vector of Q and p_i be the i-th row vector of Q. First, we obtain

$$\frac{D\hat{f}}{Dy_{v+i}y_{v+j}} = [b_1,\ldots,b_u,x_{u+1},\ldots,x_s](q_{v+i} \oplus q_{v+j}) . \qquad (5)$$

From condition (2) of this theorem, $W(q_{v+i} \oplus q_{v+j}) \geq k+1$. On the other hand, $u \leq k$. Therefore, the right hand side of (5) is a non-constant affine function. Hence, $D\hat{f}/Dy_{v+i}y_{v+j}$ is balanced from Proposition 2.

Next, for g, define

$$\hat{g}(x_{u+1},\ldots,x_s) \triangleq g(b_1,\ldots,b_u,x_{u+1},\ldots,x_s) .$$

Further, define $\dfrac{D\hat{g}}{Dx_{u+i}}$ and $\dfrac{D\hat{g}}{Dx_{u+i}x_{u+j}}$ similarly to $\hat{f}$. Then we obtain

$$\frac{D\hat{f}}{Dx_{u+i}x_{u+j}} = (p_{u+i} \oplus p_{u+j})[c_1, \ldots, c_v, y_{v+i}, \ldots, y_t]^T \oplus \frac{D\hat{g}}{Dx_{u+i}x_{u+j}} .$$

From condition (3) of this theorem, $W(p_{u+i} \oplus p_{u+j}) \geq k+1$. On the other hand, $v \leq k$. Therefore, $(p_{u+i} \oplus p_{u+j})[c_1, \ldots, c_v, y_{v+i}, \ldots, y_t]^T$ is a non-constant affine function. Hence, $D\hat{f}/Dx_{u+i}x_{u+j}$ is balanced from Proposition 3.

Finally, we have

$$\frac{D\hat{f}}{Dx_{u+i}y_{v+j}} = p_{u+i}[c_1, \ldots, c_v, y_{v+i}, \ldots, y_t]^T$$

$$\oplus [b_1, \ldots, b_u, x_{u+1}, \ldots, x_s]q_{v+j} \oplus \frac{D\hat{g}}{Dx_{u+i}} .$$

Here, $p_{u+i}[c_1, \ldots, c_v, y_{v+i}, \ldots, y_t]^T$ is a non-constant affine function since $v \leq k$ and $W(p_{u+i}) \geq k+1$. Hence, $D\hat{f}/Dx_{u+i}y_{v+j}$ is balanced from Proposition 3.

Thus, we have proved that $\hat{f}(x) \oplus \hat{f}(x \oplus \alpha)$ is balanced for any α such that $W(\alpha) = 2$. Similarly, we can show that it is balanced for $W(\alpha) = 1$. Consequently, f satisfies PC(2) of order k. $\qquad\square$

3.2 How to Find Q

This subsection shows that the matrix Q of Theorem 16 can be obtained by using generator matrices of error correcting codes.

Definition 17. A linear $[N, h, d]$ code is a binary linear code of length N, dimension h and the minimum Hamming distance at least d.

Definition 18. The dual code $C^\perp$ of a linear code C is defined as

$$C^\perp \triangleq \{u \mid u \cdot v = 0 \text{ for all } v \in C\} .$$

The dual minimum Hamming distance of C is defined as the minimum Hamming distance of $C^\perp$.

Theorem 19. *Let G_1 be a generator matrix of a linear $[t, h, d_1]$ code C_1 with the dual minimum Hamming distance d'_1. Let G_2 be a generator matrix of a linear $[s, h, d_2]$ code C_2 with the dual minimum Hamming distance d'_2. Let*

$$Q \triangleq G_2^T G_1 .$$

Then Q satisfies the conditions of Theorem 16 for

$$l = \min(d'_1, d'_2) - 1$$
$$k = \min(d_1, d_2) - 1 .$$

Proof. We first show that Q satisfies condition (2) of Theorem 16. Let γ_1 be a $t \times 1$ vector such that $1 \leq W(\gamma_1) \leq l$. γ_1 is not a codeword of $C_1^\perp$ because $W(\gamma_1) \leq l < d_1'$. Then,

$$G_1\gamma_1 \neq 0$$

because G_1 is a parity check matrix of $C_1^\perp$. Therefore,

$$Q\gamma_1 = G_2^T(G_1\gamma_1)$$

is a nonzero codeword of C_2 because G_2 is a generator matrix of C_2. Hence,

$$W(Q\gamma_1) \geq d_2 \geq k+1 \ .$$

Similarly, Q satisfies condition (3) of Theorem 16. □

By using Theorem 19, we can obtain the following results, for example.

Proposition 20. *[6, p.30] Let C be a $[2^r - 1, 2^r - 1 - r, 3]$ Hamming code. Then $C^\perp$ is a $[2^r - 1, r, 2^{r-1}]$ simplex code.*

Corollary 21. *For $r \geq 2$, there exists*

(1) a $PC(2^{r-1} - 1)$ of order 2 function such that $n = 2^{r+1} - 2$ and
(2) a $PC(2)$ of order $2^{r-1} - 1$ function such that $n = 2^{r+1} - 2$.

Proposition 22. *[6, p.31] Let C be a $[2^r, 2^r - 1 - r, 4]$ extended Hamming code. Then $C^\perp$ is a $[2^r, r+1, 2^{r-1}]$ first order Reed–Muller code.*

Corollary 23. *For $r \geq 2$, there exists*

(1) a $PC(2^{r-1} - 1)$ of order 3 function such that $n = 2^{r+1}$ and
(2) a $PC(3)$ of order $2^{r-1} - 1$ function such that $n = 2^{r+1}$.

4 Balance, Large Degree and High Nonlinearity

We can obtain "balanced" $PC(l)$ of order k functions by choosing g appropriately in Theorem 16. Large degree and high nonlinearity can also be obtained.

4.1 Balanced PC(l) of Order k

Definition 24. We say that g is balanced for a matrix Q if

$$\left|\{x \mid g(x) = 0, xQ = 0\}\right| = \left|\{x \mid g(x) = 1, xQ = 0\}\right| \ . \tag{6}$$

Theorem 25. *In (4), f is balanced if g is balanced for Q.*

Proof. Substitute $x_1 = b_1, \ldots, x_s = b_s$ into (4), where $b_1, \ldots, b_s$ are constant bits. Then we have

$$f(b_1, \ldots, b_s, y_1, \ldots, y_t) = [b_1, \ldots, b_s]Q[y_1, \ldots, y_t]^T \oplus g(b_1, \ldots, b_s) . \qquad (7)$$

If $[b_1, \ldots, b_s]Q \neq 0$, the right hand side of (7) is a non-constant affine function. Therefore, $f(b_1, \ldots, b_s, y_1, \ldots, y_t)$ is balanced from Proposition 2. For $[b_1, \ldots, b_s]$ such that $[b_1, \ldots, b_s]Q = 0$, we have

$$f(b_1, \ldots, b_s, y_1, \ldots, y_t) = g(b_1, \ldots, b_s) .$$

Then because g is balanced for Q, we see that $f(x_1, \ldots, x_s, \hat{y}_1, \ldots, \hat{y}_t)$ is balanced for Q for any fixed $(\hat{y}_1, \ldots, \hat{y}_t)$.

Consequently, $f(x_1, \ldots, x_s, y_1, \ldots, y_t)$ is balanced. $\qquad \square$

We can find such g in the following way.

Lemma 26. *Suppose that* $g(x_1, \ldots, x_n)$ *is written as*

$$g(x_1, \ldots, x_s) = a_1 x_1 \oplus \cdots \oplus a_s x_s \qquad (8)$$

if $[x_1, \ldots, x_n]Q = 0$. *Then* g *is balanced for* Q *if and only if* $[a_1, \ldots, a_s]^T$ *is linearly independent of the columns of* Q.

Proof. First, it is easy to see that g of (8) is balanced for Q if and only if there is an x such that

$$xQ = 0 \text{ but } g(x) = 1 . \qquad (9)$$

This condition is equivalent to say that the kernel (zero space) of Q^T is not contained in the zero space of the linear mapping

$$g(x) = [a_1, \ldots, a_s]x^T .$$

This holds if and only if $[a_1, \ldots, a_s]$ is linearly independent of the rows of Q^T. $\quad \square$

Corollary 27. *Let* $xQ = [h_1(x), \ldots, h_t(x)]$. *Define*

$$g(x_1, \ldots, x_s) \overset{\triangle}{=} a_1 x_1 \oplus \cdots \oplus a_s x_s \oplus h_1(x)h_2(x) \ldots h_t(x)H(x) ,$$

where $H(x)$ *is any function. Then* g *is balanced for* Q *if and only if* $[a_1, \ldots, a_s]^T$ *is linearly independent of the columns of* Q.

Another way of finding a balanced g for Q is to write its truth table.

4.2 Large Degree and High Nonlinearity

In (4), we can obtain $\deg(f) = s$ by letting

$$g(x_1, \ldots, x_s) = x_1 \ldots x_s \ .$$

Further, $PC(l)$ of order k functions which possess high nonlinearity and large degree at the same time can be obtained as follows.

Theorem 28. *There exists a $PC(l)$ of order k function f such that*

- $\deg(f) = s/2$ and $N(f) \geq 2^{t+s-1} - 2^{t+s/2-1}$ for $s = $ even.
- $\deg(f) = (s-1)/2$ and $N(f) \geq 2^{t+s-1} - 2^{t+(s-1)/2}$ for $s = $ odd,

where s and t are defined in Theorem 16.

Proof. For $s = $ even, there exists a bent function $g(x_1, \ldots, x_s)$ such that $\deg(g) = s/2$. By choosing this g in (4), we obtain $\deg(f) = s/2$. Next, we compute the distance between this f and an affine function $A(x_1, \ldots, x_s, y_1, \ldots, y_t)$. Substitute $y_1 = c_1, \ldots y_t = c_t$ into f and A, where $c_1, \ldots, c_t$ are constant bits. Let

$$f_0(x_1, \ldots, x_s) \stackrel{\triangle}{=} f(x_1, \ldots, x_s, c_1, \ldots c_t) = g(x_1, \ldots, x_s) \oplus B(x_1, \ldots, x_s)$$

$$A_0(x_1, \ldots, x_s) \stackrel{\triangle}{=} A(x_1, \ldots, x_s, c_1, \ldots c_t) \ ,$$

where

$$B(x_1, \ldots, x_s) \stackrel{\triangle}{=} [x_1, \ldots, x_s]Q[c_1, \ldots c_t]^T \ .$$

Then

$$d(f, A) = \sum_{c_1, \ldots c_t} d(f_0, A_0) = \sum_{c_1, \ldots c_t} d(g \oplus B, A_0)$$

$$= \sum_{c_1, \ldots c_t} d(g, A_0 \oplus B) \geq \sum_{c_1, \ldots c_t} N(g) = 2^t(2^{s-1} - 2^{s/2-1})$$

from Proposition 7. The above inequality holds for any affine function A. Therefore, $N(f) \geq 2^t(2^{s-1} - 2^{s/2-1})$.

For $s = $ odd, let $\hat{g}(x_1, \ldots, x_{s-1})$ be a bent function with degree $(s-1)/2$ and let $g(x_1, \ldots, x_s) = \hat{g}(x_1, \ldots, x_{s-1})$. (Bent functions exist only for $s = $ even.) $\square$

Compare Theorem 28 with Proposition 6. Then we see that the above $N(f)$ is almost the maximum if t is small. (From condition (1) of Theorem 16, $t \geq \max\{l, k+1\}$, though.)

5 Balanced SAC(k) with the Maximum Degree

Proposition 10 gives an upper bound on the degree of SAC(k) functions. In Sect. 5.2, we will show that this bound is tight for $k \leq n/2 - 1$. Further, Sect. 5.3 will show that this bound is tight even for balanced SAC(k) functions for $k \leq n/2 - 1$ and $n - k - 1 = $odd.

5.1 How to Design SAC(k) Functions

First, we can obtain SAC(k) functions as a special case of Theorem 16.

Corollary 29. *Let*

$$f(x_1, \ldots, x_n) = (x_1 \oplus \cdots \oplus x_{n-k-1})(x_{n-k} \oplus \cdots \oplus x_n) \oplus g(x_1, \ldots, x_{n-k-1}) \ , \quad (10)$$

where $g(x_1, \ldots, x_{n-k-1})$ is any function. Then f satisfies SAC(k) if $k \leq \frac{n}{2} - 1$.

Proof. In Theorem 16, let

$$Q = \text{ the } (n - k - 1) \times (k + 1) \text{ matrix whose elements are all one.} \quad (11)$$

If $n - k - 1 \geq k + 1$, Q satisfies conditions (2) and (3) of Theorem 16 for $l = 1$. $\square$

5.2 SAC(k) with the Maximum Degree

Theorem 30. *There exists an SAC(k) function $f(x_1, \ldots, x_n)$ which meets the equality of (3) for $k \leq \frac{n}{2} - 1$.*

Proof. In Corollary 29, let $g(x_1, \ldots, x_{n-k-1}) = x_1 \ldots x_{n-k-1}$. Then we obtain $\deg(f) = n - k - 1$ and the equality of (3) is satisfied. $\square$

Remark. Proposition 11 shows that Proposition 10 is tight for $k = n - 2$ and $n - 3$.

5.3 Balanced SAC(k) with the Maximum Degree

Theorem 31. *There exists a balanced SAC(k) function $f(x_1, \ldots, x_n)$ which meets the equality of (3) if $k \leq \frac{n}{2} - 1$ and $k - n - 1 = $ odd.*

Proof. In (10), let

$$g(x_1, \ldots, x_{n-k-1}) = a_1 x_1 \oplus \cdots \oplus a_{n-k-1} x_{n-k-1} \oplus x_1 \ldots x_{n-k-1} \ ,$$

where

$$[a_1, \ldots, a_{n-k-1}] \neq [0, \ldots, 0], \ [1, \ldots, 1] \ . \quad (12)$$

We show that this g is balanced for Q, where Q is given by (11). Let $x = [x_1, \ldots, x_{n-k-1}]$. Note that $x_1 \ldots x_{n-k-1} = 0$ if $W(x) < n - k - 1 = $(odd). Also, $W(x) = $ even if $xQ = 0$. Therefore, $x_1 \ldots x_{n-k-1} = 0$ if $W(x) = $even and hence if $xQ = 0$. Hence,

$$g(x_1, \ldots, x_{n-k-1}) = a_1 x_1 \oplus \cdots \oplus a_{n-k-1} x_{n-k-1}$$

if $xQ = 0$. Further, $[a_1, \ldots, a_s]$ satisfying (12) is linearly independent of the columns of Q. Then g is balanced for Q from Lemma 26.

Consequently, f of (10) is balanced from Theorem 25. $\square$

Theorem 32. *For $k - n - 1 =$ even, there exists a balanced $SAC(k)$ function such that $\deg(f) = n - k - 2$.*

Proof. Let

$$g(x_1, \ldots, x_{n-k-1}) = a_1 x_1 \oplus \cdots \oplus a_{n-k-1} x_{n-k-1} \oplus x_1 \ldots x_{n-k-2} \oplus x_2 \ldots x_{n-k-1} \; ,$$

where
$$[a_1, \ldots, a_{n-k-1}] \neq [0, \ldots, 0], \; [1, \ldots, 1]$$
We can show that g is balanced for Q, where Q is given by (11). □

It will be a further work to find a tight upper bound on $\deg(f)$ of balanced $SAC(k)$ functions for $n - k - 1 =$ even.

Remark.

(1) For balanced $SAC(n-2)$ functions, see Proposition 15.
(2) Lloyd [5] showed a condition such that $SAC(n-3)$ functions arc balanced.
(3) Balanced SAC functions with high nonlinearity were constructed by [14]. Recently, other balanced SAC functions were given by [16].

6 Extension to Vector Output Boolean Functions

In this section, we extend our technique to vector output Boolean functions.

6.1 General Results

Let F denote a mapping from $\{0, 1\}^n$ to $\{0, 1\}^m$. We say that F is uniformly distributed if
$$\left| \{x \mid F(x) = \beta\} \right| = 2^{n-m}$$
for any $\beta \in \{0, 1\}^m$.

Definition 33. We say that $F(x_1, \ldots, x_n) = [f_1, \ldots, f_m]$ is an (n, m)-SAC(k) function if any nonzero linear combination of $f_1, \ldots, f_m$ satisfies SAC(k).

Definition 34. We say that $F(x_1, \ldots, x_n) = [f_1, \ldots, f_m]$ is an (n, m)-PC(l) of order k function if any nonzero linear combination of $f_1, \ldots, f_m$ satisfies PC(l) of order k.

From Theorem 16, we obtain the following corollary.

Corollary 35. *Suppose that there exist $s \times t$ binary matrices $Q_1, \ldots, Q_m$ such that any nonzero linear combination of $Q_1, \ldots, Q_m$ satisfies the conditions of Theorem 16. For $1 \leq i \leq m$, let*

$$f_i(x_1, \ldots, x_s, y_1, \ldots, y_t) \stackrel{\triangle}{=} [x_1, \ldots, x_s] Q_i [y_1, \ldots, y_t]^T \oplus g_i(x_1, \ldots, x_s) \; ,$$

where g_i is any function. Then $F = [f_1, \ldots, f_m]$ is an $(s + t, m)$-PC(l) of order k function.

Definition 36. For $F(x_1, \ldots, x_n) = [f_1, \ldots, f_m]$, define

$$\deg(F) \stackrel{\triangle}{=} \min \deg(a_1 f_1 \oplus \cdots \oplus a_m f_m),$$
$$N(F) \stackrel{\triangle}{=} \min N(a_1 f_1 \oplus \cdots \oplus a_m f_m),$$

where min is taken over all nonzero binary vectors $[a_1, \ldots, a_m]$.

Corollary 37. *In Corollary 35,*

(1) let $g_i = x_1 \ldots x_s/x_i$. Then $\deg(F) = s - 1$ if $m \leq s$.
(2) For $s =$ even and $m \leq s/2$, let $[g_1, \ldots, g_m]$ be a vector output bent function given by [9]. Then $N(f) \geq 2^{t+s-1} - 2^{t+s/2-1}$.
(3) If $s =$ odd and $m \leq (s-1)/2$, we can obtain $N(f) \geq 2^{t+s-1} - 2^{t+(s-1)/2}$.

The following corollary is obtained from Theorem 19.

Corollary 38. *Suppose that there exist*

(1) a linear $[t, h, k+1]$ code with the dual minimum Hamming distance at least $l+1$ and
(2) m matrices $G_{2,1}, \ldots G_{2,m}$ such that any nonzero linear combination of them is a generator matrix of a linear $[s, h, k+1]$ code with the dual minimum Hamming distance at least $l+1$.

Let $Q_i \stackrel{\triangle}{=} G_{2,i}^T G_1$ for $1 \leq i \leq m$. Then $Q_1, \ldots, Q_m$ satisfy the condition of Corollary 35.

6.2 Vector Output PC($_2$) of Order k

Proposition 39. *[9] Consider a linear feedback shift register of length r and with a primitive feedback polynomial. Let D be the state transition function of such a shift register. Then D is a permutation of the space Z_2^r as well as the powers D^i of D, where*

$$D^i \stackrel{\triangle}{=} D \circ \cdots \circ D, \; i = 1, 2, \ldots \; .$$

Moreover, any nonzero linear combination of $I, D, D^2, \ldots, D^{r-1}$ is also a permutation.

Lemma 40. *For any $r \geq 2$, there exist matrices $G_{2,1}, \ldots, G_{2,r}$ such that any nonzero linear combination of them is a generator matrix of the $[2^r - 1, r, 2^{r-1}]$ simplex code.*

Proof. Let $[i_1, \ldots, i_r]$ be the binary representation of i.

(1) Let $G_{2,1}$ be a $r \times (2^r - 1)$ matrix such that the i-th column vector is $[i_1, \ldots, i_r]^T$.

(2) For $2 \le j \le r$, let $G_{2,j}$ be a $r \times (2^r - 1)$ matrix such that the i-th column vector is $D^{j-1}(i_1, \ldots, i_r)$.

Then any nonzero linear combination of $G_{2,1}, \ldots, G_{2,r}$ is a parity check matrix of a $[2^r - 1, 2^r - 1 - r, 3]$ Hamming code by Proposition 39. Equivalently, any nonzero linear combination of $G_{2,1}, \ldots, G_{2,r}$ is a generator matrix of a $[2^r - 1, r, 2^{r-1}]$ simplex code. □

Theorem 41. *For $r \ge 2$,*

(1) there exists a $(2^{r+1} - 2, r)$-PC(2) of order $2^{r-1} - 1$ function F with

$$\deg(F) = 2^r - 2 .$$

(2) there exists a $(2^{r+1} - 2, r)$-PC(2) of order $2^{r-1} - 1$ function F with

$$N(F) \ge 2^{2^{r+1}-3} - 2^{3 \cdot 2^{r-1}-2} .$$

Proof. First, there exists a $[2^r - 1, r, 2^{r-1}]$ simplex code (see Proposition 20). Next, there exist matrices $G_{2,1}, \ldots, G_{2,r}$ such that any nonzero linear combination of them is a generator matrix of a $[2^r - 1, r, 2^{r-1}]$ simplex code from Lemma 40. Finally, the dual Hamming distance of a $[2^r - 1, r, 2^{r-1}]$ simplex code is 3. Hence, the conditions of Corollary 38 are satisfied.

Finally, apply Corollary 37 with $s = t = 2^r - 1$. □

6.3 Vector Output SAC(k)

Theorem 42. *For any $s > 0$,*

(1) there exists a $(2s, s - 1)$-SAC(1) function F with $\deg(F) = s - 1$.
(2) there exists a $(2s, s - 1)$-SAC(1) function F with

$$N(F) \ge \begin{cases} 2^{2s-1} - 2^{3s/2-1} & \text{if } s = even \\ 2^{2s-1} - 2^{(3s-1)/2} & \text{if } s = odd . \end{cases}$$

Proof. Let $I = (e_1, \ldots, e_s)$ be the $s \times s$ identity matrix and let P be a permutation matrix such that $P = (e_s, e_1, e_2, \ldots, e_{s-1})$. Define

$$Q_i = P^{(i-1)}(I + P) \tag{13}$$

for $1 \le i \le s - 1$. We show that $Q_1, \ldots, Q_{s-1}$ satisfy the condition of Corollary 35, that is the conditions of Theorem 16 with $s = t$. Let

$$Q = a_1 Q_1 + \cdots + a_{s-1} Q_{s-1} ,$$

where $[a_1, \ldots, a_{s-1}] \ne [0, \ldots, 0]$. Let q_i be the i-th column vector of Q and p_i be the i-th row vector of Q. Without loss of generality, we can assume that

(1) $a_1 = \cdots = a_{s-1} = 1$ or
(2) $a_1 = \cdots = a_j = 1$ and $a_{j+1} = 0$ for some $1 \le j \le s - 2$.

In case 1,

$$Q = I + P^{s-1} \ .$$

In case 2,

$$Q = I + P^j + X \ ,$$

where X cancels no elements of $I + P^j$. In any case, $W(q_i) \geq 2$ for any i and $W(p_i) \geq 2$ for any i. Thus, the conditions of Theorem 16 are satisfied for $l = 1$.

Finally, apply Corollary 37. □

Theorem 42 can be generalized as follows.

Theorem 43. *For any $k \geq 0$ and any $s \geq k + 1$, let*

$$\gamma \triangleq \lceil (k+1)/2 \rceil \ , \quad m \triangleq \lfloor (s-k-1)/\gamma + 1 \rfloor \ .$$

Then

(1) there exists a $(2s, m)$-SAC(k) function F with $\deg(F) = s - 1$.
(2) there exists a $(2s, m)$-SAC(k) function F with

$$N(F) \geq \begin{cases} 2^{2s-1} - 2^{3s/2-1} & \text{if } s = \text{even} \\ 2^{2s-1} - 2^{(3s-1)/2} & \text{if } s = \text{odd} \ . \end{cases}$$

Remark. In [3], we showed that there exists an (n, m)-SAC(k) function F if there exists a linear $[N, m, k + 1]$ code such that

$$N = \begin{cases} n - 1 & \text{if } n \text{ is even} \\ n - 2 & \text{if } n \text{ is odd} \ . \end{cases} \tag{14}$$

In this construction,

(1) $\deg(F)$ and $N(F)$ are small. Actually, $\deg(F) = 2$.
(2) However, m can be larger than that of Theorem 42 and Theorem 43.

In other words, there is a tradeoff between the construction of [3] and Theorem 42 and Theorem 43 of this paper.

Acknowledgments

We would like to thank the anonymous referees for helpful comments. Especially, lemma 4.1 was improved.

References

1. R. Forré. The strict avalanche criterion : spectral properties of Boolean functions and an extend definition. In *Advances in Cryptology — CRYPTO '88 Proceedings, Lecture Notes in Computer Science* 403, pages 450–468. Springer-Verlag, 1990.

2. T. Jakobsen and L.R. Knudsen. The interpolation attack on block ciphers. In *Preproc. of Fast Software Encryption*, pages 28–40. January, 1997.

3. K. Kurosawa and T. Satoh. Generalization of higher order SAC to vector output Boolean functions. In *Advances in Cryptology — ASIACRYPT '96 Proceedings, Lecture Notes in Computer Science* 1163, pages 218–231. Springer-Verlag, 1996.

4. S. Lidl and Niederreiter. *Finite Fields, Encyclopedia of Mathematics and Its Applications 20*. Cambridge University Press, 1983.

5. S. Lloyd. Counting binary functions with certain cryptographic properties. *Journal of Cryptology*, 5:107–131, 1992.

6. F. J. MacWilliams and N. J. A. Sloane. *The theory of error-correcting codes.* North-Holland Publishing Company, 1977.

7. M. Matsui. Linear cryptanalysis method for DES cipher. In *Advances in Cryptology — EUROCRYPT '93 Proceedings, Lecture Notes in Computer Science* 765, pages 386–397. Springer-Verlag, 1994.

8. W. Meier and O. Staffelbach. Nonlinearity criteria for cryptographic functions. In *Advances in Cryptology — EUROCRYPT '89 Proceedings, Lecture Notes in Computer Science* 434, pages 549–562. Springer-Verlag, 1990.

9. K. Nyberg. Perfect nonlinear S-boxes. In *Advances in Cryptology — EUROCRYPT '91 Proceedings, Lecture Notes in Computer Science* 547, pages 378–386. Springer-Verlag, 1991.

10. J. Pieprzyk and G. Finkelstein. Towards effective nonlinear cryptosystem design. *IEE Proceedings Part E*, 35(6):325–335, November 1988.

11. B. Preneel, W. Van Leekwijck, L. Van Linden, R. Govaerts, and J. Vandewalle. Propagation characteristics of Boolean functions. In *Advances in Cryptology — EUROCRYPT '90 Proceedings, Lecture Notes in Computer Science* 473, pages 161–173. Springer-Verlag, 1991.

12. B. Preneel, R. Govaerts, and J. Vandewalle. Boolean functions satisfying higher order propagation criteria. In *Advances in Cryptology — EUROCRYPT '91 Proceedings, Lecture Notes in Computer Science* 547, pages 141–152. Springer-Verlag, 1991.

13. O. S. Rothaus. On bent functions. *Journal of Combinatorial Theory (A)*, 20:300–305, 1976.

14. J. Seberry and X.M. Zhang. Highly nonlinear 0-1 balanced Boolean functions satisfying strict avalanche criterion. In *Advances in Cryptology — AUSCRYPT '92 Proceedings, Lecture Notes in Computer Science* 718. Springer-Verlag, 1993.

15. A. F. Webster and S. E. Tavares. On the design of S-boxes. In *Advances in Cryptology — CRYPTO '85 Proceedings, Lecture Notes in Computer Science* 218, pages 523–534. Springer-Verlag, 1986.

16. A. M. Youssef, T. W. Cusick, P. Stănică, and S. E. Tavares. New bounds on the number of functions satisfying the strict avalanche criterion. In *Third Annual Workshop on Selected Areas in Cryptography*, 1996.

Distributed "Magic Ink" Signatures

Markus Jakobsson*   Moti Yung †

Abstract

The physical analog of "blind signatures" of Chaum is a document and a carbon paper put into an envelope, allowing the signer to transfer his signature onto the document by signing on the envelope, and without opening it. Only the receiver can present the signed document while the signer cannot "unblind" its signature and get the document signed.

When an authority signs "access tokens", "electronic coins", "credentials" or "passports", it makes sense to assume that whereas the users can typically enjoy the disassociation of the blindly signed token and the token itself (i.e. anonymity and privacy), there may be cases which require "unblinding" of a signature by the signing authority itself (to establish what is known as "audit trail" and to "revoke anonymity" in case of criminal activity).

This leads us to consider a new notion of signature with the following physical parallel: The signer places a piece of paper with a carbon paper on top in an envelope as before (but the document on the paper is not yet written). The receiver then writes the document on the envelope using *magic ink*, e.g., ink that is only visible after being "developed". Due to the carbon copy, this results in the document being written in visible ink on the internal paper. Then, the signer signs the envelope (so its signature on the document is made available). The receiver gets the internal paper and the signer retains the envelope with the magic ink copy. Should the signer need to unblind the document, he can develop the magic ink and get the document copy on the envelope. Note that the signing is not blinded forever to the signer. We call this new type of signature a *magic ink signature*.

We present an efficient method for distributively generating magic ink signatures, requiring a quorum of servers to produce a signature and a (possibly different) quorum to unblind a signature. The scheme is robust, and the unblinding is guaranteed to work even if a set of up to a threshold of signers refuses to cooperate, or actively cheats during either the signing or the unblinding protocol. We base our specific implementation on the DSS algorithm. Our construction demonstrates the extended power of distributed signing.

*Department of Computer Science and Engineering, University of California, San Diego. markus@cs.ucsd.edu

†CertCo, New York, NY. moti@cs.columbia.edu, moti@certco.com

W. Fumy (Ed.): Advances in Cryptology - EUROCRYPT '97, LNCS 1233, pp. 450-464, 1997.

1 Introduction

In recent years, various notions of distribution of cryptographic functions (signature and encryption) among independent agents were considered. The typical added functionality of such a distribution include increased security of the secret key, increased availability of service, and increased flexibility of access, the latter by requiring a quorum to access information (as in, e.g., [9, 12, 20]). All these notions are functionality of distributed computing.

In this work we suggest that the distributed signature setting also provides for extended functionality by enabling "a new notion of signature itself" which is otherwise impossible (owing to the added control in this case). The notion we specifically suggest is that of "Magic Ink Signatures". In such a signature service, the signer blindly signs a message for a receiver, while retaining the capability to "unblind" the signature (analogous to developing "Magic Ink"), at any later point. What the distribution enables us is to implement the unblinding with separation in time – i.e., allowing the development of the "Magic Ink" at some point, but not earlier. This is impossible in the centralized case (what the signer can do at some point it can do earlier if there is no limiting factor such as the "Quorum Control" in the distributed case).

Note that requiring various actions of a quorum of distributed agents regarding a specific signature value needs a careful flexible design. For example, we cannot require that in each action the same identical quorum of agents be present. Requiring this may, paradoxically, reduce the availability of the service as the distribution level grows (whereas one of the initial reasons in distributing the service was increased availability). For the same reason, and quite counter-intuitively, it may also force us to put *more* trust in individual servers with a higher degree of distribution, unless care is taken.

The magic ink signature enables the generation of blind signatures which can later be unblinded by the signer (following the physical analogue given in the abstract). This is in sharp contrast with traditional blind signatures, which are information theoretically blinded to the signer [5]. The typical application where the need for unblinding arises is for cases where "privacy of individuals" is assured until some criminal or otherwise unusual activity is detected. Upon detection, identification of the origin of a signature becomes important in identifying the source of the unwanted activity. This is applied to private access tokens, authorized anonymous accounts, and electronic-money. Regarding the later setting, Chaum, Fiat and Naor's [7] original off-line scheme (and its follow-ups) offered perfect anonymity. However, the absolute privacy feature of all these schemes is not only beneficial to honest users, but also to criminal offenders, as it makes perfect crimes possible [2, 4, 8, 11, 18, 23, 25]: Various methods for anonymity revocation are suggested in some of these works mentioned. In "fair blind signatures" [3, 8, 11, 25], a signature receiver puts a pseudonym into the signature, allowing a third party (a judge) to later unblind the signature by calculating a pseudonym from a signature or vice versa. Magic ink signatures are the "distributed cousins" of fair blind signatures, increasing the availability and lowering the amount of trust required (no need to employ a third party beyond the distributed signing agents and using quorum control to assure separation of duties). Magic ink signatures is a generic tool for blind signature generation, enabling the possibility of unblinding selected blind signatures by the signer – but only under quorum agreement to do so. The method is applied to a payment scheme with revocable privacy in [17, 19]. We note that it is easy to apply proactive methods [14, 15] to our suggested solution, for maximum security and availability.

Organization: We present a magic ink signature scheme that is robust, ensuring that as long as a quorum of (a plurality of honest) servers cooperate, they will *always* be able to unblind a given signature. Thus, we ensure availability and a high degree of distribution and reduced degree of trust required from an individual server. We first specify the notion of magic ink signatures, and the format of DSS signatures. Then, in section 3, we present the intuitive approach of magic ink DSS signatures. In section 4, we explain the model, our assumptions, and the tools we utilize. Among these is a new construction of robustness applicable to certain distributed protocols. This is followed by a protocol for magic ink generation of DSS signatures in section 5. In section 6, we elaborate on the robustness of the scheme and we claim its properties in section 7 and the Appendix.

2 Requirements and Background

Specifications: We wish to obtain a signature scheme where blind signatures can be distributively produced by a quorum of trustees, and these signatures can *always* be unblinded by a (possibly different) quorum (assuming a certain linear-fraction majority of honest trustees). We specify the following properties:

- Signatures are generated using a (t, n) threshold scheme by any t out of the n trustees. Less than t trustees cannot generate a valid signature.

- The signatures are computationally blinded to any set of less than t trustees (i.e., the signature cannot be correlated to the blinded signature or the signing session by a set of less than t trustees.)

- Valid signatures can be unblinded, i.e., signatures matched to signing session or vice versa, by any t out of the n trustees, regardless of the behavior of the other $n - t$ trustees and the signature receiver.

- Furthermore, we want signatures generated by an attacker who compromises the secret key of less than t signers (or forces these signers to sign using a protocol different than the specified) to be identifiable by any t of the signers (i.e., having an audit trail of legal signatures).

2.1 The Digital Signature Standard (DSS)

We use the DSS (described herein) as the underlying signature algorithm [21].

Note: Since we use different moduli at different times, we use $[op]_z$ to denote the operation op modulo z, where this is not clear from the context.

Key Generation. A DSS key is composed of public information p, q, g, a public key y and a secret key x, where:
1. p is a prime number of length l where l is a multiple of 64 and $512 \le l \le 1024$.
2. q is a 160-bit prime divisor of $p - 1$.
3. g is an element of order q in Z_p^*. The triple (p, q, g) is public.
4. x is the secret key of the signer, a random number $1 \le x < q$.
5. $y = [g^x]_p$ is the public verification key.

Signature Algorithm. Let $m \in Z_q$ be a hash of the message to be signed. The signer picks a random number k such that $1 \le k < q$, calculates $k^{-1} \bmod q$ (w.l.o.g. k and k^{-1} values compared to DSA description are interchanged), and sets

$$
\begin{aligned}
r &= [[g^{k^{-1}}]_p]_q \\
s &= [k(m + xr)]_q
\end{aligned}
$$

The pair (r, s) is a signature of m.

Verification Algorithm. A signature (r, s) of a message m can be publicly verified by checking that $r = [[g^{ms^{-1}} y^{rs^{-1}}]_p]_q$.

3 Single-Server (Pseudo) Magic Ink Signatures

In order to communicate the intuition of our scheme, we present a method for producing Magic Ink DSS Signatures using only one signing server (which will be able to unblind the signature at will at any time). However, when we later distribute the signature server, signing and unblinding both will require quorum agreement.

3.1 (Pseudo) magic ink generation of DSS signatures

1. The signature receiver R has a hashed message $m \in Z_q$ that he wants signed. He generates two blinding factors, $a, b \in_u Z_q$, and computes a blinding of m, $\mu = [ma]_q$. R sends μ to the signature generating server S.

2. S generates a random secret session key, $\overline{k} \in_u Z_q$, and computes $\overline{r} = [g^{\overline{k}^{-1}}]_p$, which is sent to the signature receiver R.

3. The signature receiver R computes $r = [[\overline{r}^b]_p]_q$, and computes a blinding ρ of r: $\rho = [ra]_q$. R sends ρ to the signature generating server.

4. S generates a tag tag and the DSS signature σ on the message μ, using the public session key ρ. Here, tag is calculated first (which we describe how to do below), after which σ is calculated as follows: $\sigma = [\bar{k}(\mu + x\rho)]_q$. The server sends σ to R.

5. The signature receiver R unblinds the signature: $s = [\sigma a^{-1}b^{-1}]_q$. The triple (m, r, s) is a valid DSS signature on m.

Theorem 1: The protocol produces correct DSS signatures.

Proof of Theorem 1:
Recall that $\sigma \equiv_q \bar{k}(\mu + x\rho)$, $s \equiv_q \sigma a^{-1}b^{-1}$, $m \equiv_q \mu a^{-1}$, and $y \equiv_p g^x$. We can describe r either as $r = [[g^{\bar{k}^{-1}b}]_p]_q$ (from the point of view of information going from the signer(s) to the receiver) or as $r = [\rho a^{-1}]_q$ (from the point of view of information going back from the receiver to the signer(s)). We have that
$$[g^{ms^{-1}}y^{rs^{-1}}]_p = [g^{(m+xr)s^{-1}}]_p = [g^{(\mu a^{-1}+x\rho a^{-1})s^{-1}}]_p = [g^{(\mu a^{-1}+x\rho a^{-1})(\sigma a^{-1}b^{-1})^{-1}}]_p$$
$$= [g^{a^{-1}(\mu+x\rho)((\bar{k}(\mu+x\rho))a^{-1}b^{-1})^{-1}}]_p = [g^{a^{-1}(\mu+x\rho)(\bar{k}(\mu+x\rho))^{-1}ab}]_p = [g^{\bar{k}^{-1}b}]_p \equiv_q r$$
Thus, the protocol generates valid DSS signatures. □

3.2   Generation of tags

Let us start by making the following observation:

Signature-View Invariant: Let $[mr^{-1}]_q$ identify a valid signature (m, r, s), and $\{\mu, \rho\}$ part of the the view of the signer during a signature generation session. We have: $mr^{-1} \equiv_q \mu\rho^{-1}$, since $\mu = [ma]_q$ and $\rho = [ra]_q$ for a valid signature.

Justification: We have $\sigma = [\bar{k}(\mu + x\rho)]_q$ for (μ, ρ) generated by R. Linear combinations of more than one such signature are not known to give a signature of the valid form (due to the use of different values of $\bar{k}$ of different signatures; and implied by our assumption of existential unforgeability of this type of signatures) so we will only consider operations on *one* signature. Multiplying the value σ by a coefficient can maintain a valid signature; two such manipulations are known: First, for $s = [\sigma a]_q$, and $(m, r) = ([\mu a]_q, [\rho a]_q)$ we have that (m, r, s) is still a valid signature. Second, for $s = [\sigma b]_q$, and $(m, r) = (\mu, \rho) = (\mu, [[g^{\bar{k}^{-1}b}]_p]_q)$, we have that (m, r, s) is also a valid signature. For both of these manipulations, the invariant holds, and no other applicable blinding methods are known. Therefore, any way of obtaining a valid signature (m, r, s) for which $m/r \not\equiv_q \mu/\rho$, would give a new method for blinding of signatures of this type.

Use for tagging: We will use the signature-view invariant for the production of tags, which will be (possibly distributedly stored results of) a function of $[\mu\rho^{-1}]_q$. Consider the following tagging method: The signature servers distributively generate and keep a marker (session tag) specific to the signing session, and a distributed tag (unknown to any subset of less than t servers). Together, these can be used to distributively calculate the invariant $[m/r]_q$ of the related session, which can be output and compared to a signature invariant (based on m and r) or distributively (secretly) compared to a given invariant.

3.3 Tracing

There are three types of tracing we can perform:

1. **From known signing session to signed message:** The signature invariant is calculated from the tag and the marker of a given session.

2. **From known signed message to signing session:** The given signature invariant is distributively compared to the signature invariant of each potential session, which is distributively calculated from the tag and the marker of a given session.

3. **By comparison:** The given signature invariant is distributively compared to the signature invariant of the given session.

4 Model and Tools

4.1 Communication and Threat Model

We assume the standard computational model of polynomial-time randomized Turing machines. Players are connected by an insecure broadcast medium, and an (also polynomial time limited) adversary can inject messages and eavesdrop, but not disconnect any other player from the network. Furthermore, the adversary can corrupt up to $t - 1$ of the n players in the network, and by doing so, force the corrupted players to divert from the specified protocol arbitrarily. See [12] for more details about the model.

4.2 Assumptions

We will rely on the following assumptions:

1. The Undeniable Signature Assumption [6] holds (i.e., given an input quadruple (m, s, g, y), it is hard to decide whether $log_m s = log_g y$, unless $x = log_g y$ is known.) This implies that the Discrete Log problem is not in BPP, and that Pedersen's secret sharing scheme [22] is secure on random secrets.

2. The DSS signature scheme where the signature receiver is allowed to specify the message m to be signed *after* seeing the value $[g^{k^{-1}}]_p$ is secure against a chosen message attack.

4.3 Tools

Let us briefly describe the existing tools we employ:

- **Polynomial Interpolation Secret Sharing**[24]: This is the well-known result in which a secret σ is shared by choosing at random a polynomial $f(x)$ of degree t, such that $f(0) = \sigma$.

- **Joint Random Secret Sharing**[10, 22]: In a Joint Random Secret Sharing scheme the players *collectively* choose shares corresponding to a (t, n)-secret sharing of a random value.

- **Joint Zero Secret Sharing**[1]: This protocol generates a *collective* sharing of a "secret" whose value is zero. Such a protocol is similar to the above joint random secret sharing protocol but instead of local random secrets each player deals a sharing of the value zero.

- **Computing Reciprocals**[12]: Given a secret $k \bmod q$ which is shared among players $P_1, ... P_n$, generate a sharing of the value $k^{-1} \bmod q$, without revealing information on k and k^{-1}.

- **Multiplication of Secrets**[12]: Given two secrets u and v, which are both shared among the players, compute the product uv, while maintaining both of the original values secret (aside from the obvious information which is revealed from the result).

The multiplication of two secrets easily extends to linear combinations and products of three secrets, e.g., $\overline{k}_i(\mu_i + x_i\rho_i)$ for secrets $\overline{k}_i$, μ_i, x_i, and ρ_i. This is achieved without altering the method given in [12]. We also use three new tools:

- **Comparison of Secrets**: Given two secrets u and v, which are both shared among the players (or one is shared one is known), using the above tools we can compare their equality without learning the secret values.

- **Undeniable Signature Based Robustness**: We introduce the use of the verification protocol of undeniable signatures to prove correct exponentiations.

- **Destructive Robustness**: We introduce a new method for making distributed protocols robust: Instead of verifying that each individual share of the calculation is correct, we first combine the shares and then verify that the combined result is correct. If it is not, then each share of the result is verified. A minor efficiency improvement is obtained from doing so. But more importantly, this approach allows simpler and clearer protocol design. This is because we can allow the individual correctness verification to destruct important properties of the produced transcript, which, if the combined result is not correct, is a worthless transcript anyway. Therefore, we call this type of robustness *destructive robustness*.

5 Magic Ink Signature Generation

5.1 Distributed magic ink generation of DSS signatures

Let us now consider a distributed version of the protocols previously presented. Here, let Q be a quorum of t servers in $S_1 \dots S_n$:

1. The signature receiver R has a message $m \in Z_q$ that he wants signed. He generates two blinding factors, $a, b \in_u Z_q$. He then computes a blinding of m, $\mu = [ma]_q$, and a (t, n) secret sharing $(\mu_1, \dots \mu_n)$ of μ, with public information $(g^{\mu_1} \dots g^{\mu_n})$. He sends μ_i to signature generating server S_i.

2. The set of servers $S_i | i \in Q$ distributively generate a random secret session key, $\overline{k} \in_u Z_q$, where server S_i has a share $\overline{k}_i$. Server S_i publishes $[g^{\overline{k}_i}]_p$, and using the methods for computing reciprocals in [12], the servers compute $\overline{r} = [g^{\overline{k}^{-1}}]_p$, which is sent exclusively to the signature receiver R.

3. The signature receiver R computes $r = [[\overline{r}^b]_p]_q$, and blinds this: $\rho = [ra]_q$. R computes a (t, n) secret sharing $(\rho_1, \dots \rho_n)$ of ρ, with public information $(g^{\rho_1} \dots g^{\rho_n})$. R sends ρ_i to S_i.

4. The set of servers $S_i | i \in Q$ distributively generate the tag tag and the DSS signature σ on the message μ, using the (shared) public session key ρ. Here, tag is calculated first (for which we present a robust protocol below), after which σ is calculated as follows: S_i generates $\sigma_i = [\overline{k}_i(\mu_i + x_i \rho_i)]_q$. Then, $\sigma = [\overline{k}(\mu + x\rho)]_q$ is interpolated from the σ_i's using the method for multiplication of secrets in [12]. The servers send σ to R.

5. The signature receiver R unblinds the signature: $s = [\sigma a^{-1} b^{-1}]_q$. The triple (m, r, s) is a valid DSS signature on m.

We note that the proof of correctness is identical to that of the non-distributed protocol version, given robust primitives for secret sharing (e.g., [10, 22]), for computing reciprocals (e.g., [12]) and for multiplication of secrets (e.g., [12]). Also note that we can use standard zero-knowledge techniques to force the receiver to prove that the blinding of steps 1 and 3 are consistent.

5.2 Distributed tag generation and tracing

Let us review the steps of tagging method previously outlined: At the time of signing, μ and ρ are available distributedly. The servers distributively compute $[\mu/\rho]_q$ (without revealing this value to each other). Also, they select a distributed random value $[c]_q$. The servers distributively store this value, and its inverse $[c^{-1}]_q$. They compute and publish the tag $[c(\mu/\rho)]_q$ (given that the components of the multiplication are distributed and secret, this value is random). We can now trace from a session to a signature by distributed multiplication of the tag by $[c^{-1}]_q$, and comparing the result to the public signature invariant. Given a signature invariant $[m/r]_q$, we can distributedly multiply by a value $[c]_q$; if the (distributively held) result equals the the published tag, the session and the signature indeed match. For comparison of a session to a signature, on the other hand, we do not reveal the result of the last multiplication. Rather we check distributedly and secretly for equality of the computed multiplication and the session tag. Note that the probability of collision of tags is negligible.

6 Robustness of Signature Generation

So far, we have not considered the robustness of the signature generation. We will employ *destructive robustness* in order to obtain high efficiency without sacrificing anonymity.

Destructive robustness involves two steps: (1) combination of shares of the result, and error detection, by verifying the correctness of the combined result. This check can be done either internally (i.e., by the same entities that produced the shares) or externally. Then, if the combined result is not correct, the second step is invoked: (2) error tracing, in which it is determined which server(s) have deviated from the protocol. This kind of robustness is possible in protocols where partial incorrect results can be discarded and when we can withstand delays of malicious servers revealing themselves in a slow pace.

We demonstrate an *external* method of destructive robustness for the generation of the blind signature σ on μ, using ρ as public session key:

1. **Share Combination and (External) Error Detection:**
   The signature servers send σ to R, who unblinds the result, obtaining a triple (m, r, s). If this signature is not valid, then R sends a complaint to the signature servers, invoking the next step:

2. **Error Tracing:**
   $S_i | i \in Q$ reveals μ_i. If σ_i was computed correctly, then $g^{\sigma_i} \equiv_p g^{\overline{k}_i(\mu_i + x_i \rho_i)}$ $\equiv_p ((g^{\mu_i})g^{x_i\rho_i})^{\overline{k}_i} \equiv_p ((g^{\mu_i})y_i^{\rho_i})^{\overline{k}_i}$. Using a verification protocol for undeniable signatures, S_i proves that for some $I = (g^{\mu_i})y_i^{\rho_i}$ it is true that $log_I g^{\sigma_i} = log_g(g^{\overline{k}_i})$. He then proves that $log_{y_i}(I(g^{\mu_i})^{-1}) = log_g(g^{\rho_i})$. A server S_i is declared a cheater if he refuses to reveal the information, if the

information is not consistent with the public shares of the secret sharing schemes, or if the share s_i sent out earlier was incorrectly computed.

We see that the above method assures that cheating servers are caught, and that no transcript properties are lost when no complaint is filed. Also note that if R files a unjustified complaint, then *this* will be established, since it will be found that no server cheated. Finally, note that no secret information of honest servers will be leaked to R if R receives an invalid signature transcript. R has no motivation to complain about a good signature; this results in early "unblinding". Each time a threshold is used and opened, the misbehaving processors are eliminated and the process start afresh (to avoid leaking information) based on new random choices. This may result in a delay of at most t times, but enables t to be a maximal minority $n = t/2 + 1$. Note that the method is applicable due to the probabilistic nature of the computation and the care in opening erroneous results.

7 Correctness Claims

We claim that the scheme satisfies the specification of Magic Ink Signature schemes. More specifically, we claim that

- We generate correct DSS signatures in a robust way, using a (t, n) threshold scheme (t from [12]).

- It is not possible for less than t out of n signature servers to correlate a signed message to its blinded withdrawal session.

- It is always possible for t out of n signature servers to correlate a signed message to its blinded withdrawal session.

- It is always possible for t out of n signature servers to distinguish messages they signed from messages signed by an attacker who compromised their secret key (or forced them to produce a signature in a fully blinded manner.)

The claims are shown to hold in the appendix.

Finally, we note that key exchange can be reduced to magic-ink signatures (one party playing the receiver and the other party plays all signers, and the message being the key). Due to blinding, the message (key) is hidden from eavesdroppers, but not from the party playing the signers (since it can perform "unblinding" internally). This implies the difficulty of designing magic-ink signature merely based on the existence of a general one-way permutations [16].

8   Acknowledgments

Thanks to Russell Impagliazzo for numerous discussions, to Jan Camenisch and Markus Stadler for pointing out corrections to the initial draft, and to Markus Michels, Tal Rabin and Rebecca Wright for helpful comments and remarks.

References

[1] M. Ben-Or, S. Goldwasser, A. Wigderson, "Completeness Theorems for Non-cryptographic Fault-Tolerant Distributed Computations," STOC '88, pp. 1-10.

[2] E. Brickell, P. Gemmell, D. Kravitz, "Trustee-based Tracing Extensions to Anonymous Cash and the Making of Anonymous Change," Proc. 6th Annual ACM-SIAM Symposium on Discrete Algorithms (SODA), 1995, pp. 457-466.

[3] J. Camenisch, U. Maurer, M. Stadler, "Digital Payment Systems with Passive Anonymity-Revoking Trustees," Computer Security - ESORICS 96, volume 1146, pp. 33-43.

[4] J. Camenisch, J-M. Piveteau, M. Stadler, "An Efficient Fair Payment System," 3rd ACM Conf. on Comp. and Comm. Security, 1996, pp. 88-94.

[5] D. Chaum, "Blind Signatures for Untraceable Payments," Advances in Cryptology - Proceedings of Crypto '82, 1983, pp. 199-203.

[6] D. Chaum, H. Van Antwerpen, "Undeniable Signatures," Advances in Cryptology - Proceedings of Crypto '89, pp. 212-216.

[7] D. Chaum, A. Fiat and M. Naor, "Untraceable Electronic Cash," Advances in Cryptology - Proceedings of Crypto '88, pp. 319-327.

[8] G.I. Davida, Y. Frankel, Y. Tsiounis, and M. Yung, "Anonymity Control in E-Cash Systems," Financial Cryptography 97.

[9] Y. Desmedt, Y. Frankel, "Threshold Cryptosystems," Advances in Cryptology - Proceedings of Crypto '89.

[10] P. Feldman, "A Practical Scheme for Non-Interactive Verifiable Secret Sharing" FOCS '87, pp. 427-437.

[11] Y. Frankel, Y. Tsiounis, and M. Yung, "Indirect Discourse Proofs: Achieving Efficient Fair Off-Line E-Cash," Advances in Cryptology - Proceedings of Asiacrypt 96, pp. 286-300.

[12] R. Gennaro, S. Jarecki, H. Krawczyk, T. Rabin, "Robust Threshold DSS Signatures", Advances in Cryptology - Proceedings of Eurocrypt '96, pp. 354-371.

[13] S. Goldwasser and S. Micali, "Probabilistic Encryption". J. Comp. Sys. Sci. 28, pp 270-299, 1984.

[14] A. Herzberg, M. Jakobsson, S. Jarecki, H. Krawczyk, M. Yung, "Proactive Public Key and Signature Systems," 4th ACM Conf. on Comp. and Comm. Security, 1997.

[15] A. Herzberg, S. Jarecki, H. Krawczyk, M. Yung, "Proactive Secret Sharing, or How to Cope with Perpetual Leakage," Advances in Cryptology - Proceedings of Crypto '95.

[16] R. Impagliazzo and S. Rudich, Limits on the Provable Consequences of One-way Permutations, STOC '89.

[17] M. Jakobsson, "Privacy vs. Authenticity," PhD Thesis, University of California, San Diego, Department of Computer Science and Engineering, 1997. Available at http://www-cse.ucsd.edu/users/markus/.

[18] M. Jakobsson and M. Yung, "Revocable and Versatile Electronic Money," 3rd ACM Conference on Comp. and Comm. Security, 1996, pp. 76-87.

[19] M. Jakobsson and M. Yung, "Applying Anti-Trust Policies to Increase Trust in a Versatile E-Money System," Financial Cryptography '97.

[20] S. Micali, "Fair Cryptosystems," Advances in Cryptology - Proceedings of Crypto '92.

[21] National Institute for Standards and Technology, "Digital Signature Standard (DSS)," Federal Register Vol 56(169), Aug 30, 1991.

[22] T.P. Pedersen, "Distributed Provers with Applications to Undeniable Signatures," Advances in Cryptology - Proceedings of Eurocrypt '91, pp. 221-242.

[23] S. von Solms and D. Naccache, "On Blind Signatures and Perfect Crimes," Computers and Security, 11 (1992) pp. 581-583.

[24] A. Shamir, "How to Share a Secret," CACM, V. 22, 1979, pp. 612-613.

[25] M. Stadler, J-M. Piveteau, J. Camenisch, "Fair Blind Signatures," Advances in Cryptology - Proceedings of Eurocrypt '95, 1995.

9 Appendix: Correctness and Security

The magic ink signature generation is correct, as shown in the proof of the single-server version (which generates the signature the same way) in section 3, and its simulation in the distributed setting (based on [12]). The robustness of the signature generation depends on our destructive robustness method for random signatures (on top of the non-robust threshold DSS), and the soundness of the composed undeniable signature verification.

Let us next sketch the proof of the additional required properties: that the original signature is blinded and that it can be unblinded as well.

Theorem 2: A coalition of less than t cheating servers cannot, with a non-negligible advantage over guessing, correlate a signature to a signing session.

Proof of Theorem 2: (Sketch)
Let V_1 and V_2 be the view of a coalition of less than t signing servers for two different signing sessions. Let (m, r, s) be a signed message, created in either one of these signing sessions, and assume, in order to reach a contradiction, that the signed message can be correctly matched to either V_1 or V_2 with probability $\frac{1}{2} + \varepsilon$, where ε is polynomial in the size of the security parameters.
We will show that this is not possible by demonstrating that, unless the undeniable signatures assumption is invalid, a polynomial time limited adversary will not be able to tell the transcript parts given from random strings.
We will therefore perform the following thought experiment: We assume that we have a random string of the size of the signers' view during a signature generation phase, where each individual part of the string (corresponding to a transcript part (communication step) of the generation) is selected from the same distribution as its corresponding actual transcript part. Then, we will replace the individual parts of the random string with part of a real transcript one by one. For each step, we will show that it is not possible for a non-quorum of signers to distinguish which of the strings corresponds to a string before or after the last replacement. (This is the walking argument on a random variable [13]). This shows that given a generated signature, its generation view and a random string of the same size and same public distribution, a non-quorum of signers will not be able to match the signature to the the generation view with more than a negligible probability.
We divide the information into two sets, (a) the view of less than a threshold of signer servers, and (b) the signed message. The view of server S_i consists of the public information $\{g^{\mu_i}\}$, $\{g^{\overline{k}_i}\}$, $\overline{r} = [g^{\overline{k}^{-1}}]$, $\{g^{\rho_i}\}$ $\{\sigma_i\}$, σ. We also have the value tag, and intermediary results in the generation; we will consider this later. We have private information $\overline{k}_i$, μ_i, ρ_i and x_i. Since the private information are random shares of $\overline{k}$, μ, ρ and x; and σ_i is just a combination of the random shares and public information, these (or less than t of these sets) cannot help us to correlate the view to the signed message. Therefore, we focus only on

the public information and the signed message, and prove that these cannot be correlated by a non-threshold of signature servers.

Let us consider what meaningful information can be calculated from the public view and the signed message: The signed message is of the form (m, r, s), where $r = g^{\overline{k}^{-1}b}$ and s is such that $r \equiv_q g^{ms^{-1}}y^{rs^{-1}}$. Given the structure of the tag, we need to learn something about $[\mu/\rho]_q$ in order to trace (which we will show to be hard). Without loss of generality, let us consider real transcripts parts of V_1, and the following order of substituting correct transcript parts with random transcript parts in the list of random transcript parts and a potential triple (m, r, s). The following are ideas regarding the implications of possible distinguishability at each substitution stage:

1. Substitute in g^μ: It is not possible to distinguish this step, since (m, r, s) is statistically uncorrelated to μ (given that a is chosen uniformly at random, $\mu = [ma]_q$, r is not related to μ, and s is uncorrelated from μ by b, which is chosen uniformly at random.)

2. Substitute in $g^{\overline{k}}$, $\overline{r} = g^{\overline{k}^{-1}}$: It is not possible to distinguish this step either, since (m, r, s, μ) is statistically uncorrelated to $\overline{k}$. It cannot be correlated to m or μ since these are not related, and not to r since b is chosen uniformly at random and $r = [[g^{\overline{k}^{-1}b}]_p]_q$. It cannot be correlated to s, which is a linear combination of $\overline{k}$, x (both unknown) and μ, r (both in the set of potential transcript parts), or: given the linear combination, and known $\mu, \overline{r}, \sigma, \rho$ we would be able to decide the undeniable signature $(g, g^x, \overline{r}, \overline{r}^x)$, where $\overline{r}^x = (g^\sigma \overline{r}^{-\mu})^{1/\rho}$. The same argument holds for substituting in $g^{\overline{k}^{-1}}$.

3. Substitute in g^ρ: We see that ρ is unrelated to m, and $g^{\overline{k}}$. If we can correlate it to r or $\overline{r}$, this gives us an algorithm for deciding the undeniable signature (g, g^b, r, r^b) (assuming a is known); s is just a linear combination of the above, and the previous argument for linear combinations holds. It is not possible to produce a known function $[\mu/\rho]_q$ from g^μ and g^ρ, or the Diffie–Hellman assumption breaks.

4. Substitute in σ_i (or with the same argument: σ): Follows the linear combination argument above.

Next, based on the above ideas, since none of the substitutions can be distinguished from a random string, it is not possible to match with related non-negligible probability one signed message to one out of two signing views (by the triangle inequality). Let us consider *tag* now: The tag generation protocol outlined is specified so that it hides the participants inputs (guaranteed by the properties of the protocol for multiplication of secrets and inverting a secret.) In fact the public tag is a random element mod q (for each tag and signature invariant there is an element that matches it with the signature). It is therefore not possible to match a signed message to one out of two possible signing views.

It is also true, then, that it is not possible to match a signed message to one out of n signing views. Otherwise we simply would get a contradiction by constructing $n - 2$ additional signing views, none of which matches the signed message, and then match this to one of the remaining two views. $\square$

Theorem 3: A quorum of t servers will always succeed in unblinding a signature in either of the three directions given.

Proof of Theorem 3: (Sketch)
Given that $[m/r]_q \equiv_q [\mu/\rho]$ is always true for a signature generation session in which a valid signature (m, r, s) is generated (this can be guaranteed using zero-knowledge proofs if a new blinding methods is suspected to exist), we have that the tag will always be retrievable given robust protocols for tag generation and tracing. The robustness of these follows from the robustness of methods for multiplication and inversion of secrets. $\square$

An audit trail of legal signatures:
In addition to correct checking and tracing of existing signatures, we have a built-in fraud detection mechanism. Since only signatures that were generated in the proper manner by the signature servers will have a tag, it will be possible to distinguish such signatures from signatures generated by an attacker who compromised the secret key of the signer but has no access to the tags. The signature servers can compare tag by tag to the signature (using the third tracing option,) and if no tag matches, then it is invalid. This feature may provide an "audit trail" for sensitive services.

Efficient and Generalized Group Signatures

Jan Camenisch

Department of Computer Science
ETH Zurich
CH-8092 Zurich, Switzerland
camenisch@inf.ethz.ch

Abstract. The concept of group signatures was introduced by Chaum et al. at Eurocrypt '91. It allows a member of a group to sign messages anonymously on behalf of the group. In case of a later dispute a designated group manager can revoke the anonymity and identify the originator of a signature. In this paper we propose a new efficient group signature scheme. Furthermore we present a model and the first realization of generalized group signatures. Such a scheme allows to define coalitions of group members that are able to sign on the group's behalf.

1 Introduction

In [6] Chaum and van Heyst proposed a new type of signature scheme for a group of entities, called *group signatures*. Such a scheme allows a group-member to sign a message on the group's behalf such that everybody can verify the signature but no one can find out which group member provided it. However, there is a trusted third party, called the group manager, who can in case of a later dispute reveal the identity of the originator of a signature. The group manager can either be a single entity or a number of coalitions of several entities (e.g. group members). This concept can be generalized to allow defined subsets of all group members to jointly sign a message on behalf of the group.

An application of group signature schemes is a company needing a corporate identity. Members of the company can sign contracts with customers such that a customer does not know who actually signed the contract. If a problem with a particular contract occurs later, the company can find out which employee is to be held responsible.

1.1 Related Work

There exist several other group-oriented concepts for signature schemes. The most important ones are multi-signatures [3,9,15] and proxy signatures [14]. Multi-signatures can be seen as generalized group signature without the ability of "opening" signatures, while proxy signatures are group signatures that do not provide anonymity.

Solutions for group signature schemes were first presented in [6] and later in [7]. We discuss these schemes briefly. In [6] four different schemes were proposed. Three of them require the group manager to contact each group member

W. Fumy (Ed.): Advances in Cryptology - EUROCRYPT '97, LNCS 1233, pp. 465-479, 1997.
© Springer-Verlag Berlin Heidelberg 1997

in order to find out who signed a message. These scheme provide computational anonymity, whereas the forth scheme provides information theoretical anonymity. For two of the schemes it is not possible to add a new member after the scheme is set up (including the scheme giving information theoretical anonymity). In none of the proposed schemes it is possible to distribute the functionality of the group manager efficiently.

Later, Chen and Pedersen proposed two new schemes in [7] providing information theoretical anonymity and computational anonymity, respectively. These schemes allow to add new members after the setup of the system and to distribute the functionality of the group manager. They are based on proofs of knowledge of one out of several discrete logarithms, each being the secret key of a group member. The proofs they apply have the special property that when knowing all secret keys, one can tell which one was used in the proof. To realize the group manager's ability to open signatures, two such proofs of knowledge must be used in parallel, where for one the manager is told the secret keys of all group members. However, this solution has the drawback that the group manager can falsely accuse a group member of having signed a message: she therefore computes one of the proofs of knowledge using the known secret key of the member she wants to accuse. This risk can be weakened, but not prevented, by sharing the functionality of the group manager. To solve this problem, some kind of disavowal protocol would be needed.

1.2   Our Results

In this paper we propose a group signature scheme where the manager cannot falsely accuse group members (even if she is also a group member) and which is also more efficient than all the previously proposed schemes. Furthermore, this scheme is extended to a generalized group signature scheme that is also presented. In both schemes, the functionality of the group manager can be shared such that the identity of a signer can still be revealed efficiently. Both schemes allow to add (or remove) group members after the initial setup. They provide computational anonymity which we believe is satisfactory because the security of the signature scheme itself is also computational (as is the case for all signature schemes).

The paper is structured as follows. In the next section we formalize the concept of (generalized) group signatures schemes. The preliminaries are given in Section 3, and in Section 4 we formalize different protocols for proving knowledge about discrete logarithms. This formalization allows a compact and comprehensive description of the new group signature schemes in Section 5. An example of a generalized group signature scheme is also given. In Section 6 we present extensions to the scheme, such as distributing of the functionality of the group manager.

2 Defining Group Signature Schemes

In this section we define the generalized concept of group signature schemes. Let $\mathcal{P} = \{P_1, ..., P_n\}$ be a set of *group members* and M be a designated entity, called *group manager*. The set of all authorized coalitions of group members $\Gamma \subseteq 2^\mathcal{P}$ is called *authority structure*. The structure must be *monotone*, i.e., for two sets S and $S' \in 2^\mathcal{P}$, if $S \in \Gamma$ and $S' \supseteq S$, then also $S' \in \Gamma$. If $\Gamma = \{\{P_1\}, \{P_2\}, ..., \{P_n\}\}$, we call the group signature scheme *simple* (this is the only authority structure we do not require to be monotone).

A (generalized) group signature scheme for $\mathcal{P}$ and M with respect to Γ consists of four procedures:

setup: On input Γ this multi-party protocol between all members in $\mathcal{P}$ and M outputs the group public key $\mathcal{Y}$, to each group member $P_i \in \mathcal{P}$ a secret key x_i, and an *opening secret key* ω to the group manager M.

sign: On input a message m, the group public key $\mathcal{Y}$, the structure Γ, the coalition S, and the corresponding secret keys x_i, this multi-party protocol between members in some $S \in \Gamma$ outputs a signature s on m.

verify: On input a message m, the group public key $\mathcal{Y}$, the structure Γ, and a signature s, this algorithm outputs **yes** if and only if the signature is correct.

open: On input a message m, the group public key $\mathcal{Y}$, the structure Γ, a signature s and the opening secret key ω, the algorithm outputs $S \in \Gamma$ (i.e., the set of group members that signed m) and a proof that S indeed signed m.

In the procedures being multi-party protocols the private inputs of the different parties must of course remain secret during and after the execution. The requirement that **open** also outputs a proof is often omitted but is essential if the trust to be put into the group manager is to be minimized.

The group publishes its public key $\mathcal{Y}$, the authority structure Γ, and some system parameters. A group signature scheme must satisfy the following properties:

1. Only authorized coalitions S of group members, i.e., $S \in \Gamma$, can sign. The correctness of a signature can be publicly verified using $\mathcal{Y}$ and Γ.
2. It is not possible to find out which coalition $S \in \Gamma$ signed a message (anonymity) or whether two different signature are signed by the same coalition (unlinkability).
3. In case of dispute, the group manager can open a signature, i.e., find out which coalition signed a message, by running the algorithm **open**
4. The group manager must only be involved in the procedures **setup** and **open**.

These properties are demanded in all previous papers and further properties follow from them, for instance the property that a coalition must not be able so sign in the name of another coalition. However, the following natural properties should also be satisfied by a group signature scheme. The property 5 was formulated as an open problem in [6] and achieved first in [7].

5. To decrease the trust to put in the group manager, it should be possible to distribute her role among a set of entities such as the members of the group.

6. The group manager is only trusted not to open signatures at will and is not trusted with regard to anything else.

When considering the efficiency of a scheme, the following parameters are of particular interest: the amount of computation in the algorithms setup, sign, verify, and open, the size of the group public key, and the length of signatures. The possibility of adding (or removing) new group members after the initial setup falls also in this category, namely in the efficiency of the algorithms setup (i.e., whether it is possible to run it incremental or not).

3 Preliminaries

In this section a variation of the ElGamal encryption scheme is described. This variation is used as a building block for both group signature schemes we present. We give a formal definition of secret sharing schemes and describe an example. Secret sharing is used for constructing the generalized group signature scheme.

3.1 ElGamal Encryption Variant

The original encryption scheme was proposed by ElGamal [10]. In this paper we interchange the role of the base and the public key and get the following scheme with the same security properties. Let G be a finite cyclic group of prime order q and let $g \in G$ be a generator of G such that computing discrete logarithms to the base g is infeasible. In order to encrypt a message m for an entity with public key $z = g^x$, one first chooses α randomly in $\mathbb{Z}_q$ and then encrypts m by computing the pair $(A, B) = (z^\alpha, g^\alpha m)$. The entity knowing the secret key x can decrypt the message m by calculating

$$\frac{B}{A^{x^{-1}}} = \frac{g^\alpha m}{g^{x\alpha x^{-1}}} = m .$$

3.2 Secret Sharing

A secret sharing scheme is a method for distributing a *secret* σ among a set of n participants $\mathcal{P} = \{P_1, ..., P_n\}$. Each participant P_i obtains a *share* ς_i of the secret σ such that every *qualified* subset $\mathcal{S}$ of $\mathcal{P}$ can reconstruct σ by using algorithm Υ_Γ. The following must hold:

$$\forall \mathcal{S} \in \Gamma : \quad \sigma = \Upsilon_\Gamma(\mathcal{S}, \{\varsigma_i | P_i \in \mathcal{S}\}) .$$

The union of all qualified subsets $\Gamma \subseteq 2^{\mathcal{P}}$ is called the *access structure* and is required to be monotone. A common special case is a *threshold* structure where for a threshold k the access structure Γ is defined as $\{\mathcal{S} \subseteq 2^{\mathcal{P}} \mid |\mathcal{S}| \geq k\}$. Every access structure Γ has a natural *dual* access structure Γ^*:

$$\mathcal{S} \in \Gamma^* \iff \bar{\mathcal{S}} \notin \Gamma ,$$

where $\bar{\mathcal{S}}$ denotes the complement of $\mathcal{S}$ in $\mathcal{P}$. If Γ is monotone, then Γ^* is also monotone and we have $(\Gamma^*)^* = \Gamma$. If Γ is a threshold structure, then so is Γ^*. A secret sharing scheme is called *perfect* if the participants forming a non-qualified subset of $\mathcal{P}$ are not able to obtain any information on σ. A secret sharing scheme is *ideal* if it is perfect and the secret and the shares are of the same length.

To construct the shares for a given secret σ, we employ a nonstandard algorithm that, given the shares of a non-qualified set, outputs the shares for the remaining participants. Formally, the algorithm Ψ which takes as inputs the access structure Γ, a non-qualified set of participants $\mathcal{N} \notin \Gamma$, the set $\{\varsigma_i | P_i \in \mathcal{N}\}$ of their shares, and the secret σ and outputs the set $\{\varsigma_j | P_j \in \bar{\mathcal{N}}\}$, i.e.,

$$\Psi(\Gamma, \mathcal{N}, \{\varsigma_i | P_i \in \mathcal{N}\}, \sigma) = \{\varsigma_j | P_j \in \bar{\mathcal{N}}\} .$$

The algorithm relies on the fact that given the secret and the shares of a non-qualified set participants $\mathcal{N}$, it is possible to construct a complete set of shares.

As an example of a threshold secret sharing scheme with n participants and threshold k, we present Shamir's scheme [18]. A secret σ (an element of a finite field $GF(q)$, with $q > n$) is shared by randomly choosing the coefficients $\alpha_1, ..., \alpha_{k-1} \in GF(q)$ of the polynomial

$$f(X) = \alpha_{k-1} X^{k-1} + ... + \alpha_1 X + \sigma \pmod{q} .$$

The share for participant P_i is then calculated as $\varsigma_i = f(p_i)$, where p_i is a publicly known element of $GF(q)$ associated with participant P_i, e.g. $p_i = i$. Given k or more shares the function f and thus σ can be found by Lagrange interpolation on the points (p_i, ς_i). This scheme is ideal.

4 Proving Knowledge of Discrete Logarithms

In this section we define and formalize the building blocks for our scheme. They are based on different interactive proofs of knowledge of discrete logarithms that are made non-interactive using the techniques of [17]. To avoid confusion with the terminology of non-interactive proofs of knowledge, we call these building blocks *signatures of knowledge*.

The algebraic setting is as follows. Let G be a finite cyclic group of prime order q and let $g, g_1, ..., g_n \in G$ be generators of G such that computing discrete logarithms to any of the bases is infeasible. A *public key* y_i is constructed by computing $y_i = g^{x_i}$ with the *secret key* x_i chosen at random from $\mathbb{Z}_q$. The symbol $\|$ denotes the concatenation of two binary strings (or of the binary representation of group elements and integers). Finally, let $\mathcal{H} : \{0,1\}^* \to \{0,1\}^{\ell}$ ($\ell \approx 128$) denote a one-way hash function.

The first building block we define is a signature of knowledge of the discrete logarithm of a public key y to the base g.

Definition 1. A pair (c, s) satisfying

$$c = \mathcal{H}(g\|y\|g^s y^c\|m)$$

is a signature of knowledge of the discrete logarithm of a group element y to the base g for the message m and is denoted by $SKDL(g, y, m)$.

Basically, such a signature of knowledge is a Schnorr signature (see [17]) with a slightly different argument to the hash function. A $SKDL$ can be computed only if the secret key x is known, by choosing r at random from $\mathbb{Z}_q$ and computing c and s according to

$$c = \mathcal{H}(g\|y\|g^r\|m)$$

and

$$s = r - cx \pmod{q}.$$

The values g^r, c, and s are often called *commitment*, *challenge*, and *response*, respectively, although the "proof" is non-interactive. If the context is clear then the hashing of bases and public keys could be omitted.

Another building block we use is a signature of knowledge of the discrete logarithm of one out of several public keys y_i without revealing which one. Such proof-systems were first introduced in [8].

Definition 2. A $2n$-tuple $(c_1, ..., c_n, s_1, ..., s_n)$ satisfying

$$\sum_{i=1}^{n} c_i = \mathcal{H}(g\|y_1\|...\|y_n\|g^{s_1} y_1^{c_1}\|...\|g^{s_n} y_n^{c_n}\|m) \pmod{q}$$

is a signature of knowledge of the discrete logarithm of one group element out of the list $\{y_1, ..., y_n\}$ to the base g for the message m and is denoted by $SKDL\begin{bmatrix} n \\ 1 \end{bmatrix}(g, y_1, ..., y_n, m)$.

A $SKDL\begin{bmatrix} n \\ 1 \end{bmatrix}(g, y_1, ..., y_n, m)$ can only be given if at least one of the secret keys is known. We now show how to compute such a signature. Assume that the known secret key is x_1. The prover chooses $r, s_2..., s_n, c_2, ..., c_n$ randomly in $\mathbb{Z}_q$ and computes $t_1 = g^r$ and $t_i = g^{s_i} y_i^{c_i}$ for $i = 2, ..., n$. Then he computes c_1 and s_1 according to

$$c_1 = \mathcal{H}(g\|y_1\|...\|y_n\|t_1\|...\|t_n\|m) - \sum_{i=2}^{n} c_i \pmod{q}$$

and

$$s_1 = r - x_1 c_1 \pmod{q}.$$

The prover has thereby computed $SKDL\begin{bmatrix} n \\ 1 \end{bmatrix}(g, y_1, ..., y_n, m) = (c_1, ..., c_n, s_1, ..., s_n)$.

The idea behind this is the fact that a *SKDL* can be forged if the challenge c is known before the computation of the commitment t. The verification condition of $SKDL[\begin{smallmatrix}n\\1\end{smallmatrix}]$ is a linear equation over the c_i's and therefore all but one c_i can be chosen before computing the commitments. It follows that at least for one y_i the discrete logarithm must be known and one of the partial *SKDL*'s must be true.

In [8] such proof systems were generalized to proof systems for proving the knowledge of all discrete logarithms of one out of several defined subsets of the set of public keys $\mathcal{Y} = \{y_1, ..., y_n\}$ without revealing any further information. Formally, let Γ denote a monotone set of subsets of $\mathcal{Y}$, i.e., $\Gamma \subseteq 2^{\mathcal{Y}}$. By combining n signatures of knowledge $SKDL(g, y_i)$ and a secret sharing system with access structure Γ^*, it is possible to construct a system for proving the knowledge of the discrete logarithms of all $y_i \in S$ for some $S \in \Gamma$, without saying which subset S.

Definition 3. A $2n$-tuple $(c_1, ..., c_n, s_1, ..., s_n)$ satisfying

$$\forall S' \in \Gamma^* : \quad \mathcal{H}(g\|y_1\|...\|y_n\|g^{s_1}y_1^{c_1}\|...\|g^{s_n}y_n^{c_n}\|m) = \Upsilon_{\Gamma^*}(S', \{c_i \mid y_i \in S'\})$$

is a signature of knowledge of the discrete logarithm of all $y_i \in S$ $\{y_1, ..., y_n\}$ to the base g for some $S \in \Gamma$ for the message m. Such a signature is denoted by $SKDL[\Gamma](g, y_1, ..., y_n, m)$.

This signature system is similar to the one in Definition 2; here the secret sharing scheme implies conditions on the (partial) challenges c_i by interpreting them also as shares, whereas in Definition 2 we have only one condition (i.e., a linear equation) on the challenges. If the challenges and the shares do not have the same domain, a mapping must be introduced (for further technical details see [8]). Let us show how such a signature of knowledge $(c_1, ..., c_n, s_1, ..., s_n)$ can be computed. Assume that $x_1, ..., x_j$ are the known secret keys and that $S = \{P_1, ..., P_j\} \in \Gamma$. The prover chooses $r_1, ..., r_j, s_{j+1}..., s_n, c_{j+1}, ..., c_n$ randomly in $\mathbb{Z}_q$ and computes

$$\sigma = \mathcal{H}(g\|y_1\|...\|y_n\|g^{r_1}\|...\|g^{r_j}\|g^{s_{j+1}}y_{j+1}^{c_{j+1}}\|...\|g^{s_n}y_n^{c_n}\|m) ,$$

$$\{c_1, ..., c_j\} = \Psi(\Gamma^*, \{P_{j+1}, ..., P_n\}, \{c_{j+1}, ..., c_n\}, \sigma) , \text{ and}$$

$$s_k = r_k - c_k x_k \pmod{q} \qquad \text{for } k = 1, ..., j .$$

For the definition of the function Ψ see Section 3.2.

Another primitive often used in cryptography (e.g. [5]) is a signature that the logarithms of two group elements with respect to two different bases are the same. Such a signature also implies the knowledge of these logarithms.

Definition 4. A pair (c, s) satisfying

$$c = \mathcal{H}(h\|g\|z\|y\|h^s z^c\|g^s y^c\|m)$$

is signature of equality of the discrete logarithm of the group element z with respect to the base h and the discrete logarithm of the group element y with respect to the base g for the message m. It is denoted by $SEQDL(h, g, z, y, m)$.

This signature of equality can be seen as two parallel signatures of knowledge $SKDL(h, z, m)$ and $SKDL(g, y, m)$ where the exponent for the commitment, the challenges, and the responses are the same. By using several $SKEQ$ in parallel and implying conditions on their commitments (similar as in the Definitions 2 and 3), one obtains the signature systems $SEQDL[{n \atop 1}](h, g, z_1, y_1, ..., z_n, y_n, m)$ and $SEQDL[\Gamma](h, g, z_1, y_1, ..., z_n, y_n, m)$, respectively.

Our last building block are signatures of knowledge of a representation. The respective proof systems were first introduced in [4]. Let $y = \prod_{i=1}^{n} g_i^{x_i}$ for some $x_1, ..., x_n \in \mathbb{Z}_q$.

Definition 5. A $(n+1)$-tuple $(c, s_1, ..., s_n)$ satisfying

$$c = \mathcal{H}(g_1 \| ... \| g_n \| y \| y^c \prod_{i=1}^{n} g_i^{s_i} \| m)$$

is a signature of knowledge of a representation of a group element y with respect to the bases $g_1, ..., g_n$ for the message m. It is denoted by $SKREP(g_1, ..., g_n, y, m)$.

We now show how this signature of knowledge of a representation can be calculated from $x_1, ..., x_n$. The prover chooses $r_1, ..., r_n$ at random from $\mathbb{Z}_q$, computes $t = \prod_{i=1}^{n} g_i^{r_i}$,

$$c = \mathcal{H}(g_1 \| ... \| g_n \| y \| t \| m) \quad (\text{mod } q),$$

and

$$s_i = r_i - x_i c \quad (\text{mod } q) \qquad \text{for } i = 1, ..., n$$

and thus obtains an $SKREP(g_1, ..., g_n, y, m) = (c, s_1, ..., s_n)$. If the bases g_i are chosen in a random or pseudo-random manner, computation of another than the known representation is believed to be as hard as the discrete logarithm problem and is called the *representation problem*. For further discussion see [4].

5 Construction of a Group Signature Scheme

In this section an efficient simple group signature scheme and a generalized group signature scheme are proposed. They are based on the signature systems $SEQDL[{n \atop 1}]$ and $SEQDL[\Gamma]$, respectively. These underlying systems already fulfill the properties of a group signature scheme except those related to the group manager's capability of "opening" a signature.

In the following we present efficient solutions to achieve the missing properties by using a variation of the ElGamal encryption scheme (see Section 3) and the techniques discussed in the previous section. The solutions further allow a simple way of distributing the functionality of the group manager, as will be shown in Section 6.

5.1 An Efficient Simple Group Signature Scheme

The algebraic setting is the same as in Section 4. In addition, let $z = g^\omega$ denote the public key of the group manager and ω her secret key. Each group member P_i chooses his secret key x_i randomly in $\mathbb{Z}_q$ and computes the public key $y_i = g^{x_i}$. The group's public key consists the list of all members' public keys $\mathcal{Y} = (y_1, ..., y_n)$ and is published together with the manager's public key and the system parameters.

The idea behind the scheme is that in order to sign a message, a group member encrypts one of the public keys of $\mathcal{Y} = \{y_1, ..., y_n\}$ with the public key of the group manager and proves that

- he encrypted one of the y_i's and that
- he actually knows the discrete logarithm of the encrypted key.

From this follows, that the group member must have encrypted his public key. More formally, to generate a signature of a message m, the group member P_j executes the following steps:

1. choose a randomly in $\mathbb{Z}_q$
2. encrypt y_j by computing $A = z^a$ and $B = y_j g^a$
3. calculate $(c_1, ..., c_n, s_1, ..., s_n) = SEQDL[^n_1](z, g, A, \frac{B}{y_1}, ..., A, \frac{B}{y_n}, m)$
4. calculate $(\tilde{c}, \tilde{s}) = SKDL(g, B, m)$

The computed group signature is the tuple $(A, B, c_1, ..., c_n, s_1, ..., s_n, \tilde{c}, \tilde{s})$ and can be verified by checking the correctness of $SEQDL[^n_1](z, g, A, \frac{B}{y_1}, ..., A, \frac{B}{y_n}, m)$ and $SKDL(g, B, m)$.

The first signature assures that (A, B) is the encryption of an element of the list $\mathcal{Y}$ and the second signature guarantees that the signer actually knows the discrete logarithm of the public key encrypted in (A, B). The signer thus proves indirectly his knowledge of the discrete logarithm of an element of $\mathcal{Y}$ and therefore that he is a member of the group $\mathcal{P}$. It can easily be seen that only group members can sign messages.

To open a valid signature the group manager decrypts (A, B) and immediately obtains the public key of the signer. Assume that the group member P_j has signed. By computing the signature of equality

$$SEQDL(g, z, B/(y_j), A, P_j)$$

the group manager can assure that she opened the signature correctly and that indeed P_j has issued this signature.

5.2 A generalized Group Signature Scheme

The system parameters are the same as for the simple group signature scheme. In addition to all public keys and to the system parameters, an authority structure Γ must be published.

The idea of the generalized scheme is similar to the one of the simple scheme. To sign a message m all members of an authorized coalition prove that each of them encrypted an element of $\mathcal{Y} = \{y_1, ... y_n\}$ and that they know the discrete logarithms of the encrypted values. Furthermore, they must also prove that the encrypted elements are all different. The problem with this approach is that the number of encryptions equals the size of the coalition, which should be kept secret. Therefore, the coalition must also encrypt some dummy values in order to provide n encryptions.

More formally, to generate a signature of a message m, the group members forming an authorized set $S \in \Gamma$ execute together the following steps:

1. – choose $a_1, ..., a_n$, and b_i for all i with $y_i \notin S$ randomly in $\mathbb{Z}_q$

   – for all $y_j \in S$ encrypt y_j: $A_j = z^{a_j}$, $B_j = y_j g^{a_j}$

   – for all $y_i \notin S$ encrypt g^{b_i}: $A_i = z^{a_i}$, $B_i = g^{b_i} g^{a_i}$

2. calculate $(c_1, ..., c_n, s_1, ..., s_n) = SEQDL[\Gamma](z, g, A_1, \frac{B_1}{y_1}, ..., A_n, \frac{B_n}{y_n}, m)$

3. calculate $(\tilde{c}_i, \tilde{s}_i) = SKDL(g, B_i, m||c_1||...||c_n||s_1||...||s_n)$ for $i = 1, ..., n$

Member P_j must calculate the signature $SKDL(g, B_j, m)$ and also parts of the signature in Step 2 alone in order to hide his secret key from the other members. All other computations should be performed by all group members on their own in order to assure themselves of the correctness of the outcome. The random choices in these common computations must be agreed upon by the group members in advance, for instance by choosing a random string each, committing to the string by hashing it, exchanging these commitments, then exchanging the random strings, and finally taking the XOR of all these random strings. The resulting group signature is the tuple $(A_1, B_1, ..., A_n, B_n, c_1, ..., c_n, s_1, ..., s_n, \tilde{c}, \tilde{s})$ and can be verified by checking the correctness of the signatures of knowledge $SEQDL[\Gamma](z, g, A_1, \frac{B_1}{y_1}, ..., A_n, \frac{B_n}{y_n}, m)$ and $SKDL(g, B_i, m)$ for all i.

The first signature assures that the list $((A_1, B_1), ..., (A_n, B_n))$ contains the encryptions of some $y_j \in \mathcal{Y}$ such that the corresponding P_j's form an authorized coalition. The signatures generated in Step 3 assure that the authorized coalition was really involved, i.e., that the discrete logarithms of the encrypted y_j's are known. Here, the signature of Step 2 is appended to the message in order to bind the two steps together. This prevents the reuse of a $SKDL$ in another run of the scheme.

Again, it is easy to see that the group manager can find out which coalition provided the signature by checking the validity of the signature and decrypting all pairs (A_j, B_j). Note that a coalition cannot encrypt a public key of a member $P_i \notin S$ not participating in the signing because then they could not provide the corresponding signature in Step 3 and therefore the group signature would not be valid. By computing the signatures of equality

$$SEQDL(g, z, B/(y_j), A, P_j)$$

for all P_j having participated in the signing, the group manager can assure that she opened the signature correctly.

Remark. The signature can be made shorter if all $\tilde{c}_i$ are the same, i.e., all signatures $SKDL(g, B_i, m)$ are merged and are verified simultaneously by checking the equation

$$\tilde{c} = \mathcal{H}(g\|B_1\|...\|B_n\|g^{\tilde{s}_1} B_1^{\tilde{c}}\|...\|g^{\tilde{s}_n} B_n^{\tilde{c}}\|m) \ .$$

Of course, the signatures must then be computed in parallel and $\tilde{c}$ calculated accordingly. This choice also binds Steps 2 and 3 together, i.e., the concatenation of the first signature to the message is not needed in this case. This is applied in the following example.

5.3 An Example for a Threshold Group Signature Scheme

In this section we give an example for a generalized group signature scheme with a threshold authority structure. Let k be the minimum number of members that must cooperate in order to sign and let $f(x) = \sum_{i=0}^{k-1} \alpha_i x^i$ denote the polynomial of a secret sharing scheme with threshold k as described in Section 3.2. To generate a signature of a message m, the group members forming an authorized set $\mathcal{S}$, i.e., $|\mathcal{S}| \geq k$, execute the steps below. In Step 2 it is indicated when the calculations must be performed by a specific member of the coalition, whereas in Step 3, all calculations for a specific j must be performed by member P_j for $P_j \in \mathcal{S}$. All other computations should by done by the coalition members on their own using the agreed-on random string.

1.   – choose $a_1, ..., a_n$, and b_i for all i with $y_i \notin \mathcal{S}$ randomly in $\mathbb{Z}_q$

    – for all $y_j \in \mathcal{S}$, member P_j encrypts y_j: $A_j = z^{a_j}$, $B_j = y_j g^{a_j}$

    – for all $y_i \notin \mathcal{S}$ encrypt g^{b_i}: $A_i = z^{a_i}$, $B_i = g^{b_i} g^{a_i}$

2. compute $SEQDL[\Gamma](z, g, A_1, \frac{B_1}{y_1}, ..., A_n, \frac{B_n}{y_n}) = (\alpha_0, ..., \alpha_{k-1}, s_1, ..., s_n, m)$:

    – for all $y_j \in \mathcal{S}$, member P_j chooses r_j randomly in $\mathbb{Z}_q$ and calculates $t_{z,j} = z^{r_j}$ and $t_{g,j} = g^{r_j}$

    – for all $y_i \notin \mathcal{S}$ choose r_i and c_i randomly in $\mathbb{Z}_q$ and compute $t_{z,i} = z^{r_i} A_i^{c_i}$ and $t_{g,i} = g^{r_i} (\frac{B_i}{y_i})^{c_i}$

    – $c = \mathcal{H}(z\|g\|A_1\|\frac{B_1}{y_1}\|...\|A_n\|\frac{B_n}{y_n}\|t_{z,1}\|t_{g,1}\|...\|t_{z,n}\|t_{g,n}\|m)$

    – choose $\alpha_0, ..., \alpha_{k-1}$ such that $f(i) = c_i \pmod q$ for all $i|y_i \notin \mathcal{S}$ and $f(0) = c \pmod q$

    – for all $y_j \in \mathcal{S}$, member P_j computes $s_j = r_j - f(j)a_j \pmod q$

    – for all $y_i \notin \mathcal{S}$ set $s_i = r_i$

3. calculate the combined signatures $SKDL(g, B_i, m) = (\tilde{c}, \tilde{s}_1, ..., \tilde{s}_n)$:

    – for $i = 1, ..., n$ choose $\tilde{r}_i$ randomly in $\mathbb{Z}_q$

    – for $i = 1, ..., n$ compute $\tilde{t}_i = g^{\tilde{r}_i}$

    – $\tilde{c} = \mathcal{H}(g\|B_1\|...\|B_n\|\tilde{t}_1\|...\|\tilde{t}_n\|m)$

$$- \text{ for } i = 1, ..., n \text{ compute } \tilde{s}_i = \begin{cases} \tilde{r}_i - \tilde{c}(x_i + a_i) & (\text{mod } q) & \text{if } y_i \in S \\ \tilde{r}_i - \tilde{c}(b_i + a_i) & (\text{mod } q) & \text{if } y_i \notin S \end{cases}$$

The group signature of m is the tuple $(\alpha_0, ..., \alpha_{k-1}, s_1, ..., s_n, \tilde{c}, \tilde{s}_1, ..., \tilde{s}_n)$. Note that instead of all c_i's, the values $\alpha_0, ..., \alpha_{k-1}$ are included in the signature. This makes the signature shorter but not less secure because c and all c_i's are uniquely determined by $\alpha_0, ..., \alpha_{k-1}$.

The group signature can be verified by checking the following equations:

$$\alpha_0 = \mathcal{H}\left(z\|g\|A_1\left\|\frac{B_1}{y_1}\right\|...\|A_n\left\|\frac{B_n}{y_n}\right\|z^{s_1}A_1^{f(1)}\left\|g^{s_1}\left(\frac{B_1}{y_1}\right)^{f(1)}\right\|...\right.$$
$$\left....\left\|z^{s_n}A_n^{f(n)}\right\|g^{s_n}\left(\frac{B_n}{y_n}\right)^{f(n)}\right\|m\right)$$

and

$$\tilde{c} = \mathcal{H}(g\|B_1\|...\|B_n\|g^{\tilde{s}_1}B_1^{\tilde{c}}\|...\|g^{\tilde{s}_n}B_n^{\tilde{c}}\|m)$$

where

$$f(x) = \alpha_0 + \alpha_1 x + ... + \alpha_{k-1}x^{k-1} \quad (\text{mod } q).$$

5.4 Security and Efficiency Considerations

Let us shortly discuss the security properties of the generalized group signature scheme (which hold also for the simple scheme).

Non-members cannot sign: If a non-member would be able to forge a group signature, he would also be able to forge Schnorr signature.

Signatures are unlinkable and anonymous: Unlinkability follows from the properties of $SEQDL[\Gamma]$ and from the fact that the y_i's are randomly encrypted, which also guarantees anonymity.

Authorized coalitions cannot sign on behalf of another coalition: Clearly, a coalition cannot sign on behalf of a coalition that includes members that are not included in itself. If a coalition contains an true authorized subset, some members try to make it appear as if they were not involved in the signing. This attack is prevented by the mutually agreed random string.

The group manager cannot falsely accuse members: This is assured by the proof the group manager must provide as evidence in the procedure **open**.

With regard to efficiency, all algorithms except **open** have efficiency linear in the number of group members. The size of the group's public key and the length of signatures are also linear in the number of group members. The algorithm **open** is independent of the group's size (however, finding the identity of a signer given his key requires a look up in a database).

Comparing the second scheme of [7] and our simple group signature scheme, it turns out, that our scheme is approximately four times more efficient in terms of computations of the signer and signatures are about the same ratio shorter. Furthermore, in [7] the algorithm **open** has an efficiency that is linear in the group's size.

6 Extensions

In this section we show how the functionality of the group manager can be shared among several parties (e.g. among the group members) and present a method for reducing the size of the group's public key.

6.1 Sharing the Functionality of the Group Manager

To obtain higher security against fraudulent opening of signatures, the capability of the group manager can be shared among several managers according to an access structure such that only predefined subsets of the managers are able to cooperatively open a signature.

To achieve this, the group manager's secret key ω must be shared among the managers and exponentiation with ω^{-1} must be possible in a distributed manner without leaking information about the shares.

For an access structure with threshold t and k managers, a realization is based on Shamir's secret sharing scheme [18] and Feldman's verifiable secret sharing scheme [11]. A solution to powering with ω^{-1} is described in [12] for the case $t < k/2$ if all managers are honest and for the case $t < k/3$ if up to t of the managers may be actively cheating.

More general access structures are possible if exponentiation with ω^{-1} is avoided, i.e., if signatures are opened as follows. Compute B^ω/A and then compare the result with the list $\{y_1^\omega, ..., y_n^\omega\}$. This list can be (pre-)computed (without revealing ω) during the setup of the system[1]. Then, for instance the monotone circuit construction of Benaloh and Leichter [1] can be applied over $GF(q)$ and powering B with ω can be achieved by multiplying all B^{ω_j}, where ω_j denotes the share of a manager in a qualified set.

6.2 Reducing the Size of the Group's Public Key

The size of the group's public key can be reduced using a technique proposed by Blom for public key distribution [2]. Let Φ be a publicly known generator matrix of an (n, k) MDS code over $\mathbb{Z}_q$. The group's public key now becomes $\{y_1, ..., y_k\}$. The public key of member P_j is then computed as

$$\tilde{y}_j = \prod_{i=1}^{k} y_i^{\phi_{ij}},$$

where ϕ_{ij} denotes the element of Φ in row i and column j. These public keys are then used in Step 2 of the signature generating procedure. The secret keys

---

[1] The computation of such a list can be avoided if normal ElGamal encryption is used in our group signature scheme. Then the signature systems in Step 2 and 3 must be adjusted: in Step 2 the A_i's instead of the B_i's must be divided by the respective y_i's. and in Step 3 the signatures SKDL(g, B_i, m) must be replaced by signatures SKREP(g, z, B_i, m). This change would make the signatures somewhat longer, but the public key of the signer could be computed directly as A_i/B_i^ω.

of the individual group members are computed similarly. This method has the disadvantages that a trusted third party is needed to compute the group's public and secret keys, and that if more than k group members collude, they can find out all secret keys and therefore sign on behalf of any authorized set. Hence there exists a trade-off between the size of the group's public key and the security.

7 Open Problems

In all previously proposed schemes, as well as in our scheme, the size of the group's public key is linear in the number of group members. It is an open problem to construct a group signature scheme where the size of the public key and the amount of computation for signing and verifying does not depend on the size of the group (the only proposed schemes [13,16] with fixed size public keys were broken).

Acknowledgments

It is has been a pleasure to discuss group signatures and the results of this paper with Christian Cachin, Ronald Cramer, Ueli Maurer, and Markus Stadler. These discussions greatly improved the paper. The comments of the anonymous referees were also welcomed.

The author is supported by the Swiss Commission for Technology and Innovation (KTI) and by the Union Bank of Switzerland.

References

1. J. Benaloh and J. Leichter. Generalized secret sharing and monotone functions. In S. Goldwasser, editor, *Advances in Cryptology — CRYPTO '88*, volume 403 of *Lecture Notes in Computer Science*, pages 27–35. Springer-Verlag, 1990.
2. R. Blom. An optimal class of symmetric key generation systems. *Proc. EUROCRYPT'84, Lecture Notes in Comp. Sc., vol. 209, New York, NY: Springer Verlag*, pages 335–338, 1985.
3. C. Boyd. Digital multisignatures. In H. J. Beker and F. Piper, editors, *Cryptography and Coding*, pages 241–246. The Institute of Mathematics and its Applications Conference Series, Oxford Science Publications, 1989.
4. S. Brands. An efficient off-line electronic cash system based on the representation problem. Technical Report CS-R9323, CWI, Apr. 1993.
5. D. Chaum and T. Pedersen. Wallet databases with observers. In E. F. Brickell, editor, *Advances in Cryptology — CRYPTO '92*, volume 740 of *Lecture Notes in Computer Science*, pages 89–105. Springer-Verlag, 1993.
6. D. Chaum and E. van Heyst. Group signatures. In D. W. Davies, editor, *Advances in Cryptology — EUROCRYPT '91*, volume 547 of *Lecture Notes in Computer Science*, pages 257–265. Springer-Verlag, 1991.
7. L. Chen and T. P. Pedersen. New group signature schemes. In A. D. Santis, editor, *Advances in Cryptology — EUROCRYPT '94*, volume 950 of *Lecture Notes in Computer Science*, pages 171–181. Springer-Verlag, 1995.

8. R. Cramer, I. Damgård, and B. Schoenmakers. Proofs of partial knowledge and simplified design of witness hiding protocols. In Y. G. Desmedt, editor, *Advances in Cryptology - CRYPTO '94*, volume 839 of *Lecture Notes in Computer Science*, pages 174–187. Springer Verlag, 1994.

9. R. Croft and S. Harris. Public key cryptography and re-usable shared secrets. In H. J. Beker and F. Piper, editors, *Cryptography and Coding*, pages 189–201. The Institute of Mathematics and its Applications Conference Series, Oxford Science Publications, 1989.

10. T. ElGamal. A public key cryptosystem and a signature scheme based on discrete logarithms. In G. R. Blakley and D. Chaum, editors, *Advances in Cryptology - CRYPTO '84*, volume 196 of *Lecture Notes in Computer Science*, pages 10–18. Springer Verlag, 1985.

11. P. Feldman. A practical scheme for non-interactive verifiable secret sharing. In *Proc. 28th IEEE Symp. Found. Comp. Sc.*, pages 427–437, 1987.

12. R. Gennaro, S. Jarecki, H. Krawczyk, and T. Rabin. Robust threshold DSS signatures. In U. Maurer, editor, *Advances in Cryptology — EUROCRYPT '96*, volume 1070 of *Lecture Notes in Computer Science*, pages 354–371. Springer Verlag, 1996.

13. S. J. Kim, S. J. Park, and D. H. Won. Convertible group signatures. In K. Kim and T. Matsumoto, editors, *Advances in Cryptology — ASIACRYPT '96*, volume 1163 of *Lecture Notes in Computer Science*, pages 311–321. Springer Verlag, 1996.

14. M. Mambo, K. Usuda, and E. Okamoto. Proxy signatures for delegating signing operation. In *3rd ACM Conference on Computer and Communicatons Security*, pages 48–57, New Delhi, Mar. 1996. acm press.

15. K. Ohta and T. Okamoto. A digital multisignature scheme based on the Fiat-Shamir scheme. In H. Imai, R. L. Rivest, and T. Matsumoto, editors, *Advances in Cryptology — ASIACRYPT '91*, volume 739 of *Lecture Notes in Computer Science*, pages 139–148. Springer-Verlag, 1993.

16. S. J. Park, I. S. Lee, and D. H. Won. A practical group signature. In *Proceedings of the 1995 Japan-Korea Workshop on Information Security and Cryptography*, pages 127–133, Jan. 1995.

17. C. P. Schnorr. Efficient signature generation for smart cards. *Journal of Cryptology*, 4(3):239–252, 1991.

18. A. Shamir. How to share a secret. *Commun. ACM*, 22(11):612–613, Nov. 1979.

Collision-Free Accumulators and Fail-Stop Signature Schemes Without Trees*

Niko Barić[1] and Birgit Pfitzmann[2]

[1] dvg Hannover, Postfach 91 02 40, D-30422 Hannover, Germany
[2] Universität Dortmund, Informatik 6, D-44221 Dortmund, Germany;
email pfitzb@ls6.informatik.uni-dortmund.de

Abstract. One-way accumulators, introduced by Benaloh and de Mare, can be used to accumulate a large number of values into a single one, which can then be used to authenticate every input value without the need to transmit the others. However, the one-way property does is not sufficient for all applications.

In this paper, we generalize the definition of accumulators and define and construct a collision-free subtype. As an application, we construct a fail-stop signature scheme in which many one-time public keys are accumulated into one short public key. In contrast to previous constructions with tree authentication, the length of both this public key and the signatures can be independent of the number of messages that can be signed.

1 Introduction

The security of digital signature schemes depends on so-called computational assumptions, e.g., the factoring assumption. If somebody can break the assumption on which the system is based, and if he can therefore get the private key of the signer, he can construct signatures on messages chosen by himself. The signer cannot prove that she did not sign those messages herself.

This disadvantage was overcome with the introduction of "fail-stop" signature schemes, e.g., [WaPf90, PfWa90, HePe93, PePf97]. With these schemes, the signer can produce a so-called proof of forgery to demonstrate that she did not sign a message. This proof shows that the computational assumption has been broken (fail) and that the system should therefore not be used any longer (stop).

Most of the currently known basic constructions of fail-stop signature schemes (FSS schemes) can only be used to sign one single message. FSS schemes for more than one message have been constructed based on these one-time FSS schemes by using tree authentication to authenticate the public one-time keys. Consequently, the length of signatures in such a scheme grows logarithmically in the number of messages that can be signed. The question whether this can be

---

* Work done while both authors were at the University of Hildesheim. Supported by the DFG (German Research Foundation). A preliminary version was available as [Pfit94], more details can be found in [Bari96].

W. Fumy (Ed.): Advances in Cryptology - EUROCRYPT '97, LNCS 1233, pp. 480–494, 1997.

avoided was also the main gap between known lower and upper bounds on the complexity of fail-stop signature schemes [HePP93].

The accumulators presented in [BeMa94] seem to be a solution to this problem: a large number of values is accumulated into one value z. Later on, for authentication of one of those values, y, an additional value is computed that will authenticate y with respect to z. The length of z and the additional value can be independent of the number of values to be accumulated. If we use this for FSS schemes and accumulate all the public one-time keys, the length of the resulting public key and the signatures can be independent of the number of messages.

The accumulators defined in [BeMa94] have only a one-way property, i.e., given an output, it is hard to find a suitable input. Unfortunately, this is not enough for an FSS scheme, because the adversary may be able to choose the one-time public keys (i.e., the values to be accumulated), and thus to some extent the accumulated output, himself. Therefore we define and construct collision-free accumulators. We take the opportunity to generalize the accumulators defined in [BeMa94] to contain only those properties that are needed for our purpose and also to include newer accumulators from [Nybe96a, Nybe96b]. The new collision-free accumulators are then included into a modular FSS scheme. Thus now we really have a scheme where the length of both the public key and the signatures is independent of the number of messages.

The goal of constructing schemes without trees is similar to recent efforts with non-fail-stop provably secure signatures to shorten the signatures by flat trees [DwNa94, CrDa96], but the measures developed there cannot be used for FSS schemes.

1.1 Organization of this Paper

In Section 2, we present our definitions and constructions of accumulators. In Section 3, we describe conversion algorithms as an interface between the one-time FSS scheme and the accumulator which is used to authenticate the individual public one-time keys. The general construction of our accumulator FSS scheme is given in Section 4. Two example accumulator FSS schemes follow in Section 5.

2 Accumulators

Accumulators were introduced in [BeMa94] as a new way of "summarizing" a large number N of values in one value. The accumulators as defined in [BeMa94] have some properties we do not need for our purposes, so we generalize their definition a little. Then we define some subtypes of accumulators with different levels of security. Nevertheless, the accumulators as given in [BeMa94] are an important subtype that we call *elementary accumulators* (see Section 2.2).

2.1 General Accumulators

Definition 1. A *family a of accumulators* has the following components:

– Sets $accu\_keys(k, N)$, which contain all possible keys for the security parameter k and the number N of values to be accumulated, and a probabilistic polynomial-time algorithm $accu\_gen(k, N)$ that chooses an accumulator key n from $accu\_keys(k, N)$. If the choice is uniformly random, we often simply write $\in_R$.

  In our examples, $accu\_keys(k, N)$ is independent of N.

– Sets Y_n containing the suitable inputs for an accumulator key n.

– A probabilistic polynomial-time algorithm $accu\_eval$ which, on input an accumulator key n and N values $y_1, \ldots, y_N \in Y_n$, outputs a value z and an auxiliary value aux, which will be used by the other algorithms.

  We write $a_n(y_1, \ldots, y_N)$ instead of $accu\_eval(n, y_1, \ldots, y_N)$.

  Every execution of $accu\_eval$ with the same input $(n, y_1, \ldots, y_N)$ must yield the same output z.

– A probabilistic polynomial-time algorithm $auth$ that, on input n, y_i, and aux, computes a value $accu_i$ from a set $Accu_n$, which is needed to authenticate y_i.

  We write $auth_n(y_i, aux)$ instead of $auth(n, y_i, aux)$.

– A polynomial-time algorithm $authentic$ which, on input $(n, z, y_i, accu_i)$, checks whether $y_i \in Y_n$ together with $accu_i \in Accu_n$ is authenticated by z. If so, the output is ok, otherwise not_ok.

  We write $authentic_n(z, y_i, accu_i)$ instead of $authentic(n, z, y_i, accu_i)$.

Additionally, there must be two polynomial-time algorithms: one that, on input n and y, checks whether $y \in Y_n$, and one that, on input n and $accu$, checks whether $accu \in Accu_n$. Finally, we require that every y_i in the input of a_n can be authenticated by the output of a_n, formally:

$$
\begin{aligned}
&\forall k \, \forall N \, \forall n \in accu\_keys(k, N) \, \forall (y_1, \ldots, y_N) \in Y_n^N: \\
&\quad \text{If } (z, aux) \leftarrow a_n(y_1, \ldots, y_N) \\
&\quad \text{then } \forall i \in \{1, \ldots, N\}: \\
&\qquad authentic_n\big(z, y_i, auth_n(y_i, aux)\big) = ok.
\end{aligned}
$$

In [BeMa94], a one-way property is defined for accumulators. Generalized to our definition, it means that it is hard for an adversary who is given values $(y_1, \ldots, y_N)$, their accumulation result z, and another value y' to find a value $accu'$ that authenticates y' with respect to z.

  That article also informally considers a slightly stronger property that we call *strongly one-way*. It means that given only $(y_1, \ldots, y_N)$ and z, it is hard to find a pair $(y', accu')$ such that $authentic_n(z, y', accu') = ok$ with $y' \notin \{y_1, \ldots, y_N\}$. I.e., now the attacker can choose the value y' himself. The importance of a strong one-way property was also recognized in [Nybe96a].

  For our accumulator FSS scheme, we need an even stronger property, because the adversary might be able to choose all the public one-time keys that are to be accumulated, i.e., not even the values $y_1, \ldots, y_N$ are now given.

Definition 2. A family a of accumulators is *N-times collision-free* for $N \geq 1$ if it is hard to find $y_1, \ldots, y_N$, another value y' and $accu'$ such that y' is authenticated by $accu'$ and $a_n(y_1, \ldots, y_N)$: for all probabilistic polynomial-time adversaries $\tilde{A}$, all $c > 0$, and all sufficiently large k:

$$P\Big(authentic_n(z, y', accu') = ok \wedge y' \notin \{y_1, \ldots, y_N\}$$
$$\wedge\, y', y_1, \ldots, y_N \in Y_n \wedge accu' \in Accu_n ::$$
$$n \leftarrow accu\_gen(k, N);$$
$$(accu', y', y_1, \ldots, y_N) \leftarrow \tilde{A}(k, N, n);$$
$$(z, aux) \leftarrow a_n(y_1, \ldots, y_N)\Big) \leq \frac{1}{k^c}.$$

Definition 3. A family a of accumulators is *collision-free* if a is N-times collision-free for all $N \geq 1$.

2.2 Elementary Accumulators

In [BeMa94], accumulators were defined as functions $h_n \colon X_n \times Y_n \to X_n$, where n is again an accumulator key. With repeated use as in

$$z = h_n\Big(\cdots h_n\big(h_n(x, y_1), y_2\big), \cdots y_N\Big),$$

where the result of one application of h_n is inserted as the first argument in the next application of h_n, all $y_1, \ldots, y_N \in Y_N$ are accumulated to a value $z \in X_n$ given an initial value $x \in Start_n \subseteq X_n$.

With such a function h_n, we can create an accumulator $a_{(n,x)}$ according to the general definition, where the initial value x is part of the key of a, as follows:

$$(z, aux) = a_{(n,x)}(y_1, \ldots, y_N)$$

with z as above and $aux = (x, y_1, \ldots, y_N)$. We use $(x, y_1, \ldots, y_N)$ as the auxiliary output, so that we can use it for the computation of the values $accu_i$.

In [BeMa94], such a function h_n has to be *quasi-commutative*, i.e.,

$$h_n\big(h_n(x, y_1), y_2\big) = h_n\big(h_n(x, y_2), y_1\big) \text{ for all } x \in X_n \text{ and } y_1, y_2 \in Y_n.$$

We do not need this property for our accumulator FSS scheme, but if one has a function with this property, one can easily construct algorithms to create and verify the values $accu_i$ [BeMa94]:

$$auth_n\big(y_i, (x, y_1, \ldots, y_N)\big) = h_n\Big(\ldots h_n\big(h_n(\ldots h_n(x, y_1), \ldots y_{i-1}), y_{i+1}\big), \ldots y_N \Big)$$

and

$$authentic_n(z, y_i, accu_i) = ok \text{ iff } z = h_n(accu_i, y_i).$$

In this case, the list of all values $accu_i$ can be computed with $\mathcal{O}(N \cdot \log_2 N)$ applications of h_n with a tree-like evaluation. This can be done offline after z has been published.

2.3 Examples

In the following two subsections, we give two examples of accumulators. Both are based on the elementary accumulator given in [BeMa94], but with some modifications to fulfill the collision-freeness needed for the accumulator FSS scheme.

Another elementary and strongly one-way accumulator is described in [Nybe96a, Nybe96b]. In short, it uses a hash function h that generates a long random output o_i of fixed length $r \cdot d$ for every input y_i, where r and d are two security parameters. Then o_i is transformed into a bitstring b_i of length r that has far more 1's than 0's. To accumulate the values $(y_1, \ldots, y_N)$, the corresponding strings b_i are multiplied modulo 2 coordinatewise. In the result, a bit can be zero only if at least one b_i has a zero bit at the same place. The main advantage of this accumulator is the absence of any trapdoor information, whereas in the following accumulators based on the RSA assumption, someone knows the factors of the RSA modulus. A disadvantage is its long output, too long for the public key of an FSS scheme.

RSA Accumulator Without Random Oracle. The first example is almost the same accumulator as presented in [BeMa94], based on the elementary accumulator function $h_n(x, y) = x^y \bmod n$.

Definition 4. The following family a^{RSA} is called *RSA accumulator without random oracle*:

- $accu\_keys^{\text{RSA}}(k, N) := \{(n, x) \mid n \in RSA\_Mod(k) \wedge x \in \mathbb{Z}_n\}$
- $Y_{(n,x)}^{\text{RSA}} := \{y \mid y < n \wedge y \text{ prime}\}$
- $a_{(n,x)}^{\text{RSA}}(y_1, \ldots, y_N) := x^{y_1 \cdots y_N} \bmod n$
- $auth_{(n,x)}^{\text{RSA}}(y_i, (x, y_1, \ldots, y_N)) := x^{y_1 \cdots y_{i-1} y_{i+1} \cdots y_N} \bmod n$
- $authentic_{(n,x)}^{\text{RSA}}(z, y, accu) := ok$ iff $accu^y \equiv z \pmod{n}$

Here, $RSA\_Mod(k)$ is the set of RSA moduli of length k [RSA78]. The difference to the original accumulator is the restriction of the input domain to prime numbers. In addition, to prove collision-freeness, we have to make a stronger RSA assumption.

Assumption (strong RSA assumption). For all probabilistic polynomial-time algorithms $\tilde{A}$, all $c > 0$, and all sufficiently large k,

$$P\big(y^e \equiv x \pmod{n} \wedge e \text{ prime} \wedge e < n ::$$
$$n \in_R RSA\_Mod(k); x \in_R \mathbb{Z}_n; (y, e) \leftarrow \tilde{A}(n, x)\big) \leq k^{-c}.$$

Thus the adversary $\tilde{A}$ is given n and x as in a usual RSA assumption, but he may choose the exponent e for which he extracts the root. We are neither aware of any corroboration that it should be hard, nor can we break it. Four obvious attacks do not work, i.e., they are equivalent to breaking some other problem believed to be hard:

- If the adversary chooses a random e first, he has to break RSA.
- If he chooses a random y first, he has to compute a discrete logarithm.
- If he tries to find d and e with $y = x^d$ and $(x^d)^e = x$, then $\mathrm{ord}(x)$ divides $f := de - 1$, where $\mathrm{ord}(x)$ is the smallest $i > 0$ with $x^i \equiv 1 \pmod{n}$. He can then also break RSA for the same n and x: Let a random public exponent e' be given. It is sufficient to consider the case where e' is prime and no factor of f. Then we set $d' := e'^{-1} \bmod f$ and obtain $(x^{d'})^{e'} \equiv x \pmod{n}$ because $\mathrm{ord}(x)$ divides $d'e' - 1$.
- The attacker could try to choose special values e for which RSA would be easier to break. However, no such exponents seem to be known. There are attacks for short secret exponents [Wien90], but our e corresponds to the public exponent, and we see no way for the attacker to influence the corresponding secret exponent. A well-known attack on short public exponents [Håst86] only applies to situations where the attacker sees several messages encrypted with that exponent using different moduli. Similarly, the new class of attacks on short public exponents in [CFPR96] only applies to situations where the attacker sees the ciphertexts of several messages with a known polynomial relationship, encrypted using the same modulus.

Theorem 5. *Under the strong RSA assumption, a^{RSA} is collision-free.*

Proof sketch. An adversary who finds a collision in a^{RSA} for given n, x, i.e., who finds $y_1, \ldots, y_N$, y', and $accu'$ with

$$accu'^{y'} \equiv x^{y_1 \cdots y_N} \pmod{n},$$

can break the strong RSA assumption as follows: Let $e := y'$ and $r := y_1 \cdots y_N$. Now the e-th root y of x can be constructed as in [Sham83, BeMa94]: Compute $a, b \in \mathbb{Z}$ with $ar + by' = 1$ with the extended Euclidean algorithm (this is possible because y' is prime) and let $y := accu'^a x^b$. Thus

$$y^e \equiv accu'^{ay'} x^{by'} \equiv x^{ra + by'} \equiv x \pmod{n}.$$

$\square$

RSA Accumulator With Random Oracle. The second example uses, as the name of the first suggests, a random oracle Ω [BeRo93]. Whenever asked to compute $\Omega(y)$ for a new value y, the oracle generates a random number r as its answer, and it stores all previous pairs (y, r) so that it answers with the same r if asked the same y again.

In practice, one replaces the random oracle by an efficient hash function. Of course, this replacement is only a heuristic.

By using a random oracle, we can construct an accumulator that is collision-free under the normal RSA assumption. The elementary accumulator uses the function

$$h_{(n, \Omega, l)}^{\mathrm{RSA}\,\Omega}\big(x, (y, dist)\big) := x^{2^l \Omega(y) + dist} \bmod n.$$

We do not use $\Omega(y)$ directly, because in the proof we will need that the exponents are prime numbers. So we append l bits such that $2^l \Omega(y) + dist$ is prime. Of course, this might not be possible for all values of y, so we accept only those y's as input for which a suitable $dist$ exists.

Definition 6. Let a family M of sets M_k be given where membership is decidable in polynomial time. It contains the values that we really want to accumulate for each security parameter k. The following family is called *RSA accumulator with random oracle* (for M):

- $accu\_keys^{RSA\Omega}(k, N) := \{(n, \Omega, l, x) \mid n \in RSA\_Mod(k + l)$
  $\wedge\ \Omega \in \{f \mid f\colon M \to \mathbb{Z}_{n\,\mathrm{div}\,2^l}\} \wedge l = \lceil \log_2 2k \rceil \wedge x \in \mathbb{Z}_n\}$
- $Y^{RSA\Omega}_{(n,\Omega,l,x)} = \{(y, dist) \mid y \in M_k \wedge dist \in \mathbb{Z}_{2^l} \wedge 2^l \Omega(y) + dist \text{ prime}\}$, i.e.,
  the values that we actually accumulate are pairs of a value that we want to accumulate and a suffix that turns its hash value into a prime number.
- $a^{RSA\Omega}_{(n,\Omega,l,x)}((y_1, dist_1), \ldots, (y_N, dist_N)) :=$
  $x^{(2^l \Omega(y_1) + dist_1) \cdots (2^l \Omega(y_N) + dist_N)} \bmod n$
- $auth^{RSA\Omega}_{(n,\Omega,l,x)}\big((y_i, dist_i), (x, (y_1, dist_1), \ldots, (y_N, dist_N))\big) :=$
  $x^{(2^l \Omega(y_1) + dist_1) \cdots (2^l \Omega(y_{i-1}) + dist_{i-1}) \cdot (2^l \Omega(y_{i+1}) + dist_{i+1}) \cdots (2^l \Omega(y_N) + dist_N)} \bmod n$
- $authentic^{RSA\Omega}_{(n,\Omega,l,x)}(z, (y, dist), accu) := ok$ iff $accu^{2^l \Omega(y) + dist} \equiv z \pmod{n}$

Theorem 7. *This accumulator is collision-free under the normal RSA assumption.*

Proof sketch. We have to show that for all N, all probabilistic polynomial-time algorithms $\widetilde{A}$, all $c > 0$, and all sufficiently large k,

$$P\Big(accu'^{2^l \Omega(y') + dist'} \equiv x^{(2^l \Omega(y_1) + dist_1) \cdots (2^l \Omega(y_N) + dist_N)} \pmod{n}$$
$$\wedge (y', dist') \notin \{(y_1, dist_1), \ldots, (y_N, dist_N)\}$$
$$\wedge (y', dist'), (y_1, dist_1), \ldots, (y_N, dist_N) \in \{(y, dist) \mid y \in M_k$$
$$\wedge\ dist \in \{0, \ldots, 2^l - 1\} \wedge 2^l \Omega(y) + dist \text{ prime}\}$$
$$\wedge\ accu' \in \mathbb{Z}_n ::$$
$$l := \lceil \log_2 2k \rceil; n \in_R RSA\_Mod(k + l);$$
$$\Omega \in_R \{f \mid f\colon M_k \to \mathbb{Z}_{n\,\mathrm{div}\,2^l}\}; x \in_R \mathbb{Z}_n;$$
$$\big(accu', (y', dist'), (y_1, dist_1), \ldots, (y_N, dist_N)\big) \leftarrow \widetilde{A}^\Omega(k, N, n, l, x)\Big)$$
$$\leq \frac{1}{k^c},$$

where $\widetilde{A}^\Omega$ means $\widetilde{A}$ with access to the oracle Ω. Assume that an algorithm $\widetilde{A}$ contradicts this inequality for some N. We can then construct an algorithm $\widetilde{A}^\Omega_1$ that calls $\widetilde{A}^\Omega$ and, whenever that is successful, sets

$$r' := 2^l \Omega(y') + dist' \text{ and}$$
$$r_i := 2^l \Omega(y_i) + dist_i \text{ for } i = 1, \ldots, N,$$

and computes the r'-th root of x using the extended Euclidean algorithm for these values as in the proof of the previous theorem. The only exception is if r' equals one of the r_i's. Then an oracle collision has been found, which can only happen with very small probability. Hence it is sufficient to prove for all probabilistic polynomial-time algorithms $\widetilde{A}_1$, all $c > 0$, and all sufficiently large k,

$$
P\Big(y^{r'} = x \wedge r' \text{ prime} \wedge r' < n \wedge dist' < 2^l ::
$$
$$
l := \lceil \log_2 2k \rceil; n \in_R RSA\_Mod(k + l);
$$
$$
\Omega \in_R \{f \mid f \colon M_k \to \mathbb{Z}_{n \text{ div } 2^l}\}; x \in_R \mathbb{Z}_n;
$$
$$
(y, y', dist') \leftarrow \widetilde{A}_1^\Omega(k, N, n, l, x); r' := 2^l \Omega(y') + dist'\Big) \leq \frac{1}{k^c}.
$$

Without loss of generality, we can assume that $\widetilde{A}_1$ has asked the oracle for $\Omega(y')$. The number of values that $\widetilde{A}_1$ asks for is bounded by a polynomial $Q(k)$. Whatever strategy $\widetilde{A}_1$ uses in choosing its oracle queries, it amounts to the same thing as if it were given a list of $Q(k)$ random numbers ρ and had to select r' among the numbers $2^l \rho + dist$. Thus this new adversary $\widetilde{A}_2$ is given a list of $Q(k) \cdot 2^l$ exponents and has to extract a root for at least one of them. If this were possible with non-negligible probability, it would also be possible to extract an e-th root for one given random e. For this, a new adversary $\widetilde{A}_3$, given e, inserts $(e \text{ div } 2^l)$ at a random place into a list of $Q(k) - 1$ random numbers and appends the values $dist$. $\widetilde{A}_3$ calls $\widetilde{A}_2$, and with a probability smaller by the factor $Q(k) \cdot 2^l$ it gets the e-th root of x (recall that $2^l \approx k$). $\qquad \Box$

The proof also shows another result that is interesting in practice, where the function used instead of the oracle is not perfect: To find an accumulator collision, one at least either has to either find a collision of this function (where collision-freeness is a much weaker requirement than "being like an oracle") or to break the strong RSA assumption.

3 Conversion Algorithm

We want to use collision-free accumulators as defined in the previous section to accumulate the public one-time keys in an FSS scheme. But what if the public one-time keys are not suitable as input for the accumulator? For example, the RSA accumulator without random oracle as defined in Section 2.3 needs prime numbers as input, and none of the known FSS schemes uses prime numbers as public one-time keys. Hence one has to convert the public one-time keys to prime numbers that can then be accumulated by the accumulator.

Of course, such a conversion could be done within the underlying one-time FSS scheme or within the accumulator. But then one has to prove their security again. Thus it seems better to use a simple conversion algorithm that has no effect on the security as an interface between the FSS scheme and the accumulator. In this way, we get a general modular construction for which one can use any collision-free accumulator and any one-time FSS scheme provided that one finds a conversion algorithm for them. As examples, we present two instantiations in Section 5. For this purpose, we use a family Λ of conversion algorithms, which has the following components:

- A function *calc_pars* that computes the security parameters k' for the accumulator and (k^*, σ^*) for the underlying FSS scheme if given as input (k, σ, N), the security parameters of the desired accumulator FSS scheme and the number of messages to be signed. The output must fulfill

$$k', k^* \geq k \text{ and } \sigma^* \geq \sigma.$$

- A polynomial-time algorithm Λ_gen which, on input k^*, σ^* and an accumulator key n, computes a key *par* specifying an individual member of Λ.
- A probabilistic polynomial-time algorithm Λ_eval which, on input a conversion key *par* and a public one-time key pk_i, outputs either a value $\widehat{pk}_i \in Y_n$ (a suitable input for the accumulator with the key n) or *"unsuitable"*. The success probability should at least be the inverse of some polynomial; in the examples, it will be at least constant.
  We write $\Lambda_{par}(pk_i)$ instead of $\Lambda\_eval(par, pk_i)$.
- A polynomial-time inversion algorithm, abbreviated Λ_{par}^{-1}, with $\Lambda_{par}^{-1}(\Lambda_{par}(pk_i)) = pk_i$ for all $\Lambda_{par}(pk_i) \neq$ *"unsuitable"*.

Note that the conversion of a one-time key is not necessarily deterministic, but the inversion has to be. So it is possible to include some random bits in the output of Λ_{par} that are needed for an accumulator, but the result of Λ_{par}^{-1} is always unique.

We now show the core of a simple example Λ_{prim}, which we will use in Section 5. It converts input numbers into prime numbers, if possible, using the same idea as in Section 2.3: The parameter *par* is a small integer l. On input $x \in \mathbb{N}$, the algorithm $\Lambda_{\text{prim},l}$ checks for $dist = 1, 3, \ldots, 2^l - 1$ whether the number $2^l x + dist$ is prime. If so, it returns $2^l x + dist$, otherwise *"unsuitable"*. To get x back from the output $\hat{x}$, the inversion algorithm simply cuts off the l least significant bits.

Another example of a conversion algorithm is of course the identity function, which can be used whenever no conversion is necessary.

4 Accumulator FSS Scheme

In this section, we describe the accumulator FSS scheme. It is based on

- a one-time FSS scheme with prekey and parameters (k^*, σ^*),
- a family of collision-free accumulators with parameters (k', N), and
- a family of conversion algorithms for the one-time FSS scheme and the accumulator.

4.1 One-time FSS Scheme with Prekey

We use so-called one-time FSS schemes with prekey, e.g., [PePf97]. This prekey is generated by a center trusted by all recipients and verified by the signer, who need not trust the center. The center is used instead of the recipients themselves for simplicity. Based on this prekey, the signer can generate as many one-time key pairs as she wants. Among the two security parameters, σ^* is chosen by the signer for her information-theoretical security, whereas k^* is chosen by the center for the computational security of the recipients.

For simplicity, we only consider schemes that fulfil the simplified security criteria for schemes with prekey from [Pfit96, Theorem 7.34]. First, this means that proofs of forgery only depend on the prekey. This is natural because only the prekey is *not* chosen by the signer, i.e., a proof of forgery has to show a secret hidden in the prekey. Secondly, it is required that for every good prekey (one that the signer accepts with significant probability), for *every* one-time key pair based on it and every forgery, the probability that the forgery cannot be proved is at most $2^{-\sigma^*}$.

4.2 Construction

Key generation. The accumulator FSS scheme gets only (k, σ, N) as input. The remaining security parameters are calculated with

$$(k', k^*, \sigma^*) := calc\_pars(k, \sigma, N).$$

The center generates

- a prekey, using the algorithm $gen(k^*, \sigma^*)$ of the one-time FSS scheme.
- an accumulator key n with $n \leftarrow accu\_gen(k', N)$.
- the parameter for the conversion algorithm as $par := \Lambda\_gen(k^*, \sigma^*, n)$.

The signer verifies the prekey. She need not verify the accumulator key because it has no effect on her security. A weak accumulator key may make it easier for an adversary to find an accumulator collision and forge a signature. But this is no problem for the signer because she can show the collision as a proof of forgery. All these global values are part of the signer's public key, but for readability we omit them in the following.

The signer now chooses N key pairs (sk_i, pk_i) of the underlying one-time FSS scheme, based on the given prekey.

She computes $\widehat{pk}_i := \Lambda_{par}(pk_i)$ for $i = 1, \ldots, N$. If there is any $\widehat{pk}_i \notin Y_n$, i.e., $\widehat{pk}_i =$ "unsuitable", she has to generate a new key pair (sk_i, pk_i) and to repeat the computation of $\widehat{pk}_i$.

Finally, the signer computes the main public key pk of the accumulator FSS scheme by accumulating the $\widehat{pk}_i$'s:

$$(pk, aux) \leftarrow a_n(\widehat{pk}_1, \ldots, \widehat{pk}_N).$$

She publishes pk and stores aux for later use. Formally, her secret key sk contains not only the secret one-time keys $sk_1, \ldots, sk_N$, but also the converted public one-time keys $\widehat{pk}_1, \ldots, \widehat{pk}_N$ and the auxiliary output aux.

Signing. The signature on the i-th message, m_i, is

$$s := (s_i, \widehat{pk}_i, accu_i),$$

where s_i is the one-time signature on this message with the one-time key sk_i, and $\widehat{pk}_i$ and $accu_i$ are needed for the authentication of the one-time public key pk_i. The value $accu_i$ is computed using

$$accu_i \leftarrow auth_n(\widehat{pk}_i, aux).$$

Since $accu_i$ is independent of the message, it can be precomputed when the computer is idle.

Testing. A value $s = (s_i, \widehat{pk}_i, accu_i)$ is an *acceptable signature* on the message m_i iff

1. s_i is an acceptable one-time signature on m_i with respect to $pk_i = \Lambda_{par}^{-1}(\widehat{pk}_i)$,
2. $\widehat{pk}_i \in Y_n$,
3. $accu_i \in Accu_n$, and
4. pk authenticates $\widehat{pk}_i$, i.e., $authentic_n(pk, \widehat{pk}_i, accu_i) = ok$.

Proving Forgeries. If $(s', \widehat{pk}', accu')$ is an acceptable signature on a message m' not previously signed by the signer, she can generate a proof of forgery as follows:

1. If $pk' = \Lambda_{par}^{-1}(\widehat{pk}') \in \{pk_1, \ldots, pk_N\}$, she tries to generate a proof of forgery in the one-time FSS scheme.
2. Otherwise, she shows the accumulator collision

$$proof := \left((\widehat{pk}_1, \ldots, \widehat{pk}_N), (\widehat{pk}', accu')\right).$$

This proof shows that the assumption on which the accumulator is based has been broken.

Verifying Proofs of Forgery.

1. If *proof* is said to be a proof of forgery in the one-time FSS scheme, one verifies that.
2. Otherwise *proof* is accepted iff it fulfills the following conditions:
   (a) $\widehat{pk}' \notin \{\widehat{pk}_1, \ldots, \widehat{pk}_N\}$,
   (b) $\widehat{pk}_1, \ldots, \widehat{pk}_N, \widehat{pk}' \in Y_n$,
   (c) $accu' \in Accu_n$ and
   (d) $authentic_n(pk, \widehat{pk}', accu') = ok$ with $(pk, aux) \leftarrow a_n(\widehat{pk}_1, \ldots, \widehat{pk}_N)$.

4.3 Security

Theorem 8. *The accumulator FSS scheme as defined in the previous section is secure for both the signer and the recipients as defined in [PfWa90, PePf97].*

Proof sketch. For the information-theoretic security of the signer, we first show that any forgery that is not a forgery in the one-time FSS scheme, i.e., that does not fulfil the condition of Item 1 in "Proving Forgeries", is provable with probability 1: If $pk' \notin \{pk_1, \dots, pk_N\}$, then $\widehat{pk}' \notin \{\widehat{pk}_1, \dots, \widehat{pk}_N\}$ because the inversion Λ_{par}^{-1} is deterministic. Thus the value the signer computes in Item 2 is indeed an accumulator collision.

If the forgery *is* in the underlying one-time scheme, the signer can prove it with an error probability less than $2^{-\sigma^*}$, and thus less than $2^{-\sigma}$ (given that the prekey is good), because

  – with probability 1, she finds the one-time key pair (sk_i, pk_i) whose public one-time key the forger has used,
  – for *every* generated one-time key pair, the probability is at most $2^{-\sigma^*}$ that no proof of forgery can be found in the underlying FSS scheme, independent of the number of *"unsuitable"* public one-time keys generated before, and
  – the forger gains no information about sk_i by the accumulation.

The recipients want to be secure that no signatures they have accepted become invalid. Thus it should not be possible that

  – an adversary computes an acceptable signature that will be (correctly) proven to be forged by the signer, and that
  – the signer can (incorrectly) deny a previously generated signature using a proof of forgery.

Hence it is sufficient to show that no proof of forgery can be computed. This is (computationally) true because a proof of forgery of the new scheme implies either a successful proof of forgery in the underlying one-time FSS scheme or a collision of the utilized accumulator. Since for both parts the security parameter is at least k (guaranteed by the function *calc_pars*), neither should be possible for a polynomially restricted forger. That some key pairs are thrown away during key generation does not help the adversary, because the proof is based on the prekey alone.   □

5 Examples

We construct two examples of accumulator FSS schemes, using the two accumulators from Section 2.3. As the underlying one-time FSS scheme, we choose the one described in [HePe93]. It is based on the Discrete Logarithm assumption. Its public keys are pairs (a, b) of elements of the group where computing discrete logarithms is assumed to be hard; let their length in bits be the security parameter k^*. The algorithms of the accumulator FSS schemes are clear from the previous section as soon as we fix the conversion algorithms.

The first examples uses the accumulator a^{RSA}. It needs prime numbers as inputs, so we convert the one-time public keys (a, b) with Λ_{prim}, interpreting (a, b) as one $2k^*$-bit number.

The security parameters for the one-time FSS scheme and the accumulator are calculated by

$$(k', k^*, \sigma^*) = calc\_pars(k, \sigma, N) := (2k + \lceil \log_2 2k \rceil + 1, k, \sigma),$$

and the key of the conversion algorithm by

$$l = \Lambda\_gen(k^*, \sigma^*, n) := \lceil \log_2 2k^* \rceil.$$

These functions guarantee that the converted public one-time keys are in the domain of the accumulator: The parameters for the one-time FSS scheme are simply the given k and σ. The parameter k' for the accumulator is set such that the RSA modulus is longer than a one-time FSS key and the appended value $dist$. The length l of $dist$ is a somewhat arbitrary value ensuring that a prime will typically be found in the search interval.

The second example is based on the RSA accumulator with a random oracle assumption. This accumulator needs pairs $(pk_i, dist_i)$ as input, so the conversion algorithm is similar to $\Lambda_{\mathrm{prim},l}$, but returns $(pk_i, dist_i)$ instead of $2^l \Omega(pk_i) + dist_i$ if that value is prime. The security parameters are computed with

$$(k', k^*, \sigma^*) = calc\_pars(k, \sigma, N) := (k, k, \sigma)$$

and the key of the conversion algorithm is

$$\Lambda\_gen(k^*, \sigma^*, (n, \Omega, l, x)) := (l, \Omega).$$

Concretely, this means that the length of the RSA modulus used for the accumulator is independent of the length of the one-time keys, because only oracle outputs with appended values $dist$ are accumulated, and the length of the oracle output is adapted accordingly.

6 Conclusion

We have presented a generalized definition of accumulators and the definition of a collision-free subtype. We constructed two collision-free accumulators, one based on a stronger RSA assumption than usual, the other based on a random oracle and the normal RSA assumption. We remind the reader that no new assumption in cryptology should be trusted, i.e., we certainly do not recommend the first version for use in practice for quite some time. These accumulators can be used to construct fail-stop signature schemes in which the length of the public key and of the signatures is independent of the number N of messages that can be signed, while the additional cost for signing is small, especially because most of the signature can be computed and sent before the message is known.

Key generation, however, takes significantly longer than in constructions with trees. To avoid the precomputation of a very long secret key, one can combine the constructions with top-down tree authentication. In this way, we get flat trees similar to those in [DwNa94]. For instance, one might use accumulation for 1024 pairs (sk_i, pk_i) each, form a tree with two levels of such structures, and generate the structures of the lower level on demand, signing their "public" keys with the secret keys of the upper level. Thus one can sign one million messages with one public key. A complete signature consists of the accumulation result z of one lower-level structure and two accumulator FSS signatures as described in Section 4.

Acknowledgments

We thank Michael Waidner, Joachim Biskup, Andreas Pfitzmann, and Ute von Jan for helpful comments on this paper.

References

[Bari96]   NIKO BARIĆ: *Digitale Signaturen mit Fail-stop Sicherheit ohne Baumauthentifizierung*. Diplomarbeit, Institut für Informatik, Universität Hildesheim, July 1996.

[BeMa94]   JOSH BENALOH and MICHAEL DE MARE: *One-Way Accumulators: A Decentralized Alternative to Digital Signatures*. In *Advances in Cryptology — EUROCRYPT '93*, LNCS 765, pages 274–285. Springer-Verlag, Berlin, 1994.

[BeRo93]   MIHIR BELLARE and PHILLIP ROGAWAY: *Random Oracles are Practical: A Paradigm for Designing Efficient Protocols*. In *1st ACM Conference on Computer and Communications Security, November 1993*, pages 62–73. acm press, New York, 1993.

[CFPR96]   DON COPPERSMITH, MATTHEW FRANKLIN, JACQUES PATARIN, and MICHAEL REITER: *Low-Exponent RSA with Related Messages*. In *Advances in Cryptology — CRYPTO '96*, LNCS 1070, pages 1–9. Springer-Verlag, Berlin, 1996.

[CrDa96]   RONALD CRAMER and IVAN B. DAMGÅRD: *New Generation of Secure and Practical RSA-Based Signatures*. In *Advances in Cryptology — CRYPTO '96*, LNCS 1109. Springer-Verlag, Berlin, 1996.

[DwNa94]   CYNTHIA DWORK and MONI NAOR: *An Efficient Existentially Unforgeable Signature Scheme and its Application*. In *Advances in Cryptology — CRYPTO '94*, LNCS 839, pages 234–246. Springer-Verlag, Berlin, 1994.

[Håst86]   JOHAN HÅSTAD: *On Using RSA with Low Exponent in a Public Network*. In *Advances in Cryptology — CRYPTO '85*, LNCS 218, pages 403–408. Springer-Verlag, Berlin, 1986.

[HePe93]   EUGÈNE VAN HEYST and TORBEN P. PEDERSEN: *How to Make Efficient Fail-stop Signatures*. In *Advances in Cryptology — EUROCRYPT '92*, LNCS 658, pages 366–377. Springer-Verlag, Berlin, 1993.

[HePP93] EUGÈNE VAN HEIJST, TORBEN P. PEDERSEN, and BIRGIT PFITZMANN: *New Constructions of Fail-Stop Signatures and Lower Bounds*. In *Advances in Cryptology — CRYPTO '92*, LNCS 740, pages 15–30. Springer-Verlag, Berlin, 1993.

[Nybe96a] KAISA NYBERG: *Commutativity in Cryptography*. In *Proceedings of the First International Workshop on Functional Analysis at Trier University*, pages 331–342. Walter de Gruyter, Berlin, 1996.

[Nybe96b] KAISA NYBERG: *Fast Accumulated Hashing*. In *3rd Fast Software Encryption Workshop*, LNCS 1039, pages 83–87. Springer-Verlag, Berlin, 1996.

[PePf97] TORBEN P. PEDERSEN and BIRGIT PFITZMANN: *Fail-Stop Signatures*. to appear in SIAM Journal on Computing, 26(2):291–330, April 1997.

[Pfit94] BIRGIT PFITZMANN: *Fail-Stop Signatures Without Trees*. Hildesheimer Informatik-Berichte 16/94, ISSN 0941-3014, Institut für Informatik, Universität Hildesheim, June 1994.

[Pfit96] BIRGIT PFITZMANN: *Digital Signature Schemes — General Framework and Fail-Stop Signatures*. LNCS 1100. Springer-Verlag, Berlin, 1996.

[PfWa90] BIRGIT PFITZMANN and MICHAEL WAIDNER: *Formal Aspects of Fail-stop Signatures*. Interner Bericht 22/90, Fakultät für Informatik, Universität Karlsruhe, December 1990.

[RSA78] RONALD L. RIVEST, ADI SHAMIR, and LEONARD ADLEMAN: *A Method for Obtaining Digital Signatures and Public-Key Cryptosystems*. Communications of the ACM, 21(2):120–126, February 1978.

[Sham83] ADI SHAMIR: *On the Generation of Cryptographically Strong Pseudorandom Sequences*. ACM Transaction on Computer Systems, 1(1):38–44, February 1983.

[WaPf90] MICHAEL WAIDNER and BIRGIT PFITZMANN: *The Dining Cryptographers in the Disco: Unconditional Sender and Recipient Untraceability with Computationally Secure Serviceability*. In *Advances in Cryptology — EUROCRYPT '89*, LNCS 434, page 690. Springer-Verlag, Berlin, 1990.

[Wien90] MICHAEL J. WIENER: *Cryptanalysis of Short RSA Secret Exponents*. IEEE Transactions on Information Theory, 36(3):553–558, May 1990.

Selective Forgery of RSA Signatures Using Redundancy

Marc Girault
marc.girault@francetelecom.fr

Jean-François Misarsky
jean-francois.misarsky@francetelecom.fr

CNET CAEN
42, rue des Coutures
B.P. 6243
FR-14066 CAEN Cedex

Abstract: We show the weakness of several RSA signature schemes using redundancy (i.e. completing the message to be signed with some additional bits which are fixed or message-dependent), by exhibiting chosen-message attacks based on the multiplicative property of RSA signature function. Our attacks, which largely extend those of De Jonge and Chaum [DJC], make extensive use of an affine variant of Euclid's algorithm, due to Okamoto and Shiraishi [OS]. When the redundancy consists of appending any fixed bits to the message m to be signed (more generally when redundancy takes the form of an affine function of m), then our attack is valid if the redundancy is less than half the length of the public modulus. When the redundancy consists in appending to m the remainder of m modulo some fixed value (or, more generally, any function of this remainder), our attack is valid if the redundancy is less than half the length of the public modulus minus the length of the remainder. We successfully apply our attack to a scheme proposed for discussion inside ISO.

1 Introduction

Let (P, S) be a RSA [RSA] key pair, where P is the public function and S the secret one. It is well known that the "reciprocal property" (the fact that $P \circ S = S \circ P = Id$, the identity function) and the "multiplicative property" (the fact that $S(xy) = S(x)S(y)$) of RSA lead to potential weaknesses, especially when used for signatures.

The reciprocal property trivially allows to perform an existential forgery: just choose Σ at random and compute $m = P(\Sigma)$; then the pair (m, Σ) is an apparently authentic signed message. The multiplicative property allows a selective forgery by performing a 2-chosen-message attack, i.e. a chosen-message attack requiring two messages. Let m be the message to be signed, choose x as you like in $[1, n\text{-}1]$ and compute

W. Fumy (Ed.): Advances in Cryptology - EUROCRYPT '97, LNCS 1233, pp. 495-507, 1997.
© Springer-Verlag Berlin Heidelberg 1997

$y = m/x \bmod n$ where n is the public modulus; obtain the signatures of x and y and compute the signature of m as the product $S(m) = S(x)S(y) \bmod n$.

Different ways exist to eliminate these potential weaknesses. We can either add some redundancy to the message to be signed [ISO1], or use a hash-function in the signature scheme [ISO2], [BR]. The present paper is related to the redundancy solution. This solution is of particular interest when the message is short, because it prevents from specifying and implementing a hash-function (a rather delicate cryptographic challenge), and it allows to construct very compact signed messages, since messages can be recovered from the signatures themselves (and hence need not any longer be transmitted or stored). More precisely, let R be the (invertible) redundancy function. The signature of m is $\Sigma(m) = S[R(m)]$, and the signer only sends $\Sigma(m)$ to the receiver. The latter applies P to $\Sigma(m)$, and verifies that the result complies with the redundancy rule, i.e. is an element of the image set of R. Then he recovers m by discarding the redundancy (i.e. by applying R^{-1}) to this result.

But it has been shown in the past [DJC] that too simple redundancy does not avoid all the chosen-message attacks. For instance, the redundancy defined by appending trailing '0' bits to the message is insufficient because it remains possible, for any m, to construct two integers x and y such that $(m\|0..0) = (x\|0..0)(y\|0..0) \bmod n$ (implying $S(m\|0..0) = S(x\|0..0)S(y\|0..0) \bmod n$) by using Euclid's algorithm. In the standard ISO/IEC 9796 Part 1 [ISO1], a redundancy function is described, the security of which is assessed as very good. But its expansion rate (at least two) is too high in many applications, e.g. public key certification. As a consequence, there remains a need for a simple/short redundancy function providing adequate security.

The main goal of this paper is to show that a number of attractive redundancy functions, some of which proposed here and there, are subject to a 2-chosen-message attack. It is organized as follows: in section 2, we summarize our results, in section 3, we describe the mathematical tools used by our attacks, in section 4, attacks on valid messages with fixed redundancy, in section 5, attacks on valid messages with fixed and modular redundancy, in section 6, some applications including an attack on a scheme proposed for discussion inside ISO. We explain how to defeat this forgery in section 7 and we conclude in section 8.

Throughout this paper, we call valid message any message m completed with redundancy (i.e. any integer in the form $R(m)$), and bitlength (or length in short) of an integer the number of bits of its binary representation. We denote by $|m|$ the bitlength of m. We also define mb as the maximum bitlength of message accepted in a signature scheme.

2 Our Results

First, we extend the results of De Jonge and Chaum [DJC]: if the redundancy consists in appending any fixed bits to m to be signed, or more generally if redundancy takes the form of an affine function of m, that is when the signature $\Sigma(m)$ of m is computed as $\Sigma(m) = S(\omega m + a)$, for any constant a, any constant ω and message m, then the signature scheme is subject to a chosen-message attack, provided the redundancy is less than half the length of the public modulus used by S and P. De Jonge and Chaum exhibited similar attacks only in the cases when $a = 0$ (with the same amount of redundancy) or when $\omega = 1$ (with a smaller amount of redundancy).

Next, we study the case of the redundancy obtained in appending to m the remainder of m modulo some fixed value. Then, the signature scheme is still subject to a 2-chosen-message attack, provided the redundancy is less than half the length of the public modulus minus the length of the remainder. In a particular case, it even works when the redundancy is up to half the length of the modulus.

Here, the term "chosen-message attack" means the following: for any arbitrary message m it is possible to construct two messages m_1 and m_2 such that $\Sigma(m_2) / \Sigma(m_1) = \Sigma(m)$ modulo the RSA-modulus used by S. Therefore, by obtaining the signatures of m_1 and m_2, an enemy can forge the signature of m. It must be stressed that m can be entirely selected by the enemy; so this forgery is selective, not only existential.

All the attacks make extensive use of an affine variant of Euclid's algorithm, due to Okamoto and Shiraishi [OS], which is described in the coming section.

3   Basic Tools

In all our attacks, we will face the following problem:

Let n be a positive integer and d, z_0, X, Y, with X and Y "small", four positive integers less than n. Find solutions x and y to:

$$(S) \quad \begin{cases} dx = y + z_0 \pmod{n} \\ |x| < X \\ |y| < Y \end{cases}$$

3.1   Case of $z_0 = 0$

W. De Jonge and D. Chaum solved this problem [DJC]. There is at least one solution not equal to $(0, 0)$ if $XY > n$. Demonstration of this result uses the "pigeon-hole principle". It is useful to remark [GTV] that finding small x and y satisfying (S) comes to finding a good approximation of the fraction d/n. So, we find such a solution by developing it in continued fractions i.e. applying extended Euclidean algorithm to d and n.

Algorithm EE
- *Input*: n, d, X, Y (with $XY > n$)
- *Output*: nothing or some x such that $|x| < X$ and $|dx \pmod{n}| < Y$
- *Method*: apply extended Euclidean algorithm to d and n; one obtains coefficients l_i and m_i such that:

$$l_i n + m_i d = r_i \tag{1}$$

where the r_i are the successive remainders; output the smallest (in absolute value) m_i such that $n/Y < |m_{i-1}|$ (the case "such an m_i does not exist" is very rare).

- **Proof:** the fractions $|l_i/m_i| = -l_i/m_i$ are in fact the convergents of the development of d/n in continued fractions; hence:

$$|d/n + l_i/m_i| \leq 1/|m_i m_{i-1}| \implies |dm_i + nl_i| \leq n/|m_{i+1}|$$
$$\implies |dm_i \pmod n| < Y$$

Moreover, $(|m_i| \leq n/Y$ and $XY > n)$ implies $|m_i| < X$

3.2 Case of $z_0 \neq 0$

Okamoto and Shiraishi provide in [OS] an extension of extended Euclidean algorithm which very often solves this problem. We use a version of this algorithm to generate solutions.

Algorithm OS

- **Input:** n, d, X, Y (with $XY > n$), z_0
- **Output:** nothing or some x such that $|x| < X$ and $|dx - z_0 \pmod n| < Y$
- **Method:** apply extended Euclidean algorithm to d and n; introduce a sequence y_i whose first term y_0 is z_0 and following ones are defined by:

$$y_i = y_{i-1} - q_i' r_i \tag{2}$$

where q_i' is the quotient in the division of y_{i-1} by r_i; introduce also the sequence k_i whose first term k_0 is zero and the following ones are defined by:

$$k_i = k_{i-1} + q_i' m_i \tag{3}$$

Output k_i such that $n/Y < |k_i| < X$ and $|k_i| y_i \leq n$

- **Proof:** let the sequence h_i whose first term h_0 is zero and following ones defined by:

$$h_i = h_{i-1} + q_i' l_i \tag{4}$$

Then,

$$h_i n + k_i d = (h_{i-1} + q_i' l_i)n + (k_{i-1} + q_i' m_i)d$$
$$= h_{i-1} n + k_{i-1} d + q'(l_i n + m_i d)$$

(1) and (2) imply:

$$h_i n + k_i d = h_{i-1} n + k_{i-1} d + (y_{i-1} - y_i)$$

Then,

$$h_i n + k_i d = 0 + (y_0 - y_1) + (y_1 - y_2) + \ldots + (y_{i-1} - y_i)$$
$$= \qquad y_0 - y_i \qquad \implies k_i d \pmod n = z_0 - y_i$$

By taking output's conditions on k_i into account, we have:

$$|k_i| < X \text{ and } y_i \leq n/|k_i| < Y$$

Remark: to increase the number of solutions when $z_0 \neq 0$, you can combine one solution (x, y) found by algorithm OS with a solution (x', y') given by algorithm EE for the same system with $z_0 = 0$.

4   Valid Messages with Fixed Redundancy

Recall that redundancy function takes the form of an affine function of message. The signature $\Sigma(m)$ of m is computed as $\Sigma(m) = S(\omega m + a)$ for any constant a, any constant ω and message m. De Jonge and Chaum already studied multiplicative attacks on schemes using fixed redundancy. But their results were restricted to $a = 0$ (and any value of ω) or $\omega = 1$ (and any value of a). Moreover, their attack is valid if the redundancy takes up less than half of the bits in the modulus n when $a = 0$, and otherwise if the redundancy takes up less than one third of the bits in the modulus n.

Our method extend this results: the signature scheme is subject to a chosen-message attack for any value of a and ω, provided that the redundancy takes up less than half of the bits in a valid message.

In this section, we describe our attack on right-padded redundancy scheme, left-padded redundancy scheme, then on a more general scheme. Proof and efficiency are only given in the general case.

4.1   Right-Padded Redundancy Scheme

Let a be a fixed pattern of bits, and $\omega = 2^{|a|}$.
We denote by $\mathscr{E}$ the set of messages:
$$\mathscr{E} = \{\text{integers } m \text{ such that } 0 \leq m < n/\omega\}$$
and by $\mathscr{E}'$ the set of valid messages:
$$\mathscr{E}' = \{\omega m + a \text{ such that } m \in \mathscr{E}\}$$
Example: an element of $\mathscr{E}'$ has this form:

Message m	...1010010100001110101

Attack:
• Choose a message $m \in \mathscr{E}$ of which you want to forge a signature.
• Set

$$z_0 = \frac{a}{\omega}[1 - (\omega m + a)](\bmod n) \qquad (5)$$

• Solve

$$(\omega m + a)x = y + z_0 (\bmod n) \qquad (6)$$

with x and y elements of $\mathscr{E}$ by using algorithm OS. You obtain, very often, a solution if the range of m is larger than $\sqrt{n}$ (i.e. the number of bits of redundancy a is less than half of the bits of modulus n). See 4.3 for more details.
• By replacing z_0 by its expression (5) in the latter equation (6), you can easily prove that $(\omega m + a)(\omega x + a) = (\omega y + a) \pmod n$. If you get signatures of y and x (i.e. if you get $S(\omega y + a)$ and $S(\omega x + a)$), then you deduce the signature of m by dividing $S(\omega y + a)$ by $S(\omega x + a)$ modulo n.

4.2 Left-Padded Redundancy Scheme

Let a' be a fixed pattern of bits, and $\beta = 2^{mb}$. We denote by $\mathscr{E}$ the set of messages:
$$\mathscr{E} = \{\text{integers } m \text{ such that } 0 \le m < \beta\}$$
and by $\mathscr{E}'$ the set of valid messages:
$$\mathscr{E}' = \{m + a'\beta \text{ such that } m \in \mathscr{E}\}$$
Example: an element of $\mathscr{E}'$ has this form :

01001001010101111...	Message m

Attack:
- Choose a message $m \in \mathscr{E}$ of which you want to forge a signature.
- Set

$$z_0 = a'\beta[1 - (m + a'\beta)](\mathrm{mod}\, n) \tag{7}$$

- Solve

$$(m + a'\beta)x = y + z_0 (\mathrm{mod}\, n) \tag{8}$$

with x and y elements of $\mathscr{E}$ by using algorithm OS. You obtain, very often, a solution if the range of m is larger than $\sqrt{n}$ (i.e. the number of bits of redundancy a is less than half of the bits of modulus n). See 4.3 for more details.

- By replacing z_0 by its expression (7) in the latter equation (8), you can easily prove that $(m + a'\beta)(x + a'\beta) = (y + a'\beta)$ (mod n). If you get signatures of y and x (i.e. if you get $S(y + a'\beta)$ and $S(x + a'\beta)$), then you deduce the signature of m by dividing $S(y + a'\beta)$ by $S(x + a'\beta)$ modulo n.

4.3 Generalization

Let a be the lower bound to a valid message, b be the upper bound to a valid message ($a \le m < b < n$), ω a multiplicative constant. Consequently, we can define $\mathscr{E}$ as the set of messages:
$$\mathscr{E} = \{\text{integers } m \text{ such that } 0 \le m < (b - a)/\omega\}$$
and $\mathscr{E}'$ as the set of valid messages:
$$\mathscr{E}' = \{\omega m + a \text{ such that } m \in \mathscr{E}\}$$

Attack:
- Choose a message $m \in \mathscr{E}$ of which you want to forge a signature.
- Set

$$z_0 = \frac{a}{\omega}[1 - (\omega m + a)](\mathrm{mod}\, n) \tag{9}$$

- Solve

$$(\omega m + a)x = y + z_0 (\mathrm{mod}\, n) \tag{10}$$

with x and y elements of $\mathscr{E}$ by using algorithm OS. You obtain, very often, a solution if the range of m is larger than $\sqrt{n}$ (i.e. the number of bits of redundancy, multiplicative and additive, is less than half of the bits of modulus n).

- By replacing z_0 by its expression (9) in the latter equation (10), you can easily prove that $(\omega m + a)(\omega x + a) = (\omega y + a) \pmod{n}$. If you get signatures of y and x (i.e. if you get $S(\omega y + a)$ and $S(\omega x + a))$, then you deduce the signature of m by dividing $S(\omega y + a)$ by $S(\omega x + a)$ modulo n.

Proof: let x and y be a couple of solutions:

$$(\omega m + a)x \quad = \quad y + z_c \qquad \pmod{n}$$

$$(\omega m + a)\omega x \quad = \quad \omega y + \omega\left[\frac{a}{\omega}\left[1 - (\omega m + a)\right]\right] \quad \pmod{n}$$

$$(\omega m + a)\omega x \quad = \quad \omega y + a - a(\omega m + a) \qquad \pmod{n}$$

$$(\omega m + a)(\omega x + a) \quad = \quad (\omega y + a) \qquad \pmod{n}$$

Efficiency: algorithm OS gives a solution if $XY > n$ (see 3.2.), i.e. if:

$$\frac{(b-a)}{\omega}\frac{(b-a)}{\omega} > n$$

Thus, a solution is obtained when the range of m, i.e. $\dfrac{(b-a)}{\omega}$,is larger than $\sqrt{n}$ or when:

$$\log_2\left(\frac{b-a}{\omega}\right) \quad > \quad \log_2\left(\sqrt{n}\right)$$

$$\log_2(n) - \log_2\left(\frac{b-a}{\omega}\right) \quad < \quad \frac{1}{2}\log_2(n)$$

i.e. the number of bits of redundancy, multiplicative and additive redundancy, is less than half of the bits of modulus n.

Remarks :
- If ω is a power of two upper than $2^{|a|}$ then it is the right-padded redundancy scheme (see 3.1).
- If $\omega = 1$ and a is a multiple of 2^{mb} then it is the left-padded redundancy scheme (see 3.2).
- Note that with an appropriate choice of ω and a, it is a scheme with the message in the middle :

1101010101...	Message m	...00010101101

5   Valid Messages With Fixed And Modular Redundancy

The expression "modular redundancy" is used to indicate a redundancy obtained with a modular operation. We denote this modular redundancy by the function $H(x)$. In this section, we consider a modular redundancy of u bits in length.

We consider three cases: first of all, the particular case $H(m) = m \pmod{2^u + 1}$, a modular redundancy of u bits (except if $H(m) = 2^u$, an event of probability nearly equal to 0). Next $H(m) = m \pmod{2^u + v}$ where v is a negative integer greater than or equal to -2^{u-1}, and last $H(m) = (m \pmod{2^u + v})) \oplus Mask$ where v is a negative integer greater than or equal to -2^{u-1} and $Mask$ is a u-bit fixed string. We denote the message m concatenated with $H(m)$ by:

$$\Phi(m) = m \parallel H(m) \tag{11}$$

Let a and ω be integers less than n, α the length of message, and $\mathscr{E}$ the set of messages:

$$\mathscr{E} = \{m \text{ such that } 0 \le m < 2^\alpha\}$$

Then, the set of valid messages is:

$$\mathscr{E}' = \{\omega\Phi(m) + a, \text{ with } m \in \mathscr{E}\}$$

Example: if ω is a power of two, then an element of $\mathscr{E}'$ has this form :

01011....	Message m (α bits)	$H(m)$ (u bits)	...0110

5.1 $H(m) = m \pmod{2^u + 1}$

We can also write

$$m = q(2^u + 1) + r \tag{12}$$

with q the quotient and r the remainder of Euclidean division of m by $(2^u + 1)$.
Hence $\Phi(m) = [q(2^u + 1) + r] \, 2^u + r$ and finally we obtain:

$$\Phi(m) = \psi(m)(2^u + 1) \tag{13}$$

with

$$\psi(m) = q2^u + r \tag{14}$$

Consequently, a new definition of the set of valid message is possible :

$$\mathscr{E}' = \{\omega'\psi(m) + a \text{ with } m \in \mathscr{E}\}$$

with $\omega' = \omega(2^u + 1)$.
Our attack uses this new definition.

Attack:
• Choose a message m of which you want to forge a signature.
• Set

$$z_0 = \frac{a}{\omega'}\left[1 - (\omega'\psi(m) + a)\right] \pmod{n} \tag{15}$$

• Solve

$$(\omega'\psi(m) + a)x = y + z_0 \pmod{n} \tag{16}$$

with x and y positive integers less than $2^{\alpha+u} / (2^u + 1)$ by using algorithm OS. You obtain, very often, a solution if the number of bits of the message, α, is upper than half of the length of modulus n.
• By replacing z_0 by its expression (15) in the latter equation (16), you can easily prove that:

$$(\omega'\psi(m) + a) (\omega'x + a) = (\omega'y + a) \pmod{n}$$

But the definition of function ψ, (13), and the fact that $\Phi(m) < 2^{\alpha+u}$, imply the existence of a message m s.t. $\psi(m) = t$ when t is less than $2^{\alpha+u} / (2^u + 1)$. Consequently, there are two messages m_1 and m_2 such that $\psi(m_1) = x$ and $\psi(m_2) = y$. Finally, if you get signatures of m_1 and m_2 (i.e. if you get $S(\omega'\psi(m_2) + a)$ and $S(\omega'\psi(m_1) + a)$), then you deduce the signature of m by dividing $S(\omega'\psi(m_2) + a)$ by $S(\omega'\psi(m_1) + a)$ modulo n.

5.2 $H(m) = m \pmod{2^u + v}$

Let

$$m = q(2^u + v) + r \tag{17}$$

where q and r are respectively the quotient and the remainder of the Euclidean division of m by $(2^u + v)$. Thus:

$$\Phi(m) = q(2^u + v)2^u + r(2^u + 1) \tag{18}$$

Given that $v \neq 1$, it follows that we cannot apply the latter method (5.1) to reduce the number of variables. Consequently, we will rather fix the value of either the quotient or the remainder. We choose to fix r because its range is shorter than the range of q. Hence, the modular redundancy is fixed as well.

Attack:
- Choose a message m of which you want to forge a signature.
- Choose r_1 and r_2 two positive integers less than $2^u + v$.
- Set

$$a_1 = r_1(2^x + 1)\omega + a$$
$$a_2 = r_2(2^x + 1)\omega + a \tag{19}$$
$$z_0 = \frac{1}{\omega 2^x(2^u + v)}\left[(\omega\Phi(m) + a)a_1 - a_2\right]$$

- Solve

$$(\omega\Phi(m) + a)q_1 = q_2 \cdot z_0 \pmod{n} \tag{20}$$

with q_1 and q_2 positive integers less than, respectively, $(2^\alpha - r_1)/(2^u + v)$ and $(2^\alpha - r_2)/(2^u + v)$, by using algorithm OS. You obtain, very often, a solution if the number of bits of the message, α, minus the number of bits of redundancy, u, is upper than half of the length of modulus n.
- Set

$$m_1 = q_1(2^u + v) + r_1 \tag{21}$$

and

$$m_2 = q_2(2^u + v) + r_2 \tag{22}$$

The set of possible values of q_1, r_1, q_2, r_2, implies that $m_1 \in M$ and $m_2 \in M$. By replacing z_0, a_1, a_2, by their expressions (19) in the solved equation (20), you obtain, after a brief calculation :

$$(\omega\Phi(m) + a)(\omega\Phi(m_1) + a) = (\omega\Phi(m_2) + a) \pmod{n}$$

Finally, you deduce the signature of m by dividing $S(\omega\Phi(m_2) + a)$ by $S(\omega\Phi(m_1) + a)$ modulo n.

5.3 $H(m) = (m \pmod{2^u + v}) \oplus Mask$

We denote by *Mask* a u-bit fixed string and by $\oplus$ the function exclusive OR.

We apply the same method as previously, but we introduce a new function:

$$C(r) = r2^u + (r \oplus Mask) \tag{23}$$

Thus we obtain:

$$\Phi(m) = q(2^u + v)2^u + C(r) \tag{24}$$

Since during the development of the attack the two remainders r_1 and r_2 are fixed, $C(r_1)$ and $C(r_2)$ are also fixed and the mask does not generate any extra difficulty.

Attack:
- Choose a message m of which you want to forge a signature.
- Choose r_1 and r_2 two positive integers such that they are less than $2^u + v$.
- Set

$$a_1 = C(r_1)\omega + a$$
$$a_2 = C(r_2)\omega + a \tag{25}$$
$$z_0 = \frac{1}{\omega 2^u(2^u + v)}\left[(\omega\varphi(m) + a)a_1 - a_2\right]$$

- Solve

$$(\omega\Phi(m) + a)q_1 = q_2 - z_0 \pmod{n} \tag{26}$$

with q_1 and q_2 positive integers less than, respectively, $(2^\alpha - r_1)/(2^u + v)$ and $(2^\alpha - r_2)/(2^u + v)$, by using algorithm OS. You obtain, very often, a solution if the number of bits of the message, α, minus the number of bits of redundancy, u, is upper than half of the length of modulus n.

- Set

$$m_1 = q_1(2^u + v) + r_1 \tag{27}$$

and

$$m_2 = q_2(2^u + v) + r_2 \tag{28}$$

The set of possible values of q_1, r_1, q_2, r_2, implies that $m_1 \in M$ and $m_2 \in M$. By replacing z_0, a_1, a_2, by their expressions (25) in the solved equation (26), you obtain, after a brief calculation :

$$(\omega\Phi(m) + a)(\omega\Phi(m_1) + a) = (\omega\Phi(m_2) + a) \pmod{n}$$

Finally, you deduce the signature of m by dividing $S(\omega\Phi(m_2) + a)$ by $S(\omega\Phi(m_1) + a)$ modulo n.

Remark: since this attack does not depend on the exact expression of $C(r)$, it can be performed against any modular redundancy in the form:

$$H(m) = H'[m \pmod{2^u + v}], \text{ for any function } H'.$$

6   Applications

We applied our results to a part of the project on digital signature schemes giving message recovery ISO/IEC JTC 1/SC 27 [ISO]. It was a Working Draft (WD), i.e. one of the first stages of the development of International Standards. After, when the working group is satisfied with the specified solution, the next step is the Committee Draft (CD), which is submitted to a ballot. Successive Committee Drafts may be considered until consensus is reached on the technical content. Once consensus has been attained, the text is finalized for submission as a Draft International Standard (DIS). Once a DIS has been approved, the final text is published as an International Standard (IS).

Part 2 of this project aims at defining a signature scheme allowing short certificates, which is convenient for smart cards. Like ISO/IEC 9796 [ISO], it is supposed to avoid the known attacks against RSA [GQLS]. In a particular case, this project uses a simplified hash-function $H(m) = 2(m \pmod{2^{79}+1})$ to define the modular redundancy. Structure of a valid message :

Adaptation bits	More-data bit	Padding Field	Data Field	Check Field	Adaptation nibble
Fixed: 2 bits	Fixed: 1 bit	Variable: 1 or more bits	Variable	Fixed: 80 bits	Fixed: 4 bits
01	0	0, 1 or more bits set to 0 followed by 1 bit set to 1	Message	Modular redundancy	0110

We implemented algorithms OS and EE in C-language on a PC computer to obtain our results. With a message m of 384 bits, $H(m) = 2(m \pmod{2^{79}+1})$, and a 512-bit RSA-modulus to define this scheme, we found nearly 40 solutions with algorithm OS and nearly 4000 solutions by the means of a simple combination with results of algorithm EE. This result can certainly be improved if all possible combinations are considered. When the length of message is 425-bit long, we found 60 or so with OS and about 8800 with OS combined with EE.

We have modified the function $H(m)$ to study the efficiency of our algorithm. With $H(m) = Mask \oplus 2(m \pmod{2^{79}+1})$ and $Mask = $ BBBBBBBBBBBBBBBBBBBB, we found, when the length of message is 384 bits nearly 16 solutions with OS and nearly 670 with OS and EE. When the length of message is 425 bits, we found 23 or so with OS and about 1720 with OS combined with EE. As previously, the number of solutions can certainly be expanded.

Remark: in the first case, we obtain more solutions than in the second one because the redundancy is not fixed. In fact, using $H(m) = 2(m \pmod{2^{79}+1})$ is like using the particular modular redundancy defined in 5.1. Here $u = 80$ and

$$\Phi(m) = [q(2^{79} + 1) + r]\, 2^{80} + 2r$$

with q the quotient and r the remainder of Euclidean division of m by $(2^{79} + 1)$. Finally we obtain:

$$\Phi(m) = \psi(m)(2^{80} + 2) \text{ with } \psi(m) = q2^{79} + r$$

and the attack described in 5.1 can be applied.

7 How To Defeat This Forgery

At Eurocrypt'96 Rump Session, we proposed three solutions to repair the previous schemes :

- Introduce the quotient q of Euclidean division of m by $(2^u + v)$

$$H(m) = r \times q \pmod{2^u + v}$$

This definition of H implies that we cannot isolate q and r in the expression of m concatenated with $H(m)$. The principle of our attack cannot be used here.

- Append to m its remainders modulo two different values, $2^{w/2} + v$ and $2^{w/2} + w$ with $v \neq w$. Two different moduli increase the link between message and redundancy, there is an interdependence between the different quotients and remainders. One of them cannot be fixed to use our attack. Simple values can be chosen, e.g. $v = -1$ and $w = 0$.

- Split the message into different parts and keep a simple redundancy. This method increases the number of variables and OS cannot be used to solve $mx = y \pmod{n}$. The latter solution is used in ISO/IEC 9796-3 [ISO3], Working Draft, December 1996, which replaces ISO/IEC JTC 1/SC 27 [ISO].

Remark: one of the authors has recently discovered a multiplicative attack using lattice basis reduction and only the first solution is valid.

8 Conclusion

We have shown the weakness of many attractive redundancy functions for the purpose of RSA digital signatures. We successfully applied our attack to an ISO Working Draft [ISO] and a modified version using a redundancy function with mask. Thus, we showed that some redundancy function may be inappropriate, even when it is message-dependent and even when it involves non-arithmetic operations. Afterwards, we have proposed new redundancy functions, which apparently cannot be attacked by our techniques. Nevertheless a further research showed that two of them can be attacked by a LLL-based method.

Acknowledgments

We would like to thank Louis Guillou for many fruitful discussions about RSA signature schemes and for stimulating this research. We are grateful to Luc Vallée for help on the C-language and for lending of his big number library. We also thank the referees for their useful comments on the previous version of the paper, which helped improve the quality of this paper.

References

[BR]   M. Bellare, P. Rogaway, "The Exact Security of Digital Signatures - How to Sign with RSA and Rabin", Eurocrypt'96 Proceedings, Lecture Notes In Computer Science, Vol.1070, U. Maurer ed., Springer-Verlag, 1996.

[DJC]   W. De Jonge, D. Chaum, "Attacks on some RSA Signatures", Advances in Cryptology, Crypto'85 Proceedings, Lecture Notes In Computer Science, Vol.218, Springer-Verlag, Berlin, 1986, pp. 18-27.

[GQLS]   L.C. Guillou, J.J. Quisquater, P. Landrock, C. Shaer, "Precautions taken against various potential attacks in ISO/IEC DIS 9796, Digital signature scheme giving message recovery", Eurocrypt'90 Proceedings, Lecture Notes in Computer Science, Vol.473, Springer-Verlag, pp 465-473.

[GTV]   M. Girault, P. Toffin, B. Vallée. "Computation of approximation L-th roots modulo n and application to cryptography", Proc. of Crypto'88, LNCS 403, Springer-Verlag, 1988, pp.100-117.

[ISO]   ISO/IEC JTC 1/SC 27, "Digital signature schemes giving message recovery; Part 2: Mechanisms using a hash function", Working Draft, January 1996.

[ISO1]   ISO/IEC 9796-1, "Digital signature schemes giving message recovery; Part 1: Mechanisms using redundancy".

[ISO2]   ISO/IEC 9796-2, "Digital signature schemes giving message recovery; Part 2: Mechanisms using a hash-function".

[ISO3]   ISO/IEC 9796-3, "Digital signature schemes giving message recovery; Part 3: Mechanisms using a check-function".

[OS]   T. Okamoto and A. Shiraishi, "A fast signature scheme based on quadratic inequalities", Proc. of the 1985 Symposium on Security and Privacy, Apr.1985, Oakland. CA.

[RSA]   R.L. Rivest, A. Shamir and L. Adleman, "A method for obtaining digital signatures and public-key cryptosystems", CACM, Vol. 21, n°2, Feb. 1978, pp. 120-126.

Author Index